SIMULATING THE PHYSICAL WORLD

The simulation of physical systems requires a simplified, hierarchical approach, which models each level from the atomistic to the macroscopic scale. From quantum mechanics to fluid dynamics, this book systematically treats the broad scope of computer modeling and simulations, describing the fundamental theory behind each level of approximation. Berendsen evaluates each stage in relation to their applications giving the reader insight into the possibilities and limitations of the models. Practical guidance for applications and sample programs in Python are provided. With a strong emphasis on molecular models in chemistry and biochemistry, this book will be suitable for advanced undergraduate and graduate courses on molecular modeling and simulation within physics, biophysics, physical chemistry and materials science. It will also be a useful reference to all those working in the field. Additional resources for this title including solutions for instructors and programs are available online at www.cambridge.org/9780521835275.

HERMAN J. C. BERENDSEN is Emeritus Professor of Physical Chemistry at the University of Groningen. His research focuses on biomolecular modeling and computer simulations of complex systems. He has taught hierarchical modeling worldwide and is highly regarded in this field.

SIMULATING THE PHYSICAL WORLD

Hierarchical Modeling from Quantum Mechanics to Fluid Dynamics

HERMAN J. C. BERENDSEN

Emeritus Professor of Physical Chemistry,
University of Groningen, the Netherlands

CAMBRIDGE
UNIVERSITY PRESS

CAMBRIDGE UNIVERSITY PRESS
Cambridge, New York, Melbourne, Madrid, Cape Town, Singapore, São Paulo

Cambridge University Press
The Edinburgh Building, Cambridge CB2 2RU, UK

Published in the United States of America by Cambridge University Press, New York

www.cambridge.org
Information on this title: www.cambridge.org/9780521835275

First published 2007

Printed in the United Kingdom at the University Press, Cambridge

A catalog record for this publication is available from the British Library

ISBN-13   978-0-521-83527-5 hardback
ISBN-10   0-521-83527-5 hardback

Contents

v

Preface

This book was conceived as a result of many years research with students and postdocs in molecular simulation, and shaped over several courses on the subject given at the University of Groningen, the Eidgenössische Technische Hochschule (ETH) in Zürich, the University of Cambridge, UK, the University of Rome (La Sapienza), and the University of North Carolina at Chapel Hill, NC, USA. The leading theme has been the truly interdisciplinary character of molecular simulation: its gamma of methods and models encompasses the sciences ranging from advanced theoretical physics to very applied (bio)technology, and it attracts chemists and biologists with limited mathematical training as well as physicists, computer scientists and mathematicians. There is a clear hierarchy in models used for simulations, ranging from detailed (relativistic) quantum dynamics of particles, via a cascade of approximations, to the macroscopic behavior of complex systems. As the human brain cannot hold all the specialisms involved, many practical simulators specialize in their niche of interest, adopt – often unquestioned – the methods that are commonplace in their niche, read the literature selectively, and too often turn a blind eye on the limitations of their approaches.

This book tries to connect the various disciplines and expand the horizon for each field of application. The basic approach is a physical one, and an attempt is made to rationalize each necessary approximation in the light of the underlying physics. The necessary mathematics is not avoided, but hopefully remains accessible to a wide audience. It is at a level of abstraction that allows compact notation and concise reasoning, without the burden of excessive symbolism. The book consists of two parts: Part I follows the hierarchy of models for simulation from relativistic quantum mechanics to macroscopic fluid dynamics; Part II reviews the necessary mathematical, physical and chemical concepts, which are meant to provide a common background of knowledge and notation. Some of these topics may be superfluous to physicists or mathematicians, others to chemists. The chapters of Part II could be useful in courses or for self-study for those who have missed certain topics in their education; for this purpose exercises are included. Answers and further information are available on the book's website.

The subjects treated in this book, and the depth to which they are explored, necessarily reflect the personal preference and experience of the author. Within this subjective selection the literature sources are restricted to the period before January 1, 2006. The overall emphasis is on simulation of large molecular systems, such as biomolecular systems where function is related to structure and dynamics. Such systems are in the middle of the hierarchy of models: very fast motions and the fate of electronically excited states require quantum-dynamical treatment, while the sheer size of the systems and the long time span of events often require severe approximations and coarse-grained approaches. Proper and efficient sampling of the configurational space (e.g., in the prediction of protein folding and other rare events) poses special problems and requires innovative solutions. The fun of simulation methods is that they may use physically impossible pathways to reach physically possible states; thus they allow a range of innovative phantasies that are not available to experimental scientists.

This book contains sample programs for educational purposes, but it contains no programs that are optimized to run on large or complex systems. For real applications that require molecular or stochastic dynamics or energy minimization, the reader is referred to the public-domain program suite GROMACS (http://www.gromacs.org), which has been described by Van der Spoel *et al.* (2005).

Programming examples are given in Python, a public domain interpretative object-oriented language that is both simple and powerful. For those who are not familiar with Python, the example programs will still be intelligible, provided a few rules are understood:

- Indentation is essential. Consecutive statements at the same indentation level are considered as a block, as if – in C – they were placed between curly brackets.
- Python comes with many *modules*, which can be imported (or of which certain elements can be imported) into the main program. For example, after the statement *import math* the *math* module is accessible and the sine function is now known as *math.sin*. Alternatively, the sine function may be imported by *from math import sin*, after which it is known as *sin*. One may also import all the methods and attributes of the *math* module at once by the statement *from math import ∗*.

- Python variables need not be declared. Some programmers don't like this feature as errors are more easily introduced, but it makes programs a lot shorter and easier to read.

- Python knows several types of sequences or *lists*, which are very versatile (they may contain a mix of different variable types) and can be manipulated. For example, if $x = [1, 2, 3]$ then $x[0] = 1$, etc. (indexing starts at 0), and $x[0 : 2]$ or $x[: 2]$ will be the list $[1, 2]$. $x + [4, 5]$ will concatenate x with $[4, 5]$, resulting in the list $[1, 2, 3, 4, 5]$. $x * 2$ will produce the list $[1, 2, 3, 1, 2, 3]$. A multidimensional list, as $x = [[1, 2], [3, 4]]$ is accessed as $x[i][j]$, e.g., $x[0][1] = 2$. The function $range(3)$ will produce the list $[0, 1, 2]$. One can run over the elements of a list x by the statement *for i in range(len(x)):* ...

- The extra package *numpy* (numerical python) which is not included in the standard Python distribution, provides (multidimensional) arrays with fixed size and with all elements of the same type, that have fast methods or functions like matrix multiplication, linear solver, etc. The easiest way to include *numpy* and – in addition – a large number of mathematical and statistical functions, is to install the package *scipy* (scientific python). The function *arange* acts like *range*, but defines an array. An array element is accessed as $x[i, j]$. Addition, multiplication etc. now work element-wise on arrays. The package defines the very useful *universal functions* that also work on arrays. For example, if $x = array([1, 2, 3])$, $\sin(x * pi/2)$ will be $array([1., 0., -1.])$.

The reader who wishes to try out the sample programs, should install in this order: a recent version of Python (http://www.python.org), *numpy* and *scipy* (http://www.scipy.org) on his system. The use of the IDLE Python shell is recommended. For all sample programs in this book it is assumed that *scipy* has been imported:

```
from scipy import *
```

This imports universal functions as well, implying that functions like *sin* are known and need not be imported from the *math* module. The programs in this book can be downloaded from the Cambridge University Press website (http://www.cambridge.org/9780521835275) or from the author's website (http://www.hjcb.nl). These sites also offer additional Python modules that are useful in the context of this book: *plotps* for plotting data, producing postscript files, and *physcon* containing all relevant physical constants in SI units. Instructions for the installation and use of Python are also given on the author's website.

This book could not have been written without the help of many former students and collaborators. It would never have been written without the stimulating scientific environment in the Chemistry Department of the University of Groningen, the superb guidance into computer simulation methods by Aneesur Rahman (1927–1987) in the early 1970s, the pioneering atmosphere of several interdisciplinary CECAM workshops, and the fruitful collaboration with Wilfred van Gunsteren between 1976 and 1992. Many ideas discussed in this book have originated from collaborations with colleagues, often at CECAM, postdocs and graduate students, of whom I can only mention a few here: Andrew McCammon, Jan Hermans, Giovanni Ciccotti, Jean-Paul Ryckaert, Alfredo DiNola, Raúl Grigera, Johan Postma, Tjerk Straatsma, Bert Egberts, David van der Spoel, Henk Bekker, Peter Ahlström, Siewert-Jan Marrink, Andrea Amadei, Janez Mavri, Bert de Groot, Steven Hayward, Alan Mark, Humberto Saint-Martin and Berk Hess. I thank Frans van Hoesel, Tsjerk Wassenaar, Farid Abraham, Alex de Vries, Agur Sevink and Florin Iancu for providing pictures.

Finally, I thank my wife Lia for her endurance and support; to her I dedicate this book.

Symbols, units and constants

Symbols

The typographic conventions and special symbols used in this book are listed in Table 1; Latin and Greek symbols are listed in Tables 2, 3, and 4. Symbols that are listed as vectors (bold italic, e.g., $\boldsymbol{r}$) may occur in their roman italic version ($r = |\boldsymbol{r}|$) signifying the norm (absolute value or magnitude) of the vector, or in their roman bold version ($\mathbf{r}$) signifying a one-column matrix of vector components. The reader should be aware that occasionally the same symbol has a different meaning when used in a different context. Symbols that represent general quantities as a, unknowns as x, functions as $f(x)$, or numbers as i, j, n are not listed.

Units

This book adopts the SI system of units (Table 5). The SI units (Système International d'Unités) were agreed in 1960 by the CGPM, the Conférence Générale des Poids et Mesures. The CGPM is the general conference of countries that are members of the *Metre Convention*. Virtually every country in the world is a member or associate, including the USA, but not all member countries have strict laws enforcing the use of SI units in trade and commerce.[1] Certain units that are (still) popular in the USA, such as inch (2.54 cm), Ångström (10^{-10} m), kcal (4.184 kJ), dyne (10^{-5} N), erg (10^{-7} J), bar (10^5 Pa), atm (101 325 Pa), electrostatic units, and Gauss units, in principle have no place in this book. Some of these, such as the Å and bar, which are decimally related to SI units, will occasionally be used. Another exception that will occasionally be used is the still popular Debye for dipole moment ($10^{-29}/2.997\,924\,58$ Cm); the Debye relates decimally

---

[1] A European Union directive on the enforcement of SI units, issued in 1979, has been incorporated in the national laws of most EU countries, including England in 1995.

to the obsolete electrostatic units. Electrostatic and electromagnetic equations involve the vacuum permittivity (now called the *electric constant*) ε_0 and vacuum permeability (now called the *magnetic constant*) μ_0; the velocity of light does not enter explicitly into the equations connecting electric and magnetic quantities. The SI system is *rationalized*, meaning that electric and magnetic potentials, but also energies, fields and forces, are derived from their sources (charge density ρ, current density $\boldsymbol{j}$) with a multiplicative factor $1/(4\pi\varepsilon_0)$, resp. $\mu_0/4\pi$:

$$\Phi(r) \;=\; \frac{1}{4\pi\varepsilon_0} \int \frac{\rho(\boldsymbol{r}')}{|\boldsymbol{r}-\boldsymbol{r}'|} \, d\boldsymbol{r}', \tag{1}$$

$$\boldsymbol{A}(\boldsymbol{r}) = \frac{\mu_0}{4\pi} \int \frac{\boldsymbol{j}(\boldsymbol{r}')}{|\boldsymbol{r}-\boldsymbol{r}'|} \, d\boldsymbol{r}', \tag{2}$$

while in differential form the 4π vanishes:

$$\operatorname{div} \boldsymbol{E} \;=\; -\operatorname{div} \operatorname{\mathbf{grad}} \Phi = \rho/\varepsilon_0, \tag{3}$$

$$\operatorname{\mathbf{curl}} \boldsymbol{B} \;=\; \operatorname{\mathbf{curl}} \operatorname{\mathbf{curl}} \boldsymbol{A} = \mu_0 \boldsymbol{j}. \tag{4}$$

In *non-rationalized* systems without a multiplicative factor in the integrated forms (as in the obsolete electrostatic and Gauss systems, *but also in atomic units*), an extra factor 4π occurs in the integrated forms:

$$\operatorname{div} \boldsymbol{E} \;=\; 4\pi\rho, \tag{5}$$

$$\operatorname{\mathbf{curl}} \boldsymbol{B} \;=\; 4\pi\boldsymbol{j}. \tag{6}$$

Consistent use of the SI system avoids ambiguities, especially in the use of electric and magnetic units, but the reader who has been educated with *non-rationalized* units (electrostatic and Gauss units) should not fall into one of the common traps. For example, the magnetic susceptibility χ_m, which is the ratio between induced magnetic polarization $\boldsymbol{M}$ (dipole moment per unit volume) and applied magnetic intensity $\boldsymbol{H}$, is a dimensionless quantity, which nevertheless differs by a factor of 4π between rationalized and non-rationalized systems of units. Another quantity that may cause confusion is the *polarizability* $\boldsymbol{\alpha}$, which is a tensor defined by the relation $\boldsymbol{\mu} = \boldsymbol{\alpha}\boldsymbol{E}$ between induced dipole moment and electric field. Its SI unit is $\mathrm{F\,m^2}$, but its non-rationalized unit is a volume. To be able to compare α with a volume, the quantity $\alpha' = \alpha/(4\pi\varepsilon_0)$ may be defined, the SI unit of which is $\mathrm{m^3}$.

Technical units are often based on the force exerted by standard gravity $(9.806\,65\ \mathrm{m\,s^{-2}})$ on a mass of a kilogram or a pound avoirdupois [lb = $0.453\,592\,37$ kg (exact)], yielding a kilogramforce (kgf) = $9.806\,65$ N, or a poundforce (lbf) = $4.448\,22$ N. The US technical unit for pressure psi (pound per square inch) amounts to 6894.76 Pa. Such non-SI units are avoided in this book.

When dealing with electrons, atoms and molecules, SI units are not very practical. For treating quantum problems with electrons, as in quantum chemistry, *atomic units* (a.u.) are often used (see Table 7). In a.u. the electron mass and charge and Dirac's constant all have the value 1. For treating molecules, a very convenient system of units, related to the SI system, uses nm for length, u (unified atomic mass unit) for mass, and ps for time. We call these *molecular units* (m.u.). Both systems are detailed below.

SI Units

SI units are defined by the basic units *length, mass, time, electric current, thermodynamic temperature, quantity of matter* and *intensity of light*. Units for angle and solid angle are the dimensionless *radian* and *steradian*. See Table 5 for the defined SI units. All other units are derived from these basic units (Table 6).

While the *Système International* also defines the *mole* (with unit *mol*), being a number of entities (such as molecules) large enough to bring its total mass into the range of grams, one may express quantities of molecular size also per mole rather than per molecule. For macroscopic system sizes one then obtains more convenient numbers closer to unity. In chemical thermodynamics molar quantities are commonly used. Molar constants as the Faraday F (molar elementary charge), the gas constant R (molar Boltzmann constant) and the molar standard ideal gas volume V_m (273.15 K, 10^5 Pa) are specified in SI units (see Table 9).

Atomic units

Atomic units (a.u.) are based on electron mass $m_e = 1$, Dirac's constant $\hbar = 1$, elementary charge $e = 1$ and $4\pi\varepsilon_0 = 1$. These choices determine the units of other quantities, such as

$$\text{a.u. of length (Bohr radius) } a_0 = \frac{4\pi\varepsilon_0\hbar^2}{m_e e^2} = \frac{\hbar}{\alpha m_e c}, \tag{7}$$

$$\text{a.u. of time } = \frac{(4\pi\varepsilon_0)^2\hbar^3}{m_e e^4} = \frac{m_e a_0^2}{\hbar}, \tag{8}$$

$$\text{a.u. of velocity } = \hbar/(m_e a_0) = \alpha c, \tag{9}$$

$$\text{a.u. of energy (hartree)} \quad E_{\mathrm{h}} = \frac{m_e e^4}{(4\pi\varepsilon_0)^2\hbar^2} = \frac{\alpha^2 c^2 m_e}{\hbar^2}. \quad (10)$$

Here, $\alpha = e^2/(4\pi\varepsilon_0\hbar c)$ is the dimensionless *fine-structure constant*. The system is *non-rationalized* and in electromagnetic equations $\varepsilon_0 = 1/(4\pi)$ and $\mu_0 = 4\pi\alpha^2$. The latter is equivalent to $\mu_0 = 1/(\varepsilon_0 c^2)$, with both quantities expressed in a.u. Table 7 lists the values of the basic atomic units in terms of SI units.

These units employ physical constants, which are not so constant as the name suggests; they depend on the definition of basic units and on the improving precision of measurements. The numbers given here refer to constants published in 2002 by CODATA (Mohr and Taylor, 2005). Standard errors in the last decimals are given between parentheses.

Molecular units

Convenient units for molecular simulations are based on nm for length, u (unified atomic mass units) for mass, ps for time, and the elementary charge e for charge. The unified atomic mass unit is defined as $1/12$ of the mass of a ^{12}C atom, which makes 1 u equal to 1 gram divided by Avogadro's number. The unit of energy now appears to be 1 kJ/mol $= 1$ u nm^2 ps^{-2}. There is an *electric factor* $f_{\mathrm{el}} = (4\pi\varepsilon_0)^{-1} = 138.935\,4574(14)$ kJ mol^{-1} nm e^{-2} when calculating energy and forces from charges, as in $V_{\mathrm{pot}} = f_{\mathrm{el}}\,q^2/r$. While these units are convenient, the unit of pressure (kJ mol^{-1} nm^{-3}) becomes a bit awkward, being equal to $1.666\,053\,886(28)$ MPa or $16.66\ldots$ bar.

Warning: One may not change kJ/mol into kcal/mol and nm into Å (the usual units for some simulation packages) without punishment. When keeping the u for mass, the unit of time then becomes $0.1/\sqrt{4.184}$ ps $= 48.888\,821\ldots$ fs. Keeping the e for charge, the electric factor must be expressed in kcal mol^{-1} Å e^{-2} with a value of $332.063\,7127(33)$. The unit of pressure becomes $69\,707.6946(12)$ bar! These units also form a consistent system, but we do not recommend their use.

Physical constants

In Table 9 some relevant physical constants are given in SI units; the values are those published by CODATA in 2002.[2] The same constants are given in Table 10 in atomic and molecular units. Note that in the latter table

---

[2] See Mohr and Taylor (2005) and
http://physics.nist.gov/cuu/. A Python module containing a variety of physical constants, *physcon.py*, may be downloaded from this book's or the author's website.

molar quantities are not listed: It does not make sense to list quantities in molecular-sized units per mole of material, because values in the order of 10^{23} would be obtained. The whole purpose of atomic and molecular units is to obtain "normal" values for atomic and molecular quantities.

Table 1 *Typographic conventions and special symbols*

Element	Example	Meaning
$*$	c^*	complex conjugate $c^* = a - bi$ if $c = a + bi$
$\ddagger$	$\Delta G^{\ddagger}$	transition state label
hat	$\hat{H}$	operator
overline	$\bar{u}$	(1) quantity per unit mass, (2) time average
dot	$\dot{v}$	time derivative
$\langle\,\rangle$	$\langle x \rangle$	average over ensemble
bold italic (l.c.)	$\boldsymbol{r}$	vector
bold italic (u.c.)	$\boldsymbol{Q}$	tensor of rank ≥ 2
bold roman (l.c.)	$\mathbf{r}$	one-column matrix, e.g., representing vector components
bold roman (u.c.)	$\mathbf{Q}$	matrix, e.g., representing tensor components
overline	$\bar{u}$	quantity per unit mass
overline	$\overline{M}$	multipole definition
superscript T	$\mathbf{b}^{\mathsf{T}}$	transpose of a column matrix (a row matrix)
	$\mathbf{A}^{\mathsf{T}}$	transpose of a rank-2 matrix $(A^{\mathsf{T}})_{ij} = A_{ji}$
superscript $\dagger$	$\mathbf{H}^{\dagger}$	Hermitian conjugate $(\mathbf{H}^{\dagger})_{ij} = H_{ji}^*$
d	df/dx	derivative function of f
∂	$\partial f/\partial x$	partial derivative
D	D/Dt	Lagrangian derivative $\partial/\partial t + \boldsymbol{u} \cdot \boldsymbol{\nabla}$
δ	$\delta A/\delta \rho$	functional derivative
centered dot	$\boldsymbol{v} \cdot \boldsymbol{w}$	dot product of two vectors $\mathbf{v}^{\mathsf{T}}\mathbf{w}$
$\times$	$\boldsymbol{v} \times \boldsymbol{w}$	vector product of two vectors
∇		nabla vector operator $(\partial/\partial x, \partial/\partial y, \partial/\partial z)$
grad	$\nabla \phi$	gradient $(\partial\phi/\partial x, \partial\phi/\partial y, \partial\phi/\partial z)$
div	$\nabla \cdot \boldsymbol{v}$	divergence $(\partial v_x/\partial x + \partial v_y/\partial y + \partial v_z/\partial z)$
grad	$\nabla \boldsymbol{v}$	gradient of a vector (tensor of rank 2) $(\nabla\boldsymbol{v})_{xy} = \partial v_y/\partial x$
curl	$\nabla \times \boldsymbol{v}$	$\mathbf{curl}\,\boldsymbol{v}$; $(\nabla \times \boldsymbol{v})_x = \partial v_z/\partial y - \partial v_y/\partial z$
∇^2	$\nabla^2 \Phi$	Laplacian: nabla-square or Laplace operator $(\partial^2\Phi/\partial x^2 + \partial^2\Phi/\partial y^2 + \partial^2\Phi/\partial z^2)$
$\nabla\nabla$	$\nabla\nabla\Phi$	Hessian (tensor) $(\nabla\nabla\Phi)_{xy} = \partial^2\Phi/\partial x\partial y$
tr	$\mathrm{tr}\,\mathbf{Q}$	trace of a matrix (sum of diagonal elements)
calligraphic	$\mathcal{C}$	set, domain or contour
$\mathbb{Z}$		set of all integers $(0, \pm 1, \pm 2, \ldots)$
$\mathbb{R}$		set of all real numbers
$\mathbb{C}$		set of all complex numbers
$\Re$	$\Re z$	real part of complex z
$\Im$	$\Im z$	imaginary part of complex z
$\mathbf{1}$		diagonal unit matrix or tensor

Table 2 *List of lower case Latin symbols*

symbol	meaning
a	activity
a_0	Bohr radius
c	(1) speed of light, (2) concentration (molar density)
d	infinitesimal increment, as in dx
e	(1) elementary charge, (2) number 2.1828 ...
f_{el}	electric factor $(4\pi\varepsilon_0)^{-1}$
g	metric tensor
h	(1) Planck's constant, (2) molar enthalpy
$\hbar$	Dirac's constant $(h/2\pi)$
i	$\sqrt{-1}$ (j in Python programs)
$\boldsymbol{j}$	current density
k	(1) rate constant, (2) harmonic force constant
$\boldsymbol{k}$	wave vector
k_B	Boltzmann's constant
n	(1) total quantity of moles in a mixture, (2) number density
m	mass of a particle
p	(1) pressure, (2) momentum, (3) probability density
$\boldsymbol{p}$	(1) n-dimensional generalized momentum vector, (2) momentum vector $m\boldsymbol{v}$ (3D or 3N-D)
q	(1) heat, mostly as dq, (2) generalized position, (3) charge
$[q]$	$[q_0, q_1, q_2, q_3] = [q, \boldsymbol{Q}]$ quaternions
$\boldsymbol{q}$	n-dimensional generalized position vector
$\boldsymbol{r}$	cartesian radius vector of point in space (3D or 3N-D)
s	molar entropy
t	time
u	molar internal energy
u	symbol for unified atomic mass unit (1/12 of mass ^{12}C atom)
$\boldsymbol{u}$	fluid velocity vector (3D)
v	molar volume
$\boldsymbol{v}$	cartesian velocity vector (3D or 3N-D)
w	(1) probability density, (2) work, mostly as dw
z	ionic charge in units of e
$\mathbf{z}$	point in phase space $\{\boldsymbol{q}, \boldsymbol{p}\}$

Table 3 *List of upper case Latin symbols*

Symbol	Meaning
A	Helmholtz function or Helmholtz free energy
$\boldsymbol{A}$	vector potential
B_2	second virial coefficient
$\boldsymbol{B}$	magnetic field vector
D	diffusion coefficient
$\boldsymbol{D}$	dielectric displacement vector
E	energy
$\boldsymbol{E}$	electric field vector
F	Faraday constant ($N_A\,e = 96\,485$ C)
$\boldsymbol{F}$	force vector
G	(1) Gibbs function or Gibbs free energy, (2) Green's function
H	(1) Hamiltonian, (2) enthalpy
$\boldsymbol{H}$	magnetic intensity
$\boldsymbol{I}$	moment of inertia tensor
J	Jacobian of a transformation
$\boldsymbol{J}$	flux density vector (quantity flowing through unit area per unit time)
K	kinetic energy
L	Onsager coefficients
$\mathcal{L}$	(1) Liouville operator, (2) Lagrangian
$\boldsymbol{L}$	angular momentum
M	(1) total mass, (2) transport coefficient
$\boldsymbol{M}$	(1) mass tensor, (2) multipole tensor
	(3) magnetic polarization (magnetic moment per unit volume)
N	number of particles in system
N_A	Avogadro's number
P	probability density
$\boldsymbol{P}$	(1) pressure tensor,
	(2) electric polarization (dipole moment per unit volume)
Q	canonical partition function
$\boldsymbol{Q}$	quadrupole tensor
R	gas constant ($N_A\,k_B$)
$\mathbf{R}$	rotation matrix
S	(1) entropy, (2) action
$d\boldsymbol{S}$	surface element (vector perpendicular to surface)
$\mathbf{S}$	overlap matrix
T	absolute temperature
$\boldsymbol{T}$	torque vector
U	(1) internal energy, (2) interaction energy
V	(1) volume, (2) potential energy
W	(1) electromagnetic energy density
$W_\rightarrow$	transition probability
$\boldsymbol{X}$	thermodynamic driving force vector

Table 4 *List of Greek symbols*

Symbol	Meaning
α	(1) fine structure constant, (2) thermal expansion coefficient, (3) electric polarizability
α'	polarizability volume $\alpha/(4\pi\varepsilon_0)$
β	(1) compressibility, (2) $(k_\mathrm{B}T)^{-1}$
γ	(1) friction coefficient as in $\dot{v} = -\gamma v$, (2) activity coefficient
Γ	interfacial surface tension
δ	(1) delta function, (2) Kronecker delta: δ_{ij}
Δ	small increment, as in Δx
ε	(1) dielectric constant, (2) Lennard Jones energy parameter
ε_0	vacuum permittivity
ε_r	relative dielectric constant $\varepsilon/\varepsilon_0$
η	viscosity coefficient
ζ	(1) bulk viscosity coefficient, (2) friction coefficient
κ	(1) inverse Debye length, (2) compressibility
λ	(1) wavelength, (2) heat conductivity coefficient, (3) coupling parameter
μ	(1) thermodynamic potential, (2) magnetic permeability, (3) mean of distribution
$\boldsymbol{\mu}$	dipole moment vector
μ_0	vacuum permeability
ν	(1) frequency, (2) stoichiometric coefficient
π	number $\pi = 3.1415\ldots$
Π	product over terms
$\boldsymbol{\Pi}$	momentum flux density
ρ	(1) mass density, (2) number density, (3) charge density
σ	(1) Lennard–Jones size parameter, (2) variance of distribution (3) irreversible entropy production per unit volume
$\boldsymbol{\sigma}$	stress tensor
$\sum$	sum over terms
$\boldsymbol{\Sigma}$	Poynting vector (wave energy flux density)
τ	generalized time
$\boldsymbol{\tau}$	viscous stress tensor
ϕ	wave function (generally basis function)
Φ	(1) wave function, (2) electric potential, (3) delta-response function
ψ	wave function
Ψ	wave function, generally time dependent
χ	susceptibility: electric (χ_e) or magnetic (χ_m)
χ^2	chi-square probability function
Ξ	(1) grand-canonical partition function, (2) virial
ω	angular frequency ($2\pi\nu$)
$\boldsymbol{\omega}$	angular velocity vector
Ω	microcanonical partition function

Table 5 *Defined SI units*

Quantity	Name	Symbol	Definition (year adopted by CGPM)
length	**meter**	m	distance traveled by light in vacuum in 1/299 792 458 s (1983)
mass	**kilogram**	kg	mass of international prototype kilogram in Paris (1889)
time	**second**	s	duration of 9 192 631 770 periods of hyperfine transition in ^{133}Cs atoms [at rest at 0 K, in zero magnetic field] (1967)
current	**ampere**	A	current in two infinitely long and thin conductors at 1 m distance that exert a mutual force of 2×10^{-7} N/m (1948)
temperature	**kelvin**	K	1/273.16 of thermodynamic temperature of triple point of water (1967)
quantity	**mole**	mol	quantity of matter with as many specified elementary entities as there are atoms in 0.012 kg pure ^{12}C (1971)
light intensity	**candela**	cd	intensity of light source emitting 1/683 W/sr radiation with frequency 540×10^{12} Hz (1979)

Table 6 *Derived named SI units*

Quantity	Symbol	Name	Unit
planar angle	$\alpha, \dots$	**radian**	rad (circle $= 2\pi$)
solid angle	ω, Ω	**steradian**	sr (sphere$= 4\pi$)
frequency	ν, f	**hertz**	$\mathrm{Hz} = \mathrm{s}^{-1}$
force	F	**newton**	$\mathrm{N} = \mathrm{kg\,m\,s}^{-2}$
pressure	p	**pascal**	$\mathrm{Pa} = \mathrm{N/m}^2$
energy	E, U, w	**joule**	$\mathrm{J} = \mathrm{N\,m} = \mathrm{kg\,m}^2\,\mathrm{s}^{-2}$
power	P, W	**watt**	$\mathrm{J\,s} = \mathrm{kg\,m}^2\,\mathrm{s}^{-1}$
charge	q, Q	**coulomb**	$\mathrm{C} = \mathrm{A\,s}$
electric potential	V, Φ	**volt**	$\mathrm{V} = \mathrm{J/C}$
capacity	C	**farad**	$\mathrm{F} = \mathrm{C/V}$
resistance	R	**ohm**	$\Omega = \mathrm{V/A}$
conductance	G	**siemens**	$\mathrm{S} = \Omega^{-1}$
inductance	L	**henry**	$\mathrm{H} = \mathrm{Wb/A}$
magnetic flux	Φ	**weber**	$\mathrm{Wb} = \mathrm{V\,s}$
magnetic field	B	**tesla**	$\mathrm{T} = \mathrm{Wb/m}^2$

Table 7 *Atomic units (a.u.)*

Quantity	Symbol	Value in SI unit
mass	m_e	$9.109\,3826(16) \times 10^{-31}$ kg
length	a_0	$5.291\,772\,108(18) \times 10^{-11}$ m
time	$m_e a_0^2/\hbar$	$2.418\,884\,326505(16) \times 10^{-17}$ s,
velocity	αc	$2.187\,691\,2633(73) \times 10^6$ m/s
energy	$\hbar^2/(m_e a_0^2)$	$4.359\,744\,17(75) \times 10^{-18}$ J
	(E_h)	$= 27.211\,3845(23)$ eV
	(hartree)	$= 2\,625.499\,63(45)$ kJ/mol
		$= 627.509\,47(11)$ kcal/mol
force	E_h/a_0	$8.238\,7225(14) \times 10^{-8}$ N
charge	e	$1.602\,176\,53(14) \times 10^{-19}$ C,
current	a.u.	$6.623\,617\,82(57) \times 10^{-3}$ A
electric potential	a.u.	$27.211\,3845(23)$ V
electric field	a.u.	$5.142\,206\,42(44) \times 10^{11}$ V/m
electric field gradient	a.u.	$9.717\,361\,82(83) \times 10^{21}$ V m^{-2}
dipole moment	a.u.	$8.478\,353\,09(73) \times 10^{-30}$ C m
		$= 2.541\,746\,31(22)$ Debye
quadrupole moment	a.u.	$4.486\,551\,24(39) \times 10^{-40}$ C m^2
electric polarizability	a.u.	$1.648\,777\,274(16) \times 10^{-41}$ F m^2
$\alpha' = \alpha/(4\pi\varepsilon_0)$	a.u.	$a_0^3 = 1.481\,847\,114(15) \times 10^{-31}$ m^3

Table 8 *Molecular units (m.u.)*

quantity	symbol	value in SI unit
mass	u	$1.66053886(28) \times 10^{-27}$ kg
length	nm	1×10^{-9} m
time	ps	1×10^{-12} s,
velocity	nm/ps	1000 m/s
energy	kJ/mol	$1.660\,538\,86(28) \times 10^{-21}$ J
		$= 0.010\,364\,268\,99(85)$ eV
		$= 0.239\,005\,736\ldots$ kcal/mol
force	kJ mol^{-1} nm^{-1}	$1.660\,538\,86(28) \times 10^{-12}$ N
charge	e	$1.602\,176\,53(14) \times 10^{-19}$ C,
current	e/ps	$1.602\,176\,53(14) \times 10^{-7}$ A
electric potential	kJ mol^{-1} e^{-1}	$0.010\,364\,268\,99(85)$ V
electric field	kJ mol^{-1} e^{-1} nm^{-1}	$1.036\,426\,899(85) \times 10^7$ V/m
electric field gradient	kJ mol^{-1} e^{-1} nm^{-2}	$1.036\,426\,899(85) \times 10^{16}$ V m^{-2}
dipole moment	e nm	$1.602\,176\,53(14) \times 10^{-28}$ C m
		$= 48.032\,0440(42)$ Debye
quadrupole moment	e nm^2	$1.602\,176\,53(14) \times 10^{-37}$ C m^2
electric polarizability	e^2 nm^2 kJ^{-1} mol	$1.545\,865\,44(26) \times 10^{-35}$ F m^2
$\alpha' = \alpha/(4\pi\varepsilon_0)$	nm^3	1×10^{-27} m^3

Table 9 *Some physical constants in SI units (CODATA 2002)*

Constant		Equivalent	Value in SI units
magnetic constant[a]	μ_0		$4\pi \times 10^{-7}$ (ex) N/A^2
electric constant[b]	ε_0	$(\mu_0 c^2)^{-1}$	$8.854\,187\,818... \times 10^{-12}$ F/m
electric factor[c]	f_{el}	$(4\pi\varepsilon_0)^{-1}$	$8.987\,551\,787... \times 10^9$ m/F
velocity of light	c	def	$299\,792\,458$(ex) m/s
gravitation constant[d]	G	fund	$6.6742(10) \times 10^{-11}$ m^3 kg^{-1} s^{-1}
Planck constant	h	fund	$6.626\,0693(11) \times 10^{-34}$ J s
Dirac constant	$\hbar$	$h/2\pi$	$1.054\,571\,68(18) \times 10^{-34}$ J s
electron mass	m_e	fund	$9.109\,3826(16) \times 10^{-31}$ kg
elementary charge	e	fund	$1.602\,176\,53(14) \times 10^{-19}$ C
unified a.m.u.[e]	u	fund	$1.660\,53886(28) \times 10^{-27}$ kg
proton mass	m_p	fund	$1.672\,621\,71(29) \times 10^{-27}$ kg
neutron mass	m_n	fund	$1.674\,927\,28(29) \times 10^{-27}$ kg
deuteron mass	m_d	fund	$3.343\,583\,35(57) \times 10^{-27}$ kg
muon mass	m_μ	fund	$1.883\,531\,40(33) \times 10^{-28}$ kg
^{1}H atom mass	m_H	fund	$1.673\,532\,60(29) \times 10^{-27}$ kg
fine-structure const.	α	$e^2/(2\varepsilon_0 hc)$	$7.297\,352\,568(24) \times 10^{-3}$
—, inverse	α^{-1}	$2\varepsilon_0 hc/e^2$	$137.035\,999\,11(46)$
Bohr radius	a_0	$\hbar/(\alpha c m_e)$	$5.291\,772\,108(18) \times 10^{-11}$ m
Rydberg constant[f]	R_∞	$\alpha^2 m_e c/2h$	$1.097\,373\,156\,8525(73) \times 10^7$ m^{-1}
Bohr magneton	μ_B	$e\hbar/2m_e$	$9.274\,009\,49(80) \times 10^{-24}$ J/T
Boltzmann constant	k_B		$1.380\,6505(24) \times 10^{-23}$ J/K
ideal gas volume[g]	v_m^0	$k_B T^0/p^0$	$3.771\,2467(66) \times 10^{-26}$ m^3
Avogadro constant	N_A	$0.001\,\text{kg}/u$	$6.022\,1415(10) \times 10^{23}$ mol^{-1}
Faraday constant	F	$N_A e$	$96\,485.3383(83)$ C/mol
molar gas constant	R	$N_A k_B$	$8.314\,472(15)$ J mol^{-1} K^{-1}
molar gas volume[h]	V_m^0	RT^0/p^0	$22.710\,981(40) \times 10^{-3}$ m^3/mol

[a] also called *vacuum permeability.*
[b] also called *vacuum permittivity* or *vacuum dielectric constant.*
[c] as in $F = f_{el} q_1 q_2/r^2$.
[d] as in $F = G m_1 m_2/r^2$.
[e] atomic mass unit, defined as 1/12 of the mass of a ^{12}C atom
[f] very accurately known: relative uncertainty is 6.6×10^{-12}.
[g] volume per molecule of an ideal gas at a temperature of $T^0 = 273.15$ K and a pressure of $p^0 = 10^5$ Pa. An alternative, but now outdated, standard pressure is $101\,325$ Pa.
[h] volume per mole of ideal gas under standard conditions; see previous note.

Table 10 *Physical constants in atomic units and "molecular units"*

Symbol	Value in a.u.	Value in m.u.
μ_0	$6.691\,762\,564(44) \times 10^{-4}$	$1.942\,591\,810(19) \times 10^{-8}$
ε_0	$1/(4\pi)$	$5.727\,657\,506(58) \times 10^{-4}$
f_{el}	$1(\mathrm{ex})$	$138.935\,4574(14)$
c	$137.035\,99911(46)$	$299\,792.458(\mathrm{ex})$
G	$4.222\,18(63) \times 10^{-32}$	$1.108\,28(17) \times 10^{-34}$
h	2π	$0.399\,031\,2716(27)$
$\hbar$	$1(\mathrm{ex})$	$0.063\,507\,799\,32(43)$
m_{e}	$1(\mathrm{ex})$	$5.485\,799\,0945(24) \times 10^{-4}$
e	$1(\mathrm{ex})$	$1(\mathrm{ex})$
u	$1\,822.888\,484\,93(80)$	$1(\mathrm{ex})$
m_{p}	$1\,836.152\,672\,61(85)$	$1.007\,276\,46688(13)$
m_{n}	$1\,838.683\,6598(13)$	$1.008\,664\,915\,60(55)$
m_{d}	$3\,670.482\,9652(18)$	$2.013\,553\,212\,70(35)$
m_{μ}	$206.768\,2838(54)$	$0.113\,428\,9264(30)$
m_{H}	$1\,837.152\,645\,89(85)$	$1.007\,825\,032\,13(13)$
α	$7.297\,352\,568(24) \times 10^{-3}$	$7.297\,352\,568(24) \times 10^{-3}$
α^{-1}	$137.035\,999\,11(46)$	$137.035\,999\,11(46)$
a_0	$1\ (\mathrm{ex})$	$5.291\,772\,108(18) \times 10^{-2}$
R_{∞}	$0.5(\mathrm{ex})$	$0.010\,973\,731\,568\,525(73)$
μ_{B}	$0.5(\mathrm{ex})$	$57.883\,818\,04(39)$
k_{B}	$3.166\,8154(55) \times 10^{-6}$	$0.008\,314\,472(15)$
v_{m}^0	$254\,496.34(44)$	$37.712\,467(66)$

Part I

A Modeling Hierarchy for Simulations

1

Introduction

1.1 What is this book about?

1.1.1 Simulation of real systems

Computer simulations of real systems require a *model* of that reality. A model consists of both *a representation of the system* and *a set of rules* that describe the behavior of the system. For dynamical descriptions one needs in addition a specification of the *initial state* of the system, and if the response to external influences is required, a specification ofthe *external influences*.

Both the model and the method of solution depend on the *purpose* of the simulation: they should be *accurate* and *efficient*. The model should be chosen accordingly. For example, an accurate quantum-mechanical description of the behavior of a many-particle system is not efficient for studying the flow of air around a moving wing; on the other hand, the Navier–Stokes equations – efficient for fluid motion – cannot give an accurate description of the chemical reaction in an explosion motor. Accurate means that the simulation will reliably (within a required accuracy) predict the real behavior of the real system, and efficient means "feasible with the available technical means." This combination of requirements rules out a number of questions; whether a question is answerable by simulation depends on:

- the state of theoretical development (models and methods of solution);
- the computational capabilities;
- the possibilities to implement the methods of solution in algorithms;
- the possibilities to validate the model.

Validation means the assessment of the accuracy of the model (compared to physical reality) by critical experimental tests. Validation is a crucial part of modeling.

1.1.2 System limitation

We limit ourselves to models of the real world around us. This is the realm of chemistry, biology and material sciences, and includes all industrial and practical applications. We do not include the formation of stars and galaxies (*stellar dynamics*) or the physical processes in hot plasma on the sun's surface (*astrophysics*); neither do we include the properties and interactions of elementary particles (*quantum chromodynamics*) or processes in atomic nuclei or neutron stars. And, except for the purposes of validation and demonstration, we shall not consider unrealistic models that are only meant to test a theory. To summarize: we shall look at literally "down-to-earth" systems consisting of atoms and molecules under non-extreme conditions of pressure and temperature.

This limits our discussion in practice to systems that are made up of interacting *atomic nuclei*, which are specified by their mass, charge and spin, *electrons*, and *photons* that carry the electromagnetic interactions between the nuclei and electrons. Occasionally we may wish to add gravitational interactions to the electromagnetic ones. The internal structure of atomic nuclei is of no consequence for the behavior of atoms and molecules (if we disregard radioactive decay): nuclei are so small with respect to the spatial spread of electrons that only their *monopole* properties as total charge and total mass are important. Nuclear excited states are so high in energy that they are not populated at reasonable temperatures. Only the spin degeneracy of the nuclear ground state plays a role when nuclear magnetic resonance is considered; in that case the nuclear magnetic dipole and electric quadrupole moment are important as well.

In the normal range of temperatures this limitation implies a practical division between electrons on the one hand and nuclei on the other: while all particles obey the rules of quantum mechanics, the quantum character of electrons is essential but the behavior of nuclei approaches the classical limit. This distinction has far-reaching consequences, but it is rough and inaccurate. For example, protons are light enough to violate the classical rules. The validity of the classical limit will be discussed in detail in this book.

1.1.3 Sophistication versus brute force

Our interest in *real* systems rather than simplified model systems is consequential for the kind of methods that can be used. Most real systems concern some kind of condensed phase: they (almost) never consist of isolated molecules and can (almost) never be simplified because of inherent symmetry. Interactions between particles can (almost) never be described by mathematically simple forms and often require numerical or tabulated descriptions. Realistic systems usually consist of a very large number of interacting particles, embedded in some kind of environment. Their behavior is (almost) always determined by statistical averages over ensembles consisting of elements with random character, as the random distribution of thermal kinetic energy over the available degrees of freedom. That is why statistical mechanics plays a crucial role in this book.

The complexity of real systems prescribes the use of methods that are easily extendable to large systems with many degrees of freedom. Physical theories that apply to simple models only, will (almost) always be useless. Good examples are the very sophisticated statistical-mechanical theories for atomic and molecular fluids, relating fluid structural and dynamic behavior to interatomic interactions. Such theories work for atomic fluids with simplified interactions, but become inaccurate and intractable for fluids of polyatomic molecules or for interactions that have a complex form. While such theories thrived in the 1950s to 1970s, they have been superseded by accurate simulation methods, which are faster and easier to understand, while they predict liquid properties from interatomic interactions much more accurately. Thus sophistication has been superseded by brute force, much to the dismay of the sincere basic scientist.

Many mathematical tricks that employ the simplicity of a toy model system cannot be used for large systems with realistic properties. In the example below the brute-force approach is applied to a problem that has a simple and elegant solution. To apply such a brute-force method to a simple problem seems outrageous and intellectually very dissatisfying. Nevertheless, the elegant solution cannot be readily extended to many particles or complicated interactions, while the brute-force method can. Thus not only sophistication in physics, but also in mathematics, is often replaced by brute force methods. There is an understandable resistance against this trend among well-trained mathematicians and physicists, while scientists with a less elaborate training in mathematics and physics welcome the opportunity to study complex systems in their field of application. The field of simulation has made theory much more widely applicable and has become accessible to a much wider range of scientists than before the "computer age." Simulation has become a "third way" of doing science, not instead of, but in addition to theory and experimentation.

There is a danger, however, that applied scientists will use "standard" simulation methods, or even worse use "black-box" software, without realizing on what assumptions the methods rest and what approximations are

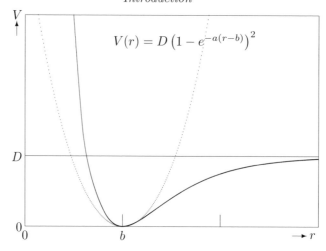

Figure 1.1 Morse curve with $a = 2/b$ (solid curve). Dotted curve: parabola with same curvature as Morse curve at $r = b$: $V = Da^2(r - b)^2$.

implied. This book is meant to provide the necessary scientific background and to promote awareness for the limitations and inaccuracies of simulating the "real world".

Example: An oscillating bond
In this example we use brute-force simulation to attack a problem that could be approached analytically, albeit with great difficulty. Consider the classical bond length oscillation of a simple diatomic molecule, using the molecule hydrogen fluoride (HF) as an example. In the simplest approximation the potential function is a parabola:

$$V(r) = \tfrac{1}{2}k(r - b)^2, \tag{1.1}$$

with r the H–F distance, k the force constant and b the equilibrium distance. A better description of the potential function is the *Morse function* (see Fig. 1.1)

$$V(r) = D\left(1 - e^{-a(r-b)}\right)^2, \tag{1.2}$$

where D is the dissociation energy and a is a constant related to the steepness of the potential. The Morse curve is approximated near the minimum at $r = b$ by a parabola with force constant $k = 2Da^2$.

The Morse curve (Morse, 1929) is only a convenient analytical expression that has some essential features of a diatomic potential, including a fairly good agreement with vibration spectra of diatomic molecules, but there is no theoretical justification for this particular form. In many occasions we may not even have an analytical form for the potential, but know the potential at a number of discrete points, e.g., from quantum-chemical calculations. In that case the best way to proceed is to construct the potential function from *cubic spline interpolation* of the computed points. Be-

Table 1.1 *Data for hydrogen fluoride*

mass H	m_{H}	1.0079	u
mass F	m_{F}	18.9984	u
dissocation constant	D	569.87	kJ/mol
equilibrium bond length	b	0.09169	nm
force constant	k	5.82×10^5	$\mathrm{kJ\,mol^{-1}\,nm^{-2}}$

cause cubic splines (see Chapter 19) have continuous second derivatives, the forces will behave smoothly as they will have continuous first derivatives everywhere.

A little elementary mechanics shows that we can split off the translational motion of the molecule as a whole, and that – in the absence of rotational motion – the bond will vibrate according to the equation of motion:

$$\mu\ddot{r} = -\frac{dV}{dr}, \tag{1.3}$$

where $\mu = m_{\mathrm{H}}m_{\mathrm{F}}/(m_{\mathrm{H}} + m_{\mathrm{F}})$ is the reduced mass of the two particles. When we start at time $t = 0$ with a displacement Δr and zero velocity, the solution for the harmonic oscillator is

$$r(t) = b + \Delta r \cos \omega t, \tag{1.4}$$

with $\omega = \sqrt{k/\mu}$. So the analytical solution is simple, and we do not need any numerical simulation to derive the frequency of the oscillator. For the Morse oscillator the solution is not as straightforward, although we can predict that it should look much like the harmonic oscillator with $k = 2Da^2$ for small-amplitude vibrations. But we may expect anharmonic behavior for larger amplitudes. Now numerical simulation is the easiest way to derive the dynamics of the oscillator. For a spline-fitted potential we *must* resort to numerical solutions. The extension to more complex problems, like the vibrations of a molecule consisting of several interconnected harmonic oscillators, is quite straightforward in a simulation program, while analytical solutions require sophisticated mathematical techniques.

The reader is invited to write a simple molecular dynamics program that uses the following very general routine `mdstep` to perform one dynamics step with the *velocity-Verlet* algorithm (see Chapter 6, (6.83) on page 191). Define a function `force(r)` that provides an array of forces $\boldsymbol{F}$, as well as the total potential energy V, given the coordinates $\boldsymbol{r}$, both for the harmonic and the Morse potential. You may start with a one-dimensional version. Try out a few initial conditions and time steps and look for energy conservation and stability in long runs. As a rule of thumb: start with a time step such that the fastest oscillation period contains 50 steps (first compute what the oscillation period will be). You may generate curves like those in Fig. 1.2. See what happens if you give the molecule a rotational velocity! In this case you of course need a two- or three-dimensional version. Keep to "molecular units": *mass*: u, *length*: nm, *time*: ps, *energy*: kJ/mol. Use the data for hydrogen fluoride from Table 1.1.

The following function performs one velocity-Verlet time step of MD on a system of n particles, in m (one or more) dimensions. Given initial positions r, velocities v and forces F (at position r), each as arrays of shape (n, m), it returns r, v, F and

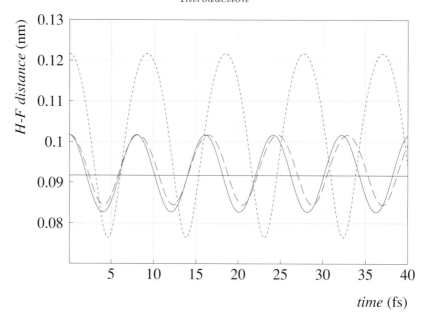

Figure 1.2 Oscillation of the HF bond length, simulated with the harmonic oscillator (solid curve) and the Morse curve (long dash), both with initial deviation from the equilibrium bond length of 0.01 nm, Dotted curve: Morse oscillator with initial deviation of 0.03 nm, showing increased anharmonic behavior. Note that the frequency of the Morse oscillator is lower than that of the harmonic oscillator. A time step of 0.2 fs was used; the harmonic oscillator simulation is indistinguishable from the analytical solution.

the potential energy V one time step later. For convenience in programming, the inverse mass should be given as an array of the same shape (n, m) with repeats of the same mass for all m dimensions. In Python this $n \times m$ array `invmass` is easily generated from a one-dimensional array `mass` of arbitrary length n:

`invmass=reshape(repeat(1./mass,m),(alen(mass),m)),`

or equivalently

`invmass=reshape((1./mass).repeat(m),(alen(mass),m))`

An external function `force(r)` must be provided that returns $[F, V]$, given r. V is not actually used in the time step; it may contain any property for further analysis, even as a list.

PYTHON PROGRAM 1.1 **mdstep(invmass,r,v,F,force,delt)**
General velocity-Verlet Molecular Dynamics time step

```
01  def mdstep(invmass,r,v,F,force,delt):
02  # invmass: inverse masses [array (n,m)] repeated over spatial dim. m
03  # r,v,F: initial coordinates, velocities, forces [array (n,m)]
04  # force(r): external routine returning [F,V]
05  # delt: timestep
```

```
06  # returns [r,v,F,V] after step
07    v=v+0.5*delt*invmass*F
08    r=r+v*delt
09    FV=force(r)
10    v=v+0.5*delt*invmass*FV[0]
11    return [r,v,FV[0],FV[1]]
```

Comments
As mentioned in the *Preface* (page xiii), it is assumed that *scipy* has been imported. The initial values of r, v, F, V are valid at the time before the step, and normally available from the output of the previous step. To start the run, the routine `force(r)` must have been called once to initiate F. The returned values are valid at the end of the step. The arguments are not modified in place.

1.2 A modeling hierarchy

The behavior of a system of particles is in principle described by the rules of relativistic quantum mechanics. This is – within the limitation of our system choices – the highest level of description. We shall call this *level 1*. All other levels of description, such as considering atoms and molecules instead of nuclei and electrons, classical dynamics instead of quantum dynamics, or continuous media instead of systems of particles, represent approximations to level 1. These approximations can be ordered in a hierarchical sense from fine atomic detail to coarse macroscopic behavior. Every lower level loses detail and loses applicability or accuracy for a certain class of systems and questions, but gains applicability or efficiency for another class of systems and questions. The following scheme lists several levels in this hierarchy.

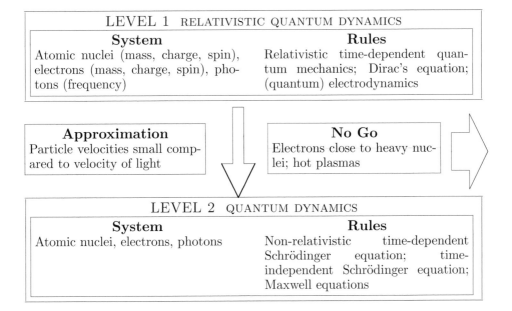

<table>
<tr><td>

Approximation
Born–Oppenheimer approx.: electrons move much faster than nuclei

</td><td>

</td><td>

No Go
Electron dynamics (e.g., in semiconductors); fast electron transfer processes; dynamic behavior of excited states

</td><td>

</td></tr>
</table>

LEVEL 3 ATOMIC QUANTUM DYNAMICS

System	**Rules**
Atoms, ions, molecules, (photons)	Atoms move in effective potential due to electrons; atoms may behave according to time-dependent Schrödinger equation

<table>
<tr><td>

Approximation
Atomic motion is classical

</td><td>

</td><td>

No Go
Proton transfer; hydrogen and helium at low temperatures; fast reactions and high-frequency motions

</td><td>

</td></tr>
</table>

LEVEL 4 MOLECULAR DYNAMICS

System	**Rules**
Condensed matter: (macro)molecules, fluids, solutions, liquid crystals, fast reactions	Classical mechanics (Newton's equations); statistical mechanics; molecular dynamics

<table>
<tr><td>

Approximation
Reduce number of degrees of freedom

</td><td>

</td><td>

No Go
Details of fast dynamics, transport properties

</td><td>

</td></tr>
</table>

LEVEL 5 GENERALIZED LANGEVIN DYNAMICS ON REDUCED SYSTEM

System	**Rules**
Condensed matter: large molecular aggregates, polymers, defects in solids, slow reactions	Superatoms, reaction coordinates; averaging over local equilibrium, constraint dynamics, free energies and potentials of mean force.

<table>
<tr><td>

Approximation
Neglect time correlation and/or spatial correlation in fluctuations

</td><td>

</td><td>

No Go
Correlations in motion, short-time accuracy

</td><td>

</td></tr>
</table>

LEVEL 6   SIMPLE LANGEVIN DYNAMICS

System	**Rules**
"Slow" dynamic (non-equilibrium) processes and reactions	Accelerations given by systematic force, friction, and noise; Fokker–Planck equations

Approximation	**No Go**
Neglect inertial terms: coarse graining in time	Dynamic details

LEVEL 7   BROWNIAN DYNAMICS

System	**Rules**
Coarse-grained non-equilibrium processes; colloidal systems; polymer systems	Velocities given by force and friction, plus noise; Brownian (diffusive) dynamics; Onsager flux/force relations

Approximation	**No Go**
Reduce description to continuous densities of constituent species	Details of particles

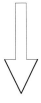

LEVEL 8   MESOSCOPIC DYNAMICS

System	**Rules**
As for level 7: self-organizing systems; reactive non-equilibrium systems	Density description: mass conservation plus dynamic flux equation, with noise.

Approximation	**No Go**
Average over "infinite" number of particles	Spontaneous structure formation driven by fluctuations

LEVEL 9   REACTIVE FLUID DYNAMICS

System	**Rules**
Non-equilibrium macroscopic mixture of different species (as the atmosphere for weather forecasting	Energy, momentum and mass conservation; reactive fluxes

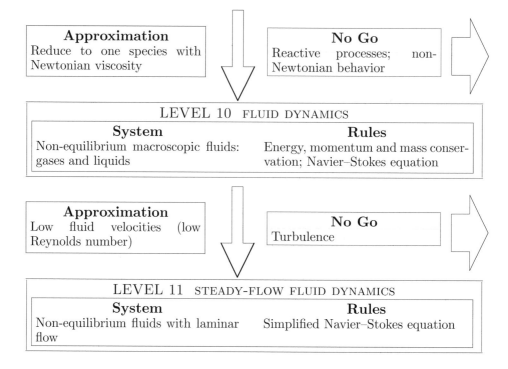

From level 5 onward, not all atomic details are included in the system description: one speaks of *coarse graining in space*. From level 6 onward dynamic details on a short time scale are disregarded by *coarse graining in time*.

In the last stages of this hierarchy (levels 8 to 11), the systems are not modeled by a set of particles, but rather by properties of a *continuum*. Equations describe the time evolution of the continuum properties. Usually such equations are solved on a spatial grid using finite difference or finite elements methods for discretizing the continuum equations. A different approach is the use of particles to represent the continuum equations, called *dissipative particle dynamics* (DPD). The particles are given the proper interactions representing the correct physical properties that figure as parameters in the continuum equations.

Note that we have considered dynamical properties at all levels. Not all questions we endeavor to answer involve dynamic aspects, such as the prediction of static equilibrium properties (e.g., the binding constant of a ligand to a macromolecule or a solid surface). For such static questions the answers may be found by sampling methods, such as Monte Carlo simulations, that generate a representative statistical ensemble of system configurations rather than a trajectory in time. The ensemble generation makes use of random displacements, followed by an acceptance or rejection based on a probabilistic criterion that ensures *detailed balance* between any pair of configurations: the ratio of forward and backward transition probabilities is made equal to the ratio of the required probabilities of the two configurations. In this book the emphasis will be on dynamic methods; details on Monte Carlo methods can be found in Allen and Tildesley (1987) or Frenkel and Smit (2002) for chemically oriented applications and in Binder and Heermann (2002) or Landau and Binder (2005) for physically oriented applications.

1.3 Trajectories and distributions

Dynamic simulations of many-particle systems contain fluctuations or stochastic elements, either due to the irrelevant particular choice of initial conditions (as the exact initial positions and velocities of particles in a classical simulation or the specification of the initial wave function in a quantum-dynamical simulation), or due to the "noise" added in the method of solution (as in Langevin dynamics where a stochastic force is added to replace forces due to degrees of freedom that are not explicitly represented). Fluctuations are implicit in the dynamic models up to and including level 8.

The precise details of a particular trajectory of the particles have no relevance for the problem we wish to solve. What we need is always an average over many trajectories, or at least an average property, such as the average or the variance of a single observable or a correlation function, over one long trajectory. In fact, an individual trajectory may even have *chaotic* properties: two trajectories with slightly different initial conditions may deviate drastically after a sufficiently long time. However, the average behavior is deterministic for most physical systems of interest.

Instead of generating distribution functions and correlation functions from trajectories, we can also try to define equations, such as the Fokker–Planck equation, for the distribution functions (probability densities) or correlation functions themselves. Often the latter is very much more difficult than generating the distribution functions from particular trajectories. An exception is the generation of equilibrium distributions, for which Monte Carlo methods are available that circumvent the necessity to solve specific equations for the distribution functions. Thus the simulation of trajectories is often the most efficient – if not the only possible – way to generate the desired average properties.

While the notion of a trajectory as the time evolution of positions and velocities of all particles in the system is quite valid and clear in classical mechanics, there is no such notion in quantum mechanics. The description of a system in terms of a wave function Ψ is by itself a description in terms of a *probability density*: $\Psi^*\Psi(\boldsymbol{r}_1, \ldots, \boldsymbol{r}_n, t)$ is the probability density that the particles $1, \ldots, n$ are at positions $\boldsymbol{r}_1, \ldots, \boldsymbol{r}_n$ at time t. Even if the initial state is precisely defined by a sharp wave function, the wave function evolves under the quantum-dynamical equations to yield a probability distribution rather than a precise trajectory. From the wave function evolution expectation values (i.e., average properties over a probability distribution) of physical observables can be obtained by the laws of quantum mechanics, but the wave function cannot be interpreted as the (unmeasurable) property of a single particle.

Such a description fits in well with equations for the evolution of probability distributions in classical systems, but it is not compatible with descriptions in terms of classical trajectories. This fundamental difference in interpretation lies at the basis of the difficulties we encounter if we attempt to use a hybrid quantum/classical description of a complex system. If we insist on a trajectory description, the quantum-dynamical description should be reformulated by some kind of contraction and sampling to yield trajectories that have the same statistical properties as prescribed by the quantum evolution. It is for the same reason of incompatibility of quantum descriptions and trajectories that quantum corrections to classical trajectories cannot be unequivocally defined, while quantum corrections to equilibrium probability distributions can be systematically derived.

1.4 Further reading

While Part I treats most of the theoretical models behind simulation and Part II provides a fair amount of background knowledge, the interested reader may feel the need to consult standard texts on further background material, or consult books on aspects of simulation and modeling that are not treated in this book. The following literature may be helpful.

1 S. Gasiorowicz, *Quantum Physics* (2003) is a readable, over 30 years old but updated, textbook on quantum physics with a discussion of the limits of classical physics.

2 L. I. Schiff, *Quantum Mechanics* (1968). A compact classic textbook, slightly above the level of Gasiorowicz.

3 E. Merzbacher, *Quantum Mechanics* (1998) is another classic textbook with a complete coverage of the main topics.

4 L. D. Landau and E.M. Lifshitz, *Quantum Mechanics (Non-relativis-*

tic Theory) (1981). This is one volume in the excellent series "Course of Theoretical Physics." Its level is advanced and sophisticated.

5 P. A. M. Dirac, *The Principles of Quantum Mechanics* (1958). By one of the founders of quantum mechanics: advisable reading only for the dedicated student.

6 F. S. Levin, *An Introduction to Quantum Theory* (2002) introduces principles and methods of basic quantum physics at great length. It has a part on "complex systems" that does not go far beyond two-electron atoms.

7 A. Szabo and N. S. Ostlund, *Modern Quantum Chemistry* (1982) is a rather complete textbook on quantum chemistry, entirely devoted to the solution of the time-independent Schrödinger equation for molecules.

8 R. McWeeny, *Methods of Molecular Quantum Mechanics* (1992) is the classical text on quantum chemistry.

9 R. G. Parr and W. Yang, *Density Functional Theory* (1989). An early, and one of the few books on the still-developing area of density-functional theory.

10 F. Jensen, *Introduction to Computational Chemistry* (2006). First published in 1999, this is a modern comprehensive survey of methods in computational chemistry including a range of ab initio and semi-empirical quantum chemistry methods, but also molecular mechanics and dynamics.

11 H. Goldstein, *Classical Mechanics* (1980) is the classical text and reference book on mechanics. The revised third edition (Goldstein *et al.*, 2002) has an additional chapter on chaos, as well as other extensions, at the expense of details that were present in the first two editions.

12 L. D. Landau and E. M. Lifshitz, *Mechanics* (1982). Not as complete as Goldstein, but superb in its development of the theory.

13 L. D. Landau and E. M. Lifshitz, *Statistical Physics* (1996). Basic text for statistical mechanics.

14 K. Huang, *Statistical Mechanics* (2nd edn, 1987). Statistical mechanics textbook from a physical point of view, written before the age of computer simulation.

15 T. L. Hill, *Statistical Mechanics* (1956). A classic and complete, but now somewhat outdated, statistical mechanics textbook with due attention to chemical applications. Written before the age of computer simulation.

16 D. A. McQuarrie, *Statistical Mechanics* (1973) is a high-quality text-book, covering both physical and chemical applications.

17 M. Toda, R. Kubo and N. Saito, *Statistical Physics. I. Equilibrium Statistical Mechanics* (1983) and R. Kubo, M. Toda and N. Hashit-sume *Statistical Physics. II. Nonequilibrium Statistical Mechanics* (1985) emphasize physical principles and applications. These texts were originally published in Japanese in 1978. Volume II in particular is a good reference for linear response theory, both quantum-mechanical and classical, to which Kubo has contributed significantly. It describes the connection between correlation functions and macro-scopic relaxation. Not recommended for chemists.

18 D. Chandler, *Introduction to Modern Statistical Mechanics* (1987). A basic statistical mechanics textbook emphasizing fluids, phase transitions and reactions, written in the age of computer simulations.

19 B. Widom, *Statistical Mechanics, A Concise Introduction for Chemists* (2002) is what it says: an introduction for chemists. It is well-written, but does not reach the level to treat the wonderful inventions in computer simulations, such as particle insertion methods, for which the author is famous.

20 M. P. Allen and D. J. Tildesley, *Computer Simulation of Liquids* (1987). A practical guide to molecular dynamics simulations with emphasis on the methods of solution rather than the basic underlying theory.

21 D. Frenkel and B. Smit, *Understanding Molecular Simulation* (2002). A modern, instructive, and readable book on the principles and practice of Monte Carlo and molecular dynamics simulations.

22 D. P. Landau and K. Binder, *A Guide to Monte Carlo Simulations in Statistical Physics* (2005). This book provides a detailed guide to Monte Carlo methods with applications in many fields, from quantum systems to polymers.

23 N. G. van Kampen, *Stochastic Processes in Physics andChemistry* (1981) gives a very precise and critical account of the use of stochastic and Fokker–Planck type equations in (mostly) physics and (a bit of) chemistry.

24 H. Risken, *The Fokker–Planck equation* (1989) treats the evolution of probability densities.

25 C. W. Gardiner, *Handbook of Stochastic Methods for Physics, Chemistry and the Natural Sciences* (1990) is a reference book for modern developments in stochastic dynamics. It treats the relations between stochastic equations and Fokker–Planck equations.

26 M. Doi and S. F. Edwards, *The Theory of Polymer Dynamics* (1986) is the already classic introduction to mesoscopic treatment of polymers.

27 L. D. Landau and E. M. Lifshitz, *Fluid Mechanics* (1987) is an excellent account of the physics behind the equations of fluid dynamics.

28 T. Pang, *Computational Physics* (2006). First published in 1997, this is a modern and versatile treatise on methods in computational physics, covering a wide range of applications. The emphasis is on the computational aspects of the methods of solution, not on the physics behind the models.

29 F. J. Vesely, *Computational Physics, An Introduction* (2nd ed., 2001) is an easily digestable treatment of computational problems in physics, with emphasis on mathematical and computational methodology rather than on the physics behind the equations.

30 M. Griebel, S. Knapek, G. Zumbusch and A. Caglar, *Numerische Simulation in der Moleküldynamik* (2003) gives many advanced details on methods and algorithms for dynamic simulation with particles. The emphasis is on computational methods including parallelization techniques; programs in C are included. Sorry for some readers: the text is in German.

31 D. Rapaport, *The Art of Molecular Dynamics Simulation* (2004) is the second, reworked edition of a detailed, and readable, account of classical molecular dynamics methods and applications.

32 M. M. Woolfson and G. J. Pert, *An Introduction to Computer Simulation* (1999) is not on models but on methods, from solving partial differential equations to particle simulation, with accessible mathematics.

33 A. R. Leach, *Molecular Modelling, Principles and Applications* (1996) aims at the simulation of molecular systems leading up to drug discovery. Starting with quantum chemistry, the book decribes energy minimization, molecular dynamics and Monte Carlo methods in detail.

34 C. J. Cramer, *Essentials of Computational Chemistry* (2004) is the second edition of a detailed textbook of modern computational chemistry including quantum methods, simulation, optimization and reaction dynamics.

2

Quantum mechanics: principles and relativistic effects

Readers who are not sensitive to the beauty of science can skip this entire chapter, as nothing is said that will help substantially to facilitate the solution of practical problems!

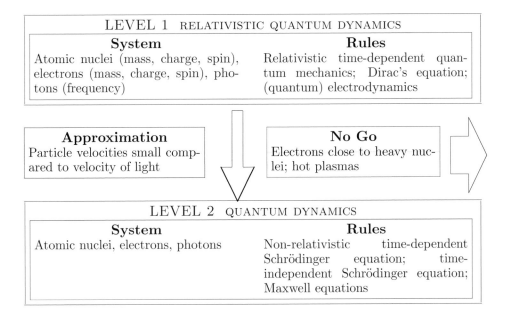

2.1 The wave character of particles

Textbooks on quantum mechanics abound, but this is not one of them. Therefore, an introduction to quantum mechanics is only given here as a guideline to the approximations that follow. Our intention is neither to be complete nor to be rigorous. Our aim is to show the beauty and simplicity of the basic quantum theory; relativistic quantum theory comprises such subtleties as electron spin, spin-orbit and magnetic interactions in a natural way. For practical reasons we must make approximations, but by descending down the hierarchy of theoretical models, we unfortunately lose the beauty of the higher-order theories. Already acquired gems, such as electron spin, must be re-introduced at the lower level in an ad hoc fashion, thus muting their brilliance.

Without going into the historical development of quantum mechanics, let us put two classes of observations at the heart of quantum theory:

- Particles (such as electrons in beams) show diffraction behavior as if they are waves. The wavelength λ appears to be related to the momentum $p = mv$ of the particle by $\lambda = h/p$, where h is Planck's constant. If we define $\boldsymbol{k}$ as the *wave vector* in the direction of the velocity of the particle and with absolute value $k = 2\pi/\lambda$, then

$$\boldsymbol{p} = \hbar\boldsymbol{k} \tag{2.1}$$

(with $\hbar = h/2\pi$) is a fundamental relation between the momentum of a particle and its wave vector.

- Electromagnetic waves (such as monochromatic light) appear to consist of packages of energy of magnitude $h\nu$, where ν is the frequency of the (monochromatic) wave, or $\hbar\omega$, where $\omega = 2\pi\nu$ is the angular frequency of the wave. Assuming that particles have a wave character, we may generalize this to identify the frequency of the wave with the energy of the particle:

$$E = \hbar\omega. \tag{2.2}$$

Let us further define a *wave function* $\Psi(\boldsymbol{r}, t)$ that describes the wave. A homogeneous plane wave, propagating in the direction of $\boldsymbol{k}$ with a phase velocity ω/k is described by

$$\Psi(\boldsymbol{r}, t) = c \exp[i(\boldsymbol{k} \cdot \boldsymbol{r} - \omega t)]$$

where c is a complex constant, the absolute value of which is the amplitude of the wave, while its argument defines the phase of the wave. The use of complex numbers is a matter of convenience (restriction to real numbers would require two amplitudes, one for the sine and one for the cosine constituents of the wave; restriction to the absolute value would not enable us to describe interference phenomena). In general, a particle may be described by a superposition of many (a continuum of) waves of different wave vector and frequency:

$$\Psi(\boldsymbol{r}, t) = \int d\boldsymbol{k} \int d\omega G(\boldsymbol{k}, \omega) \exp[i(\boldsymbol{k} \cdot \boldsymbol{r} - \omega t)], \tag{2.3}$$

where G is a distribution function of the wave amplitude in $\boldsymbol{k}, \omega$ space. Here we recognize that $\Psi(\boldsymbol{r}, t)$ and $G(\boldsymbol{k}, \omega)$ are each other's Fourier transform, although the sign conventions for the spatial and temporal transforms differ. (See Chapter 12 for details on Fourier transforms.) Of course, the transform can also be limited to the spatial variable only, yielding a time-dependent distribution in $\boldsymbol{k}$-space (note that in this case we introduce a factor of $(2\pi)^{-3/2}$ for symmetry reasons):

$$\Psi(\boldsymbol{r}, t) = (2\pi)^{-3/2} \int d\boldsymbol{k} \, g(\boldsymbol{k}, t) \exp[i(\boldsymbol{k} \cdot \boldsymbol{r})]. \tag{2.4}$$

The inverse transform is

$$g(\boldsymbol{k}, t) = (2\pi)^{-3/2} \int d\boldsymbol{r} \, \Psi(\boldsymbol{r}, t) \exp(-i\boldsymbol{k} \cdot \boldsymbol{r}). \tag{2.5}$$

The next crucial step is one of interpretation: we interpret $\Psi^*\Psi(\boldsymbol{r}, t)$ as the *probability density* that the particle is at $\boldsymbol{r}$ at time t. Therefore we require for a particle with continuous existence the probability density to be normalized at all times:

$$\int d\boldsymbol{r} \, \Psi^*\Psi(\boldsymbol{r}, t) = 1, \tag{2.6}$$

where the integration is over all space. Likewise g^*g is the probability density in $\boldsymbol{k}$-space; the normalization of g^*g is automatically satisfied (see Chapter 12):

$$\int d\boldsymbol{k} \, g^*g(\boldsymbol{k}, t) = 1 \tag{2.7}$$

The *expectation value*, indicated by triangular brackets, of an observable $f(\boldsymbol{r})$, which is a function of space only, then is

$$\langle f(\boldsymbol{r}) \rangle(t) = \int d\boldsymbol{r} \, \Psi^*\Psi(\boldsymbol{r}, t) f(\boldsymbol{r}) \tag{2.8}$$

and likewise the expectation value of a function of $\boldsymbol{k}$ only is given by

$$\langle f(\boldsymbol{k}) \rangle(t) = \int d\boldsymbol{k} \, g^*g(\boldsymbol{k}, t) f(\boldsymbol{k}). \tag{2.9}$$

If we apply these equations to define the expectation values of the *variances* of one coordinate x and its conjugate $k = k_x$:

$$\sigma_x^2 = \langle (x - \langle x \rangle)^2 \rangle, \tag{2.10}$$
$$\sigma_k^2 = \langle (k - \langle k \rangle)^2 \rangle, \tag{2.11}$$

we can show (see Chapter 12) that

$$\sigma_x \sigma_k \geq \tfrac{1}{2}, \tag{2.12}$$

which shows that two conjugated variables as x and k (that appear as product ikx in the exponent of the Fourier transform) cannot be simultaneously sharp. This is Heisenberg's *uncertainty relation*, which also applies to t and ω. Only for the Gaussian function $\exp(-\alpha x^2)$ the product of variances reaches the minimal value.

As shown in Chapter 12, averages over $\boldsymbol{k}$ and powers of k can be rewritten in terms of the spatial wave function Ψ:

$$\langle \boldsymbol{k} \rangle(t) = \int d\boldsymbol{r}\ \Psi^*(-i\boldsymbol{\nabla})\Psi(\boldsymbol{r},t), \tag{2.13}$$

$$\langle k^2 \rangle(t) = \int d\boldsymbol{r}\ \Psi^*(-\boldsymbol{\nabla}^2)\Psi(\boldsymbol{r},t). \tag{2.14}$$

Thus, the expectation of some observable A, being either a function of $\boldsymbol{r}$ only, or being proportional to $\boldsymbol{k}$ or to k^2, can be obtained from

$$\langle A \rangle(t) = \int d\boldsymbol{r}\ \Psi^*(\boldsymbol{r},t)\hat{A}\Psi(\boldsymbol{r},t), \tag{2.15}$$

where $\hat{A}$ is an *operator* acting on Ψ, and

$$\hat{\boldsymbol{k}} = -i\boldsymbol{\nabla}, \tag{2.16}$$

$$\hat{k}^2 = -\boldsymbol{\nabla}^2. \tag{2.17}$$

Similarly (but with opposite sign due to the opposite sign in ωt), the expectation value of the angular frequency ω is found from equation by using the operator

$$\hat{\omega} = i\frac{\partial}{\partial t}. \tag{2.18}$$

The identifications $\boldsymbol{p} = \hbar\boldsymbol{k}$ (2.1) and $E = \hbar\omega$ (2.2) allow the following operator definitions:

$$\hat{\boldsymbol{p}} = -i\hbar\boldsymbol{\nabla}, \tag{2.19}$$

$$\hat{p}^2 = -\hbar^2\boldsymbol{\nabla}^2, \tag{2.20}$$

$$\hat{E} = i\hbar\frac{\partial}{\partial t}. \tag{2.21}$$

From these relations and expression of the energy as a function of momenta and positions, the equations of motion for the wave function follow.

2.2 Non-relativistic single free particle

In principle we need the relativistic relations between energy, momentum and external fields, but for clarity we shall first look at the simple non-relativistic case of a single particle in one dimension without external interactions. This will allow us to look at some basic propagation properties of wave functions.

Using the relation

$$E = \frac{p^2}{2m}, \tag{2.22}$$

then (2.21) and (2.20) give the following equations of motion for the wave function:

$$i\hbar \frac{\partial \Psi(x,t)}{\partial t} = -\frac{\hbar^2}{2m} \frac{\partial^2 \Psi}{\partial x^2}, \tag{2.23}$$

or

$$\frac{\partial \Psi}{\partial t} = \frac{i\hbar}{2m} \frac{\partial^2 \Psi}{\partial x^2}. \tag{2.24}$$

This is in fact the time-dependent Schrödinger equation. This equation looks much like Fick's diffusion equation, with the difference that the diffusion constant is now imaginary (or, equivalently, that the diffusion takes place in imaginary time). If you don't know what that means, you are in good company.

If we choose an initial wave function $\Psi(x,0)$, with Fourier transform $g(k,0)$, then the solution of (2.24) is simply

$$\Psi(x,t) = \frac{1}{\sqrt{2\pi}} \int_{-\infty}^{\infty} dk\, g(k,0) \exp[ikx - i\omega(k)t], \tag{2.25}$$

$$\omega(k) = \frac{\hbar k^2}{2m} \tag{2.26}$$

The angular frequency corresponds to the energy:

$$E = \hbar\omega = \frac{(\hbar k)^2}{2m} = \frac{p^2}{2m}, \tag{2.27}$$

as it should. If ω had been just proportional to k (and not to k^2) then (2.25) would represent a wave packet traveling at constant velocity without any change in the form of the packet. But because of the k^2-dependence the wave packet slowly broadens as it proceeds in time. Let us assume that $g(k,0)$ is a narrow distribution around a constant k_0, and write

$$k^2 = k_0^2 + 2k_0\Delta k + (\Delta k)^2, \tag{2.28}$$

$$\hbar k_0 = mv. \tag{2.29}$$

In these terms the wave function can be written as

$$\Psi(x,t) = \frac{1}{\sqrt{2\pi}} \exp[ik_0(x - \tfrac{1}{2}vt)]$$

$$= \int_{-\infty}^{\infty} d\Delta k\, g(\Delta k, t) \exp[i\Delta k(x - vt)], \tag{2.30}$$

$$g(k,t) = g(k,0)\exp[-i\frac{\hbar(\Delta k)^2 t}{2m}]. \tag{2.31}$$

The factor in front of the integral is a time-dependent phase factor that is irrelevant for the shape of the density distribution since it cancels in $\Psi^*\Psi$. The packet shape (in space) depends on $x' = x - vt$ and thus the packet travels with the *group velocity* $v = d\omega/dk$. However, the packet changes shape with time. In fact, the package will always broaden unless it represents a stationary state (as a standing wave), but the latter requires an external confining potential.

Let us take a Gaussian packet with initial variance (of $\Psi^*|\Psi\rangle$) of σ_0^2 and with velocity v (i.e., $\langle k \rangle = k_0$) as an example. Its initial description (disregarding normalizing factors) is

$$\Psi(x,0) \propto \exp\left[-\frac{x^2}{4\sigma_0^2} + ik_0 x\right], \tag{2.32}$$

$$g(k,0) \propto \exp[-\sigma_0^2(\Delta k)^2]. \tag{2.33}$$

The wave function $\Psi(x' = x - vt, t)$ is, apart from the phase factor, equal to the inverse Fourier transform in Δk of $g(k,t)$ of (2.31):

$$g(k,t) \propto \exp\left[-\left(\sigma_0^2 + i\frac{\hbar t}{2m}\right)(\Delta k)^2\right], \tag{2.34}$$

which works out to

$$\Psi(x,t) \propto \exp\left[-\frac{x'^2}{4(\sigma_0^2 + i\hbar t/2m)}\right]. \tag{2.35}$$

By evaluating $\Psi^*\Psi$, we see that a Gaussian density is obtained with a variance $\sigma(t)$ that changes in time according to

$$\sigma^2(t) = \sigma_0^2\left(1 + \frac{\hbar^2 t^2}{4m^2\sigma_0^4}\right). \tag{2.36}$$

The narrower the initial package, the faster it will spread. Although this seems counterintuitive if we think of particles, we should remember that the wave function is related to the probability of finding the particle at a certain place at a certain time, which is all the knowledge we can possess. If the initial wave function is narrow in space, its momentum distribution is broad; this implies a larger uncertainty in position when time proceeds.

Only free particles broaden beyond measure; in the presence of confining potentials the behavior is quite different: stationary states with finite width emerge.

Because the packet becomes broader in space, it seems that the Heisenberg uncertainty relation would predict that it therefore becomes sharper in momentum distribution. This, however, is an erroneous conclusion: the broadening term is imaginary, and Ψ is not a pure real Gaussian; therefore the relation $\sigma_x \sigma_k = 1/2$ is not valid for $t > 0$. In fact, the width in k-space remains the same.

2.3 Relativistic energy relations for a free particle

The relation between energy and momentum (for a free particle) that we used in the previous section (2.22) is incorrect for velocities that approach the speed of light. In non-relativistic physics we assume that the laws of physics are invariant for a translation of the spatial and time origins of our coordinate system and also for a rotation of the coordinate system; this leads to fundamental conservation laws for momentum, energy, and angular momentum, respectively.[1] In the theory of special relativity the additional basic assumption is that the laws of physics, including the velocity of light, are also invariant if we transform our coordinate system to one moving at a constant speed with respect to the original one. Where for normal rotations in 3D-space we require that the square of length elements $(d\boldsymbol{r})^2$ is invariant, the requirement of the constant speed of light implies that for transformations to a moving frame $(c\,d\tau)^2 = (c\,dt)^2 - (d\boldsymbol{r})^2$ is invariant. For $1+1$ dimensions where we transform from (x,t) to x', t' in a frame moving with velocity v, this leads to the *Lorentz transformation*

$$\begin{pmatrix} x' \\ ct' \end{pmatrix} = \begin{pmatrix} \gamma & -\gamma v/c \\ -\gamma v/c & \gamma \end{pmatrix} \begin{pmatrix} x \\ ct \end{pmatrix}, \tag{2.37}$$

where

$$\gamma = \frac{1}{\sqrt{1 - v^2/c^2}}. \tag{2.38}$$

In *Minkovsky space* of $1+3$ dimensions $(ct, x, y, z) = (ct, \boldsymbol{r})$ vectors are *four-vectors* $v_\mu = (v_0, \boldsymbol{v})(\mu = 0, 1, 2, 3)$ and we define the scalar or inner product

---

[1] Landau and Lifschitz (1982) give a lucid derivation of these laws.

of two four-vectors as

$$v_\mu w_\mu \overset{\text{def}}{=} v_0 w_0 - v_1 w_1 - v_2 w_2 - v_3 w_3 = v_0 w_0 - \boldsymbol{v} \cdot \boldsymbol{w}. \tag{2.39}$$

The notation $v_\mu w_\mu$ uses the Einstein summation convention ($\sum_{\mu=0}^{3}$ over repeating indices is assumed, taking the signs into account as in (2.39)).[2] The square magnitude or length of a four-vector is the scalar product with itself; note that such a square length may be positive or negative. Lorentz transformations are all transformations in Minkowski space that leave $dx_\mu dx_\mu = (c\, d\tau)^2$ invariant; they of course include all space-like rotations for which $d\tau = 0$. Vectors that represent physical quantities are invariant for Lorentz transformations, and hence their scalar products and square magnitudes are constants.

Without any derivation, we list a number of relevant physical four-vectors, as they are defined in relativistic mechanics:

- coordinates: $x_\mu = (ct, \boldsymbol{r})$;
- wave vector: $k_\mu = (\omega/c, \boldsymbol{k})$;
- velocity: $u_\mu = (\gamma c, \gamma \boldsymbol{v})$;
- momentum: $p_\mu = m u_\mu = (\gamma m c, \gamma m \boldsymbol{v})$.

Here m is the (rest) mass of the particle. The first component of the momentum four-vector is identified with the energy E/c, so that $E = \gamma m c^2$. Note the following constant square lengths:

$$u_\mu u_\mu = c^2, \tag{2.40}$$

$$p_\mu p_\mu = \frac{E^2}{c^2} - \boldsymbol{p}^2 = m^2 c^2, \tag{2.41}$$

or

$$E^2 = m^2 c^4 + \boldsymbol{p}^2 c^2. \tag{2.42}$$

This is the relation between energy and momentum that we are looking for. From the quadratic form it is immediately clear that E will have equivalent positive and negative solutions, one set around $+mc^2$ and the other set around $-mc^2$. Only the first set corresponds to the solutions of the non-relativistic equation.

---

[2] We use subscripts exclusively and do not use general tensor notation which distinguishes covariant and contravariant vectors and uses a metric tensor to define vector products. We note that the "Einstein summation convention" in non-relativistic contexts, for example in matrix multiplication, is meant to be simply a summation over repeated indices.

Now identifying E with $i\hbar\partial/\partial t$ and $\boldsymbol{p}$ with $-i\hbar\boldsymbol{\nabla}$, we obtain the *Klein–Gordon equation*

$$\left[\left(-\frac{\partial^2}{c^2\partial t^2}+\boldsymbol{\nabla}^2\right)-\left(\frac{mc}{\hbar}\right)^2\right]\Psi=0. \tag{2.43}$$

This equation has the right relativistic symmetry (which the Schrödinger equation does not have), but unfortunately no solutions with real scalar densities $\Psi^*\Psi$ exist.

Dirac devised an ingeneous way to linearize (2.42). Let us first consider the case of one spatial dimension, where motion is allowed only in the x-direction, and angular momentum cannot exist. Instead of taking a square root of (2.42), which would involve the square root of the operator $\hat{p}$, one can devise a two-dimensional *matrix equation* which in fact equals a set of equations with multiple solutions:

$$i\hbar\frac{\partial\Psi}{\partial t}=c(\boldsymbol{\alpha}\hat{p}+\boldsymbol{\beta}mc)\Psi=\hat{H}\Psi, \tag{2.44}$$

where Ψ is a two-component vector, and $\boldsymbol{\alpha}$ and $\boldsymbol{\beta}$ are dimensionless Hermitian 2×2 matrices, chosen such that (2.42) is satisfied for all solutions of (2.44):

$$(\boldsymbol{\alpha}\hat{p}+\boldsymbol{\beta}mc)^2=(\hat{p}^2+m^2c^2)\mathbf{1}. \tag{2.45}$$

This implies that

$$\boldsymbol{\alpha}^2\hat{p}^2+(\boldsymbol{\alpha}\boldsymbol{\beta}+\boldsymbol{\beta}\boldsymbol{\alpha})mc\hat{p}+\boldsymbol{\beta}^2m^2c^2=(\hat{p}^2+m^2c^2)\mathbf{1}, \tag{2.46}$$

or

$$\boldsymbol{\alpha}^2=\mathbf{1},\quad \boldsymbol{\beta}^2=\mathbf{1},\quad \boldsymbol{\alpha}\boldsymbol{\beta}+\boldsymbol{\beta}\boldsymbol{\alpha}=\mathbf{0}. \tag{2.47}$$

In other words, $\boldsymbol{\alpha}$ and $\boldsymbol{\beta}$ are Hermitian, anticommuting, and unitary matrices.[3] The trivial solutions of the first two equations: $\boldsymbol{\alpha}=\pm\mathbf{1}$ and/or $\boldsymbol{\beta}=\pm\mathbf{1}$ do not satisfy the third equation.

There are many solutions to all three equations (2.47). In fact, when a matrix pair $\boldsymbol{\alpha},\boldsymbol{\beta}$ forms a solution, the matrix pair $\mathbf{U}\boldsymbol{\alpha}\mathbf{U}^\dagger,\mathbf{U}\boldsymbol{\beta}\mathbf{U}^\dagger$, constructed by a unitary transformation $\mathbf{U}$, forms a solution as well. A simple choice is

$$\boldsymbol{\alpha}=\begin{pmatrix}0&1\\1&0\end{pmatrix};\quad \boldsymbol{\beta}=\begin{pmatrix}1&0\\0&-1\end{pmatrix}. \tag{2.48}$$

---

[3] *Hermitian* ($\boldsymbol{\alpha}^\dagger=\boldsymbol{\alpha}$) because the eigenvalues must be real, *anticommuting* because $\boldsymbol{\alpha}\boldsymbol{\beta}+\boldsymbol{\beta}\boldsymbol{\alpha}=\mathbf{0}$, *unitary* because $\boldsymbol{\alpha}^2=\boldsymbol{\alpha}^\dagger\boldsymbol{\alpha}=\mathbf{1}$.

Inserting this choice into (2.44) yields the following matrix differential equation:

$$i\hbar\frac{\partial}{dt}\begin{pmatrix} \Psi_L \\ \Psi_S \end{pmatrix} = c\begin{pmatrix} mc & \hat{p} \\ \hat{p} & -mc \end{pmatrix}\begin{pmatrix} \Psi_L \\ \Psi_S \end{pmatrix}. \tag{2.49}$$

We see that in a coordinate frame moving with the particle ($p = 0$) there are two solutions: Ψ_L corresponding to particles (electrons) with positive energy $E = mc^2$; and Ψ_S corresponding to *antiparticles* (positrons) with negative energy $E = -mc^2$. With non-relativistic velocities $p \ll mc$, the wave function Ψ_S mixes slightly in with the particle wave function Ψ_L (hence the subscripts L for "large" and S for "small" when we consider particles). The eigenfunctions of the Hamiltonian matrix are

$$\hat{H} = \pm c(m^2c^2 + \hat{p}^2)^{1/2}, \tag{2.50}$$

which gives, after expanding the square root to first order in powers of p/mc, the particle solution

$$i\hbar\frac{\partial\Psi}{\partial t} \approx (mc^2 + \frac{\hat{p}^2}{2m})\Psi, \tag{2.51}$$

in which we recognize the Schrödinger equation for a free particle, with an extra constant, and irrelevant, zero-energy term mc^2.

In the case of three spatial dimensions, there are three α-matrices for each of the spatial components; i.e., they form a vector $\boldsymbol{\alpha}$ of three matrices $\boldsymbol{\alpha}_x, \boldsymbol{\alpha}_y, \boldsymbol{\alpha}_z$. The simplest solution now requires four dimensions, and Ψ becomes a four-dimensional vector. The Dirac equation now reads

$$i\hbar\frac{\partial\boldsymbol{\Psi}}{\partial t} = c(\boldsymbol{\alpha}\cdot\hat{\boldsymbol{p}} + \boldsymbol{\beta}mc)\boldsymbol{\Psi} = \hat{H}\boldsymbol{\Psi}, \tag{2.52}$$

where $\boldsymbol{\alpha}_x, \boldsymbol{\alpha}_y, \boldsymbol{\alpha}_z$ and $\boldsymbol{\beta}$ are mutually anti-commuting 4×4 matrices with their squares equal to the unit matrix. One choice of solutions is:

$$\boldsymbol{\alpha} = \begin{pmatrix} \mathbf{0} & \boldsymbol{\sigma} \\ \boldsymbol{\sigma} & \mathbf{0} \end{pmatrix}, \tag{2.53}$$

$$\sigma_x = \begin{pmatrix} 0 & 1 \\ 1 & 0 \end{pmatrix}, \quad \sigma_y = \begin{pmatrix} 0 & -i \\ i & 0 \end{pmatrix}, \quad \sigma_z = \begin{pmatrix} 1 & 0 \\ 0 & -1 \end{pmatrix}, \tag{2.54}$$

while $\boldsymbol{\beta}$ is a diagonal matrix $\{1, 1, -1, -1\}$ that separates two solutions around $+mc^2$ from two solutions around $-mc^2$. The wave function now also has four components, which refer to the two sets of solutions (electrons and positrons) each with two spin states. Thus spin is automatically introduced; it gives rise to an angular momentum $\boldsymbol{S}\hbar$ and an extra quantum number $S = 1/2$. By properly incorporating electromagnetic interactions, the small *spin-orbit interaction*, arising from magnetic coupling between the electron spin S and angular orbital momentum $L\hbar$, is included in the solution of the Dirac equation. This term makes it impossible to exactly separate the spin and orbital momenta; in fact there is one quantum number for the total angular momentum.

Let us now look at the relativistic effects viewed as a perturbation of the non-relativistic Schrödinger equation. We may first remark that spin can be separately and ad hoc introduced into the non-relativistic case as a new degree of freedom with two states. Each electron spin has associated with it an angular momentum $S\hbar$ and a *magnetic moment* $\mu = -\gamma_e S\hbar$, where γ_e is the electron's gyromagnetic ratio. The spin-orbit interaction term can then be computed from the classical interaction of the electron magnetic moment with the magnetic field that arises at the electron due to its orbital motion around a charged nucleus.

The relativistic effects arising from the high velocity of the electron can be estimated from a Taylor expansion of the positive solution of (2.42):

$$\frac{E}{c} = \sqrt{m^2 c^2 + p^2}, \tag{2.55}$$

$$E = mc^2 \left(1 + \frac{p^2}{2m^2 c^2} - \frac{p^4}{8m^4 c^4} + \cdots \right), \tag{2.56}$$

$$= mc^2 + \frac{p^2}{2m} - \frac{p^4}{8m^3 c^2} + \cdots \tag{2.57}$$

The first term is an irrelevant zero-point energy, the second term gives us the non-relativistic Schrödinger equation, and the third term gives a relativistic correction. Let us estimate its magnitude by a classical argument.

Assume that the electron is in a circular orbital at a distance r to a nucleus with charge Ze. From the balance between the nuclear attraction and the centrifugal force we conclude that

$$p^2 = -2mE, \tag{2.58}$$

where E is the (negative) total energy of the electron, not including the term mc^2 (this also follows from the virial

equation valid for a central Coulombic field: $E_{\text{pot}} = -2E_{\text{kin}}$, or $E = -E_{\text{kin}}$). For the expectation value of the first relativistic correction we find a lower bound

$$\frac{\langle p^4 \rangle}{8m^3 c^2} \geq \frac{\langle p^2 \rangle^2}{8m^3 c^2} = \frac{E^2}{2mc^2}. \tag{2.59}$$

The correction is most important for 1s-electrons near highly charged nuclei; since $-E$ is proportional to Z^2, the correction is proportional to Z^4. For the hydrogen atom $E = -13.6$ eV while $mc^2 = 511$ keV and hence the correction is 0.18 meV or 17 J/mol; for germanium (charge 32) the effect is expected to be a million times larger and be in the tens of MJ/mol. Thus the effect is not at all negligible and a relativistic treatment for the inner shells of heavy atoms is mandatory. For molecules with first-row atoms the relativistic correction to the total energy is still large (-146 kJ/mol for H_2O), but the effects on binding energies and on equilibrium geometry are small (dissociation energy of H_2O into atoms: -1.6 kJ/mol, equilibrium OH distance: -0.003 pm, equilibrium angle: -0.08 deg).[4]

In addition to the spin and energetic effects, the 1s-wave functions contract and become "smaller"; higher s-wave functions also contract because they remain orthogonal to the 1s-functions. Because the contracted s-electrons offer a better shielding of the nuclear charge, orbitals with higher angular momentum tend to expand. The effect on outer shell behavior is a secondary effect of the perturbation of inner shells: therefore, for quantum treatments that represent inner shells by *effective core potentials*, as in most practical applications of density functional theory, the relativistic corrections can be well accounted for in the core potentials without the need for relativistic treatment of the outer shell electrons.

Relativistic effects show up most clearly in the properties of heavy atoms, such as gold (atom number 79) and mercury (80). The fact that gold has its typical color, in contrast to silver (47) which has a comparable electron configuration, arises from the relatively high energy of the highest occupied d-orbital (due to the expansion of $5d_{3/2}$-orbital in combination with a high spin-orbit coupling) and the relatively low energy of the s-electrons in the conduction band (due to contraction of the 6s-orbitals), thus allowing light absorption in the visible region of the spectrum. The fact that mercury is a liquid (in contrast to cadmium (48), which has a comparable electron configuration) arises from the contraction of the 6s-orbitals, which are doubly occupied and so localized and "buried" in the electronic structure that they contribute little to the conduction band. Mercury atoms therefore resemble noble gas atoms with weak interatomic interactions. Because the enthalpy of fusion (2.3 kJ/mol) is low, the melting point (234 K) is low. For cadmium the heat of fusion is 6.2 kJ/mol and the melting point is 594 K. For the same reason mercury is a much poorer conductor (by a factor of 14) than cadmium. For further reading on this subject the reader is referred to Norrby (1991) and Pyykkö (1988).

---

[4] Jensen (1999), p. 216.

2.4 Electrodynamic interactions

¿From the relation $E = p^2/2m$ and the correspondence relations between energy or momentum and time or space derivatives we derived the non-relativistic Schrödinger equation for a non-interacting particle (2.24). How is this equation modified if the particle moves in an external potential?

In general, what we need is the operator form of the *Hamiltonian H*, which for most cases is equivalent to the total kinetic plus potential energy. When the potential energy in an external field is a function $V(\boldsymbol{r})$ of the coordinates only,such as produced by a stationary electric potential, it is simply added to the kinetic energy:

$$i\hbar\frac{\partial\Psi}{\partial t} = -\frac{\hbar^2}{2m}\boldsymbol{\nabla}^2\Psi + V(\boldsymbol{r})\Psi. \qquad (2.60)$$

In fact, electrons feel the environment through electromagnetic interactions, in general with both an electric and a magnetic component. If the electric field is not stationary, there is in principle always a magnetic component. As we shall see, the magnetic component acts through the *vector potential* that modifies the momentum of the particle. See Chapter 13 for the basic elements of electromagnetism.

In order to derive the proper form of the electromagnetic interaction of a particle with charge q and mass m, we must derive the *generalized momentum* in the presence of a field. This is done by the Lagrangian formalism of mechanics, which is reviewed in Chapter 15. The Lagrangian $L(\boldsymbol{r}, \boldsymbol{v})$ is defined as $T - V$, where T is the kinetic energy and V is the potential energy. In the case of an electromagnetic interaction, the electrical potential energy is modified with a velocity-dependent term $-q\boldsymbol{A} \cdot \boldsymbol{v}$, where $\boldsymbol{A}$ is the *vector potential* related to the magnetic field $\boldsymbol{B}$ by

$$\boldsymbol{B} = \operatorname{\mathbf{curl}} \boldsymbol{A}, \qquad (2.61)$$

in a form which is invariant under a Lorentz transformation:

$$V(\boldsymbol{r}, \boldsymbol{v}) = q\phi - q\boldsymbol{A} \cdot \boldsymbol{v}. \qquad (2.62)$$

Thus the Lagrangian becomes

$$L(\boldsymbol{r}, \boldsymbol{v}) = \tfrac{1}{2}m\boldsymbol{v}^2 - q\phi + q\boldsymbol{A} \cdot \boldsymbol{v}. \qquad (2.63)$$

The reader should verify that with this Lagrangian the Euler–Lagrange equations of motion for the components of coordinates and velocities

$$\frac{d}{dt}\left(\frac{\partial L}{\partial v_i}\right) = \frac{\partial L}{\partial x_i} \qquad (2.64)$$

lead to the common Lorentz equation for the acceleration of a charge q in an electromagnetic field

$$m\dot{\boldsymbol{v}} = q(\boldsymbol{E} + \boldsymbol{v} \times \boldsymbol{B}), \tag{2.65}$$

where

$$\boldsymbol{E} \stackrel{\text{def}}{=} -\boldsymbol{\nabla}\phi - \frac{\partial \boldsymbol{A}}{\partial t} \tag{2.66}$$

(see Chapter 13). The generalized momentum components p_i are defined as (see Chapter 15)

$$p_i = \frac{\partial L}{\partial v_i}, \tag{2.67}$$

and hence

$$\boldsymbol{p} = m\boldsymbol{v} + q\boldsymbol{A}, \tag{2.68}$$

or

$$\boldsymbol{v} = \frac{1}{m}(\boldsymbol{p} - q\boldsymbol{A}). \tag{2.69}$$

For the Schrödinger equation we need the Hamiltonian H, which is defined as (see Chapter 15)

$$H \stackrel{\text{def}}{=} \boldsymbol{p} \cdot \boldsymbol{v} - L = \frac{1}{2m}(\boldsymbol{p} - q\boldsymbol{A})^2 + q\phi. \tag{2.70}$$

Thus the non-relativistic Schrödinger equation of a particle with charge q and mass m, in the presence of an electromagnetic field, is

$$i\hbar\frac{\partial \Psi}{\partial t} = \hat{H}\Psi = \left[-\frac{\hbar^2}{2m}\left(\boldsymbol{\nabla} - \frac{iq\boldsymbol{A}}{\hbar} \right)^2 + q\phi(\boldsymbol{r}) \right]\Psi. \tag{2.71}$$

Being non-relativistic, this description ignores the magnetic effects of spin and orbital momentum, i.e., both the spin-Zeeman term and the spin-orbit interaction, which must be added ad hoc if required.[5]

The magnetic field component of the interaction between nuclei and electrons or electrons mutually is generally ignored so that these interactions are described by the pure Coulomb term which depends only on coordinates and not on velocities. If we also ignore magnetic interactions with external fields ($\boldsymbol{A} = 0$), we obtain for a N-particle system with masses m_i and

---

[5] The Dirac equation (2.52) in the presence of an external field ($\boldsymbol{A}, \phi$) has the form:

$$i\hbar\frac{\partial \Psi}{\partial t} = [c\boldsymbol{\alpha} \cdot (\hat{\boldsymbol{p}} - q\boldsymbol{A}) + \boldsymbol{\beta}mc^2 + q\phi\mathbf{1}]\Psi = \hat{H}\Psi.$$

This equation naturally leads to both orbital and spin Zeeman interaction with a magnetic field and to spin-orbit interaction. See Jensen (1999).

charges q_i the time-dependent Schrödinger equation for the wave function $\Psi(\boldsymbol{r}_1, \ldots, \boldsymbol{r}_N, t)$:

$$i\hbar \frac{\partial \Psi}{\partial t} = \hat{H}\Psi$$

$$= [-\sum_i \frac{\hbar^2}{2m_i} \boldsymbol{\nabla}_i^2 + \frac{1}{2}\frac{1}{4\pi\varepsilon_0} \sum_{i,j}' \frac{q_i q_j}{r_{ij}} + V_{\text{ext}}(\boldsymbol{r}_1, \ldots, \boldsymbol{r}_N, t)]\Psi, \quad (2.72)$$

where the $1/2$ in the mutual Coulomb term corrects for double counting in the sum, the prime on the sum means exclusion of $i = j$, and the last term, if applicable, represents the energy in an external field.

Let us finally derive simplified expressions in the case of external electromagnetic fields. If the external field is a "slow" electric field, (2.72) suffices. If the external field is either a "fast" electric field (that has an associated magnetic field) or includes a separate magnetic field, the nabla operators should be modified as in (2.71) to include the vector potential:

$$\boldsymbol{\nabla}_i \to \boldsymbol{\nabla}_i - \frac{iq}{\hbar}\boldsymbol{A}(\boldsymbol{r}_i). \quad (2.73)$$

For simplicity we now drop the particle index i, but note that for final results summation over particles is required. Realizing that

$$\boldsymbol{\nabla} \cdot (\boldsymbol{A}\Psi) = (\boldsymbol{\nabla} \cdot \boldsymbol{A})\Psi + \boldsymbol{A} \cdot (\boldsymbol{\nabla}\Psi), \quad (2.74)$$

the kinetic energy term reduces to

$$\left(\boldsymbol{\nabla} - \frac{iq}{\hbar}\boldsymbol{A}\right)^2 = \boldsymbol{\nabla}^2 - \frac{iq}{\hbar}(\boldsymbol{\nabla} \cdot \boldsymbol{A}) - \frac{2iq}{\hbar}\boldsymbol{A} \cdot \boldsymbol{\nabla} - \frac{q^2}{\hbar^2}\boldsymbol{A}^2. \quad (2.75)$$

Let us consider two examples: a stationary homogeneous magnetic field $\boldsymbol{B}$ and an electromagnetic plane wave.

2.4.1 Homogeneous external magnetic field

Consider a constant and homogeneous magnetic field $\boldsymbol{B}$ and let us find a solution $\boldsymbol{A}(\boldsymbol{r})$ for the equation $\boldsymbol{B} = \text{curl}\,\boldsymbol{A}$. There are many solutions (because any gradient field may be added) and we choose one for which $\boldsymbol{\nabla} \cdot \boldsymbol{A} = 0$ (the *Lorentz convention* for a stationary field, see Chapter 13):

$$\boldsymbol{A}(\boldsymbol{r}) = \tfrac{1}{2}\boldsymbol{B} \times \boldsymbol{r}. \quad (2.76)$$

The reader should check that this choice gives the proper magnetic field while the divergence vanishes. The remaining terms in (2.75) are a linear term in $\boldsymbol{A}$:

$$\boldsymbol{A} \cdot \boldsymbol{\nabla} = \tfrac{1}{2}(\boldsymbol{B} \times \boldsymbol{r}) \cdot \boldsymbol{\nabla} = \tfrac{1}{2}\boldsymbol{B} \cdot (\boldsymbol{r} \times \boldsymbol{\nabla}), \quad (2.77)$$

which gives a term in the Hamiltonian that represents the Zeeman interaction of the magnetic field with the orbital magnetic moment:

$$\hat{H}_{\text{zeeman}} = \frac{e}{2m} \boldsymbol{B} \cdot \hat{\boldsymbol{L}} \hbar, \tag{2.78}$$

$$\hat{\boldsymbol{L}} \hbar = \boldsymbol{r} \times \hat{\boldsymbol{p}} = -i\hbar \, \boldsymbol{r} \times \boldsymbol{\nabla}, \tag{2.79}$$

where $\hat{\boldsymbol{L}}$ is the dimensionless orbital angular momentum operator, and a quadratic term in $\boldsymbol{A}$ that is related to magnetic susceptibility.[6]

The Zeeman interaction can be considered as the energy $-\boldsymbol{\mu} \cdot \boldsymbol{B}$ of a dipole in a field; hence the (orbital) magnetic dipole operator equals

$$\hat{\boldsymbol{\mu}} = -\frac{e\hbar}{2m} \hat{\boldsymbol{L}} = -\mu_{\text{B}} \hat{\boldsymbol{L}}, \tag{2.80}$$

where $\mu_{\text{B}} = e\hbar/2m$ is the *Bohr magneton*. In the presence of spin this modifies to

$$\hat{\boldsymbol{\mu}} = -g\mu_{\text{B}} \hat{\boldsymbol{J}}, \tag{2.81}$$

where

$$\hat{\boldsymbol{J}} = \hat{\boldsymbol{L}} + \hat{\boldsymbol{S}}, \tag{2.82}$$

and g is the *Lande g-factor*, which equals 1 for pure orbital contributions, 2.0023 for pure single electron-spin contributions, and other values for mixed states. The total angular momentum $\hat{\boldsymbol{J}}\hbar$ is characterized by a quantum number J and, if the spin-orbit coupling is small, there are also meaningful quantum numbers L and S for orbital and spin angular momentum. The g-factor then is approximately given by

$$g = 1 + \frac{J(J+1) + S(S+1) - L(L+1)}{2J(J+1)}. \tag{2.83}$$

2.4.2 Electromagnetic plane wave

In the case of perturbation by an electromagnetic wave (such as absorption of light) we describe for simplicity the electromagnetic field by a linearly polarized monochromatic plane wave in the direction $\boldsymbol{k}$ (see Chapter 13):

$$\boldsymbol{E} = \boldsymbol{E}_0 \exp[i(\boldsymbol{k} \cdot \boldsymbol{r} - \omega t)], \tag{2.84}$$

$$\boldsymbol{B} = \frac{1}{c}\left(\frac{\boldsymbol{k}}{k} \times \boldsymbol{E}\right), \tag{2.85}$$

$$\omega = kc. \tag{2.86}$$

---

[6] For details see Jensen (1999).

These fields can be derived from the following potentials:

$$A = \frac{i}{\omega}E, \tag{2.87}$$

$$\phi = 0, \tag{2.88}$$

$$\nabla \cdot A = 0. \tag{2.89}$$

Note that physical meaning is attached to the real parts of these complex quantities.

As in the previous case, the Hamiltonian with (2.75) has a linear and a quadratic term in A. The quadratic term is related to dynamic polarization and light scattering and (because of its double frequency) to "double quantum" transitions. The linear term in A is more important and gives rise to first-order dipole transitions to other states (absorption and emission of radiation). It gives the following term in the Hamiltonian:

$$\hat{H}_{\text{dip}} = \frac{i\hbar q}{m}A \cdot \nabla = -\frac{q}{m}A \cdot \hat{p}. \tag{2.90}$$

If the wavelength is large compared to the size of the interacting system, the space dependence of A can be neglected, and A can be considered as a spatially constant vector, although it is still time dependent. Let us consider this term in the Hamiltonian as a perturbation and derive the form of the interaction that will induce transitions between states. In first-order perturbation theory, where the wave functions $\Psi_n(r, t)$ are still solutions of the unperturbed Hamiltonian $\hat{H}_0$, transitions from state n to state m occur if the frequency of the perturbation $\hat{H}_1$ matches $|E_n - E_m|/h$ and the corresponding matrix element is nonzero:

$$\langle m|\hat{H}_1|n\rangle \overset{\text{def}}{=} \int_{-\infty}^{\infty} \psi_m^* \hat{H}_1 \psi_n \, dr \neq 0. \tag{2.91}$$

Thus we need the matrix element $\langle m|\hat{p}|n\rangle$, which can be related to the matrix element of the corresponding coordinate:

$$\langle m|\hat{p}|n\rangle = \frac{m}{i\hbar}(E_m - E_n)\langle m|r|n\rangle \tag{2.92}$$

(the proof is given at the end of this section). The matrix element of the perturbation $\hat{H}_{\text{dip}}$ (see (2.90)), summed over all particles, is

$$\langle m|\hat{H}_{\text{dip}}|n\rangle = \frac{i}{\hbar}(E_m - E_n)A \cdot \langle m|\sum_i q_i r_i|n\rangle$$

$$= -\frac{(E_m - E_n)}{\omega_{mn}}E_0 \cdot \mu_{mn}, \tag{2.93}$$

where we have made use of (2.87). The term between angular brackets is the *transition dipole moment* $\boldsymbol{\mu}_{mn}$, the matrix element for the dipole moment operator. Note that this dipolar interaction is just an approximation to the total electromagnetic interaction with the field.

Proof We prove (2.92). We first show that

$$\hat{\boldsymbol{p}} = \frac{m}{i\hbar}[\hat{H}_0, \boldsymbol{r}], \tag{2.94}$$

which follows (for one component) from

$$[\hat{H}_0, x] = -\frac{1}{2m}[\hat{p}^2, x],$$

and

$$[p^2, x] = ppx - xpp = ppx - pxp + pxp - xpp = 2p[p, x] = -2i\hbar p,$$

because $[p, x] = -i\hbar$.

Next we compute the matrix element for $[\hat{H}_0, \boldsymbol{r}]$ (for one component):

$$\langle m|[\hat{H}_0, x]|n\rangle = \langle m|\hat{H}_0 x|n\rangle - \langle m|x\hat{H}_0|n\rangle.$$

The last term is simply equal to $E_n\langle m|x|n\rangle$. The first term rewrites by using the Hermitian property of $\hat{H}_0$:

$$\int \psi_m^* \hat{H}_0(x\psi_n)\, d\boldsymbol{r} = \int (\hat{H}_0^* \psi_m^*)x\psi_n\, d\boldsymbol{r} = E_m\langle m|x|n\rangle.$$

Collecting terms, (2.92) is obtained.   □

2.5 Fermions, bosons and the parity rule

There is one further basic principle of quantum mechanics that has far-reaching consequences for the fate of many-particle systems. It is the rule that *particles have a definite parity*. What does this mean?

Particles of the same type are in principle indistinguishable. This means that the exchange of two particles of the same type in a many-particle system should not change any observable, and therefore should not change the probability density $\Psi^*\Psi$. The wave function itself need not be invariant for particle exchange, because any change of phase $\exp(i\phi)$ does not change the probability distribution. But if we exchange two particles twice, we return exactly to the original state, so the phase change can only be $0°$ (no change) or $180°$ (change of sign). This means that the *parity* of the wave function (the change of sign on exchange of two particles) is either positive (even) or negative (odd).

The parity rule (due to Wolfgang Pauli) says that the parity is a basic, invariant, property of a particle. Thus there are two kinds of particle: *fermions* with odd parity and *bosons* with even parity. Fermions are particles with half-integral spin quantum number; bosons have integral spins. Some examples:

- *fermions* (half-integral spin, odd parity): electron, proton, neutron, muon, positron, ^{3}He nucleus, ^{3}He atom, D atom;
- *bosons* (integral spin, even parity): deuteron, H-atom, ^{4}He nucleus, ^{4}He atom, H_2 molecule.

The consequences for electrons are drastic! If we have two one-electron orbitals (including spin state) φ_a and φ_b, and we put two non-interacting electrons into these orbitals (one in each), the odd parity prescribes that the total two-particle wave function must have the form

$$\psi(1,2) \propto \varphi_a(1)\varphi_b(2) - \varphi_a(2)\varphi_b(1). \tag{2.95}$$

So, if $\varphi_a = \varphi_b$, the wave function cannot exist! Hence, *two (non-interacting) electrons (or fermions in general) cannot occupy the same spin-orbital.* This is Pauli's exclusion principle. Note that this exclusion has nothing to do with the energetic (e.g., Coulomb) interaction between the two particles.

Exercises

2.1   Derive (2.30) and (2.31).

2.2   Show that (2.35) is the Fourier transform of (2.34). See Chapter 12.

2.3   Show that the width of g^*g does not change with time.

2.4   Show that $c^2(dt')^2 - (dx')^2 = c^2(dt)^2 - (dx)^2$ when dt' and dx' transform according to the Lorentz transformation of (2.37).

3

From quantum to classical mechanics: when and how

3.1 Introduction

In this chapter we shall ask (and possibly answer) the question how quantum mechanics can produce classical mechanics as a limiting case. In what circumstances and for what kind of particles and systems is the classical approximation valid? When is a quantum treatment mandatory? What errors do we make by assuming classical behavior? Are there indications from experiment when quantum effects are important? Can we derive quantum corrections to classical behavior? How can we proceed if quantum mechanics is needed for a specific part of a system, but not for the remainder? In the following chapters the quantum-dynamical and the mixed quantum/classical methods will be worked out in detail.

The essence of quantum mechanics is that particles are represented by a wave function and have a certain width or uncertainty in space, related to an uncertainty in momentum. By a handwaving argument we can already judge whether the quantum character of a particle will play a dominant role or not. Consider a (nearly) classical particle with mass m in an equilibrium system at temperature T, where it will have a Maxwellian velocity distribution (in each direction) with $\langle p^2 \rangle = mk_{\mathrm{B}}T$. This uncertainty in momentum implies that the particle's *width* σ_x, i.e., the standard deviation of its wave function distribution, will exceed the value prescribed by Heisenberg's uncertainty principle (see Chapter 2):

$$\sigma_x \geq \frac{\hbar}{2\sqrt{mk_{\mathrm{B}}T}}. \tag{3.1}$$

There will be quantum effects if the forces acting on the particle vary appreciably over the width[1] of the particle. In condensed phases, with interparticle

---

[1] The width we use here is proportional to the *de Broglie wavelength* $\Lambda = h/\sqrt{2\pi mk_{\mathrm{B}}T}$ that figures in statistical mechanics. Our width is five times smaller than Λ.

Table 3.1 *The minimal quantum width in Å of the electron and some atoms at temperatures between 10 and 1000 K, derived from Heisenberg's uncertainty relation. All values above 0.1 Å are given in bold type*

	m(u)	10 K	30 K	100 K	300 K	1000 K
e	0.000545	**47**	**27**	**15**	**8.6**	**4.7**
H	1	**1.1**	**0.64**	**0.35**	**0.20**	**0.11**
D	2	**0.78**	**0.45**	**0.25**	**0.14**	0.078
C	12	**0.32**	**0.18**	**0.10**	0.058	0.032
O	16	**0.28**	**0.16**	0.087	0.050	0.028
I	127	0.098	0.056	0.031	0.018	0.010

distances of a few Å, this is the case when the width of the particle exceeds, say, 0.1 Å. In Table 3.1 the particle widths are given for the electron and for several atoms for temperatures between 10 and 1000 K.

It is clear that electrons are fully quantum-mechanical in all cases (except hot, dilute plasmas with interparticle separations of hundreds of Å). Hydrogen and deuterium atoms are suspect at 300 K but heavier atoms will be largely classical, at least at normal temperatures. It is likely that quantum effects of the heavier atoms can be treated by quantum corrections to a classical model, and one may only hope for this to be true for hydrogen as well. There will be cases where the intermolecular potentials are so steep that even heavy atoms at room temperature show essential quantum effects: this is the case for most of the bond vibrations in molecules. The criterion for classical behavior here is that vibrational frequencies should not exceed $k_B T/h$, which at $T = 300$ K amounts to about 6 THz, or a wave number of about 200 cm^{-1}.

We may also consider experimental data to judge the importance of quantum effects, at least for systems in thermal equilibrium. In classical mechanics, the excess free energy (excess with respect to the ideal gas value) of a conservative system depends only on the potential energy $V(\boldsymbol{r})$ and not on the mass of the particles (see Chapter 17):

$$A = A^{\mathrm{id}} - k_B T \ln V^{-N} \int e^{-\beta V(\boldsymbol{r})} \, d\boldsymbol{r}. \qquad (3.2)$$

Since the ideal gas pressure at a given molar density does not depend on atomic mass either, the phase diagram, melting and boiling points, critical constants, second virial coefficient, compressibility, and several molar properties such as density, heat capacity, etc. do not depend on isotopic composition for a classically behaving substance. Neither do equilibrium

Table 3.2 *Critical point characteristic for various isotopes of helium,*
hydrogen and water

	$T_c(\mathrm{K})$	$p_c(\mathrm{bar})$	$V_c\ (\mathrm{cm^3\,mol^{-1}})$
$^4\mathrm{He}$	5.20	2.26	57.76
$^3\mathrm{He}$	3.34	1.15	72.0
$\mathrm{H_2}$	33.18	12.98	66.95
HD	35.9	14.6	62.8
$\mathrm{D_2}$	38.3	16.3	60.3
$\mathrm{H_2O}$	647.14	220.64	56.03
$\mathrm{D_2O}$	643.89	216.71	56.28

constants as dissociation or association constants or partition coefficients
depend on isotopic composition for classical substances. If such properties
appear to be dependent on isotopic composition, this is a sure sign of the
presence of quantum effects on atomic behavior.

Look at a few examples. Table 3.2 lists critical constants for different
isotopes of helium, hydrogen and water. Table 3.3 lists some equilibrium
properties of normal and heavy water. It is not surprising that the properties of helium and hydrogen at (very) low temperatures are strongly isotope
dependent. The difference between H_2O and D_2O is not negligible: D_2O
has a higher temperature of maximum density, a higher enthalpy of vaporization and higher molar heat capacity; it appears more "structured" than
H_2O. The most likely explanation is that it forms stronger hydrogen bonds
as a result of the quantum-mechanical zero-point energy of the intermolecular vibrational and librational modes of hydrogen-bonded molecules. Accurate simulations must either incorporate this quantum behavior, or make
appropriate corrections for it.

It is instructive to see how the laws of classical mechanics, i.e., Newton's
equations of motion, follow from quantum mechanics. We consider three
different ways to accomplish this goal. In Section 3.2 we derive equations of
motion for the *expectation values* of position and velocity, following Ehrenfest's arguments of 1927. A formulation of quantum mechanics which is
equivalent to the Schrödinger equation but is more suitable to approach the
classical limit, is Feynman's *path integral formulation*. We give a short introduction in Section 3.3. Then, in Section 3.4, we consider a formulation
of quantum mechanics, originally proposed by Madelung and by de Broglie
in 1926/27, and in 1952 revived by Bohm, which represents the evolution

Table 3.3 *Various properties of normal and heavy water*

	H_2O	D_2O
melting point (°C)	0	3.82
boiling point (°C)	100	101.4
temperature of maximum density (°C)	3.98	11.19
vaporization enthalpy at 3.8 °C (kJ/mol)	44.8	46.5
molar volume at 25 °C (cm^3/mol)	18.07	18.13
molar heat capacity at 25 °C ($J K^{-1} mol^{-1}$)	74.5	83.7
ionization constant $-\log[K_w/(mol^2 dm^{-6})]$ at 25 °C	13.995	14.951

of the wave function by a fluid of particles which follow trajectories guided by a special quantum force. The application of quantum corrections to equilibrium properties computed with classical methods, and the actual incorporation of quantum effects into simulations, is the subject of following chapters.

3.2 From quantum to classical dynamics

In this section we ask the question: *Can we derive classical equations of motion for a particle in a given external potential V from the Schrödinger equation?* For simplicity we consider the one-dimensional case of a particle of mass m with position x and momentum $p = m\dot{x}$. The classical equations of Newton are

$$\frac{dx}{dt} = \frac{p}{m}, \tag{3.3}$$

$$\frac{dp}{dt} = -\frac{dV(x)}{dx}. \tag{3.4}$$

Position and momentum of a quantum particle must be interpreted as the *expectation* of x and p. The classical force would then be the value of the gradient of V taken at the expectation of x. So we ask whether

$$\frac{d\langle x \rangle}{dt} \ ? =? \ \frac{\langle p \rangle}{m}, \tag{3.5}$$

$$\frac{d\langle p \rangle}{dt} \ ? =? \ -\left(\frac{dV}{dx}\right)_{\langle x \rangle}. \tag{3.6}$$

We follow the argument of Ehrenfest (1927). See Chapter 14 for details of the operator formalism and equations of motion.

Recall that the expectation $\langle A \rangle$ of an observable A over a quantum system with wave function $\Psi(x,t)$ is given by

$$\langle A \rangle = \int \Psi^* \hat{A} \Psi \, dx, \tag{3.7}$$

where $\hat{A}$ is the operator of A. From the time-dependent Schrödinger equation the equation of motion (14.64) for the expectation of A follows:

$$\frac{d\langle A \rangle}{dt} = \frac{i}{\hbar} \langle [\hat{H}, \hat{A}] \rangle. \tag{3.8}$$

Here $[\hat{H}, \hat{A}]$ is the *commutator* of $\hat{H}$ and $\hat{A}$:

$$[\hat{H}, \hat{A}] = \hat{H}\hat{A} - \hat{A}\hat{H}. \tag{3.9}$$

We note that the Hamiltonian is the sum of kinetic and potential energy:

$$\hat{H} = \hat{K} + \hat{V} = \frac{\hat{p}^2}{2m} + V(x), \tag{3.10}$$

and that $\hat{p}$ commutes with $\hat{K}$ but not with $\hat{V}$, while $\hat{x}$ commutes with $\hat{V}$ but not with $\hat{K}$. We shall also need the commutator

$$[\hat{p}, \hat{x}] = \hat{p}\hat{x} - \hat{x}\hat{p} = \frac{\hbar}{i}. \tag{3.11}$$

This follows from inserting the operator for p:

$$\frac{\hbar}{i} \frac{\partial(x\psi)}{\partial x} - x \frac{\hbar}{i} \frac{\partial \psi}{\partial x} = \frac{\hbar}{i} \psi. \tag{3.12}$$

Now look at the first classical equation of motion (3.5). We find using (3.8) that

$$\frac{d\langle x \rangle}{dt} = \frac{i}{2m\hbar} \langle [\hat{p}^2, \hat{x}] \rangle = \frac{\langle p \rangle}{m}, \tag{3.13}$$

because

$$[\hat{p}^2, \hat{x}] = \hat{p}\hat{p}\hat{x} - \hat{x}\hat{p}\hat{p} = \hat{p}\hat{p}\hat{x} - \hat{p}\hat{x}\hat{p} + \hat{p}\hat{x}\hat{p} - \hat{x}\hat{p}\hat{p}$$
$$= \hat{p}[\hat{p}, \hat{x}] + [\hat{p}, \hat{x}]\hat{p} = \frac{2\hbar}{i} \hat{p}. \tag{3.14}$$

Hence the first classical equation of motion is always valid.

The second equation of motion (3.6) works out as follows:

$$\frac{d\langle p \rangle}{dt} = \frac{i}{\hbar} \langle [\hat{H}, \hat{p}] \rangle = \frac{i}{\hbar} \left\langle [\hat{V}, \hat{p}] \right\rangle$$
$$= \frac{i}{\hbar} \left\langle V \frac{\hbar}{i} \frac{\partial}{\partial x} \right\rangle - \frac{i}{\hbar} \left\langle \frac{\hbar}{i} \frac{\partial}{\partial x} V \right\rangle = - \left\langle \frac{dV}{dx} \right\rangle. \tag{3.15}$$

This is the *expectation* of the force over the wave function, not the force at the expectation of x! When the force is constant, there is no difference between the two values and the motion is classical, as far as the expectations of x and p are concerned. This is even true if the force depends linearly on x:

$$F(x) = F_0 + F'(x - \langle x \rangle),$$

where $F_0 = F(\langle x \rangle)$ and F' is a constant, because

$$\langle F \rangle = F_0 + F'\langle x - \langle x \rangle \rangle = F_0.$$

Expanding the force (or potential) in a Taylor series, we see that the leading correction term on the force is proportional to the second derivative of the force times the variance of the wave packet:

$$\left\langle \frac{dV}{dx} \right\rangle = \left(\frac{dV}{dx} \right)_{\langle x \rangle} + \frac{1}{2!} \left(\frac{d^3 V}{dx^3} \right)_{\langle x \rangle} \langle (x - \langle x \rangle)^2 \rangle + \cdots. \tag{3.16}$$

The motion is classical if the force (more precisely, the gradient of the force) does not vary much over the *quantum width* of the particle. This is true even for electrons in macroscopic fields, as they occur in accelerators and in dilute or hot plasmas; this is the reason that hot plasmas can be treated with classical equations of motion, as long as the electromagnetic interactions are properly incorporated. For electrons near point charges the force varies enormously over the quantum width and the classical approximation fails completely.

It is worth mentioning that a *harmonic oscillator* moves in a potential that has no more than two derivatives, and – as given by the equations derived above – *moves according to classical dynamics*. Since we know that a quantum oscillator behaves differently from a classical oscillator (e.g., it has a zero-point energy), this is surprising at first sight! But even though the classical equations do apply for the expectation values of position and momentum, a quantum particle is not equal to a classical particle. For example, $\langle p^2 \rangle \neq \langle p \rangle^2$. A particle at rest, with $\langle p \rangle = 0$, can still have a kinetic and potential energy. Thus, for classical behavior it is not enough that the expectation of x and p follow classical equations of motion.

3.3 Path integral quantum mechanics

3.3.1 Feynman's postulate of quantum dynamics

While the Schrödinger description of wave functions and their evolution in time is adequate and sufficient, the Schrödinger picture does not connect smoothly to the classical limit. In cases that the particles we are interested in are nearly classical (this will often apply to atoms, but not to electrons) the *path integral formulation* of quantum mechanics originating from Feynman[2] can be more elucidating. This formulation renders a solution to the propagation of wave functions equivalent to that following from Schrödinger's equation, but has the advantage that the classical limit is more naturally obtained as a limiting case. The method allows us to obtain quantum corrections to classical behavior. In particular, corrections to the classical partition function can be obtained by numerical methods derived from path integral considerations. These *path integral Monte Carlo and molecular dynamics methods*, PIMC and PIMD, will be treated in more detail in Section 3.3.9.

Since the Schrödinger equation is linear in the wave function, the time propagation of the wave function can be expressed in terms of aGreen's function $G(\boldsymbol{r}_f, t_f; \boldsymbol{r}_0, t_0)$, which says how much the amplitude of the wave function at an initial position $\boldsymbol{r}_0$ at time t_0 contributes to the amplitude of the wave function at a final position $\boldsymbol{r}_f$ at a later time t_f. All contributions add up to an (interfering) total wave function at time t_f:

$$\Psi(\boldsymbol{r}_f, t_f) = \int d\boldsymbol{r}_0 \, G(\boldsymbol{r}_f, t_f; \boldsymbol{r}_0, t_0) \Psi(\boldsymbol{r}_0, t_0). \quad (3.17)$$

The Green's function is the *kernel* of the integration.

In order to find an expression for G, Feynman considers all possible paths $\{\boldsymbol{r}(t)\}$ that run from position $\boldsymbol{r}_0$ at time t_0 to position $\boldsymbol{r}_f$ at time t_f. For each path it is possible to compute the mechanical *action* S as an integral of the Lagrangian $\mathcal{L}(\boldsymbol{r}, \dot{\boldsymbol{r}}, t) = K - V$ over that path (see Chapter 15):

$$S = \int_{t_0}^{t_f} \mathcal{L}(\boldsymbol{r}, \dot{\boldsymbol{r}}, t) \, dt. \quad (3.18)$$

Now define G as the sum over all possible paths of the function $\exp(iS/\hbar)$, which represents a *phase* of the wave function contribution:

$$G(\boldsymbol{r}_f, t_f; \boldsymbol{r}_0, t_0) \stackrel{\text{def}}{=} \sum_{\text{all paths}} e^{iS/\hbar}. \quad (3.19)$$

This, of course, is a mathematically dissatisfying definition, as we do not know how to evaluate "all possible paths" (Fig. 3.1a). Therefore we first approximate a path as a contiguous sequence of *linear* paths over small time steps τ, so a path is defined by straight line segments between the

---

[2] Although Feynman's ideas date from 1948 (Feynman, 1948) and several articles on the subject are available, the most suitable original text to study the subject is the book by Feynman and Hibbs (1965).

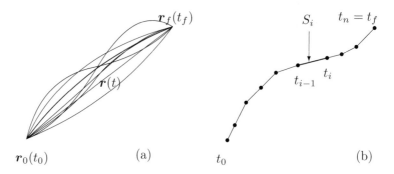

Figure 3.1 (a) Several paths connecting r_0 at time t_0 with r_f at time t_f. One path (thick line) minimizes the action S and represents the path followed by classical mechanics. (b) One path split up into many linear sections (r_{i-1}, r_i) with actions S_i.

initial point r_0 at time t_0, the intermediate points $r_1, \ldots, r_{n-1}$ at times $\tau, 2\tau, \ldots, (n-1)\tau$, and the final point $r_n = r_f$ at time $t_n = t_0 + n\tau$, where $n = (t_f - t_0)/\tau$ (Fig. 3.1b). Then we construct the sum over all possible paths of this kind by integrating $r_1, \ldots, r_{n-1}$ over all space. Finally, we take the limit for $\tau \to 0$:

$$G(r_f, t_f; r_0, t_0) = \lim_{\tau \to 0} C(\tau) \int dr_1 \cdots \int dr_{n-1} \exp\left(\sum_{k=1}^{n} iS_k/\hbar\right), \quad (3.20)$$

where

$$S_k = \int_{t_{k-1}}^{t_k} \mathcal{L}(r, \dot{r}, t) \, dt, \quad (3.21)$$

over the linear path from r_{i-1} to r_i. Note that a normalizing constant $C(\tau)$ is incorporated which takes care of the normalizing condition for G, assuring that the wave function remains normalized in time. This normalizing constant will depend on τ: the more intermediate points we take, the larger the number of possible paths becomes.

Equations (3.20) and (3.21) properly define the right-hand side (3.19). For small time intervals S_k can be approximated by

$$S_k \approx \tau \mathcal{L}(r, \dot{r}, t), \quad (3.22)$$

with $r, \dot{r}$ and t evaluated somewhere in – and with best precision precisely halfway – the interval (t_{k-1}, t_k). The velocity then is

$$\dot{r} = \frac{r_k - r_{k-1}}{\tau}. \quad (3.23)$$

3.3.2 Equivalence with the Schrödinger equation

Thus far the path integral formulation for the Green's function has been simply stated as an alternative postulate of quantum mechanics. We must still prove that this postulate leads to the Schrödinger equation for the wave function. Therefore, we must prove that the time derivative of Ψ, as given by the path integral evolution over an infinitesimal time interval, equals $-i/\hbar$ times the Hamiltonian operator acting on the initial wave function. This is indeed the case, as is shown in the following proof for the case of a single particle in cartesian coordinate space in a possibly time-dependent external field. The extension to many particles is straightforward, as long as the symmetry properties of the total wave function are ignored, i.e., exchange is neglected.

Proof Consider the wave evolution over a small time step τ, from time t to time $t + \tau$:

$$\Psi(\boldsymbol{r}, t + \tau) = \int d\boldsymbol{r}_0 \, G(\boldsymbol{r}, t + \tau; \boldsymbol{r}_0, t) \Psi(\boldsymbol{r}_0, t). \tag{3.24}$$

Now, for convenience, change to the integration variable $\boldsymbol{\delta} = \boldsymbol{r}_0 - \boldsymbol{r}$ and consider the linear path from $\boldsymbol{r}_0$ to $\boldsymbol{r}$. The one-particle Lagrangian is given by

$$\mathcal{L} = \frac{m\boldsymbol{\delta}^2}{2\tau^2} - V(\boldsymbol{r}, t). \tag{3.25}$$

The action over this path is approximated by

$$S \approx \frac{m\boldsymbol{\delta}^2}{2\tau} - V(\boldsymbol{r}, t)\tau, \tag{3.26}$$

which leads to the following evolution of Ψ:

$$\Psi(\boldsymbol{r}, t + \tau) \approx C(\tau) \int d\boldsymbol{\delta} \exp \frac{im\boldsymbol{\delta}^2}{2\hbar\tau} \exp\left(-\frac{i}{\hbar} V(\boldsymbol{r}, t)\tau\right) \Psi(\boldsymbol{r} + \boldsymbol{\delta}, t). \tag{3.27}$$

We now expand both sides to first order in τ, for which we need to expand $\Psi(\boldsymbol{r} + \boldsymbol{\delta}, t)$ to second order in $\boldsymbol{\delta}$. The exponent with the potential energy can be replaced by its first-order term. We obtain

$$\Psi + \tau \frac{\partial \Psi}{\partial t} \approx C(\tau) \left(1 - \frac{i}{\hbar} V\tau\right)$$
$$\times \left[\Psi \int \exp\left(\frac{im\boldsymbol{\delta}^2}{2\hbar\tau}\right) d\boldsymbol{\delta} + \frac{1}{2} \frac{\partial^2 \Psi}{\partial x^2} \int \exp\left(\frac{im\boldsymbol{\delta}^2}{2\hbar\tau}\right) \delta_x^2 \, d\boldsymbol{\delta} + \cdots (y, z)\right] \tag{3.28}$$

where Ψ, its derivatives and V are to be taken at $(\boldsymbol{r}, t)$. The first-order term in δ and the second-order terms containing mixed products as $\delta_x \delta_y$

cancel because they occur in odd functions in the integration. The first integral evaluates[3] to $(ih\tau/m)^{3/2}$, which must be equal to the reciprocal of the normalization constant C, as the zeroth-order term in Ψ must leave Ψ unchanged. The second integral evaluates to $(i\hbar\tau/m)(ih\tau/m)^{3/2}$ and thus the right-hand side of (3.28) becomes

$$\left(1 - \frac{i}{\hbar}V\tau\right)\left(\Psi + \frac{1}{2}\nabla^2\Psi\frac{i\hbar\tau}{m}\right),$$

and the term proportional to τ yields

$$\frac{\partial\Psi}{\partial t} = -\frac{i}{\hbar}\left(-\frac{\hbar^2}{2m}\nabla^2\Psi + V\Psi\right), \qquad (3.29)$$

which is the Schrödinger equation. $\qquad\qquad\square$

3.3.3 The classical limit

From (3.20) we see that different paths will in general contribute widely different phases, when the total actions differ by a quantity much larger than $\hbar$. So most path contributions will tend to cancel by destructive interference, except for those paths that are near to the *path of minimum action* $S_{\min}$. Paths with $S - S_{\min} \approx \hbar$ or smaller add up with roughly the same phase. In the classical approximation, where actions are large compared to $\hbar$, only paths very close to the path of minimum action survive the interference with other paths. So in the classical limit particles will follow the path of minimum action. This justifies the *postulate* of classical mechanics, that the path of minimum action prescribes the equations of motion (see Chapter 15). Perturbations from classical behavior can be derived by including paths close to, but not coinciding with, the classical trajectory.

3.3.4 Evaluation of the path integral

When the Lagrangian can be simply written as

$$\mathcal{L}(\boldsymbol{r}, \dot{\boldsymbol{r}}, t) = \tfrac{1}{2}m\dot{\boldsymbol{r}}^2 - V(\boldsymbol{r}, t), \qquad (3.30)$$

the action S_k over the short time interval (t_{k-1}, t_k) can be approximated by

$$S_k = \frac{m(\boldsymbol{r}_k - \boldsymbol{r}_{k-1})^2}{2\tau} - V(\boldsymbol{r}_k, t_k)\tau, \qquad (3.31)$$

---

[3] This is valid for one particle in three dimensions; for N particles the 3 in the exponent is replaced by $3N$. The use of Planck's constant $h = 2\pi\hbar$ is not an error!

and the kernel becomes

$$G(\boldsymbol{r}_f, t_f; \boldsymbol{r}_0, t_0) = \lim_{\tau \to 0} C(\tau) \int d\boldsymbol{r}_1 \cdots \int d\boldsymbol{r}_{n-1}$$

$$\exp \frac{i}{\hbar} \sum_{k=1}^{n} \left(\frac{m(\boldsymbol{r}_k - \boldsymbol{r}_{k-1})^2}{2\tau} - V(\boldsymbol{r}_k, t_k)\tau \right). \quad (3.32)$$

The normalization constant $C(\tau)$ can be determined by considering the normalization condition for G. A requirement for every kernel is that it conserves the integrated probability density during time evolution from t_1 to t_2:

$$\int \Psi^*(\boldsymbol{r}_2. t_2)\Psi(\boldsymbol{r}_2, t_2)\, d\boldsymbol{r}_2 = \int \Psi^*(\boldsymbol{r}_1, t_1)\Psi(\boldsymbol{r}_1, t_1)\, d\boldsymbol{r}_1, \quad (3.33)$$

or, in terms of G:

$$\int d\boldsymbol{r}_2 \int d\boldsymbol{r}_1' \int d\boldsymbol{r}_1\, G^*(\boldsymbol{r}_2, t_2; \boldsymbol{r}_1', t_1)G(\boldsymbol{r}_2, t_2; \boldsymbol{r}_1, t_1)\Psi^*(\boldsymbol{r}_1', t_1)\Psi(\boldsymbol{r}_1, t_1)$$

$$= \int d\boldsymbol{r}_1\, \Psi^*(\boldsymbol{r}_1, t_1)\Psi(\boldsymbol{r}_1, t_1) \quad (3.34)$$

must be valid for any Ψ. This is only true if

$$\boxed{\int d\boldsymbol{r}_2\, G^*(\boldsymbol{r}_2, t_2; \boldsymbol{r}_1', t_1)G(\boldsymbol{r}_2, t_2; \boldsymbol{r}_1, t_1) = \delta(\boldsymbol{r}_1' - \boldsymbol{r}_1)} \quad (3.35)$$

for any pair of times t_1, t_2 for which $t_2 > t_1$. This is the *normalization condition for G*.

Since the normalization condition must be satisfied for any time step, it must also be satisfied for an infinitesimal time step τ, for which the path is linear from $\boldsymbol{r}_1$ to $\boldsymbol{r}_2$:

$$G(\boldsymbol{r}_2, t + \tau; \boldsymbol{r}_1, t) = C(\tau) \exp \left[\frac{i}{\hbar} \left(\frac{m(\boldsymbol{r}_2 - \boldsymbol{r}_1)^2}{2\tau} + V(\boldsymbol{r}, t)\tau \right) \right]. \quad (3.36)$$

If we apply the normalization condition (3.35) to this G, we find that

$$C(\tau) = \left(\frac{ih\tau}{m} \right)^{-3/2}, \quad (3.37)$$

just as we already found while proving the equivalence with the Schrödinger equation. The 3 in the exponent relates to the dimensionality of the wave function, here taken as three dimensions (one particle in 3D space). For N particles in 3D space the exponent becomes $-3N/2$.

Proof Consider N particles in 3D space. Now

$$\int G^*G \, d\mathbf{r}_2 = C^*C \int d\mathbf{r}_2 \exp\left[\frac{im}{2\hbar\tau}\{(\mathbf{r}_2 - \mathbf{r}_1)^2 - (\mathbf{r}_2 - \mathbf{r}_1')^2\}\right]$$

$$= C^*C \exp\left[\frac{im}{2\hbar\tau}(r_1^2 - r_1'^2)\right] \int d\mathbf{r}_2 \exp\left[\frac{im}{\hbar\tau}(\mathbf{r}_1' - \mathbf{r}_1) \cdot \mathbf{r}_2\right].$$

Using one of the definitions of the δ-function:

$$\int_{-\infty}^{+\infty} \exp(\pm ikx) \, dx = 2\pi\delta(k),$$

or, in $3N$ dimensions:

$$\int \exp(\pm i\mathbf{k} \cdot \mathbf{r}) \, d\mathbf{r} = (2\pi)^{3N}\delta(\mathbf{k}),$$

where the delta function of a vector is the product of delta functions of its components, the integral reduces to

$$(2\pi)^{3N}\delta\left(\frac{m(\mathbf{r}_1' - \mathbf{r}_1)}{\hbar\tau}\right) = (2\pi)^{3N}\left(\frac{\hbar\tau}{m}\right)^{3N}\delta(\mathbf{r}_1' - \mathbf{r}_1).$$

Here we have made use of the transformation $\delta(ax) = (1/a)\delta(x)$. The presence of the delta functions means that the exponential factor before the integral reduces to 1, and we obtain

$$\int G^*G \, d\mathbf{r}_2 = C^*C \left(\frac{\hbar\tau}{m}\right)^{3N}\delta(\mathbf{r}_1' - \mathbf{r}_1).$$

Thus the normalization condition (3.35) is satisfied if

$$C^*C = \left(\frac{\hbar\tau}{m}\right)^{-3N}.$$

This is a sufficient condition to keep the integrated probability of the wave function invariant in time. But there are many solutions for C, differing by an arbitrary phase factor, as long as the absolute value of C equals the square root of the right-hand side (real) value. However, we do not wish a solution for G that introduces a changing phase into the wave function, and therefore the solution for C found in the derivation of the Schrödinger equation, which leaves not only $\Psi^*\Psi$, but also Ψ itself invariant in the limit of small τ, is the appropriate solution. This is (3.37).   □

Considering that we must make $n = (t_f - t_0)/\tau$ steps to evolve the system from t_0 to $t_f = t_n$, we can rewrite (3.32) for a N-particle system as

$$G(\mathbf{r}_f, t_f; \mathbf{r}_0, t_0) = \lim_{\tau \to 0} \left(\frac{ih\tau}{m}\right)^{-3nN/2} \int d\mathbf{r}_1 \cdots \int d\mathbf{r}_{n-1}$$

$$\exp \frac{i}{\hbar} \sum_{k=1}^{n} \left(\frac{m(\mathbf{r}_k - \mathbf{r}_{k-1})^2}{2\tau} - V(\mathbf{r}_k, t_k)\tau\right). \quad (3.38)$$

Here, $\mathbf{r}$ is a $3N$-dimensional vector. Note that in the limit $\tau \to 0$, the number of steps n tends to infinity, keeping $n\tau$ constant. The potential V may still be an explicit function of time, for example due to a time-dependent source. In most cases solutions can only be found by numerical methods. In simple cases with time-independent potentials (free particle, harmonic oscillator) the integrals can be evaluated analytically.

In the important case that the quantum system is bound in space and not subjected to a time-dependent external force, the wave function at time t_0 can be expanded in an orthonormal set of eigenfunctions ϕ_n of the Hamiltonian:

$$\Psi(\mathbf{r}, t_0) = \sum_n a_n \phi_n(\mathbf{r}). \quad (3.39)$$

As the eigenfunction ϕ_n develops in time proportional to $\exp(-iE_n t/\hbar)$, we know the time dependence of the wave function:

$$\Psi(\mathbf{r}, t) = \sum_n a_n \phi_n(\mathbf{r}) \exp\left(-\frac{i}{\hbar} E_n(t - t_0)\right). \quad (3.40)$$

¿From this it follows that the kernel must have the following form:

$$\boxed{G(\mathbf{r}, t; \mathbf{r}_0, t_0) = \sum_n \phi_n(\mathbf{r})\phi_n^*(\mathbf{r}_0) \exp\left(-\frac{i}{\hbar} E_n(t - t_0)\right).} \quad (3.41)$$

This is easily seen by applying G to the initial wave function $\sum_m a_m \phi_m(\mathbf{r}_0)$ and integrating over $\mathbf{r}_0$. So in this case the path integral kernel can be expressed in the eigenfunctions of the Hamiltonian.

3.3.5 Evolution in imaginary time

A very interesting and useful connection can be made between path integrals and the canonical partition function of statistical mechanics. This connection suggests a numerical method for computing the thermodynamic properties of systems of quantum particles where the symmetry properties of wave functions and, therefore, effects of exchange can be neglected. This is usually the case when systems of atoms or molecules are considered at normal temperatures: the repulsion between atoms is such that the quantum-mechanical exchange between particles (nuclei) is irrelevant. The quantum effects due to symmetry properties (distinguishing fermions and bosons) are completely drowned in the quantum effects due to the curvature of the interatomic potentials within the de Broglie wavelength of the nuclei.

Consider the quantum-mechanical canonical partition function of an N-particle system

$$Q = \sum_n \exp(-\beta E_n), \tag{3.42}$$

where the sum is to be taken over all quantum states (not energy levels!) and β equals $1/k_{\mathrm{B}}T$. Via the free energy relation (A is the Helmholtz free energy)

$$A = -k_{\mathrm{B}}T \ln Q \tag{3.43}$$

and its derivatives with respect to temperature and volume, all relevant thermodynamic properties can be obtained. Unfortunately, with very few exceptions under idealized conditions, we cannot enumerate all quantum states and energy levels of a complex system, as this would mean the determination of all eigenvalues of the full Hamiltonian of the system.

Since the eigenfunctions ϕ_n of the system form an orthonormal set, we can also write (3.42) as

$$Q = \int d\boldsymbol{r} \sum_n \phi_n(\boldsymbol{r})\phi_n^*(\boldsymbol{r}) \exp(-\beta E_n). \tag{3.44}$$

Now compare this with the expression for the path integral kernel of (3.41). Apart from the fact that initial and final point are the same ($\boldsymbol{r}$), and the form is integrated over $d\boldsymbol{r}$, we see that instead of time we now have $-i\hbar\beta$. So the canonical partition function is closely related to a path integral over negative *imaginary time*. The exact relation is

$$Q = \int d\boldsymbol{r} G(\boldsymbol{r}, -i\hbar\beta; \boldsymbol{r}, 0), \tag{3.45}$$

with (inserting $\tau = -i\hbar\beta/n$ into (3.38))

$$G(\boldsymbol{r}, -i\hbar\beta; \boldsymbol{r}, 0) = \lim_{n \to \infty} C(n) \int d\boldsymbol{r}_1 \cdots \int d\boldsymbol{r}_{n-1}$$
$$\exp\left(-\beta \sum_{k=1}^n \left[\frac{nm}{2\hbar^2\beta^2}(\boldsymbol{r}_k - \boldsymbol{r}_{k-1})^2 + \frac{1}{n}V(\boldsymbol{r}_k)\right]\right), \tag{3.46}$$

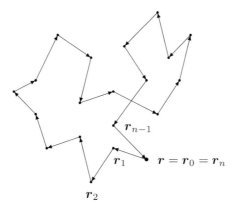

Figure 3.2 A closed path in real space and imaginary time, for the calculation of quantum partition functions.

where

$$C(n) = \left(\frac{h^2 \beta}{2\pi n m} \right)^{-3nN/2}, \tag{3.47}$$

and

$$r_0 = r_n = r.$$

Note that r stands for a $3N$-dimensional cartesian vector of all particles in the system. Also note that all paths in the path integral are closed: they end in the same point where they start (Fig. 3.2). In the multiparticle case, the imaginary time step is made for all particles simultaneously; each particle therefore traces out a three-dimensional path.

A path integral over imaginary time does not add up phases of different paths, but adds up real exponential functions over different paths. Only paths with reasonably-sized exponentials contribute; paths with highly negative exponents give negligible contributions. Although it is difficult to imagine what imaginary-time paths mean, the equations derived for real-time paths can still be used and lead to real integrals.

3.3.6 Classical and nearly classical approximations

Can we easily see what the classical limit is for imaginary-time paths? Assume that each path (which is closed anyway) does not extend very far from its initial and final point r. Assume also that the potential does not vary appreciably over the extent of each path, so that it can be taken equal to $V(\boldsymbol{r})$ for the whole path. Then we can write, instead of (3.46):

$$G(\boldsymbol{r}, -i\hbar\beta; \boldsymbol{r}, 0) = \exp(-\beta V(\boldsymbol{r})) \lim_{n\to\infty} C(n) \int d\boldsymbol{r}_1 \cdots \int d\boldsymbol{r}_{n-1}$$

$$\exp\left(-\frac{nm}{2\hbar^2\beta} \sum_{k=1}^{n} (\boldsymbol{r}_k - \boldsymbol{r}_{k-1})^2\right). \qquad (3.48)$$

The expression under the limit sign yields $(2\pi m k_{\mathrm{B}} T/h^2)^{3N/2}$, *independent of the number of nodes n.* The evaluation of the multiple integral is not entirely trivial, and the proof is given below. Thus, after integrating over $\boldsymbol{r}$ we find the *classical* partition function

$$Q = \left(\frac{2\pi m k_{\mathrm{B}} T}{h^2}\right)^{3N/2} \int e^{-\beta V(\boldsymbol{r})} \, d\boldsymbol{r}. \qquad (3.49)$$

Since the expression is independent of n, there is no need to take the limit for $n \to \infty$. Therefore the imaginary-time path integral without any intervening nodes also represents the classical limit.

Note that the integral is not divided by $N!$, since the indistinguishability of the particles has not been introduced in the path integral formalism. Therefore we cannot expect that path integrals for multiparticle systems will treat exchange effects correctly. For the application to nuclei in condensed matter, which are always subjected to strong short-range repulsion, exchange effects play no role at all.

Proof We prove that

$$C(n) \int d\boldsymbol{r}_1 \ldots \int d\boldsymbol{r}_{n-1} \exp\left(-\frac{nm}{2\hbar^2\beta} \sum_{k=1}^{n} (\boldsymbol{r}_k - \boldsymbol{r}_{k-1})^2\right) = \left(\frac{2\pi m k_{\mathrm{B}} T}{h^2}\right)^{3N/2}.$$

First make a coordinate transformation from $\boldsymbol{r}_k$ to $\boldsymbol{s}_k = \boldsymbol{r}_k - \boldsymbol{r}_0, k = 1, \ldots, n-1$. Since the Jacobian of this transformation equals one, $d\boldsymbol{r}$ can be replaced by $d\boldsymbol{s}$. Inspection of the sum shows that the integral I now becomes

$$I = \int d\boldsymbol{s}_1 \cdots \int d\boldsymbol{s}_{n-1} \exp[-\alpha\{\boldsymbol{s}_1^2 + (\boldsymbol{s}_2 - \boldsymbol{s}_1)^2 + \cdots + (\boldsymbol{s}_{n-1} - \boldsymbol{s}_{n-2})^2 + \boldsymbol{s}_{n-1}^2\}],$$

where

$$\alpha = \frac{nm}{2\hbar^2\beta}.$$

The expression between { } in $\boldsymbol{s}$ can be written in matrix notation as

$$\boldsymbol{s}^{\mathsf{T}} \mathbf{A}_n \boldsymbol{s},$$

with $\mathbf{A}_n$ a symmetric tridiagonal $(n-1) \times (n-1)$ matrix with 2 along the diagonal, -1 along both subdiagonals, and zero elsewhere. The integrand becomes a product of independent Gaussians after an orthogonal transformation that diagonalizes $\mathbf{A}_n$; thus the integral depends only on the product of eigenvalues, and evaluates to

$$I = \left(\frac{\pi}{\alpha}\right)^{3(n-1)N/2} (\det \mathbf{A}_n)^{-3N/2}.$$

The determinant can be easily evaluated from its recurrence relation,

$$\det \mathbf{A}_n = 2 \det \mathbf{A}_{n-1} - \det \mathbf{A}_{n-2},$$

and turns out to be equal to n. Collecting all terms, and replacing $C(n)$ by (3.47), we find the required result. Note that the number of nodes n cancels: the end result is valid for any n. □

In special cases, notably a free particle and a particle in an isotropic harmonic potential, analytical solutions to the partition function exist. When the potential is not taken constant, but approximated by a Taylor expansion, quantum corrections to classical simulations can be derived as perturbations to a properly chosen analytical solution. These applications will be treated in Section 3.5; here we shall derive the analytical solutions for the two special cases mentioned above.

3.3.7 The free particle

The canonical partition function of a system of N free (non-interacting) particles is a product of $3N$ independent terms and is given by

$$Q = \lim_{n \to \infty} Q^{(n)},$$

$$Q^{(n)} = (q^{(n)})^{3N},$$

$$q^{(n)} = \left(\frac{2\pi mn}{h^2 \beta}\right)^{n/2} \int dx_0 \cdots \int dx_{n-1}$$

$$\exp\left[-a \sum_{k=1}^{n}(x_k - x_{k-1})^2\right], \tag{3.50}$$

$$a = \frac{nm}{2\hbar^2 \beta}; \quad x_n = x_0. \tag{3.51}$$

The sum in the exponent can be written in matrix notation as

$$\sum_{k=1}^{n}(x_k - x_{k-1})^2 = \mathbf{x}^{\mathsf{T}}\mathbf{A}\mathbf{x} = \mathbf{y}^{\mathsf{T}}\mathbf{\Lambda}\mathbf{y}, \tag{3.52}$$

where $\mathbf{A}$ is a symmetric cyclic tridiagonal matrix:

$$\mathbf{A} = \begin{pmatrix} 2 & -1 & 0 & & & & -1 \\ -1 & 2 & -1 & & & & 0 \\ & & & \cdots & & & \\ 0 & & & & -1 & 2 & -1 \\ -1 & & & & 0 & -1 & 2 \end{pmatrix} \tag{3.53}$$

and $\mathbf{y}$ is a set of coordinates obtained by the *orthogonal* transformation $\mathbf{T}$ of $\mathbf{x}$ that diagonalizes $\mathbf{A}$ to the diagonal matrix of eigenvalues $\mathbf{\Lambda} = \mathrm{diag}\,(\lambda_0, \ldots, \lambda_{n-1})$:

$$\mathbf{y} = \mathbf{Tx}, \quad \mathbf{T}^{-1} = \mathbf{T}^\mathsf{T}, \quad \mathbf{x}^\mathsf{T}\mathbf{A}\mathbf{x} = \mathbf{y}^\mathsf{T}\mathbf{TAT}^\mathsf{T}\mathbf{y} = \mathbf{y}^\mathsf{T}\mathbf{\Lambda}\mathbf{y}. \tag{3.54}$$

There is one zero eigenvalue, corresponding to an eigenvector proportional to the sum of x_k, to which the exponent is invariant. This eigenvector, which we shall label "0," must be separated. The eigenvector $\mathbf{y}_0$ is related to the *centroid* or average coordinate x_c:

$$x_c \stackrel{\text{def}}{=} \frac{1}{n}\sum_{k=0}^{n-1} x_k, \tag{3.55}$$

$$\mathbf{y}_0 = \frac{1}{\sqrt{n}}(1,1,\ldots,1)^\mathsf{T} = \mathbf{r}_c\sqrt{n}. \tag{3.56}$$

Since the transformation is orthogonal, its Jacobian equals 1 and integration over $d\mathbf{x}$ can be replaced by integration over $d\mathbf{y}$. Thus we obtain

$$q^{(n)} = \left(\frac{2\pi mn}{h^2\beta}\right)^{n/2} n^{1/2} \int dx_c \int dy_1 \cdots \int dy_{n-1} \exp\left[-a\sum_{k=1}^{n-1}\lambda_k y_k^2\right]. \tag{3.57}$$

Thus the distribution of node coordinates (with respect to the centroid) is a multivariate Gaussian distribution. Its integral equals

$$\int dy_1 \cdots \int dy_{n-1} \exp\left[-a\sum_{k=1}^{n-1}\lambda_k y_k^2\right] = \left(\frac{\pi}{a}\right)^{(n-1)/2}\left(\Pi_{k=1}^{n-1}\lambda_k\right)^{-1/2}. \tag{3.58}$$

The product of non-zero eigenvalues of matrix $\mathbf{A}$ turns out to be equal to n^2 (valid for any n). Collecting terms we find that the partition function equals the classical partition function for a 1D free particle for any n:

$$q^{(n)} = \left(\frac{2\pi m}{h^2\beta}\right)^{1/2}\int dx_c, \tag{3.59}$$

as was already shown to be the case in (3.49).

Table 3.4 *Intrinsic variance of a discrete imaginary-time path for a one-dimensional free particle as a function of the number of nodes in the path, in units of $\hbar^2\beta/m$*

n	σ^2	n	σ^2	n	σ^2
2	0.062 500	8	0.082 031	50	0.083 300
3	0.074 074	9	0.082 305	60	0.083 310
4	0.078 125	10	0.082 500	70	0.083 316
5	0.080 000	20	0.083 125	80	0.083 320
6	0.081 019	30	0.083 241	90	0.083 323
7	0.081 633	40	0.083 281	∞	0.083 333

The variance σ^2 of the multivariate distribution is given by

$$\sigma^2 \stackrel{\text{def}}{=} \left\langle \frac{1}{n}\sum_{k=0}^{n-1}(x_k - x_c)^2 \right\rangle = \frac{1}{n}\sum_{k=1}^{n-1}\langle y_k^2\rangle = \frac{1}{2an}\sum_{k=1}^{n-1}\lambda_k^{-1} = \frac{\hbar^2\beta}{n^2 m}\sum_{k=1}^{n-1}\lambda_k^{-1}.$$
(3.60)

We shall call this the *intrinsic* variance in order to distinguish this from the distribution of the centroid itself. The sum of the inverse non-zero eigenvalues has a limit of $n^2/12 = 0.083\,33\ldots n^2$ for $n \to \infty$. In Table 3.4 the intrinsic variance of the node distribution is given for several values of n, showing that the variance quickly converges: already 96% of the limiting variance is realized by a path consisting of five nodes.

3.3.8 Non-interacting particles in a harmonic potential

The other solvable case is a system of N non-interacting particles that each reside in an external harmonic potential $V = \frac{1}{2}m\omega^2 r^2$. Again, the partition function is the product of $3N$ 1D terms:

$$Q = \lim_{n\to\infty} (q^{(n)})^{3N},$$

$$q^{(n)} = \left(\frac{2\pi mn}{h^2\beta}\right)^{n/2} \int dx_0 \cdots \int dx_{n-1}$$

$$\exp\left[-a\sum_{k=1}^{n}(x_k - x_{k-1})^2 - b\sum_{k=1}^{n}x_k^2\right],$$
(3.61)

$$b = \frac{\beta m\omega^2}{2n},$$
(3.62)

a is defined in (3.51). We can separate the centroid harmonic term as follows:

$$\frac{1}{n}\sum_{k=1}^{n} x_k^2 = x_c^2 + \frac{1}{n}\sum_{k=1}^{n}(x_k - x_c)^2 = x_c^2 + \frac{1}{n}\sum_{k=1}^{n-1} y_k^2, \tag{3.63}$$

and obtain for the 1D partition function

$$q^{(n)} = \left(\frac{2\pi mn}{h^2\beta}\right)^{n/2} n^{1/2} \int \exp[-\frac{1}{2}\beta m\omega^2 x_c^2]\, dx_c$$
$$\times \int dy_1 \cdots \int dy_{n-1} \exp\left[-a\mathbf{y}^\mathsf{T}\mathbf{B}\mathbf{y}\right], \tag{3.64}$$

where $\mathbf{B}$ is a modified version of $\mathbf{A}$ from (3.53):

$$\mathbf{B} = \mathbf{A} + \frac{(\beta\hbar\omega)^2}{n^2}\mathbf{1}. \tag{3.65}$$

The eigenvalues of $\mathbf{B}$ are equal to the eigenvalues of $\mathbf{A}$ increased with $(\beta\hbar\omega/n)^2$ and the end result is

$$q^{(n)} = \left(\frac{2\pi mn}{h^2\beta}\right)^{n/2} n^{1/2} \int \exp\left[-\frac{1}{2}\beta m\omega^2 x_c^2\right]\, dx_c$$
$$\times \Pi_{k=1}^{n-1} \int \exp[-a\lambda_k' y_k^2]\, dy_k \tag{3.66}$$
$$= \left(\frac{2\pi m}{h^2\beta}\right)^{1/2} \int_{-\infty}^{\infty} \exp\left[-\frac{1}{2}\beta m\omega^2 x_c^2\right]\, dx_c \left(\frac{1}{n^2}\Pi_{k=1}^{n-1}\lambda_k'\right)^{-1/2},$$
$$\lambda_k' = \lambda_k + \frac{(\beta\hbar\omega)^2}{n^2}. \tag{3.67}$$

So we need the product of all eigenvalues $\lambda_1', \ldots, \lambda_{n-1}'$. For $n \to \infty$ the partition function goes to the quantum partition function of the harmonic oscillator (see (17.84) on page 472), which can be written in the form

$$q^{(\infty)} = \left(\frac{2\pi m}{h^2\beta}\right)^{1/2} \int \exp\left[-\frac{1}{2}\beta m\omega^2 x_c^2\right]\, dx_c$$
$$\times \frac{\xi}{2\sinh(\xi/2)}, \quad \xi = \beta\hbar\omega. \tag{3.68}$$

The last term is an exact quantum correction to the classical partition function. From this it follows that

$$\lim_{n\to\infty} \frac{1}{n^2}\Pi_{k=1}^{n-1}\lambda_k' = \frac{\xi}{2\sinh(\xi/2)}, \tag{3.69}$$

which can be verified by direct evaluation of the product of eigenvalues for large n. Figure 3.3 shows the temperature dependence of the free energy and

Helmholtz free energy *A/hν*

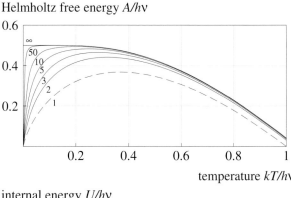

temperature *kT/hν*

internal energy *U/hν*

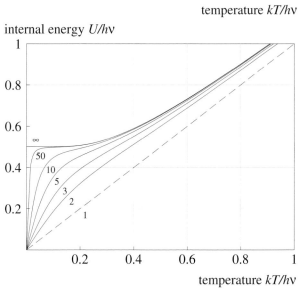

temperature *kT/hν*

Figure 3.3 Free energy A and energy U of a 1D harmonic oscillator evaluated as an imaginary-time path integral approximated by n nodes. The curves are labeled with n; the broken line represents the classical approximation $(n = 1)$; $n = \infty$ represents the exact quantum solution. Energies are expressed in units of $h\nu = \hbar\omega$.

the energy of a 1D harmonic oscillator, evaluated by numerical solution of (3.67) for several values of n. The approximation fairly rapidly converges to the exact limit, but for low temperatures a large number of nodes is needed, while the limit for $T = 0$ $(A = U = 0.5\hbar\omega)$ is never reached correctly.

The values of U, as plotted in Fig. 3.3, were obtained by numerical differentiation as $U = A - T dA/dT$. One can also obtain $U = -\partial \ln q/\partial \beta$ by differentiating q of (3.67), yielding

$$U^{(n)} = \frac{1}{\beta} + \beta \hbar^2 \omega^2 \frac{1}{n^2} \sum_{k=1}^{n-1} \frac{1}{\lambda'_k}. \qquad (3.70)$$

The first term is the internal energy of the classical oscillator while the second term is a correction caused by the distribution of
nodes relative to the centroid. Both terms consist of two equal halves representing kinetic and potential energy, respectively.

The intrinsic variance of the distribution of node coordinates, i.e., relative to the centroid, is – as in (3.60) – given by

$$\sigma_{\text{intr}}^2 = \frac{\hbar^2 \beta}{m} \frac{1}{n^2} \sum_{k=1}^{n-1} \frac{1}{\lambda_k'}. \tag{3.71}$$

We immediately recognize the second term of (3.70). If we add the "classical" variance of the centroid itself

$$\sigma_{\text{centroid}}^2 = \frac{1}{\beta m \omega^2}, \tag{3.72}$$

we obtain the total variance, which can be related to the total energy:

$$\sigma^2 = \sigma_{\text{centroid}}^2 + \sigma_{\text{intr}}^2 = \frac{U}{m\omega^2}. \tag{3.73}$$

This is compatible with U being twice the potential energy U_{pot}, which equals $0.5m\omega^2 \langle x^2 \rangle$. Its value as a function of temperature is proportional to the curve for U (case $n \to \infty$) in Fig. 3.3. As is to be expected, the total variance for $n \to \infty$ equals the variance of the wave function, averaged over the occupation of quantum states v:

$$\langle x^2 \rangle = \sum_{v=0}^{\infty} P(v) \langle x^2 \rangle_v = \sum_{v=0}^{\infty} \frac{P(v) E_v}{m\omega^2} = \frac{U^{\text{qu}}}{m\omega^2}, \tag{3.74}$$

where $P(v)$ is the probability of occurrence of quantum state v, because the variance of the wave function in state v is given by

$$\langle x^2 \rangle_v = \int \Psi_v^* x^2 \Psi \, dx = \frac{(v + \frac{1}{2})\hbar}{m\omega} = \frac{E_v}{m\omega^2}. \tag{3.75}$$

Note that this relation between variance and energy is not only valid for a canonical distribution, but for any distribution of occupancies.

We may summarize the results for $n \to \infty$ as follows: a particle can be considered as a distribution of imaginary-time closed paths around the centroid of the particle. The intrinsic (i.e., with respect to the centroid) spatial distribution for a free particle is a multivariate Gaussian with a variance of $\hbar^2 \beta/(12m)$ in each dimension. The variance (in each dimension) of the distribution for a particle in an isotropic harmonic well (with force

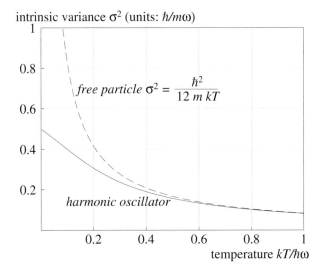

intrinsic variance σ^2 (units: $\hbar/m\omega$)

free particle $\sigma^2 = \dfrac{\hbar^2}{12\,m\,kT}$

harmonic oscillator

temperature $kT/\hbar\omega$

Figure 3.4 The intrinsic variance in one dimension of the quantum imaginary-time path distribution for the free particle and for the harmonic oscillator.

constant $m\omega^2$) is given by

$$\sigma^2_{\text{intr}} = \frac{U^{\text{qu}} - U^{\text{cl}}}{m\omega^2} = \frac{\hbar}{m\omega}\left(\frac{1}{2}\coth\frac{\xi}{2} - \frac{1}{\xi}\right), \quad \xi = \beta\hbar\omega. \tag{3.76}$$

For high temperatures (small ξ) this expression goes to the free-particle value $\hbar^2\beta/12m$; for lower temperatures the variance is reduced because of the quadratic potential that restrains the spreading of the paths; the low-temperature (ground state) limit is $\hbar/(2m\omega)$. Figure 3.4 shows the intrinsic variance as a function of temperature.

3.3.9 Path integral Monte Carlo and molecular dynamics simulation

The possibility to use imaginary-time path integrals for equilibrium quantum simulations was recognized as early as 1962 (Fosdick, 1962) and developed in the early eighties (Chandler and Wolynes, 1981; Ceperley and Kalos, 1981; and others). See also the review by Berne and Thirumalai (1986). Applications include liquid neon (Thirumalai *et al.*, 1984), hydrogen diffusion in metals (Gillan, 1988), electrons in fused salts (Parrinello and Rahman, 1984, using a molecular dynamics variant), hydrogen atoms and muonium in water (de Raedt *et al.*, 1984), and liquid water (Kuharski and Rossky, 1985; Wallqvist and Berne, 1985).

If the potential is not approximated, but evaluated for every section of the path, the expression for Q becomes

$$Q = \left(\frac{2\pi m k_B T}{h^2}\right)^{3N/2} \int d\mathbf{r} \lim_{n\to\infty} \left(\frac{2\pi m k_B T}{h^2}\right)^{3(n-1)N/2} n^{3nN/2}$$

$$\times \int d\mathbf{r}_1 \cdots \int d\mathbf{r}_{n-1} \exp\left(-\beta \sum_{k=1}^{n} \left[\frac{nm}{2\hbar^2\beta^2}(\mathbf{r}_k - \mathbf{r}_{k-1})^2 + \frac{V(\mathbf{r}_k)}{n}\right]\right), \quad (3.77)$$

with $\mathbf{r} = \mathbf{r}_n = \mathbf{r}_0$. The constant after the limit symbol exactly equals the inverse of the integral over the harmonic terms only, as was shown in the proof of the classical limit given on page 54:

$$\left(\frac{2\pi m k_B T}{h^2}\right)^{-3(n-1)N/2} n^{-3nN/2} = \quad (3.78)$$

$$\int d\mathbf{r}_1 \cdots \int d\mathbf{r}_{n-1} \exp\left(-\beta \sum_{k=1}^{n-1} \left[\frac{nm}{2\hbar^2\beta^2}(\mathbf{r}_k - \mathbf{r}_{k-1})^2\right]\right). \quad (3.79)$$

It therefore "compensates" in the partition function for the harmonic terms in the extra degrees of freedom that are introduced by the beads.

Interestingly, the expression for Q in (3.77) is proportional to the partition function of a system of particles, where each particle i is represented by a *closed string of beads* (a "necklace"), with two adjacent beads connected by a harmonic spring with spring constant

$$\kappa_i = \frac{nm_i}{\hbar^2\beta^2}, \quad (3.80)$$

and feeling $1/n$ of the interparticle potential at the position of each bead. The interaction $V(\mathbf{r}_i - \mathbf{r}_j)$ between two particles i and j acts at the k-th step between the particles positioned at the k-th node $\mathbf{r}_k$. Thus the k-th node of particle i interacts only with the k-th node of particle j (Fig. 3.5), with a strength of $1/n$ times the full interparticle interaction.

The propagator (3.48) used to derive the "string-of-beads" homomorphism, is a high-temperature free particle propagator, which – although in principle exact – converges slowly for bound particles in potential wells at low temperature. Mak and Andersen (1990) have devised a "low-temperature" propagator that is appropriate for particles in harmonic wells. It contains the resonance frequency (for example derived from the second derivative of the potential) and converges faster for bound states with similar frequencies.

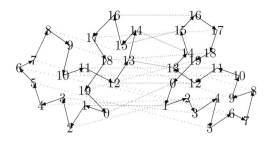

Figure 3.5 The paths of two interacting particles. Interactions act between equally-numbered nodes, with strength $1/n$.

For this propagator the Boltzmann term in (3.77) reads

$$\exp\left(-\beta\sum_{k=1}^{n-1}\left[\frac{m\omega(\boldsymbol{r}_k-\boldsymbol{r}_{k-1})^2}{2\hbar\beta\sinh(\beta\hbar\omega/n)}+\frac{2\{\cosh(\beta\hbar\omega/n)-1\}}{\beta\hbar\omega\sinh(\beta\hbar\omega/n)}V(\boldsymbol{r}_k)\right]\right).\quad(3.81)$$

The system consisting of strings of beads can be simulated in equilibrium by conventional classical Monte Carlo (MC) or molecular dynamics (MD) methods. If MD is used, and the mass of each particle is evenly distributed over its beads, the time step will become quite small. The oscillation frequency of the bead harmonic oscillators is approximately given by nk_BT/h, which amounts to about 60 THz for a conservative number of ten beads per necklace, at $T = 300$ K. Taking 50 steps per oscillation period then requires a time step as small as 0.3 fs. Such PIMC or PIMD simulations will yield a set of necklace configurations (one necklace per atom) that is representative for an equilibrium ensemble at the chosen temperature. The solution is in principle exact in the limit of an infinite number of beads per particle, if exchange effects can be ignored.

While the PIMC and PIMD simulations are valid for equilibrium systems, their use in non-equilibrium dynamic simulations is questionable. One may equilibrate the "quantum part," i.e., the necklace configurations, at any given configuration of the geometric centers of the necklaces, either by MC or MD, and compute the necklace-averaged forces between the particles. Then one may move the system one MD step ahead with those effective forces. In this way a kind of quantum dynamics is produced, with the momentum change given by quantum-averaged forces, rather than by forces evaluated at the quantum-averaged positions. This is exactly what should be done for the momentum expectation value, according to the derivation by Ehrenfest (see page 43). One may hope that this method of computing forces incorporates essential quantum effects in the dynamics of the system. However, this cannot be proven, as the wave function distribution that is generated contains no memory of the past dynamics as it should in a full quantum-dynamical treatment. Neither does this kind of "quantum dynamics" produce a bifurcation into more than one quantum state. Note that this method cannot handle exchange.[4]

In Section 3.5 we'll return to the path-integral methods and employ them to make approximate quantum corrections to molecular dynamics.

3.4 Quantum hydrodynamics

In this section a different approach to quantum mechanics is considered, which originates from Madelung (1926) and de Broglie (1927) in the 1920's, and was revived by Bohm (1952a, 1952b) in the fifties. It survived the following decades only in the periphery of the main stream of theoretical physics, but has more recently regained interest because of its applicability to simulation. The approach is characterized by the use of a classical model consisting of a system of particles or a classical fluid that – under certain modified equations of motions that include a *quantum force* – behaves according to the Schrödinger equation. Models based on deterministic fluid dynamics are known as *quantum hydrodynamics* or *Madelung fluid*. If the fluid is represented by a statistical distribution of point particles, the term *Bohmian particle dynamics* is often used. Particles constituting the quantum fluid have also been called "beables," in analogy and contrast to the "observables" of traditional quantum mechanics (Bell 1976, 1982; Vink 1993). The quantum force is supposed to originate from a wave that accompanies the quantum particles.[5]

In addition to the interpretation in terms of a fluid or a distribution of particles behaving according to *causal* relations, several attempts have been made to eliminate the quantum force and ascribe its effects to the diffusional behavior of particles that undergo stochastic forces due to some unknown external agent. Such *quantum stochastic dynamics* methods (Fényes 1952; Weizel 1954; Kershaw 1964; Nelson 1966; Guerra 1981) will not be considered further in our context as they have not yet led to useful simulation techniques.

It is possible to rewrite the time-dependent Schrödinger equation in a different form, such that the square of the wave function can be interpreted

---

[4] Exchange can be introduced into path integral methods, see Roy and Voth (1999), and should never be applied to electrons in systems with more than one electron.

[5] See for an extensive description of the particle interpretation, including a discussion of its origins, the book of Holland (1993).

as the density of a frictionless classical fluid evolving under hydrodynamic equations, with the fluid particles subjected to a quantum-modified force. The force consists of two parts: the *potential force*, which is minus the gradient of the potential V, and a *quantum force*, which is minus the gradient of a *quantum potential* Q. The latter is related to the local curvature of the density distribution. This mathematical equivalence can be employed to generate algorithms for the simulation of the evolution of wave packets, but it can also be used to evoke a new interpretation of quantum mechanics in terms of *hidden variables* (positions and velocities of the fluid particles).

Unfortunately, the hidden-variable aspect has dominated the literature since Bohm. Unfortunately, because any invocation of hidden variables in quantum mechanics is in conflict with the (usual) Copenhagen interpretation of quantum mechanics, and rejected by the main stream physicists. The Copenhagen interpretation[6] considers the wave function of a system of particles as no more than an expression from which the probability of the outcome of a measurement of an observable can be derived; it attaches no meaning to the wave function as an actual, physically real, attribute of the system. The wave function expresses all there is to know about the system from the point of view of an external observer. Any interpretation in terms of more details or hidden variables does not add any knowledge that can be subjected to experimental verification and is therefore considered by most physicists as irrelevant.

We shall not enter the philosophical discussion on the interpretation of quantum mechanics at all, as our purpose is to simulate quantum systems including the evolution of wave functions. But this does not prevent us from considering hypothetical systems of particles that evolve under specified equations of motion, when the wave function evolution can be derived from the behavior of such systems by mathematical equivalence. A similar equivalence is the path-integral Monte Carlo method to compute the evolution of ensemble-averaged quantum behavior (see Section 3.3.9), where a ring of particles interconnected by springs has a classical statistical behavior that is mathematically equivalent to the ensemble-averaged wave function evolution of a quantum particle. Of course, such equivalences are only useful when they lead to simulation methods that are either simpler or more efficient than the currently available ones. One of the reasons that interpretations of the quantum behavior in terms of distributions of particles can be quite useful in simulations is that such interpretations allow the construction of quantum trajectories which can be more naturally combined with

---

[6] Two articles, by Heisenberg (1927) and Bohr (1928), have been reprinted, together with a comment on the Copenhagen interpretation by A. Herrmann, in Heisenberg and Bohr (1963).

classical trajectories. They may offer solutions to the problem how to treat the back reaction of the quantum subsystem to the classical degrees of freedom. The ontological question of *existence* of the particles is irrelevant in our context and will not be considered.

3.4.1 The hydrodynamics approach

Before considering a quantum particle, we shall first summarize the equations that describe the time evolution of a fluid. This topic will be treated in detail in Chapter 9, but we only need a bare minimum for our present purpose.

Assume we have a fluid with mass density $m\rho(r,t)$ and fluid velocity $u(r,t)$. We shall consider a fluid confined within a region of space such that $\int \rho\, dr = 1$ at all times, so the total mass of the fluid is m. The fluid is homogeneous and could consist of a large number $N \to \infty$ of "particles" with mass m/N and number density $N\rho$, but ρ could also be interpreted as the probability density of a single particle with mass m. The velocity $u(r,t)$ then is the average velocity of the particle, averaged over the distribution $\rho(r,t)$, which in macroscopic fluids is often called the *drift velocity* of the particle. It does not exclude that the particle actual velocity has an additional random contribution. We define the *flux density* J as

$$J = \rho u. \tag{3.82}$$

Now the fact that particles are not created or destroyed when they move, or that total density is preserved, implies the *continuity equation*

$$\frac{\partial \rho}{\partial t} + \nabla \cdot J = 0. \tag{3.83}$$

This says that the outward flow $\iint J \cdot dS$ over the surface of a (small) volume V, which equals the integral of the divergence $\nabla \cdot J$ of J over that volume, goes at the expense of the integrated density present in that volume.

There is one additional equation, expressing the acceleration of the fluid by forces acting on it. This is the *equation of motion*. The local acceleration, measured in a coordinate system that *moves with the flow*, and indicated by the *material* or *Lagrangian* derivative D/Dt, is given by the force f acting per unit volume, divided by the local mass per unit volume

$$\frac{Du}{Dt} = \frac{f(r)}{m\rho}. \tag{3.84}$$

The material derivative of any attribute A of the fluid is defined by

$$\frac{DA}{Dt} \overset{\text{def}}{=} \frac{\partial A}{\partial t} + \frac{\partial A}{\partial x}\frac{dx}{dt} + \frac{\partial A}{\partial y}\frac{dy}{dt} + \frac{\partial A}{\partial z}\frac{dz}{dt} = \frac{\partial A}{\partial t} + \boldsymbol{u} \cdot \nabla A \qquad (3.85)$$

and thus (3.84) can be written as the *Lagrangian equation of motion*

$$m\rho\frac{D\boldsymbol{u}}{Dt} = m\rho\left(\frac{\partial \boldsymbol{u}}{\partial t} + (\boldsymbol{u} \cdot \nabla)\boldsymbol{u}\right) = \boldsymbol{f}(\boldsymbol{r}). \qquad (3.86)$$

The force per unit volume consists of an *external component*

$$\boldsymbol{f}^{\text{ext}} = -\rho\nabla V(\boldsymbol{r}, t) \qquad (3.87)$$

due to an external potential $V(\boldsymbol{r}, t)$, and an internal component

$$\boldsymbol{f}^{\text{int}} = \nabla \cdot \boldsymbol{\sigma}, \qquad (3.88)$$

where σ is the local *stress tensor*, which for isotropic frictionless fluids is diagonal and equal to minus the pressure (see Chapter 9 for details). This is all we need for the present purpose.

Let us now return to a system of quantum particles.[7] For simplicity we consider a single particle with wave function $\Psi(\boldsymbol{r}, t)$ evolving under the time-dependent Schrödinger equation (5.22). Generalization to the many-particle case is a straightforward extension that we shall consider later. Write the wave function in polar form:

$$\Psi(\boldsymbol{r}, t) = R(\boldsymbol{r}, t)\exp[iS(\boldsymbol{r}, t)/\hbar]. \qquad (3.89)$$

Here, $R = \sqrt{\Psi^*\Psi}$ and S are real functions of space and time. Note that R is non-negative and that S will be periodic with a period of $2\pi\hbar$, but S is undefined when $R = 0$. In fact, S can be discontinuous in nodal planes where $R = 0$; for example, for a real wave function S jumps discontinuously from 0 to $\pi\hbar$ at a nodal plane. The Schrödinger equation,

$$i\hbar\frac{\partial \Psi}{\partial t} = -\frac{\hbar^2}{2m}\nabla^2\Psi + V(\boldsymbol{r}, t)\Psi, \qquad (3.90)$$

can be split into an equation for the real and one for the imaginary part. We then straightforwardly obtain two real equations for the time dependence of R and S, both only valid when $R \neq 0$:

$$\frac{\partial R}{\partial t} = -\frac{1}{m}\nabla R \cdot \nabla S - \frac{R}{2m}\nabla^2 S, \qquad (3.91)$$

$$\frac{\partial S}{\partial t} = -\frac{1}{2m}(\nabla S)^2 - (V + Q), \qquad (3.92)$$

---

[7] We follow in essence the lucid treatment of Madelung (1926), with further interpretations by Bohm (1952a, 1952b), and details by Holland (1993).

where Q is defined as

$$Q(r) \stackrel{\text{def}}{=} -\frac{\hbar^2}{2m}\frac{\nabla^2 R}{R}. \tag{3.93}$$

The crucial step now is to identify $\nabla S/m$ as the local *fluid velocity* u:

$$u(r,t) \stackrel{\text{def}}{=} \frac{\nabla S}{m}. \tag{3.94}$$

This is entirely reasonable, since the expectation value of the velocity equals the average of $\nabla S/m$ over the distribution ρ:

$$\langle v \rangle = \frac{\hbar}{m}\langle k \rangle = \frac{\hbar^2}{im}\int \Psi^* \nabla \Psi \, dr = \frac{\hbar^2}{im}\int R(\nabla R)\, dr + \int R^2\frac{\nabla S}{m}\, dr.$$

The first term is zero because R vanishes over the integration boundaries, so that

$$\langle v \rangle = \int R^2\frac{\nabla S}{m}\, dr. \tag{3.95}$$

Applying this identification to (3.91), and writing

$$\rho(r,t) = R^2, \tag{3.96}$$

we find that

$$\frac{\partial \rho}{\partial t} + \nabla \cdot (\rho u) = 0. \tag{3.97}$$

This is a continuity equation for ρ (see (3.83))!

Equation (3.92) becomes

$$m\frac{\partial u}{\partial t} = -\nabla(\tfrac{1}{2}mu^2 + V + Q). \tag{3.98}$$

The gradient of u^2 can be rewritten as

$$\tfrac{1}{2}\nabla(u^2) = (u \cdot \nabla)u, \tag{3.99}$$

as will be shown below; therefore

$$m\left(\frac{\partial u}{\partial t} + (u \cdot \nabla)u\right) = m\frac{Du}{Dt} = -\nabla(V + Q), \tag{3.100}$$

which is the Lagrangian equation of motion, similar to (3.86). The local force per unit volume equals $-\rho\nabla(V + Q)$. This force depends on position, but not on velocities, and thus the fluid motion is frictionless, with an external force due to the potential V and an "internal force" due to the *quantum potential* Q.

Proof We prove (3.99). Since $\boldsymbol{u}$ is the gradient of S, the $\boldsymbol{u}$-field is irrotational: $\operatorname{\mathbf{curl}} \boldsymbol{u} = 0$, for all regions of space where $\rho \neq 0$. Consider the x-component of the gradient of $\boldsymbol{u}^2$:

$$\frac{1}{2}(\nabla \boldsymbol{u}^2)_x = u_x \frac{\partial u_x}{\partial x} + u_y \frac{\partial u_y}{\partial x} + u_z \frac{\partial u_z}{\partial x} = u_x \frac{\partial u_x}{\partial x} + u_y \frac{\partial u_x}{\partial y} + u_z \frac{\partial u_x}{\partial z}$$

$$= (\boldsymbol{u} \cdot \nabla) u_x, \tag{3.101}$$

because $\operatorname{\mathbf{curl}} \boldsymbol{u} = 0$ implies that

$$\frac{\partial u_y}{\partial x} = \frac{\partial u_x}{\partial y} \quad \text{and} \quad \frac{\partial u_z}{\partial x} = \frac{\partial u_x}{\partial z}.$$

□

The quantum potential Q, defined by (3.93), can also be expressed in derivatives of $\ln \rho$:

$$Q = -\frac{\hbar^2}{4m} \left(\nabla^2 \ln \rho + \tfrac{1}{2}(\nabla \ln \rho)^2 \right), \tag{3.102}$$

which may be more convenient for some applications. The quantum potential is some kind of internal potential, related to the density distribution, as in real fluids. One may wonder if a simple definition for the stress tensor (3.88) exists. It is indeed possible to define such a tensor (Takabayasi, 1952), for which

$$\boldsymbol{f}^{\text{int}} = -\rho \nabla Q = \nabla \sigma, \tag{3.103}$$

when we define

$$\sigma \stackrel{\text{def}}{=} \frac{\hbar^2}{4m} \rho \nabla \nabla \ln \rho. \tag{3.104}$$

This equation is to be read in cartesian coordinates (indexed by $\alpha, \beta, \ldots$) as

$$\sigma_{\alpha\beta} = \frac{\hbar^2}{4m} \rho \frac{\partial^2 \ln \rho}{\partial x_\alpha \partial x_\beta}. \tag{3.105}$$

3.4.2 The classical limit

In the absence of the quantum force Q the fluid behaves entirely classically; each fluid element moves according to the classical laws in a potential field $V(\boldsymbol{r})$, without any interaction with neighboring fluid elements belonging to the same particle. If the fluid is interpreted as a probability density, and the initial distribution is a delta-function, representing a point particle, then in the absence of Q the distribution will remain a delta function and follow a classical path. Only under the influence of the quantum force will the distribution change with time. So the classical limit is obtained when the quantum force (which is proportional to $\hbar^2$) is negligible compared to the interaction force. Note, however, that the quantum force will never be small for a point particle, and even near the classical limit particles will have a non-zero quantum width.

3.5 Quantum corrections to classical behavior

For molecular systems at normal temperatures that do not contain very fast motions of light particles, and in which electronically excited states play no role, classical simulations will usually suffice to obtain relevant dynamic and thermodynamic behavior. Such simulations are in the realm of *molecular dynamics* (MD),which is the subject of Chapter 6. Still, when high-frequency motions do occur or when lighter particles and lower temperatures are involved and the quantum wavelength of the particles is not quite negligible compared to the spatial changes of interatomic potentials, it is useful to introduce quantum effects as a *perturbation* to the classical limit and evaluate the first-order quantum corrections to classical quantities. This can be done most simply as a posterior correction to quantities computed from unmodified classical simulations, but it can be done more accurately by modifying the equations of motions to include quantum corrections. In general we shall be interested to preserve equilibrium and long-term dynamical properties of the real system by the classical approximation. This means that correctness of thermodynamic properties has priority over correctness of dynamic properties.

As a starting point we may either take the quantum corrections to the partition function, as described in Chapter 17, Section 17.6 on page 472, or the imaginary-time path-integral approach, where each particle is replaced by a closed string of n harmonically interacting beads (see Section 3.3 on page 44). The latter produces the correct quantum partition function. In the next section we shall start with the Feynman–Hibbs quantum-corrected pair potential and show that this potential results in corrections to thermodynamic quantities that agree with the quantum corrections known from statistical mechanics.

3.5.1 Feynman-Hibbs potential

In Sections 3.3.7 (page 55) and 3.3.8 (page 57) the intrinsic quantum widths of free and of harmonically-bound particles were derived. Both are Gaussian distributions, with variances:

$$\sigma^2 = \frac{\hbar^2}{12\,mk_{\mathrm B}T} \quad \text{(free particle)} \tag{3.106}$$

$$\sigma^2 = \frac{\hbar}{m\omega}\left[\frac{1}{2}\coth\frac{\hbar\omega}{2k_{\mathrm B}T} - \frac{kT}{\hbar\omega}\right] \quad \text{(harmonic particle)} \tag{3.107}$$

These can be used to modify pair potentials. We shall avoid the complications caused by the use of a reference potential,[8] needed when the harmonic width is used, and only use the free particle distribution. Feynman and Hibbs (1965) argued that each *pair interaction* $V_{ij}(r_{ij}) = U(r)$ between two particles i and j with masses m_i and m_j should be modified by a 3D convolution with the free-particle intrinsic quantum distribution:

$$V_{ij}^{\mathrm{FH}}(\boldsymbol{r}) = (2\pi\sigma^2)^{-3/2}\int d\boldsymbol{s}\, U(|\boldsymbol{r} + \boldsymbol{s}|)\exp\left[-\frac{s^2}{2\sigma^2}\right], \tag{3.108}$$

where

$$\boldsymbol{r} = \boldsymbol{r}_{ij} = \boldsymbol{r}_i - \boldsymbol{r}_j, \tag{3.109}$$

$$\sigma^2 = \frac{\hbar}{12\mu k_{\mathrm B}T}, \tag{3.110}$$

$$\mu = \frac{m_1 m_2}{m_1 + m_2}. \tag{3.111}$$

This is the Feynman–Hibbs potential. It can be evaluated for any well-behaved interaction function $U(r)$ from the integral (we write $z = $ cosine of the angle between $\boldsymbol{r}$ and $\boldsymbol{s}$):

$$V_{ij}^{\mathrm{FH}}(r) = (\sigma\sqrt{2\pi})^{-3}\int_0^\infty ds \int_{-1}^1 dz\, 2\pi s^2 U(\sqrt{r^2 + s^2 - 2rsz})\exp\left[-\frac{s^2}{2\sigma^2}\right]. \tag{3.112}$$

Some insight is obtained by expanding U to second order in s/r. Using

$$\sqrt{r^2 + s^2 - 2rsz} = r\left[1 - \frac{s}{r}z + \frac{1}{2}\frac{s^2}{r^2}(1 - z^2) + \mathcal{O}(\frac{s^3}{r^3})\right], \tag{3.113}$$

$U(\sqrt{r^2 + s^2 - 2rsz})$ expands as

$$U = U(r) - szU'(r) + \frac{1}{2}s^2\left[(1 - z^2)\frac{U'(r)}{r} + z^2 U''(r)\right]. \tag{3.114}$$

Evaluating the integral (3.112), we find

$$V_{ij}^{\mathrm{FH}}(r) = U(r) + \frac{\hbar^2}{24\mu k_{\mathrm B}T}\left(\frac{2U'(r)}{r} + U''(r)\right). \tag{3.115}$$

[8] See Mak and Andersen (1990) and Cao and Berne (1990) for a discussion of reference potentials.

It is left to Exercises 3.2 and 3.3 to evaluate the practical importance of the potential correction. For applications to Lennard-Jones liquids see Sesé (1992, 1993, 1994, 1995, 1996). Guillot and Guissani (1998) applied the Feynman–Hibbs approach to liquid water.

3.5.2 The Wigner correction to the free energy

The Wigner $\hbar^2$ corrections to the classical canonical partition function Q and Helmholtz free energy A are treated in Section 17.6 with the final result in terms of Q given in (17.102) on page 476. Summarizing it is found that

$$Q = Q^{\text{cl}}(1 + \langle f_{\text{cor}} \rangle), \tag{3.116}$$

$$A = A^{\text{cl}} - k_{\text{B}} T \langle f_{\text{cor}} \rangle, \tag{3.117}$$

$$f_{\text{cor}} = -\frac{\hbar^2}{12\,k_{\text{B}}^2 T^2} \sum_i \frac{1}{m_i} \left[\nabla_i^2 V - \frac{1}{2k_{\text{B}}T} (\nabla_i V)^2 \right]. \tag{3.118}$$

The two terms containing potential derivatives can be expressed in each other when averaged over the canonical ensemble:

$$\langle (\nabla_i V)^2 \rangle = k_{\text{B}} T \langle \nabla_i^2 V \rangle, \tag{3.119}$$

as we shall prove below. Realizing that the force $\boldsymbol{F}_i$ on the i-th particle is equal to $-\nabla_i V$, (3.118) can be rewritten as

$$\langle f_{\text{cor}} \rangle = -\frac{\hbar^2}{24\,k_{\text{B}}^3 T^3} \sum_i \frac{1}{m_i} \langle \boldsymbol{F}_i^2 \rangle. \tag{3.120}$$

This is a convenient form for practical use. For molecules it is possible to split the sum of squared forces into translational and rotational degrees of freedom (see Powles and Rickayzen, 1979). These are potential energy corrections; one should also be aware of the often non-negligible quantum corrections to the classical rotational partition function, which are of a kinetic nature. For formulas the reader is referred to Singh and Sinha (1987), Millot *et al.* (1998) and Schenter (2002). The latter two references also give corrections to the second virial coefficient of molecules, with application to water.

Proof We prove (3.119). Consider one particular term, say the second derivative to x_1 in ∇V:

$$\int \frac{\partial^2 V}{\partial x_1^2} e^{-\beta V} \, dx_1 d\boldsymbol{r}',$$

where the prime means integration over all space coordinates except x_1. Now, by partial integration, we obtain

$$\int \left(e^{-\beta V} \frac{\partial V}{\partial x_1} \right)_{x_1=-\infty}^{x_1=\infty} d\mathbf{r}' + \int \left(\frac{\partial V}{\partial x_1} \right)^2 e^{-\beta V} \, d\mathbf{r}.$$

The first term is a boundary term, which vanishes for a finite system where the integrand goes to zero at the boundary if the latter is taken beyond all particles. It also vanishes for a periodic system because of the equality of the integrand at periodic boundaries. Since every term in $\nabla^2 V$ can be equally transformed, (3.119) follows. $\qquad\square$

3.5.3 Equivalence between Feynman–Hibbs and Wigner corrections

We now show that application of the Feynman-Hibbs potential (3.115) yields the same partition function and free energy as application of the Wigner correction (3.118). We start by rewriting (3.118), using (3.119):

$$\langle f_{\text{cor}} \rangle = -\frac{\hbar^2}{24 \, k_{\text{B}}^2 T^2} \sum_i \frac{1}{m_i} \nabla_i^2 V. \tag{3.121}$$

Assuming V can be written as a sum of pair potentials $U(r)$:

$$V = \sum_{i<j} U(r_{ij}) = \frac{1}{2} \sum_{i,j\neq i} U(r_{ij}), \tag{3.122}$$

we can evaluate the Laplacian and arrive at

$$\langle f_{\text{cor}} \rangle = -\frac{\hbar^2}{24 \, k_{\text{B}}^2 T^2} \sum_i \frac{1}{m_i} \sum_{j\neq i} \left(U''(r_{ij}) + \frac{2U'(r_{ij})}{r_{ij}} \right). \tag{3.123}$$

Next we rewrite the total potential energy on the basis of Feynman–Hibbs pair potentials:

$$\begin{aligned} V^{\text{FH}} &= \frac{1}{2} \sum_{i,j\neq i} V_{ij}^{\text{FH}}(r_{ij}) \\ &= V^{\text{cl}} + \frac{1}{2} \sum_{i,j\neq i} \frac{\hbar^2}{24 \, k_{\text{B}} T} \left(\frac{1}{m_i} + \frac{1}{m_j} \right) \left(U''(r_{ij}) + \frac{2U'(r_{ij})}{r_{ij}} \right) \\ &= V^{\text{cl}} + \frac{\hbar^2}{24 \, k_{\text{B}} T} \sum_i \frac{1}{m_i} \sum_{j\neq i} \left(U''(r_{ij}) + \frac{2U'(r_{ij})}{r_{ij}} \right). \end{aligned} \tag{3.124}$$

Expanding $\exp(-\beta V^{\text{FH}})$ to first order in the correction term we find:

$$e^{-\beta V^{\text{FH}}} = e^{-\beta V^{\text{cl}}} \left[1 - \frac{\hbar^2}{24\, k_{\text{B}}T} \sum_i \frac{1}{m_i} \sum_{j \neq i} \left(U''(r_{ij}) + \frac{2U'(r_{ij})}{r_{ij}} \right) \right], \quad (3.125)$$

which, after integration, gives exactly the $\langle f_{\text{cor}} \rangle$ of (3.123).

3.5.4 Corrections for high-frequency oscillators

Bond oscillations are often of such a high frequency that $\hbar\omega/k_{\text{B}}T > 1$ and order-$\hbar^2$ corrections are not sufficient to describe the thermodynamics of the vibrations correctly. A good model for non-classical high-frequency vibrations is the harmonic oscillator, which is treated in Chapter 17. Figure 17.5 on page 478 shows the free energy for the harmonic oscillator for the classical case, the $\hbar^2$-corrected classical case and the exact quantum case. When $\hbar\omega/k_{\text{B}}T >\approx 5$, the bond vibrational mode is essentially in its ground state and may be considered as flexible constraint in simulations.[9] For $\hbar\omega/k_{\text{B}}T <\approx 2$, the $\hbar^2$-corrected values, which can be obtained by a proper Feynman–Hibbs potential, are quite accurate. For the difficult range $2 < \hbar\omega/k_{\text{B}}T < 5$ it is recommended to use the exact quantum corrections. At $T = 300$ K, this "difficult" range corresponds to wave numbers between 400 and 1000 cm^{-1} in which many vibrations and librations occur in molecules. One may also choose not to include the $\hbar^2$ corrections at all and apply the exact quantum corrections for the full range $\hbar\omega/k_{\text{B}}T >\approx 0.5$, i.e., all frequencies above 100 cm^{-1}.

In a condensed-phase simulation, one does not know all the normal modes of the system, from which the quantum corrections could be computed. The best way to proceed is to perform a classical simulation and compute the power spectrum of the *velocities* of the particles.[10] The power spectrum will contain prominent peaks at frequencies corresponding to normal modes. We follow the description by Berens *et al.* (1983), who applied quantum corrections to water.

First compute the total mass-weighted velocity-correlation function for an N-particle fluid:

$$C(\tau) = \sum_{i=1}^{3N} m_i \langle v_i(t) v_i(t + \tau), \rangle \qquad (3.126)$$

[9] See the treatment of flexible constraints in molecular dynamics on page 158.
[10] See Section 12.8 for the description of power spectra and their relation to correlation functions.

and from this the *spectral density of states* $S(\nu)$:

$$S(\nu) = \frac{4}{k_{\rm B}T} \int_0^\infty C(\tau) \cos 2\pi\nu\tau \, d\tau. \tag{3.127}$$

Note that the correlation function is the inverse transform of $S(\nu)$ (see Section 12.8):

$$C(\tau) = k_{\rm B}T \int_0^\infty S(\nu) \cos 2\pi\nu\tau \, d\nu, \tag{3.128}$$

with special case

$$\int_0^\infty S(\nu) \, d\nu = \frac{C(0)}{k_{\rm B}T} = 3N. \tag{3.129}$$

Now we have the classical density of states, we can compute the quantum corrections to thermodynamics quantities. The results are (Berens *et al.*, 1983):

$$A^{\rm qu} - A^{\rm cl} = k_{\rm B}T \int_0^\infty d\nu S(\nu) \left[\ln \frac{1 - e^{-\xi}}{e^{-\xi/2}} - \ln \xi \right], \tag{3.130}$$

$$U^{\rm qu} - U^{\rm cl} = k_{\rm B}T \int_0^\infty d\nu S(\nu) \left[\frac{\xi}{2} + \frac{\xi}{e^\xi - 1} - 1 \right], \tag{3.131}$$

$$S^{\rm qu} - S^{\rm cl} = k_{\rm B}T \int_0^\infty d\nu S(\nu) \left[\frac{\xi}{e^\xi - 1} - \ln \left(1 - e^{-\xi}\right) \right], \tag{3.132}$$

$$C_V^{\rm qu} - C_V^{\rm cl} = k_{\rm B}T \int_0^\infty d\nu S(\nu) \left[\frac{\xi^2 e^\xi}{(1 - e^\xi)^2} - 1 \right], \tag{3.133}$$

where

$$\xi = \frac{h\nu}{k_{\rm B}T}. \tag{3.134}$$

One should check if the computed spectral density of states integrate to $3N$.

3.5.5 The fermion–boson exchange correction

In Section 17.6 the classical approximation to quantum statistical mechanics has been derived. In Eq. (17.112) on page 479 a correction for the fermion or boson character of the particle is given in the form of a repulsive or attractive short-range potential. As shown in Fig. 17.6, the exchange correction potential for nuclei can be neglected in all but very exceptional cases.

Exercises

3.1   Check the expansions (3.113) and (3.114).

3.2   Give an analytical expression for the Feynman-Hibbs potential in the approximation of (3.115) for a Lennard-Jones interaction.

3.3   Evaluate both the full integral (3.112) and the approximation of the previous exercise for a He–He interaction at $T = 40$ K. Plot the results.

3.4   Apply (3.120) to compute the partition function and the Helmholtz free energy of a system of N non-interacting harmonic oscillators and prove the correctness of the result by expanding the exact expression (from (17.84) on page 472).

4

Quantum chemistry: solving the time-independent Schrödinger equation

4.1 Introduction

As has become clear in the previous chapter, electrons (almost) always behave as quantum particles; classical approximations are (almost) never valid. In general one is interested in the *time-dependent* behavior of systems containing electrons, which is the subject of following chapters.

The time-dependent behavior of systems of particles spreads over very large time ranges: while optical transitions take place below the femtosecond range, macroscopic dynamics concerns macroscopic times as well. The light electrons move considerably faster than the heavier nuclei, and collective motions over many nuclei are slower still. For many aspects of long-time behavior the motion of electrons can be treated in an environment considered stationary. The electrons are usually in *bound states*, determined by the positions of the charged nuclei in space, which provide an *external field* for the electrons. If the external field is stationary, the electron wave functions are stationary oscillating functions. The approximation in which the motion of the particles (i.e., nuclei) that generate the external field, is neglected, is called the *Born–Oppenheimer approximation*. Even if the external field is not stationary (to be treated in Chapter 5), the non-stationary solutions for the electronic motion are often expressed in terms of the pre-computed stationary solutions of the Schrödinger equation. This chapter concerns the computation of such stationary solutions.

Thus, in this chapter, the Schrödinger equation reduces to a time-independent problem with a stationary (i.e., still time-dependent, but periodic) solution. Almost all of chemistry is covered by this approximation. It is not surprising, therefore, that theoretical chemistry has been almost equivalent to quantum chemistry of stationary states, at least up to the 1990s, when the scope of theory in chemistry slowly started to be broadened to include the study of more complex dynamic behavior.

For completeness, in the last section of this chapter attention will be given to the stationary quantum behavior of *nuclei*, rather than electrons. This includes the rotational and vibrational steady state behavior of molecules, which is useful in spectroscopic (infrared and Raman) studies, in the prediction of spectroscopic behavior by simulations, or in the use of spectroscopic data to evaluate force fields designed for simulations.

4.2 Stationary solutions of the TDSE

The general form of the time-dependent Schrödinger equation (TDSE) is

$$i\hbar\frac{\partial\Psi}{\partial t} = \hat{H}\Psi, \tag{4.1}$$

where the usual (already simplified!) form of the Hamiltonian is that of (2.72). If the Hamiltonian does not contain any explicit time dependence and is only a function of the particle coordinates and a *stationary* external potential, the TDSE has stationary solutions that represent bound states:

$$\Psi_n(\mathbf{r}, t) = \psi_n(\mathbf{r})\exp\left(-\frac{i}{\hbar}E_n t\right), \tag{4.2}$$

where $\psi_n(\mathbf{r})$ and E_n are solutions of the eigenvalue equation

$$\hat{H}\psi(\mathbf{r}) = E\psi(\mathbf{r}). \tag{4.3}$$

The latter is also called the *time-independent Schrödinger equation*

The spatial parts of the wave functions are stationary in time, and so is the probability distribution $\Psi_n^*\Psi_n$ for each state.

In this chapter we shall look at ways to solve the *time-independent Schrödinger equation*, (4.3), assuming stationary external fields. In chapter 5 we consider how a quantum system behaves if the external field is not stationary, for example if the nuclei move as well, or if there are external sources for fluctuating potentials.

There are several ways in which *ab initio* solutions of the time-independent Schrödinger equation can be obtained. In quantum physics the emphasis is often on the behavior of a number of quantum particles, which are either bosons, as in helium-4 liquids, or fermions as in electrons in (semi)conductors or in helium-3 liquids. In chemistry the main concern is the structure and properties of single atoms and molecules; especially large molecules with many electrons pose severe computational problems and elude exact treatment.

Before considering methods to solve the many-electron problem, we shall look into the methods that are available to find the stationary solution for one or a few interacting quantum particles. Then we consider the question whether it will be possible to separate the nuclear motion from the electronic motion in atoms and molecules: this separation is the essence of the *Born–Oppenheimer approximation*. When valid, the electronic motion can be considered in a stationary external field, caused by the nuclei, while the nuclear motion can be described under the influence of an *effective* potential caused by the electrons.

4.3 The few-particle problem

Let us first turn to simple low-dimensional cases. Mathematically, the SE is a *boundary-value* problem, with acceptable solutions only existing for discrete values of E. These are called the *eigenvalues*, and the corresponding solutions the *eigenfunctions*. The boundary conditions are generally zero values of the wave function at the boundaries,[1] and square-integrability of the function, i.e., $\int \psi^* \psi(x) \, dx$ must exist. As any multiple of a solution is also a solution, this property allows to *normalize* each solution, such that the integral of its square is equal to one.

Since any Hamiltonian is *Hermitian* (see Chapter 14), its eigenvalues E are real. But most Hamiltonians are also real, except when velocity-dependent potentials as in magnetic interactions occur. Then, when ψ is a solution, also ψ^* is a solution for the same eigenvalue, and the sum of ψ and ψ^* is a solution as well. So the eigenfunctions can be chosen as real functions. Often, however, a complex function is chosen instead for convenience. For example, instead of working with the real functions $\sin m\phi$ and $\cos m\phi$, one may more conveniently work with the complex functions $\exp(im\phi)$ and $\exp(-im\phi)$. Multiplying a wave function by a constant $\exp(ia)$ does not change any of the physical quantities derived from the wave function.

Consider a single quantum particle with mass m in a given, stationary, external potential $V(\boldsymbol{x})$. We shall not treat the analytical solutions for simple cases such as the hydrogen atom, as these can be found in any text book on quantum physics or theoretical chemistry. For the one-dimensional case there are several ways to solve the time-independent SE numerically.

---

[1] In the case of periodic boundary conditions, the wave function and its derivatives must be continuous at the boundary.

4.3.1 Shooting methods

The popular *shooting* methods integrate the second-order differential equation

$$\frac{d^2\psi(x)}{dx^2} = \frac{2m}{\hbar^2}[V(x) - E]\psi(x) \tag{4.4}$$

from one end with an estimate of the eigenvalue E and iterate over changes of E; only when E is equal to an eigenvalue will the wave function fulfill the proper boundary condition at the other end. In fact, one replaces a *boundary value problem* with an iterative *initial value problem*.[2] The *Numerov* method is recommended; it consists of solving (4.4) to fourth-order precision in the grid spacing, requiring the second derivative of the potential, which is given by (4.4) and which is discretized using three points. It is left to the reader as an exercise to derive the corresponding algorithm. Function `numerov(m,E,V)` finds the nearest eigenvalue by iteration, starting from a guessed value for E. It shoots from both sides to a common point, and after scaling, compares the first derivatives at the end points (Pang, 1997). Then E is adjusted to equalize both relative derivatives with a linear inter-/extrapolation root search method. The shooting method is in principle exact (limited only by discretization errors) and has the advantage that it can also generate excited states, but it is not very suitable for higher dimensional cases.

PYTHON PROGRAM 4.1 **numerov(m,E,V)**
Finds the nearest eigenvalue for the single-particle Schrödinger equation.

```
01   def numerov(m,E,V):
02   # m=index for matching point
03   # E=trial energy*mass*delx**2/hbar**2
04   # V=array of pot. energy*mass*delx**2/hbar**2
05   # returns  [nr of zerocrossings, difference in relat. first
derivatives]
06      E1=E; E2=E1*1.05
07      F1=shoot(m,E1,V)
08      while (abs(E2-E1)> 1.e-8):
09         nF=shoot(m,E2,V); F2=nF[1]
10         Etemp=E2
11         E2=(F1*E2-F2*E1)/(F1-F2)
12         E1=Etemp
13         F1=F2
14         print '%3d %13.10f' %(nF[0], E2)
15      return [nF[0],E2]
16
```

---

[2] The one-dimensional Schrödinger equation is a special case of the class of Sturm–Liouville problems: $\frac{d}{dx}\left(p(x)\frac{df(x)}{dx}\right) + q(x)f(x) = s(x)$. See Pang (1997) or Vesely (2001) for a discussion of such methods in a more mathematical context. Both books describe Numerov's method in detail.

```
17  def shoot(m,E,V):
18  # m=index of matching point, should be near right end
19  # E=trial energy*mass*delx**2/hbar**2
20  # V=array of pot. energy*mass*delx**2/hbar**2
21  # returns   [nr of zerocrossings, difference in first derivatives]
22  nx=len(V)
23  ypresent=0.; yafter=0.001
24  i=1
25  sign=1.
26  nz=0
27  while i <= m: # shoot from left to right
28     ybefore=ypresent; ypresent=yafter
29     gplus=1.-(V[i+1]-E)/6.
30     gmed=2.+(5./3.)*(V[i]-E)
31     gmin=1.-(V[i-1]-E)/6.
32     yafter=gmed*ypresent/gplus -gmin*ybefore/gplus
33     if (yafter*sign < 0.):
34        nz=nz+1
35        sign=-sign
36     i=i+1
37  ym=ypresent
38  forwardderiv=yafter-ybefore
39  ypresent=0.; yafter=0.001
40  i=nx-2
41  while i >= m: #shoot from right to left
42     ybefore=ypresent; ypresent=yafter
43     gplus=1.-(V[i-1]-E)/6.
44     gmed=2.+(5./3.)*(V[i]-E)
45     gmin=1.-(V[i+1]-E)/6.
46     yafter=gmed*ypresent/gplus -gmin*ybefore/gplus
47     i=i-1
48  backwardderiv=(ybefore-yafter)*ym/ypresent
49  return [nz,forwardder-backwardder]
```

Comments

Line 02: m is the point where the forward and backward "shooting" should match. It is best taken near the right border, say at 80% of the length of the vectors.

Line 06: the first two guesses are E_1 and $E_1 + 5\%$.

Line 09: this produces the next guess, using the value of E produced in line 12 of the previous step, which is based on a linear relation between E and the output of shoot (difference between derivatives at matching point).

The routine may not always converge to the expected (nearest) eigenvalue, as the output of shoot is very erratic when E deviates far from any eigenvalue.

Line 15: The routine also returns the number of nodes in the wave function, which is an indication of the eigenvalue number.

Note that numerov does not produce the wave function itself. In order to generate ϕ when E is already known, the function psi(m,E,V) can be called.

PYTHON PROGRAM 4.2 **psi(m,E,V)**

Constructs the wave function for a given exact eigenvalue from the single-particle Schrödinger equation.

```
01  def psi(m,E,V):
02  # m=index of matching point
03  # E=energy*mass*delx**2/hbar**2 must be converged for same m and V
04  # V=array of pot. energy*mass*delx**2/hbar**2
05  # returns wave function y; sum(y**2)=1
06     nx=len(V)
```

```
07    y=zeros(nx,dtype=float)
08    y[1]=0.001
09    i=1
10    while i < m: # shoot from left to right
11        gplus=1.-(V[i+1]-E)/6.
12        gmed=2.+(5./3.)*(V[i]-E)
13        gmin=1.-(V[i-1]-E)/6.
14        y[i+1]=gmed*y[i]/gplus -gmin*y[i-1]/gplus
15        i=i+1
16    ym=y[m]
17    y[-2]=0.001
18    i=nx-2
19    sign=1
20    while i > m: #shoot from right to left
21        gplus=1.-(V[i-1]-E)/6.
22        gmed=2.+(5./3.)*(V[i]-E)
23        gmin=1.-(V[i+1]-E)/6.
24        y[i-1]=gmed*y[i]/gplus -gmin*y[i+1]/gplus
25        i=i-1
26    scale=ym/y[m]
27    for i in range(m,nx): y[i]=scale*y[i]
28    y=y/sqrt(sum(y**2))
29    return y
```

Comments
This algorithm is similar to the shoot algorithm, except that an array of y-values is kept and the number of zero crossings is not monitored. Lines 26–27 scale the backward shot on the forward one; line 28 normalizes the total sum of squares to 1.

Figure 4.1 illustrates the use of `numerov` and `psi` to compute the first six eigenvalues and eigenfunctions for a double-well potential, exemplifying the proton potential energy in a symmetric hydrogen bond.[3] The potential is composed of the sum of two opposing Morse potentials, each with $D = 600$ kJ/mol and a harmonic vibration frequency of 100 ps^{-1} (3336 cm^{-1}), one with minimum at 0.1 nm and the other at 0.2 nm. The lowest two levels are *adiabatic tunneling states*, which differ only slightly in energy (0.630 kJ/mol) with wave functions that are essentially the sum and difference of the `diabatic` ground state wave functions of the individual wells. In each of the adiabatic states the proton actually oscillates between the two wells with a frequency of $(E_1 - E_0)/h = 1.579$ THz; it tunnels *through* the barrier. The excited states all lie above the energy barrier; the proton then has enough energy to move *over* the barrier.

4.3.2 Expansion on a basis set

Another class of solutions is found by expanding the unknown solution in a finite number of properly chosen basis functions ϕ_n:

---

[3] See, e.g., Mavri and Grdadolnik (2001), who fit two Morse potentials plus modifying terms to high-level quantum calculations in the case of acetyl acetone.

V,E (kJ/mol)

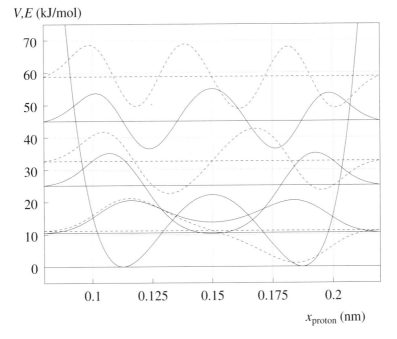

x_{proton} (nm)

Figure 4.1 Lowest six proton quantum states in a double-well potential (thick curve), typical for a proton in a symmetric hydrogen bond. Levels and wave functions were generated with the shooting method (see text). Wave functions are indicated by alternating solid and dotted thin lines. The energies of the states are (in kJ/mol): 0:10.504, 1:11.135, 2:25.102, 3:32.659, 4:45.008, 5:58.804.

$$\psi(x) = \sum_n c_n \phi_n(x). \tag{4.5}$$

The time-independent Schrödinger equation now becomes an eigenvalue equation (see Chapter ??):

$$\mathbf{Hc} = \lambda \mathbf{Sc}, \tag{4.6}$$

where $\mathbf{S}$ is the overlap matrix

$$S_{nm} = \int \phi_n^* \phi_m \, dx. \tag{4.7}$$

For an orthogonal basis set the overlap matrix is the unit matrix. The eigenvalue equation can be solved by standard methods (see, e.g., Press *et al.*, 1992). It is most convenient to diagonalize the basis set first,[4] which

---

[4] Diagonalization is not a unique procedure, as there are more unknown mixing coefficients than equations. For example, mixing two functions requires four coefficients, while there are three conditions: two normalizations and one zero overlap. One unique method is to start with

is a one-time operation after which the eigenvalue problems become much easier. Such methods do yield excited states as well, and are extendable to higher dimensions, but they are never exact: their accuracy depends on the suitability of the chosen basis set. Still, the solution of many-electron problems as in quantum chemistry depends on this approach.

An example of the use of a basis set can be found in the study of proton transfer over hydrogen bonds (Berendsen and Mavri, 1997). For solving the time-dependent SE (see Chapter 5) a description of the proton wave function for a fluctuating double-well potential in terms of a simple basis set was needed. The simplest basis set that can describe the tunneling process consists of two Gaussian functions, each resembling the diabatic ground state solution of one well (Mavri *et al.*, 1993). Many analytical theories rely on such two-state models. It turns out, however, that a reasonable accuracy can only be obtained with more than two Gaussians; five Gaussians, if properly chosen, can reproduce the first few eigenstates reasonably well (Mavri and Berendsen, 1995).

4.3.3 Variational Monte Carlo methods

The variational method consists of defining some flexible function with a number of adjustable parameters that is expected to encompass a good approximation to the true wave function. The variational principle says that the expectation of the Hamiltonian over any function $\psi(\boldsymbol{r})$ (which must be quadratically integrable) is always larger than or equal to the ground state energy with equality only when the function is identical to the ground state eigenfunction:

$$\langle E \rangle = \frac{\int \psi(\boldsymbol{r})\hat{H}\psi(\boldsymbol{r})\,d\boldsymbol{r}}{\int \psi^2(\boldsymbol{r})} = \left\langle \frac{\hat{H}\psi}{\psi} \right\rangle_{\psi^2} \geq E_0 \tag{4.8}$$

Therefore the parameter values that minimize the expectation of $\hat{H}$ yield the best approximation to the ground state wave function. For low-dimensional problems and for linear dependence on the parameters, the integral can in general be evaluated, and the minimization achieved. However, for multidimensional cases and when the integral cannot be split up into a linear combination of computable components, the multidimensional integral is better

one normalized eigenfunction, say ϕ_1, then make ϕ_2 orthogonal to ϕ_1 by mixing the right amount of ϕ_1 into it, then normalizing ϕ_2, and proceeding with ϕ_3 in a similar way, making it orthogonal to both ϕ_1 and ϕ_2, etc. This procedure is called *Schmidt orthonormalization*, see, e.g., Kyrala (1967). A more symmetrical result is obtained by diagonalizing the overlap matrix of normalized functions by an orthogonal transformation.

solved by Monte Carlo techniques. When an ensemble of configurations is generated that is representative for the probability distribution ψ^2, the integral is approximated by the ensemble average of $\hat{H}\psi/\psi$, which is a local property that is usually easy to determine. The parameters are then varied to minimize $\langle E \rangle$. The trial wave function may contain electron–electron correlation terms, for example in the *Jastrow form* of pair correlations, and preferably also three-body correlations, while it must fulfill the parity requirements for the particles studied.

The generation of configurations can be done as follows. Assume a starting configuration r_1 with $\psi^2(r_1) = P_1$ is available. The "local energy" for the starting configuration is

$$\varepsilon_1 = \frac{\hat{H}\psi(r_1)}{\psi(r_1)}. \tag{4.9}$$

(i) Displace either one coordinate at the time, or all coordinates simultaneously, with a random displacement homogeneously distributed over a given symmetric range $(-a, a)$. The new configuration is r_2.
(ii) Compute $P_2 = \psi^2(r_2)$.
(iii) If $P_2 \geq P_1$, accept the new configuration; if $P_2 < P_1$, accept the new configuration with probability P_2/P_1. This can be done by choosing a random number η between 0 and 1 and accept when $\eta \leq P_2/P_1$.
(iv) If the move is accepted, compute the "local energy"

$$\varepsilon_2 = \frac{\bar{H}\psi(r_2)}{\psi(r_2)}; \tag{4.10}$$

if the move is rejected, count the configuration r_1 and its energy ε_1 again;
(v) Repeat steps (i)–(iv) N times.
(vi) The expectation of the Hamiltonian is the average energy

$$\langle \hat{H} \rangle = \frac{\sum_{i=1}^{N} \varepsilon_i}{N}. \tag{4.11}$$

The range for the random steps should be chosen such that the acceptance ratio lies in the range 40 to 70%. Note that variational methods are not exact, as they depend on the quality of the trial function.

4.3.4 Relaxation methods

We now turn to solutions that make use of *relaxation* towards the stationary solution in time. We introduce an artificial time dependence into the wave function $\psi(x, \tau)$ and consider the partial differential equation

$$\frac{\partial \psi}{\partial \tau} = \frac{\hbar}{2m} \nabla^2 \psi - \frac{V - E}{\hbar} \psi. \tag{4.12}$$

It is clear that, if ψ equals an eigenfunction and E equals the corresponding eigenvalue of the Hamiltonian, the right-hand side of the equation vanishes and ψ will not change in time. If E differs from the eigenvalue by ΔE, the total magnitude of ψ (e.g., the integral over ψ^2) will either increase or decrease with time:

$$\frac{dI}{d\tau} = \frac{\Delta E}{\hbar} I, \quad I = \int \psi^2 \, dx. \tag{4.13}$$

So, the magnitude of the wave function is not in general conserved. If ψ is not an eigenfunction, it can be considered as a superposition of eigenfunctions ϕ_n:

$$\psi(x, \tau) = \sum_n c_n(\tau) \phi_n(x). \tag{4.14}$$

Each component will now behave in time according to

$$\frac{dc_n}{d\tau} = \frac{E - E_n}{\hbar} c_n, \tag{4.15}$$

or

$$c_n(\tau) = c_n(0) \exp\left[\frac{E - E_n}{\hbar} \tau\right]. \tag{4.16}$$

This shows that functional components with high eigenvalues will decay faster than those with lower eigenvalues; after sufficiently long time only the ground state, having the lowest eigenvalue, will survive. Whether the ground state will decay or grow depends on the value chosen for E and it will be possible to determine the energy of the ground state by monitoring the scaling necessary to keep the magnitude of ψ constant. Thus the relaxation methods will determine the *ground state* wave function and energy. Excited states can only be found by explicitly preventing the ground state to mix into the solution; e.g., if any ground state component is consistently removed during the evolution, the function will decay to the first excited state.

Comparing (4.12), setting $E = 0$, with the time-dependent Schrödinger equation (4.1), we see that these equations are equivalent if t is replaced by $i\tau$. So, formally, we can say that the relaxation equation is the TDSE in *imaginary time*. This sounds very sophisticated, but there is no deep physics behind this equivalence and its main function will be to impress one's friends!

As an example we'll generate the ground state for the Morse oscillator (see page 6) of the HF molecule. There exists an analytical solution for the Morse oscillator:[5]

$$E_n = \hbar\omega_0(n + \frac{1}{2}) - \frac{(\hbar\omega_0)^2}{4D}(n + \frac{1}{2})^2, \quad (4.17)$$

yielding

$$E_0 = \hbar\omega_0 \left(\frac{1}{2} - \frac{\hbar\omega_0}{16D} \right), \quad (4.18)$$

$$E_1 = \hbar\omega_0 \left(\frac{3}{2} - \frac{9\hbar\omega_0}{16D} \right). \quad (4.19)$$

For HF (see Table 1.1) the ground state energy is 24.7617 kJ/mol for the harmonic approximation and 24.4924 kJ/mol for the Morse potential. The first excited state is 74.2841 kJ/mol (h.o.) and 71.8633 kJ/mol (Morse). In order to solve (4.12) first discretize the distance x in a selected range, with interval Δx. The second derivative is discretized as

$$\frac{\partial^2 \psi}{\partial x^2} = \frac{\psi_{i-1} - 2\psi_i + \psi_{i+1}}{(\Delta x)^2}. \quad (4.20)$$

If we choose

$$\Delta\tau = \frac{m(\Delta x)^2}{\hbar}, \quad m \stackrel{\text{def}}{=} \frac{m_H + m_F}{m_H m_F}, \quad (4.21)$$

then we find that the second derivative leads to a computationally convenient change in ψ_i:

$$\psi_i(\tau + \Delta\tau) = \frac{1}{2}\psi_{i-1}(\tau) + \frac{1}{2}\psi_{i+1}(\tau). \quad (4.22)$$

Using a table of values for the Morse potential at the discrete distances, multiplied by $\Delta\tau/\hbar$ and denoted below by W, the following Python function will perform one step in τ.

PYTHON PROGRAM 4.3 **SRstep(n,x,y,W)**
Integrates one step of single-particle Schrödinger equation in imaginary time.

```
01   def SRstep(x,y,W):
02   # x=distance array;
03   # y=positive wave function; sum(y)=1 required; y[0]=y[1]=y[-2]=y[-1]=0.
04   # W=potential*delta tau/hbar
05   # returns wave function and energy*delta tau/hbar
06     z=concatenate(([0.],0.5*(y[2:]+y[:-2]),[0.]))
07     z[1]=z[-2]=0.
```

[5] See the original article of Morse (1929) or more recent texts as Levin (2002); for details and derivation see Mathews and Walker (1970) or Flügge (1974).

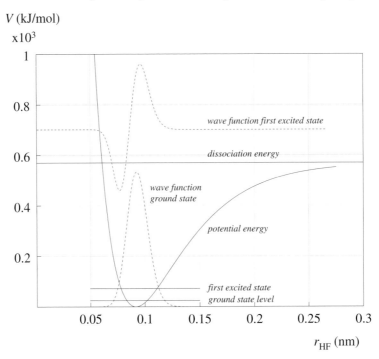

Figure 4.2 The Morse potential for the vibration of hydrogen fluoride and the solution for the ground and first excited state vibrational levels (in the absence of molecular rotation), obtained by relaxation in imaginary time.

```
08      z=z*exp(-W)
09      s=sum(z)
10      E=-log(s)
11      y=z/s
12      return [y,E]
```

Comments
Line 03: the first two and last two points are zero and are kept zero.
Line 06: diffusion step: each point becomes the average of its neighbors; zeros added at ends.
Line 07: second and before-last points set to zero.
Line 08: evolution due to potential.
Line 10: Energy $(*\Delta\tau/\hbar)$ needed to keep y normalized. This converges to the ground state energy.
Line 12: Use $y = $ SRstep[0] as input for next step. Monitor $E = $ step[1] every 100 steps for convergence. Last y is ground state wave function.

Using the values for HF given in Table 1.1, using 1000 points $(0, 3b)$ (b is the bond length), and starting with a Gaussian centered around b with $\sigma = 0.01$ nm, the energy value settles after several thousand iterations to 24.495 kJ/mol. The resulting wave function is plotted in Fig. 4.2. It very closely resembles the Gaussian expected in the harmonic approximation. The algorithm is only of second-order accuracy and the discretization error is proportional to $(\Delta x)^2$, amounting to some 3 J/mol for 1000 points.

The first excited state can be computed when the wave function is kept orthogonal to the ground state. Denoting the ground state by ψ_0, we add a proportion of ψ_0 to ψ such that

$$\int \psi\psi_0 \, dx = 0, \tag{4.23}$$

which can be accomplished by adding a line between line 8 and 9 in the program SRstep:

```
08a   z=z-sum(z*y0)*y0
```

where y0 is the converged and normalized ground state wave function. Figure 4.2 includes the first excited state, with energy 71.866 kJ/mol. This is accurate within 3 J/mol. The vibration wave numbers then are 4139.8 (harmonic oscillator) and 3959.9 (Morse function). Higher excited states can be computed as well, as long as the wave function is kept orthogonal to all lower state wave functions.

4.3.5 Diffusional quantum Monte Carlo methods

If we look carefully at (4.12), we see that the first term on the r.h.s. is a *diffusion term*, as in Fick's equation for the time dependence of the concentration c of diffusing particles:

$$\frac{\partial c}{\partial t} = D\nabla^2 c. \tag{4.24}$$

This fits with what this term does after discretization (4.22): the function splits in two halves located at the neighboring points. This is what would happen with a probability distribution of particles after each particle has made a random step over one grid element, either to the left or to the right.

This equivalence suggests a different way to solve the SE, which has been pioneered by Anderson (1975, 1976).[6] Suppose we wish to obtain the ground state of a system of n particles. Consider an *ensemble* of a large number of replicas of the system, each with its own configuration of the n particles. The members of the ensemble are called *psi-particles* (psips) or *walkers*. Then evolve each member of the ensemble as follows:

(i) Displace each of the coordinates of the particles in a "time" $\Delta\tau$ with a random displacement with variance

$$\langle(\Delta x)^2\rangle = 2D\Delta\tau = \frac{\hbar}{m}\Delta\tau. \tag{4.25}$$

---

[6] See reviews by Anderson (1995) and by Foulkes *et al.* (2001), the latter with applications to solids.

This can be done by sampling from a Gaussian distribution with that variance, or by displacing the coordinate by $\pm\sqrt{\hbar\Delta\tau/m}$.

(ii) Duplicate or annihilate the walker, according to the probability

$$P = \exp[-(V - E)\Delta\tau/\hbar] \tag{4.26}$$

in order to satisfy the second term (source term) on the r.h.s. of (4.12). This, again, can be done in several ways. One way is to let the walker survive with a probability P if $P < 1$, and let it survive but create a new walker with probability $P - 1$ if $P \geq 1$. This is accomplished (Foulkes *et al.*, 1995) by defining the new number of walkers M_{new} as

$$M_{\mathrm{new}} = integer(P + \eta), \tag{4.27}$$

where η is a uniform random number between 0 and 1. A higher-order accuracy is obtained when the potential energy for the creation/annihilation (in (4.26)) is taken as the average before and after the diffusion step:

$$V = \tfrac{1}{2}[V(\tau) + V(\tau + \Delta\tau)]. \tag{4.28}$$

Another way is to enhance or reduce a *weight* per walker, and then applying a scheme like: duplication when the weight reaches 2 or annihilation with 50% probability when the weight falls below 0.5; the surviving and new walkers will then start with weight 1. Applying weights only without creation/annihilation scheme does not work, as this produces uneven distributions with dominating walkers.

(iii) The energy E determines the net gain or loss of the number of walkers and should be adjusted to keep that number stationary.

The advantage of such a stochastic scheme above relaxation on a grid is that it can more easily be expanded to several dimensions. For example, a four-particle system in 3D space (a hydrogen molecule) involves 12 degrees of freedom, reducible to six internal degrees of freedom by splitting-off overall translation and rotation. A grid with only 100 divisions per dimension would already involve the impossible number of 10^{12} grid points, while an ensemble of 10^5 to 10^6 walkers can do the job. The method remains exact, and includes all correlations between particles. The result has a statistical error that reduces with the inverse square root of the number of steps.

Diffusional Monte Carlo methods cannot straightforwardly handle wave functions with *nodes*, which have positive as well as negative regions. Nodes occur not only for excited states, but also for ground states of systems containing more than two identical fermions, or even with two fermions with the same spin. Only two fermions with opposite spin can occupy the same all-positive ground state wave function. The diffusing particles represent either positive or negative wave functions, and when a positive walker crosses a nodal surface, it would in principle cancel out with the negative wave function on the other side. One scheme that avoids these difficulties is to keep the nodal surface fixed and annihilate the walkers that diffuse against that surface; however, by fixing the nodal surface the method is no longer exact. Schemes that use exact cancelation between positive and negative walkers, instead of fixed nodal surfaces, have been successfully applied by Anderson (1995).

The implementation of diffusional Monte Carlo methods is much more efficient when an approximate analytical wave function is used to guide the random walkers into important parts of configuration space. This *importance sampling* was introduced by Grimm and Storer (1971) and pursued by Ceperley and Alder (1980). Instead of sampling ψ, the function $f(r, \tau) = \psi(r, \tau)\psi_T(r)$ is sampled by the walkers, where ψ_T is a suitable trial function. The trial function should be close to the real solution and preferably have the same node structure as the exact solution; in that case the function f is everywhere positive. Noting that the exact energy E is the eigenvalue of $\hat{H}$ for the solution ψ at infinite time

$$\hat{H}\psi = E\psi \tag{4.29}$$

and that $\hat{H}$ is Hermitian:

$$\int \psi \hat{H}\psi_T \, dr = \int \psi_T \hat{H}\psi \, dr = E \int \psi_T \psi \, dr, \tag{4.30}$$

we can write the energy as the expectation of the *local energy for the trial wave function* (which can be evaluated for every member of the generated ensemble) over an ensemble with weight $f = \psi\psi_T$:

$$E = \frac{\int \psi \hat{H}\psi_T \, dr}{\int \psi_T \psi \, dr} = \frac{\int f \frac{\hat{H}\psi_T}{\psi_T} \, dr}{\int f \, dr} = \left\langle \frac{\hat{H}\psi_T}{\psi_T} \right\rangle_f. \tag{4.31}$$

This has the advantage that the energy follows from an ensemble average instead of from a rate of change of the number of walkers.

The time-dependent equation for f follows, after some manipulation, from the definition of f and the Schrödinger equation in imaginary time (4.12:

$$\frac{\partial f}{\partial \tau} = \frac{\hbar}{2m}[\nabla^2 f - 2\nabla \cdot (f\nabla \ln \psi_T)] - \frac{1}{\hbar}\left(\frac{\hat{H}\psi_T}{\psi_T} - E\right)f. \tag{4.32}$$

Note that in the multiparticle case the mass-containing term must be summed over all particles. The first term on the r.h.s. of this equation is a diffusion term; the second term is a *drift* term: it is (minus) the divergence of a flux density $f\boldsymbol{u}$ with a drift velocity

$$\boldsymbol{u} = \frac{\hbar}{m}\nabla \ln \psi_T. \tag{4.33}$$

The term $-\ln \psi_T$ acts as a *guiding potential*, steering the walker in the direction of large values of ψ_T. The third term replaces the strongly varying potential $V(\boldsymbol{r})$ by the much weaker varying local energy for the trial function.

QMC algorithms of this type suffer from a *time-step error* and extrapolation to zero time step is needed for a full evaluation. The accuracy can be considerably improved, allowing larger time steps, by inserting an *acceptance/rejection* step after the diffusion-drift step has been made (Reynolds *et al.*, 1982; Umrigar *et al.*, 1993). The procedure without the reactive step should lead to sampling of ψ_T^2, which can be made exact (for any time step) by accepting/rejecting or reweighting moves such as to maintain detailed balance under the known distribution ψ_T^2.

4.3.6 A practical example

Let us, for the sake of clarity, work out the programming steps for a realistic case: the helium atom. We shall use atomic units (see page xvii) for which the electron mass, elementary charge, and $\hbar$ are all unity. The helium atom consists of three particles: a nucleus with mass M (equal to 6544 for He-4), charge $+2$ and coordinates $\boldsymbol{R}$, and two electrons with mass 1, charge -1 and coordinates $\boldsymbol{r}_1, \boldsymbol{r}_2$. The Hamiltonian, with nine degrees of freedom, reads

$$\hat{H} = -\frac{1}{2M}\nabla_{\boldsymbol{R}}^2 - \tfrac{1}{2}\nabla_1^2 - \tfrac{1}{2}\nabla_2^2 - \frac{2}{|\boldsymbol{r}_1-\boldsymbol{R}|} - \frac{2}{|\boldsymbol{r}_2-\boldsymbol{R}|} + \frac{1}{r_{12}}, \tag{4.34}$$

where

$$r_{12} = |\boldsymbol{r}_1 - \boldsymbol{r}_2|. \tag{4.35}$$

In the Born–Oppenheimer approximation we can eliminate $\boldsymbol{R}$ as a variable by assuming $M = \infty$ and $\boldsymbol{R} \equiv 0$, but there is no pressing need to do so. Neither is there a need to separate center-of-mass and rotational coordinates and reduce the number of internal degrees of freedom to the possible minimum of three, and rewrite the Hamiltonian. We can simply use all nine coordinates; the effect of reduced masses using internal coordinates is implied through the first term concerning the diffusion of the heavy nucleus. The center-of-mass motion (corresponding to a free particle) will now also be included, and will consist of a simple random walk, leading to an indefinitely expanding Gaussian distribution. All relevant data will concern the relative distribution of the particles.

The attractive Coulomb terms in the Hamiltonian cause troublesome behavior when electrons move close to the nucleus. The time step should be taken very small in such configurations and it becomes virtually impossible to gather sufficient statistics. However, the use of importance sampling with a trial function that is the product of the single-electron ground-state solutions for the two electrons eliminates these attractive Coulombic terms altogether. Take as trial function

$$\psi_T = \exp[-\alpha|\boldsymbol{r}_1 - \boldsymbol{R}| - \alpha|\boldsymbol{r}_2 - \boldsymbol{R}|]. \tag{4.36}$$

Choosing

$$\alpha = \frac{2M}{M+1}, \tag{4.37}$$

we find that

$$\frac{\hat{H}\psi_T}{\psi_T} = -\frac{4M}{M+1} + \frac{1}{r_{12}}. \tag{4.38}$$

We note that this trial function is a very rough approximation to the real wave function. For realistic applications it is necessary to use much better trial functions, e.g., obtained from variational Monte Carlo or density functional theory (see Section 4.7).

The time evolution of f according to (4.32) is solved by a collection of walkers, each consisting of a nucleus and two electrons, that:

(i) diffuse with a diffusion constant $1/2M$, resp. $\frac{1}{2}$;
(ii) drift with a drift velocity (see (4.33))

$$u_0 = \frac{\alpha}{M} \left(\frac{\boldsymbol{r}_1 - \boldsymbol{R}}{|\boldsymbol{r}_1 - \boldsymbol{R}|} + \frac{\boldsymbol{r}_2 - \boldsymbol{R}}{|\boldsymbol{r}_2 - \boldsymbol{R}|} \right) \tag{4.39}$$

for the nucleus, and

$$u_i = -\alpha \frac{\boldsymbol{r}_i - \boldsymbol{R}}{|\boldsymbol{r}_i - \boldsymbol{R}|} \tag{4.40}$$

for the two electrons $i = 1, 2$;
(iii) create/annihilate according to

$$\frac{\partial f}{\partial \tau} = \left(E + \frac{4M}{M+1} - \frac{1}{r_{12}} \right) f. \tag{4.41}$$

If the creation/annihilation step is implemented at each time by a stochastic process according to (4.27), additional noise is introduced into the process. It is better to assign a *weight* to each walker and readjust the weights every step, while some form of *population control* is carried out at regular intervals.[7] The latter may involve duplication of heavy and annihilation of light walkers (according to Grassberger, 2002: if the weight exceeds a given upper threshold, then duplicate, giving each copy half the original weight; if the weight is less than a given lower threshold, draw a random number η between 0 and 1, annihilate if $\eta < \frac{1}{2}$, but keep with the double weight if $\eta \geq \frac{1}{2}$), or a complete random reassignment of walkers chosen from the weighted distribution of the existing walkers.

In the following Python program a number of functions are defined to realize the initial creation of walkers, the drift-diffusion step with readjustments of weights, and the population control. It is left to the reader to employ these functions in a simple program that computes the ground state energy of the helium-4 atom. There are two different ways to compute the energy in excess of $-4M/(M+1)$: first by monitoring the factor by which the weights must be readjusted to keep their sum constant (E), and second the average of $1/r_{12}$ over the ensemble of walkers (V). When the time step is small enough, both energies tend to be equal; their difference is a good indicator for the suitability of the time step. One may choose 1000 walkers, a time step of 0.002 and make 1000 steps with weight adjustment before the walkers are renewed. The excess energy above the value of $-3.999\,455$ hartree for $-4M/(M+1)$ should be recovered by E or V; this value should equal $+1.095\,731$, given the exact energy of the helium atom of $-2.903\,724$ (Anderson *et al.*, 1993). With this simple approach one may reach this value within 0.01 hartree, much better than the Hartree–Fock limit (see page 101) which is 0.04 hartree too high due to lack of correlation (Clementi and Roetti, 1974).

PYTHON PROGRAM 4.4 **walkers**
Three functions to be used for simple QMD of the helium atom

```
01 def initiate(N):
02 # create array of N walkers (helium atom)
03 # returns [walkers,weights]
04    walkers=zeros((N,3,3), dtype=float)
05    sigma=0.5
06    walkers[:,1,:]=sigma*randn(N,3)
07    walkers[:,2,:]=sigma*randn(N,3)
```

---

[7] See Hetherington (1984), Sorella (1998) and Assaraf *et al.* (2000) for a discussion of noise and bias related to stochastic reconfiguration.

```
08    r1=walkers[:,1]-walkers[:,0]; r2=walkers[:,2]-walkers[:,0]
09    r1sq=sum(r1**2,1); r2sq=sum(r2**2,1)
10    weights=exp(-2.*(sqrt(r1sq)+sqrt(r2sq))+0.5*(r1sq+r2sq)/sigma**2)
11    ratio=N/sum(weights)
12    weights=ratio*weights
13    return [walkers,weights]
14
15 def step(walkers,weights,delt,M):
16 # move walkers one time step delt; M=nuclear/electron mass
17    N=len(walkers)
18    r1=walkers[:,1]-walkers[:,0]; r2=walkers[:,2]-walkers[:,0]
19    r1norm=sqrt(sum(r1**2,1)); r2norm=sqrt(sum(r2**2,1))
20    for i in range(3):
21       r1[:,i]=r1[:,i]/r1norm; r2[:,i]=r2[:,i]/r2norm
22    alphadelt=2.*M/(M+1.)*delt
23    d1=-alphadelt*r1; d2=-alphadelt*r2
24    d0=-(d1+d2)/M
25    sd0=sqrt(delt/M); sd1=sqrt(delt)
26    walkers[:,0,:]=walkers[:,0,:]+d0+sd0*randn(N,3)
27    walkers[:,1,:]=walkers[:,1,:]+d1+sd1*randn(N,3)
28    walkers[:,2,:]=walkers[:,2,:]+d2+sd2*randn(N,3)
29 # adjust weights one time step
30    V=1./sqrt(sum((walkers[:,1]-walkers[:,2])**2,1))
31    weights=weights*exp(-V*delt)
32    ratio=N/sum(weights)
33    E=log(ratio)/delt
34    weights=ratio*weights
35    return [walkers,weights,E]
36
37 def renew(walkers,weights,Nnew):
38 # select Nnew new walkers with unit weight
39    wtacc=cumsum(weights)
40    s=wtacc[-1]
41    index=[]
42    for i in range(Nnew):
43       u=s*rand()
44       arg=argmax(where(greater(wtacc,u),0.,wtacc)) + 1
45       index=index+[arg]
46    wa=take(walkers,index)
47    wt=ones((Nnew))
48    return [wa,wt]
```

Comments

The coordinates of the walkers form an array **walkers**$[n, i, j]$, where n numbers the walkers, i numbers the particles ($0 = $ nucleus, $1, 2 = $ electrons) in each walker and $j = 0, 1, 2$ indicates the x, y, z coordinates of a particle.

Function **initiate** creates a number of N walkers: lines 06 and 07 assign a normal distribution to the electron coordinates (**randn** generates an array of normally distributed numbers); line 10 adjusts the weights to make the distribution exponential, and lines 11 and 12 normalize the total weight to N.

The function **step** moves the particles in lines 26–28 by a simultaneous diffusion and drift displacement through sampling a normal distribution with prescribed mean and standard deviation. Then the weights are adjusted according to the computed excess energy $1/r_{12}$ of each walker in lines 30–31, and restored in lines 32–34 to the original total weight, which yields the "energy" E.

The function **renew** reconstructs a new set of walkers each with unit weight, representing the same distribution as the old set. It first constructs the cumulative sum of the original weights (in line 39); then, for each new walker, a random number, uniformly distributed over the total weight, is selected, and the index of the original walker in whose range the random number falls, is determined. All indices are appended in one list **index** (line 45) which is used in line 46 to copy the walkers that correspond to the elements of that list.

4.3.7 Green's function Monte Carlo methods

The diffusional and drift step of a walker can be considered as random samples from the Green's function of the corresponding differential equation. The Green's function $G(\boldsymbol{r}, \tau; \boldsymbol{r}_0, 0)$ of a linear, homogeneous differential equation is the solution $\psi(\boldsymbol{r}, \tau)$ when the boundary or initial condition is given by a delta-function $\psi(\boldsymbol{r}, 0) = \delta(\boldsymbol{r} - \boldsymbol{r}_0)$; the general solution is the integral over the product of the Green's function and the full boundary function:

$$\psi(\boldsymbol{r}, \tau) = \int G(\boldsymbol{r}, \tau; \boldsymbol{r}_0, 0)\psi(\boldsymbol{r}_0, 0)\, d\boldsymbol{r}_0. \tag{4.42}$$

For the diffusion equation the Green's function is a Gaussian.

There is an alternative, iterative way to solve the time-independent Schrödinger equation by Monte Carlo moves, by iterating ψ_n according to

$$\left(-\nabla^2 - \frac{2m}{\hbar^2}E\right)\psi_{n+1} = \frac{2m}{\hbar^2}V\psi_n. \tag{4.43}$$

The function ψ_n is sampled, again, by walkers, who make a step sampled from the Greens function of the differential operator on the left-hand-side, which in this case involves a modified Bessel function of the second kind. The iterations converge to the "exact" solution. We shall not further pursue these Green's function Monte Carlo methods, which were originally described by Kalos (1962), and refer the reader to the literature.[8]

4.3.8 Some applications

Quantum Monte Carlo methods have been used to solve several few-particle and some many-particle problems.[9] Particular attention has been paid to the full potential energy surface of H_3 in the Born–Oppenheimer approximation: a nine-dimensional problem (Wu *et al.*, 1999). In such calculations involving a huge number of nuclear configurations, one can take advantage of the fact that the final distribution of walkers for one nuclear configuration is an efficient starting distribution for a different but nearby configuration. All electron correlation is accounted for, and an accuracy of better than 50 J/mol is obtained. This accuracy is of the same order as the relativistic correction to the energy, as calculated for H_2 (Wolniewicz, 1993). However, the *adiabatic error* due to the Born–Oppenheimer approximation is of the order of 1 kJ/mol and thus not all negligible.

---

[8] See also Anderson (1995); three earlier papers, Ceperley and Kalos (1979), Schmidt and Kalos (1984) and Schmidt and Ceperley (1992), give a comprehensive review of quantum Monte Carlo methods. Schmidt (1987) gives a tutorial of the Green's function Monte Carlo method.

[9] See Anderson (1976) and the references quoted in Anderson (1995).

There is room for future application of QMC methods for large systems (Foulkes *et al.*, 2001; Grossman and Mitas, 2005). Systems with up to a thousand electrons can already be treated. Trial wave functions can be obtained from density functional calculations (see below) and the QMC computation can be carried out on the fly to provide a force field for nuclear dynamics (Grossman and Mitas, 2005). Because QMC is in principle exact, it provides an *ab initio* approach to integrated dynamics involving nuclei and electrons, similar to but more exact than the "*ab initio*" molecular dynamics of Car and Parrinello (see Section 6.3.1). But even QMC is not exact as it is limited by the Born–Oppenheimer approximation, and because the nodal structure of the trial wave function is imposed on the wave function. The latter inconsistency may result in an incomplete recovery of the electron correlation energy, estimated by Foulkes *et al.* (2001) as some 5% of the total correlation energy. For elements beyond the second row of the periodic table, the replacement of core electrons by pseudopotentials becomes desirable for reasons of efficiency, which introduces further inaccuracies.

Finally, we note that a QMC program package "Zori" has been developed by a Berkeley group of scientists, which is available in the public domain (Aspuru-Guzik *et al.*, 2005).[10]

4.4 The Born–Oppenheimer approximation

The Born–Oppenheimer (B–O) approximation is an expansion of the behavior of a system of nuclei and electrons in powers of a quantity equal to the electron mass m divided by the (average) nuclear mass M. Born and Oppenheimer (1927) have shown that the expansion should be made in $(m/M)^{1/4}$; they also show that the first and third order in the expansion vanish. The zero-order approximation assumes that the nuclear mass is infinite, and therefore that the nuclei are stationary and their role is reduced to that of source of electrical potential for the electrons. This zero-order or *clamped nuclei* approximation is usually meant when one speaks of the B–O approximation *per se*. When nuclear motion is considered, the electrons adjust *infinitely fast* to the nuclear position or wave function in the zero-order B–O approximation; this is the *adiabatic* limit. In this approximation the nuclear motion causes no changes in the electronic state, and the nuclear motion – both classical and quantum-mechanical – is governed by an effective internuclear potential resulting from the electrons in their "stationary" state.

The effect of the adiabatic approximation on the energy levels of the

[10] Internet site: http://www.zori-code.com.

hydrogen atom (where effects are expected to be most severe) is easily evaluated. Disregarding relativistic corrections, the energy for a single electron atom with nuclear charge Z and mass M for quantum number $n = 1, 2, \ldots$ equals

$$E = -\frac{1}{2n^2} Z^2 \frac{\mu}{m} \quad \text{hartree},\tag{4.44}$$

where μ is the reduced mass $mM/(m + M)$. All energies (and hence spectroscopic frequencies) scale with $\mu/m = 0.999\,455\,679$. For the ground state of the hydrogen atom this means:

$$E(\text{adiabatic}) = -0.500\,000\,000 \quad \text{hartree},$$
$$E(\text{exact}) = -0.499\,727\,840 \quad \text{hartree},$$
$$\text{adiabatic error} = -0.000\,272\,160 \quad \text{hartree} \;,$$
$$= -0.714\,557 \quad \text{kJ/mol}.$$

Although this seems a sizeable effect, the effect on properties of molecules is small (Handy and Lee, 1996). For example, since the adiabatic correction to H_2 amounts to 1.36 kJ/mol, the dissociation energy D_0 of the hydrogen molecule increases by only 0.072 kJ/mol (on a total of 432 kJ/mol). The bond length of H_2 increases by 0.0004 a.u. or 0.0002 Åand the vibrational frequency (4644 cm^{-1}) decreases by about 3 cm^{-1}. For heavier atoms the effects are smaller and in all cases negligible. Handy and Lee (1996) conclude that for the motion of nuclei the *atomic* masses rather than the nuclear masses should be taken into account. This amounts to treating the electrons as "following" the nuclei, which is in the spirit of the BO-approximation.

The real effect is related to the quantum-dynamical behavior of moving nuclei, especially when there are closely spaced electronic states involved. Such effects are treated in the next chapter.

4.5 The many-electron problem of quantum chemistry

Traditionally, the main concern of the branch of theoretical chemistry that is called *quantum chemistry* is to find solutions to the stationary Schrödinger equation for a system of (interacting) electrons in a stationary external field. This describes isolated molecules in the Born–Oppenheimer approximation.

There are essentially only two radically different methods to solve Schrödinger's equation for a system of many (interacting) electrons in an external field: *Hartree–Fock* methods with refinements and *Density Functional Theory (DFT)*. Each requires a book to explain the details (see Szabo and

Ostlund, 1982; Parr and Yang, 1989), and we shall only review the principles of these methods.

Statement of the problem

We have N electrons in the field of M point charges (nuclei). The point charges are stationary and the electrons interact only with electrostatic Coulomb terms. The electrons are collectively described by a wave function, which is a function of $3N$ spatial coordinates r_i and N spin coordinates ω_i, which we combine into $4N$ coordinates $x_i = r_i, \omega_i$. Moreover, the wave function is antisymmetric for exchange of any pair of electrons (parity rule for fermions) and the wave functions are solutions of the time-independent Schrödinger equation:

$$\hat{H}\Psi = E\Psi \tag{4.45}$$

$$\Psi(x_1, \ldots, x_i, \ldots, x_j, \ldots,) = -\Psi(x_1, \ldots, x_j, \ldots, x_i, \ldots,), \tag{4.46}$$

$$\hat{H} = -\sum_{i=1}^{N} \frac{\hbar^2}{2m} \nabla_i^2 - \sum_{i=1}^{N}\sum_{k=1}^{M} \frac{z_k e^2}{4\pi\varepsilon_0 r_{ik}} + \sum_{i,j=1;i<j}^{N} \frac{e^2}{4\pi\varepsilon_0 r_{ij}}. \tag{4.47}$$

By expressing quantities in *atomic units* (see page xvii), the Hamiltonian becomes

$$\hat{H} = -\frac{1}{2}\sum_i \nabla_i^2 - \sum_i\sum_k \frac{z_k}{r_{ik}} + \sum_{i<j} \frac{1}{r_{ij}}. \tag{4.48}$$

Note that $\hat{H}$ is real, which implies that Ψ can be chosen to be real (if Ψ is a solution of (4.45), then Ψ^* is a solution as well for the same energy, and so is $\Re(\Psi)$.)

4.6 Hartree–Fock methods

The Hartree–Fock description of the wave function is in terms of products of one-electron wave functions $\psi(r)$ that are solutions of one-electron equations (what these equations are will be described later). The one-electron wave functions are built up as a linear combination of spatial basis functions:

$$\psi_i(r) = \sum_{\mu=1}^{K} c_{\mu i} \phi_\mu(r). \tag{4.49}$$

If the set of spatial basis functions would be complete (requiring an infinite set of functions), the one-electron wave function could be exact solutions of the one-electron wave equation; in practise one selects a finite number of appropriate functions, generally

"Slater-type" functions that look like the 1s, 2s, 2p, ... hydrogen atom functions, which are themselves for computational reasons often composed of several local Gaussian functions. The one-electron wave functions are therefore approximations.

The one-electron wave functions are ortho-normalized:

$$\langle \psi_i | \psi_j \rangle = \int \psi_i^*(r)\psi_j(r)\,dr = \delta_{ij}, \tag{4.50}$$

and are completed to twice as many functions χ with spin α or β:

$$\chi_{2i-1}(x) = \psi_i(r)\alpha(\omega), \tag{4.51}$$
$$\chi_{2i}(x) = \psi_i(r)\beta(\omega). \tag{4.52}$$

These χ-functions are also orthonormal, and are usually called *Hartree–Fock spin orbitals*.

In order to construct the total wave function, first the N-electron *Hartree product function* is formed:

$$\Psi^{HP} = \chi_i(x_1)\chi_j(x_2)\cdots\chi_k(x_N), \tag{4.53}$$

but this function does not satisfy the fermion parity rule. For example, for two electrons:

$$\Psi^{HP}(x_1, x_2) = \chi_i(x_1)\chi_j(x_2)$$
$$\neq -\chi_j(x_1)\chi_i(x_2),$$

while the following antisymmetrized function does:

$$\Psi(x_1, x_2) = 2^{-\frac{1}{2}}[\chi_i(x_1)\chi_j(x_2) - \chi_j(x_1)\chi_i(x_2)]$$
$$= \frac{1}{\sqrt{2}}\begin{vmatrix} \chi_i(x_1) & \chi_j(x_1) \\ \chi_i(x_2) & \chi_j(x_2) \end{vmatrix}. \tag{4.54}$$

In general, antisymmetrization is obtained by constructing the *Slater determinant*:

$$\Psi(x_1, x_2),\ldots,x_N) = \frac{1}{\sqrt{N}}\begin{vmatrix} \chi_i(x_1) & \chi_j(x_1) & \cdots & \chi_k(x_1) \\ \chi_i(x_2) & \chi_j(x_2) & \cdots & \chi_k(x_2) \\ \vdots & & & \vdots \\ \chi_i(x_N) & \chi_j(x_N) & \cdots & \chi_k(x_N) \end{vmatrix}. \tag{4.55}$$

This has antisymmetric parity because any exchange of two rows (two particles) changes the sign of the determinant. The Slater determinant is abbreviated as $\Psi = |\chi_i\chi_j\cdots\chi_k\rangle$.

Thus far we have not specified how the one-electron wave functions ψ_i, and hence χ_i, are obtained. These functions are solutions of one-dimensional eigenvalue equations with a special *Fock operator* $\hat{f}(i)$ instead of the hamiltonian:

$$\hat{f}(i)\chi(\boldsymbol{r}_i) = \varepsilon\chi(\boldsymbol{r}_i) \tag{4.56}$$

with

$$\hat{f}(i) = -\frac{1}{2}\nabla_i^2 - \sum_k \frac{z_k}{r_{ik}} + v^{\text{HF}}(i) \tag{4.57}$$

Here v^{HF} is an effective mean-field potential that is obtained from the combined charge densities of all other electrons. Thus, in order to solve this equation, one needs an initial guess for the wave functions, and the whole procedure needs an iterative approach until the electronic density distribution is consistent with the potential v (*self-consistent field*, SCF).

For solving the eigenvalue equation (4.56) one applies the *variational principle*: for any function $\psi' \neq \psi_0$, where ψ_0 is the exact ground state solution of the eigenvalue equation $\hat{H}\psi = E\psi$, the expectation value of $\hat{H}$ does not exceed the exact ground state eigenvalue E_0:

$$\frac{\int \psi'^* \hat{H}\psi' \, d\boldsymbol{r}}{\int \psi'^*\psi' \, d\boldsymbol{r}} \geq E_0 \tag{4.58}$$

The wave function χ is varied (i.e., the coefficients of its expansion in basis functions are varied) while keeping $\int \chi^*\chi \, d\boldsymbol{r} = 1$, until $\int \chi^* f \chi \, d\boldsymbol{r}$ is a minimum.

The electrons are distributed over the HF spin orbitals χ and form a *configuration*. This distribution can be done by filling all orbitals from the bottom up with the available electrons, in which case a ground state configuration is obtained. The energy of this "ground state" is called the *Hartree–Fock energy*, with the *Hartree–Fock limit* in the case that the basis set used approaches an infinite set.

But even the HF limit is not an accurate ground state energy because the whole SCF-HF procedure neglects the *correlation energy* between electrons. Electrons in the same spatial orbital (but obviously with different spin state) tend to avoid each other and a proper description of the two-electron wave function should take the electron correlation into account, leading to a lower energy. This is also the case for electrons in different orbitals. In fact, the London dispersion interaction between far-away electrons is based on electron correlation and will be entirely neglected in the HF approximation. The way out is to mix other, excited, configurations into the description of the wave function; in principle this *configuration interaction* (CI) allows for electron correlation. In practise the CI does not systematically converge and requires a huge amount of computational effort. Modern developments use a perturbative approach to the electron correlation problem, such as the popular *Møller–Plesset* (MP) perturbation theory. For further details see Jensen (1999).

4.7 Density functional theory

In SCF theory electron exchange is introduced through the awkward Slater determinant, while the introduction of electron correlation presents a major problem by itself. Density functional theory (DFT) offers a radically different approach that leads to a much more efficient computational procedure. Unfortunately it is restricted to the ground state of the system. It has one disadvantage: the functional form needed to describe exchange and correlation cannot be derived from first principles. In this sense DFT is not a pure *ab initio* method. Nevertheless: in its present form DFT reaches accuracies that can be approached by pure *ab initio* methods only with orders of magnitude higher computational effort. In addition, DFT can handle much larger systems.

The basic idea is that the *electron charge density* $\rho(\boldsymbol{r})$ determines the exact ground state wave function and energy of a system of electrons. Although the inverse of this statement is trivially true, the truth of this statement is not obvious; in fact this statement is the *first theorem of Hohenberg and Kohn* (1964). It can be rigorously proven. An intuitive explanation was once given by E. Bright Wilson at a conference in 1965:[11] assume we know $\rho(\boldsymbol{r})$. Then we see that ρ shows sharp maxima (cusps) at the positions of the nuclei. The local nuclear charge can be derived from the limit of the gradient of ρ near the nucleus, since at the nuclear position $|\nabla\rho| = -2z\rho$. Thus, from the charge density we can infer the positions and charges of the nuclei. But if we know that and the number of electrons, the Hamiltonian is known and there will be a unique ground state solution to the time-independent Schrödinger equation, specifying wave function and energy.

Thus the energy and its constituent terms are *functionals* of the density ρ:

$$E[\rho] = V_{\text{ne}}[\rho] + K[\rho] + V_{\text{ee}}[\rho] \tag{4.59}$$

where

$$V_{\text{ne}} = \int \rho(\boldsymbol{r})v_{\text{n}}(\boldsymbol{r})\,d\boldsymbol{r} \tag{4.60}$$

---

[11] Bright Wilson (1968), quoted by Handy (1996).

is the electron–nuclear interaction, with v_n the potential due to the nuclei, K is the kinetic energy of the electrons, and V_{ee} is the electron–electron interaction which includes the mutual Coulomb interaction J:

$$J[\rho] = \tfrac{1}{2} \int d\boldsymbol{r}_1 \, d\boldsymbol{r}_2 r_{12}^{-1} \rho(\boldsymbol{r}_1)\rho(\boldsymbol{r}_2), \tag{4.61}$$

as well as the exchange and correlation contributions. Now, the second theorem of Hohenberg and Kohn states that for any density distribution $\rho' \neq \rho$ (where ρ is the exact ground state density), the energy is never smaller than the true ground state energy E:

$$E[\rho'] \geq E[\rho]. \tag{4.62}$$

Thus finding ρ and E reduces to applying the variational principle to $E[\rho]$, i.e., minimizing E by varying $\rho(\boldsymbol{r})$, while keeping $\int \rho(\boldsymbol{r}) \, d\boldsymbol{r} = N$. Such a solution would provide the ground state energy and charge distribution, which is all we want to know: there is no need for knowledge of the detailed wave function. There is a slight problem, however: the functional form of the terms in (4.59) is not known!

A practical solution was provided by Kohn and Sham (1965), who considered the equations that a hypothetical system of N *non-interacting* electrons should satisfy in order to yield the same density distribution as the real system of interacting electrons. Consider N non-interacting electrons in $\tfrac{1}{2}N$ ($+\tfrac{1}{2}$ for odd N) orbitals; the total properly antisymmetrized wave function would be the Slater determinant of the occupied spin-orbitals. For this system the exact expressions for the kinetic energy and the density are

$$K_s[\rho] = \sum_{i=1} n_i \int \phi_i^*(-\tfrac{1}{2}\nabla^2)\phi_i \, d\boldsymbol{r}, \tag{4.63}$$

$$\rho[\boldsymbol{r}] = \sum_{i=1} n_i \phi_i^* \phi_i(\boldsymbol{r}). \tag{4.64}$$

Here $n_i = 1$ or 2 is the number of electrons occupying ϕ_i. The wave functions are solutions of the eigenvalue equation

$$\{-\tfrac{1}{2}\nabla^2 + v_s(\boldsymbol{r})\}\phi_i = \varepsilon_i \phi_i, \tag{4.65}$$

where $v_s(\boldsymbol{r})$ is an as yet undetermined potential. The solution can be obtained by the variational principle, e.g., by expanding the functions ϕ_i in a suitable set of basis functions. Slater-type Gaussian basis sets may be used, but it is also possible to use a basis set of simple plane waves, particularly if the system under study is periodic.

In order to find expressions for $v_s(\boldsymbol{r})$, we first note that the energy functional of the non-interacting system is given by

$$E[\rho] = K_s[\rho] + \int v_s(\boldsymbol{r})\rho(\boldsymbol{r})\,d\boldsymbol{r}. \tag{4.66}$$

The energy functional of the real interacting system is given by (4.59). Now writing the potential $v_s(\boldsymbol{r})$ in the hamiltonian for the Kohn-Sham orbitals (4.65) as

$$v_s(\boldsymbol{r}) = v_n(\boldsymbol{r}) + \int \frac{\rho(\boldsymbol{r'})}{|\boldsymbol{r}-\boldsymbol{r'}|}\,d\boldsymbol{r'} + v_{xc}(\boldsymbol{r}), \tag{4.67}$$

the Kohn–Sham wave functions (or their expansion coefficients in the chosen basis set) can be solved. In this potential the nuclear potential and the electrostatic electronic interactions are included; all other terms (due to electron correlation, exchange and the difference between the real kinetic energy and the kinetic energy of the non-interacting electrons) are absorbed in the *exchange-correlation potential* v_{xc}. The equations must be solved iteratively until self-consistency because they contain the charge density that depends on the solution. Thus the Kohn–Sham equations are very similar to the SCF equations of Hartree–Fock theory.

As long as no theory is available to derive the form of the exchange-correlation potential from first principles, approximations must be made. In its present implementation it is assumed that v_{xc} depends only on *local* properties of the density, so that it will be expressible as a function of the local density and its lower derivatives. This excludes the London dispersion interaction, which is a correlation effect due to dipole moments induced by distant fluctuating dipole moments. The first attempts to find a form for the exchange-correlation functional (or potential) started from the exact result for a uniform electron gas, in which case the exchange potential is inversely proportional to the cubic root of the local density:

$$v_x^{LDA} = -\frac{3}{4}\left(\frac{3}{\pi}\right)^{1/3}\rho^{1/3} \tag{4.68}$$

so that the exchange functional E_x equals

$$E_x^{LDA}[\rho] = -\frac{3}{4}\left(\frac{3}{\pi}\right)^{1/3}\int \rho^{4/3}(\boldsymbol{r})\,d\boldsymbol{r}. \tag{4.69}$$

This *local density approximation* (LDA) is not accurate enough for atoms and molecules. More sophisticated corrections include at least the gradient of the density, as the popular exchange functional proposed by Becke (1988, 1992). With the addition of a proper correlation functional, as the Lee,

Yang and Parr functional (Lee *et al.*, 1988), which includes both first and second derivatives of the density, excellent accuracy can be obtained for structures and energies of molecules. The combination of these exchange and correlation functionals is named the BLYP exchange-correlation functional. A further modification B3LYP exists (Becke, 1993). The functional forms can be found in Leach (2001).

4.8 Excited-state quantum mechanics

Normally, quantum-chemical methods produce energies and wave functions (or electron densities) for the electronic ground state. In many applications excited-state properties are required. For the prediction of spectroscopic properties one wishes to obtain energies of selected excited states and transition moments between the ground state and selected excited states. For the purpose of simulation of systems in which excited states occur, as in predicting the fate of optically excited molecules, one wishes to describe the potential energy surface of selected excited states, i.e., the electronic energy as a function of the nuclear coordinates. While dynamic processes involving electronically excited states often violate the Born–Oppenheimer approximation and require quantum-dynamical methods, the latter will make use of the potential surfaces of both ground and excited states, generated under the assumption of stationarity of the external potential (nuclear positions).

Within the class of Hartree–Fock methods, certain excited states, defined by the configuration of occupied molecular orbitals, can be selected and optimized. In the configuration interaction (CI) scheme to incorporate electron correlation, such excited states are considered, and used to mix with the ground state. The popular *complete active space SCF* (CASSCF) method of Roos (1980) can also be applied to specific excited-state configurations and produce excited-state potential surfaces.

Unfortunately, density-functional methods are only valid for the ground state and cannot be extended to include excited states. However, not all is lost, as *time-dependent* DFT allows the prediction of excited-state properties. The linear response of a system to a periodic perturbation (e.g., an electric field) can be computed by DFT; excited states show up by a peak in absorbance, so that at least their relative energies and transition moments can be computed. If this is done for many nuclear configurations, the excited-state energy surface can be probed. This application is not straightforward, and, thus far, DFT has not been used much for the purpose of generating excited-state energy surfaces.

4.9 Approximate quantum methods

While DFT scales more favorably with system size than extended Hartree-Fock methods, both approaches are limited to relatively small system sizes. This is particularly true if the electronic calculation must be repeated for many nuclear configurations, such as in molecular dynamics applications. In order to speed up the electronic calculation, many approximations have been proposed and implemented in widespread programs. Approximations to HF methods involve:

(i) restricting the quantum treatment to valence electrons;

(ii) restricting the shape of the atomic orbitals, generally to *Slater-type orbitals* (STO) of the form $r^{n-1} \exp(-\alpha r) Y_l^m(\theta, \phi)$;

(iii) neglecting or simplifying the overlap between neighboring atomic orbitals;

(iv) neglecting small integrals that occur in the evaluation of the Hamiltonian needed to minimize the expectation of the energy (4.58);

(v) replacing other such integrals by parameters.

Such methods require parametrization based on experimental (structural, thermodynamic and spectroscopic) data and are therefore classified as *semi-empirical* methods. This is not the place to elaborate on these methods; for a review the reader is referred to Chapter 5 of Cramer (2004). Suffice to say that of the numerous different approximations, the MNDO (modified neglect of differential overlap) and NDDO (neglect of diatomic differential overlap) methods seem to have survived. Examples of popular approaches are AM1 (Dewar *et al.*, 1985: the Austin Model 1) and the better parameterized PM3 (Stewart, 1989a, 1989b: Parameterized Model 3), which are among others available in Stewart's public domain program MOPAC 7.

Even semi-empirical methods do not scale linearly with the number of atoms in the system, and are not feasible for systems containing thousands of atoms. For such systems one looks for *linear-scaling* methods, such as the DAC ("divide-and-conquer") DFT scheme of Yang (1991a, 1991b). In this scheme the system is partitioned into local areas (groups of atoms, or even atoms themselves), where the local density is computed directly from a density functional, without evoking Kohn–Sham orbitals. One needs a local Hamiltonian which is a projection of the Hamiltonian onto the local partition. The local electron occupation is governed by a global Fermi level (electron free energy), which is determined by the total number of electrons in the system. This description has been improved by a formulation in terms of local density matrices (Yang and Lee, 1995) and promises to be applicable to very large molecules (Lee *et al.*, 1996).

For solids, an empirical approach to consider the wave function as a linear combination of atomic orbitals with fitted parameters for the interactions and overlap, is known under the name *tight-binding approximation*. The TB approximation is suitable to be combined with molecular dynamics (Laasonen and Nieminen, 1990).

4.10 Nuclear quantum states

While this chapter has so far dealt only with electronic states in stationary environments, nuclear motion, if undisturbed and considered over long periods of time, will also develop into stationary states, governed by the time-independent Schrödinger equation. The knowledge of such nuclear *rotational-vibrational states* is useful in connection with infrared and Raman spectroscopy. We shall assume the Born–Oppenheimer approximation (discussed in Section 4.4) to be valid, i.e., for each nuclear configuration the electronic states are pure solutions of the time-independent Schrödinger equation, as if the nuclei do not move, and thus the electrons provide a *potential field* for the nuclear interactions. The electrons have been factored-out of the complete nuclear-electronic wave function, and the electronic degrees of freedom do not occur in the nuclear Schrödinger equation

$$\left[-\sum_i \frac{\hbar^2}{2m_i} \nabla_i^2 + V(\mathbf{r}_1, \ldots \mathbf{r}_N) \right] \Psi = E\Psi, \tag{4.70}$$

where $i = 1, \ldots, N$ enumerates the nuclei, m_i is the nuclear mass, Ψ is a function of the nuclear coordinates and V is the interaction potential function of the nuclei, including the influence of the electrons. For every electronic state there is a different potential function and a different set of solutions.

The computation of eigenstates (energies and wave functions) is in principle not different from electronic calculations. Since there is always a strong repulsion at small distances between nuclei in molecules, exchange can be safely neglected. This considerably reduces the complexity of the solution. This also implies that the spin states of the nuclei generally have no influence on the energies and spatial wave functions of the nuclear eigenstates. However, the total nuclear wave function of a molecule containing identical nuclei must obey the symmetry properties of bosons or fermions (whichever is applicable) when two identical nuclei are exchanged. This leads to symmetry requirements implying that certain nuclear states are not allowed.

For isolated molecules or small molecular complexes, the translational motion factors out, but the rotational and vibrational modes couple into *vibrational-rotational-tunneling* (VRT) states. While the energies and wave functions for a one-dimensional oscillator can be computed easily by numerical methods, as treated in Section 4.3, in the multidimensional case the solution is expressed in a suitable set of basis functions. These are most easily expressed as functions of internal coordinates, like Euler angles and intramolecular distances and angles, taking symmetry properties into account. The use of internal coordinates implies that the kinetic energy operator must also be expressed in internal coordinates, which is not entirely trivial. We note that the much more easily obtained *classical* solution of internal vibrational modes corresponds to the quantum solution only in the case that rotational and vibrational modes are separable, and the vibration is purely harmonic.

The complete treatment of the VRT states for molecular complexes is beyond the scope of this book. The reader is referred to an excellent review by Wormer and van der Avoird (2000) describing the methods to compute VRT states in van der Waals complexes like argon-molecule clusters and hydrogen-bonded complexes like water clusters. Such weakly bonded complexes often havemultiple minima connected through relatively low saddle-point regions, thus allowing for effective tunneling between minima. The case of the water dimer, for which highly accurate low-frequency spectroscopic data are available, both for D_2O (Braly *et al.*, 2000a) and H_2O (Braly *et al.*, 2000b), has received special attention. There are eight equivalent global minima, all connected by tunneling pathways, in a six-dimensional intermolecular vibration-rotation space (Leforestier *et al.*, 1997; Fellers *et al.*, 1999). The comparison of predicted spectra with experiment provides an extremely sensitive test for intermolecular potentials.

5

Dynamics of mixed quantum/classical systems

5.1 Introduction

We now move to considering the dynamics of a system of nuclei *and* electrons. Of course, both electrons and nuclei are subject to the laws of quantum mechanics, but since nuclei are 2000 to 200 000 times heavier than electrons, we expect that classical mechanics will be a much better approximation for the motion of nuclei than for the motion of electrons. This means that we expect a level of approximation to be valid, where some of the degrees of freedom (d.o.f.) of a system behave essentially classically and others behave essentially quantum-mechanically. The system then is of a *mixed quantum/classical* nature.

Most often the quantum subsystem consists of system of electrons in a dynamical field of classical nuclei, but the quantum subsystem may also be a selection of generalized nuclear coordinates (e.g., corresponding to high-frequency vibrations) while other generalized coordinates are supposed to behave classically, or describe the motion of a proton in a classical environment.

So, in this chapter we consider the dynamics of a quantum system in a *non-stationary* potential. In Section 5.2 we consider the time-dependent potential as externally given, without taking notice of the fact that the sources of the time-dependent potential are moving nuclei, which are quantum particles themselves, feeling the interaction with the quantum d.o.f. Thus we consider the time evolution of the quantum system, which now involves mixing-in of excited states, but we completely ignore the *back reaction* of the quantum system onto the d.o.f. that cause the time-dependent potential, i.e., the moving nuclei. In this way we avoid the main difficulty of mixing quantum and classical d.o.f.: how to reconcile the evolution of probability density of the quantum particles with the evolution of the trajectory of the classical particles. The treatment in this approximation is applicable to some practical cases, notably when the energy exchange between classical and quantum part is negligible (this is the case, for example, for the motion of nuclear spins in a bath of classical particles at normal temperatures), but will fail completely when energy changes in the quantum system due to the external force are of the same order as the energy fluctuations in the classical system. How the quantum subsystem can be properly embedded in the environment, including the back reaction, is considered in Section 5.3.

As we shall see in Section 5.2, the effect of time-dependent potentials is that initially pure quantum states evolve into mixtures of different states. For example, excited states will mix in with the ground state as a result of a time-dependent potential. Such time dependence may arise from a time-dependent external field, as a radiation field that causes the system to "jump" to an excited state. It may also arise from internal interactions, such as the velocity of nuclei that determine the potential field for the quantum system under consideration, or from thermal fluctuations in the environment, as dipole fluctuations that cause a time-dependent electric field. The wave functions that result do not only represent additive mixtures of different quantum states, but the wave function also carries information on *phase coherence* between the contributing states. The mixed states will in turn relax under the influence of thermal fluctuations that cause *dephasing* of the mixed states. The occurrence of coherent mixed states is a typical quantum behavior, for which there is no classical analog. It is the cause of the difficulty to combine quantum and classical treatments, and of the difficulty to properly treat the back reaction to the classical system. The reason is that the wave function of a dephased mixed state can be viewed as the superposition of different quantum states, each with a given population. Thus the wave function does not describe one trajectory, but rather a probability distribution of several trajectories, each with its own back reaction to the classical part of the system. If the evolution is not split into several trajectories, and the back reaction is computed as resulting from the mixed state, one speaks of a *mean-field* solution, which is only an approximation.

When the quantum system is in the ground state, and all excited states have energies so high above the ground state that the motions of the "classical" degrees of freedom in the system will not cause any admixture of excited states, the system remains continuously in its ground state. The evolution now is *adiabatic*, as there is no transfer of "heat" between the classical environment and the quantum subsystem. In that case the back reaction is simply the expectation of the force over the ground-state wave function, and a consistent mixed quantum-classical dynamics is obtained.

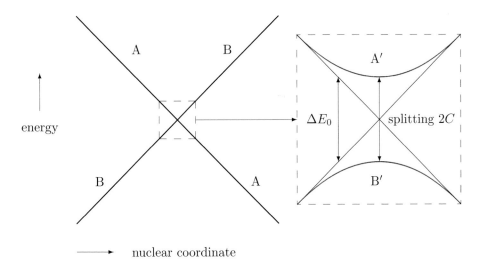

nuclear coordinate

Figure 5.1 Two crossing *diabatic* states A and B. Precise solution in the crossing region (right) yields two *adiabatic* non-crossing states A' and B'. Time dependence may cause transitions in the crossing region.

One method in this category, the *Car–Parrinello* method, also referred to as *ab initio molecular dynamics* (see Section 6.3.1), has proved to be very successful for chemically reactive systems in the condensed phase.

In cases where excited states are relevant, adiabatic dynamics is not sufficient and the separation between quantum and classical d.o.f. is no longer trivial. Now we are fully confronted with the question how to treat the evolution into a multitude of trajectories and how to evaluate the back reaction of the quantum system onto the classical d.o.f.

Consider the case that the quantum system develops into a mixture of two "pure" states. This could easily happen if the trajectory arrives at a point where the quantum system is degenerate or almost degenerate, i.e., where two states of the quantum system cross or nearly cross (see Fig. 5.1). When there is a small *coupling term* $H_{12} = H_{21} = C$ between the two states, the hamiltonian in the neighborhood of the crossing point will be:

$$\mathbf{H} = \begin{pmatrix} -\frac{1}{2}\Delta E_0 & C \\ C & \frac{1}{2}\Delta E_0 \end{pmatrix}, \tag{5.1}$$

and the wave functions will mix. The eigenvalues are

$$E_{1,2} = \pm\sqrt{\frac{1}{4}(\Delta E_0)^2 + C^2}. \tag{5.2}$$

At the crossing point ($\Delta E_0 = 0$), the adiabatic solutions are equal mixtures of both diabatic states, with a splitting of $2C$. Then there will be essentially two trajectories of the classical system possible, each related to one of the pure states. The system "choses" to evolve in either of the two branches. Only by taking the quantum character of the "classical" system into account can we fully understand the behavior of the system as a whole; the full wave function of the complete system would describe the evolution. Only that full wave function will contain the branching evolution into two states with the probabilities of each. In that full wave function the two states would still be related to each other in the sense that the wave functions corresponding to the two branches are not entirely separated; their phases remain correlated. In other words, a certain degree of "coherence" remains also after the "splitting" event, until random external disturbances destroy the phase relations, and the two states can be considered as unrelated. The coherence is related to reversibility: as long as the coherence is not destroyed, the system is time-reversible and will retrace its path if it returns to the same configuration (of the "classical" d.o.f.) where the splitting originally occurred. Such retracing may occur in small isolated systems (e.g., a diatomic molecule in the gas phase) if there is a reflection or a turning point for the classical d.o.f., as with the potential depicted in Fig. 5.2; in many-particle systems such revisiting of previously visited configurations becomes very unlikely. If in the mean time the coherence has been destroyed, the system has lost memory of the details of the earlier splitting event and will not retrace to its original starting point, but develop a new splitting event on its way back.

If we would construct only one trajectory based on the expectation value of the force, the force would be averaged over the two branches, and – assuming symmetry and equal probabilities for both branches (Fig. 5.1) after the splitting – the classical d.o.f. would feel no force and proceed with constant velocity. In reality the system develops in either branch A, continuously accelerating, or branch B, decelerating until the turning point is reached. It does not do both at the same time. Thus, the behavior based on the average force, also called the *mean-field* treatment, is clearly incorrect. It will be correct if the system stays away from regions where trajectories may split up into different branches, but cannot be expected to be correct if branching occurs.

In Section 5.3 simulations in a mixed quantum-classical system with back reaction are considered. The simplest case is the mean-field approach (Section 5.3.1), which gives consistent dynamics with proper conservation of energy and momentum over the whole system. However, it is expected to be valid only for those cases where either the back reaction does not notice-

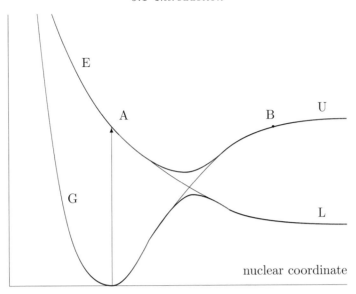

Figure 5.2 Two crossing *diabatic* states G and E, developing into two *adiabatic* states U (upper) and L (lower). After excitation from G to E (reaching point A) the system either stays on the adiabatic level U or crosses diabatically to L, depending on coupling to dynamical variables in the crossing region. If it stays in U, it reaches a turning point B and retraces its steps, if in the meantime no dephasing has taken place due to external fluctuations. With dephasing, the system may cross to G on the way back and end up vibrating in the ground state.

ably influence the classical system, or the nuclei remain localized without branching. An approximation that catches the main deficiency of the mean-field treatment is the *surface-hopping* procedure (Section 5.3.3), introduced by Tully (1990). The method introduces random hops between different potential energy surfaces, with probabilities governed by the wave function evolution. So the system can also evolve on excited-state surfaces, and incorporate non-adiabatic behavior. Apart from the rather ad hoc nature of the surface-hopping method, it is incorrect in the sense that a single, be it stochastic, trajectory is generated and the coherence between the wave functions on different trajectories is lost. Attempts to remedy this deficiency and to validate the surface-hopping method include multitrajectory methods that preserve the coherence between trajectories. Although not often applied, the mixed quantum-classical system could also be treated using the Bohmian particle description of quantum mechanics, introduced in Section 3.4. A Bohmian trajectory is a sample from the possible evolutions, and the selection of a particular trajectory allows to handle the back reaction for that particular trajectory (Section 5.3.4). In order to be able to handle the full back reaction, an ensemble of Bohmian trajectories will be needed.

The mixed quantum-classical behavior is not limited to electrons and nuclei; we can just as well treat the quantum behavior of selected nuclei (notably the proton) in a dynamical classical environment. In this case the electrons are treated in the full Born–Oppenheimer approximation; they are consistently in the electronic ground state and provide a potential field for the nuclear interactions. But the quantum nucleus has the ability to tunnel between energy wells, and move non-adiabatically with the involvement of nuclear excited states. It should be noted that the quantum effects of nuclei, being so much closer to classical behavior than electrons, can often be treated by the use of effective potentials in classical simulations, or even by quantum corrections to classical behavior. Section 3.5 is devoted to these approximate methods.

5.2 Quantum dynamics in a non-stationary potential

Assume that a quantum system of n particles $r_1, \ldots r_n$ interacts with a time-dependent Hamiltonian

$$\hat{H}(t) = \hat{K} + \hat{V}(t). \tag{5.3}$$

Then the time-dependent Schrödinger equation

$$\frac{\partial}{\partial t}\Psi(r, t) = -\frac{i}{\hbar}\hat{H}(t)\Psi(r, t) \tag{5.4}$$

can *formally* be solved as (see Chapter 14 for details of exponential operators)

$$\Psi(r, t) = \exp\left(-\frac{i}{\hbar}\int_0^t \hat{H}(t')\,dt'\right)\Psi(r, 0). \tag{5.5}$$

This way of writing helps us to be concise, but does not help very much to actually solve the time-dependent wave function.

In most applications a time-dependent solution in terms of *expansion on a basis set* is more suitable than a direct consideration of the wave function itself. If the time-dependence is weak, i.e., if the time-dependent interaction can be considered as a perturbation, or if the time dependence arises from the parametric dependence on slowly varying nuclear coordinates, the solution approaches a steady-state. Using expansion in a set of eigenfunctions of the time-independent Schrödinger equation, all the hard work is then done separately, and the time-dependence is expressed as the way the time-independent solutions mix as a result of the time-dependent perturbation.

The basis functions in which the time-dependent wave function is expanded can be either *stationary* or *time-dependent*, depending on the character of the problem. For example, if we are interested in the action of time-dependent small perturbations, resulting from fluctuating external fields, on a system that is itself stationary, the basis functions will be stationary as well. On the other hand, if the basis functions are solutions of the time-independent Schrödinger equation that contains moving parameters (as nuclear positions), the basis functions are themselves time-dependent. The equations for the mixing coefficients are different in the two cases.

In the following subsections we shall first consider direct evolution of the wave function in a time-dependent field by integration on a grid (Section 5.2.1), followed by a consideration of the evolution of the vector representing the wave function in Hilbert space. In the latter case we distinguish two cases: the basis set itself is either time-independent (Section 5.2.2) or time-dependent (Section 5.2.3).

5.2.1 Integration on a spatial grid

One way to obtain a solution is to describe the wave function on a spatial grid, and integrate values on the grid in small time steps. This actually works well for a single particle in up to three dimensions, where a grid of up to 128^3 can be used, and is even applicable for higher dimensionalities up to six, but is not suitable if the quantum system contains several particles.

For the actual solution the wave function must be sampled on a grid. Consider for simplicity the one-dimensional case, with grid points $x_0 = 0, x_1 = \delta, \ldots, x_n = n\delta, \ldots x_L = L\delta$. We assume that boundary conditions for the wave function are given (either periodic, absorbing or reflecting) and that the initial wave function $\Psi(x, t)$ is given as well. We wish to solve

$$\frac{\partial \Psi(x, t)}{\partial t} = -\frac{i}{\hbar} \left(-\frac{\hbar^2}{2m} \frac{\partial^2}{\partial x^2} + V(x, t) \right) \Psi. \tag{5.6}$$

The simplest discretization of the second spatial derivative[1] is

$$\frac{\partial^2 \Psi}{\partial x^2} = \frac{\Psi_{n-1} - 2\Psi_n + \Psi_{n+1}}{\delta^2}, \tag{5.7}$$

yielding

$$\frac{\partial \Psi(x, t)}{\partial t} = -\frac{i}{\hbar} \left[\left(\frac{\hbar^2}{m\delta^2} + V_n \right) \Psi_n - \frac{\hbar^2}{2m\delta^2} (\Psi_{n-1} + \Psi_{n+1}) \right]$$

---

[1] The error in this three-point discretization is of order δ^2. A five-point discretization with error of order δ^4 is $(1/12\delta^2)(-\Psi_{n-2} + 16\Psi_{n-1} - 30\Psi_n + 16\Psi_{n+1} - \Psi_{n+2})$. See Abramowitz and Stegun (1965) for more discretizations of partial derivatives.

$$= -i(\mathbf{H}\mathbf{\Psi})_n, \tag{5.8}$$

where $V_n = V(x_n, t)$ and

$$\mathbf{H} = \begin{pmatrix} a_0 & b & & & & & \\ b & a_1 & b & & & & \\ & \ddots & \ddots & \ddots & & & \\ & & b & a_n & b & & \\ & & & \ddots & \ddots & \ddots & \\ & & & & b & a_{L-1} & b \\ & & & & & b & a_L \end{pmatrix}. \tag{5.9}$$

Note that we have absorbed $\hbar$ into $\mathbf{H}$. The matrix elements are

$$a_n = \frac{\hbar}{m\delta^2} + \frac{V_n}{\hbar}, \qquad b = -\frac{\hbar}{2m\delta^2}. \tag{5.10}$$

We seek the propagation after one time step τ:

$$\mathbf{\Psi}(t + \tau) = e^{-i\mathbf{H}\tau}\mathbf{\Psi}(t), \tag{5.11}$$

using an approximation that preserves the unitary character of the propagator in order to obtain long-term stability of the algorithm. A popular unitary approximation is the *Crank–Nicholson* scheme:[2]

$$U_{\mathrm{CN}} = (1 + i\mathbf{H}\tau/2)^{-1}(1 - i\mathbf{H}\tau/2), \tag{5.12}$$

which is equivalent to

$$(1 + i\mathbf{H}\tau/2)\mathbf{\Psi}(t + \tau) = (1 - i\mathbf{H}\tau/2)\mathbf{\Psi}(t). \tag{5.13}$$

Both sides of this equation approximate $\mathbf{\Psi}(t + \tau/2)$, the left side by stepping $\tau/2$ backward from $\mathbf{\Psi}(t + \tau)$ and the right side by stepping $\tau/2$ forward from $\mathbf{\Psi}(t + \tau)$ (see Fig. 5.3). The evaluation of the first line requires knowledge of $\mathbf{\Psi}(t + \tau)$, which makes the Crank–Nicholson step into an *implicit* scheme:

$$[2 + i\tau a_n(t + \tau)]\Psi_n(t + \tau) - ib\tau[\Psi_{n-1}(t + \tau) + \Psi_{n+1}(t + \tau)]$$
$$= [2 - i\tau a_n(t)]\Psi_n(t) - ib\tau[\Psi_{n-1}(t) + \Psi_{n+1}(t)]. \tag{5.14}$$

Finding $\mathbf{\Psi}(t + \tau)$ requires the solution of a system of equations involving a tridiagonal matrix, which can be quickly solved. This, however is only true for one dimension; for many dimensions the matrix is no longer tridiagonal and special sparse-matrix techniques should be used.

An elegant and efficient way to solve (5.11) has been devised by de Raedt

---

[2] Most textbooks on partial differential equations will treat the various schemes to solve initial value problems like this one. In the context of the Schrödinger equation Press *et al.* (1992) and Vesely (2001) are useful references.

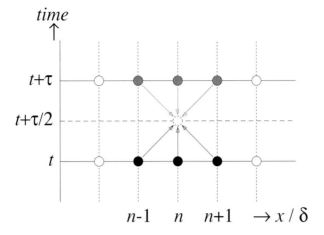

Figure 5.3 Space-time grid for the implicit, time-reversible Crank–Nicholson scheme. A virtual point at $x_n, t+\tau/2$ is approached from t and from $t+\tau$, yielding $U(-\tau/2)\Psi(t+\tau) = U(\tau/2)\Psi(t)$.

(1987, 1996), using the split-operator technique (see Chapter 14). It is possible to split the operator $\mathbf{H}$ in (5.9) into a sum of easily and exactly solvable block-diagonal matrices, such as

$$\mathbf{H} = \mathbf{H}_1 + \mathbf{H}_2,$$

with

$$\mathbf{H}_1 = \begin{pmatrix} a_0 & b & & & & \\ b & a_1/2 & & & & \\ & & a_2/2 & b & & \\ & & b & a_3/2 & & \\ & & & & \ddots & \end{pmatrix} \tag{5.15}$$

$$\mathbf{H}_2 = \begin{pmatrix} 0 & & & & & \\ & a_1/2 & b & & & \\ & b & a_2/2 & & & \\ & & & a_3/2 & b & \\ & & & b & a_4/2 & \\ & & & & & \ddots \end{pmatrix} \tag{5.16}$$

Each 2×2 exponential matrix can be solved exactly, and independently of other blocks, by diagonalization (see Chapter 14, page 388), yielding a 2×2 matrix. Then the total exponential matrix can be applied using a form of Trotter–Suzuki splitting, e.g., into the second-order product (see

Chapter 14, page 386)

$$e^{-i\mathbf{H}\tau} = e^{-i\mathbf{H}_1\tau/2}e^{-i\mathbf{H}_2\tau}e^{-i\mathbf{H}_1\tau/2}. \tag{5.17}$$

The Hamiltonian matrix can also be split up into one diagonal and two block matrices of the form

$$\exp\left[-i\tau\begin{pmatrix} 0 & b \\ b & 0 \end{pmatrix}\right] = \begin{pmatrix} \cos b\tau & -i\sin b\tau \\ -i\sin b\tau & \cos b\tau \end{pmatrix}, \tag{5.18}$$

simplifying the solution of the exponential block matrices even more. For d dimensions, the operator may be split into a sequence of operators in each dimension.[3]

While the methods mentioned above are pure *real-space* solutions, one may also split the Hamiltonian operator into the kinetic energy part $\hat{K}$ and the potential energy part $\hat{V}$. The latter is diagonal in real space and poses no problem. The kinetic energy operator, however, is diagonal in *reciprocal space*, obtained after Fourier transformation of the wave function. Thus the exponential operator containing the kinetic energy operator can be applied easily in Fourier space, and the real-space result is then obtained after an inverse Fourier transformation.

An early example of the evolution of a wave function on a grid, using this particular kind of splitting, is the study of Selloni *et al.* (1987) on the evolution of a solvated electron in a molten salt (KCl). The electron occupies local vacancies where a chloride ion is missing, similar to F-centers in solid salts, but in a very irregular and mobile way. The technique they use relies on the Trotter expansion of the exponential operator and uses repeated Fourier transforms between real and reciprocal space.

For a small time step Δt the update in Ψ is approximated by

$$\Psi(t + \Delta t) = \exp\left(-\frac{i}{\hbar}\int_t^{t+\Delta t}[\hat{K} + \hat{V}(t')]\,dt'\right)\Psi(\mathbf{r}, t)$$

$$\approx \exp\left(-\frac{i\hat{K}\Delta t}{2\hbar}\right)\exp\left(-\frac{i\hat{V}(t + \frac{1}{2}\Delta t)\Delta t}{\hbar}\right)$$

$$\times \exp\left(-\frac{i\hat{K}\Delta t}{2\hbar}\right)\Psi(\mathbf{r}, t). \tag{5.19}$$

Here we have used in the last equation the Trotter split-operator approximation for exponential operators with a sum of non-commuting exponents

---

[3] De Raedt (1987, 1996) employs an elegant notation using creation and annihilation operators to index off-diagonal matrix elements, thus highlighting the correspondence with particle motion on lattice sites, but for the reader unfamiliar with Fermion operator algebra this elegance is of little help.

(see Chapter 14), which is of higher accuracy than the product of two exponential operators. Note that it does not matter what the origin of the time-dependence of V actually is: V may depend parametrically on time-dependent coordinates of classical particles, it may contain interactions with time-dependent fields (e.g., electromagnetic radiation) or it may contain stochastic terms due to external fluctuations. The operator containing V is straightforwardly applied to the spatial wave function, as it corresponds simply to multiplying each grid value with a factor given by the potential, but the operator containing $\hat{K}$ involves a second derivative over the grid. The updating will be straightforward in reciprocal space, as $\hat{K}$ is simply proportional to k^2. Using fast Fourier transforms (FFT), the wave function can first be transformed to reciprocal space, then the operator with $\hat{K}$ applied, and finally transformed back into real space to accomplish the evolution in the kinetic energy operator. When the motion does not produce fluctuations in the potential in the frequency range corresponding to transition to the excited state, i.e., if the energy gap between ground and excited states is large compared to $k_B T$, the solvated electron behaves adiabatically and remains in the ground Born–Oppenheimer state.

5.2.2 Time-independent basis set

Let us first consider stationary basis functions and a Hamiltonian that contains a time-dependent perturbation:

$$\hat{H} = \hat{H}^0 + \hat{H}^1(t). \tag{5.20}$$

Assume that the basis functions $\phi_n(r)$ are solutions of $\hat{H}_0$:

$$\hat{H}^0 \phi_n = E_n \phi_n, \tag{5.21}$$

so that the Hamiltonian matrix $\mathbf{H}^0$ is
diagonal on that basis set (see Chapter 14). We write the solution of the time-dependent equation

$$\frac{\partial}{\partial t} \Psi(r, t) = -\frac{i}{\hbar} (\hat{H}^0 + \hat{H}^1(t)) \Psi \tag{5.22}$$

as a linear combination of time-independent basis functions:

$$\Psi(r, t) = \sum_n c_n(t) \phi_n(r). \tag{5.23}$$

Note that $c_n(t)$ contains an oscillating term $\exp(-i\omega_n t)$, where $\omega_n = E_n/\hbar$.

The time-dependent Schrödinger equation now implies

$$\frac{\partial}{\partial t} \sum_m c_m \phi_m = \sum \dot{c}_m \phi_m = -\frac{i}{\hbar} \sum_m c_m \hat{H} \phi_m, \qquad (5.24)$$

which, after left-multiplying with ϕ_n and integrating, results in

$$\dot{c}_n = -\frac{i}{\hbar} \sum_m c_m \langle \phi_n | \hat{H} | \phi_m \rangle = -\frac{i}{\hbar} (\mathbf{Hc})_n, \qquad (5.25)$$

or in matrix notation:

$$\dot{\mathbf{c}} = -\frac{i}{\hbar} \mathbf{Hc}. \qquad (5.26)$$

Since $\mathbf{H}$ is diagonal, there are two terms in the time-dependence of c_n:

$$\dot{c}_n = -i\omega_n c_n - \frac{i}{\hbar} (\mathbf{H}^1 \mathbf{c})_n. \qquad (5.27)$$

The first term simply gives the oscillating behavior $\exp(-i\omega_n t)$ of the unperturbed wave function; the second term describes the mixing-in of other states due to the time-dependent perturbation.

Often a description in terms of the density matrix ρ is more convenient, as it allows ensemble-averaging without loss of any information on the quantum behavior of the system (see Chapter 14). The density matrix is defined by

$$\rho_{nm} = c_n c_m^* \qquad (5.28)$$

and its equation of motion *on a stationary basis set* is the Liouville-Von Neumann equation:

$$\dot{\rho} = \frac{i}{\hbar}[\rho, \mathbf{H}] = \frac{i}{\hbar}[\rho, \mathbf{H}^0 + \mathbf{H}^1(t)] \qquad (5.29)$$

The diagonal terms of ρ, which do not oscillate in time, indicate how much each state contributes to the total wave function probability $\Psi^* \Psi$ and can be interpreted as a *population* of a certain state; the off-diagonal terms ρ_{nm}, which oscillate because they contain a term $\exp(-i(\omega_n - \omega_m)t)$, reveal a coherence in the phase behavior resulting from a specific history (e.g., a recent excitation). If averaged over an equilibrium ensemble, the off-diagonal elements cancel because their phases are randomly distributed. See also Section 14.8.

In Section 5.2.4 of this chapter it will be shown for a two-level system how a randomly fluctuating perturbation will cause relaxation of the density matrix.

5.2.3 Time-dependent basis set

We now consider the case that the basis functions are time-dependent themselves through the parametric dependence on nuclear coordinates $\mathbf{R}$: $\phi_n = \phi_n(\mathbf{r}; \mathbf{R}(t))$. They are eigenfunctions of the time-independent Schrödinger equation, in which the time dependence of $\mathbf{R}$ is neglected. The total wave function is expanded in this basis set:

$$\Psi = \sum_n c_n \phi_n. \tag{5.30}$$

Inserting this into the time-dependent Schrödinger equation (5.22) we find (see also Section 14.7 and (14.54))

$$\sum_m \dot{c}_m \phi_m + \dot{\mathbf{R}} \cdot \sum_m c_m \nabla_R \phi_m = -\frac{i}{\hbar} \sum_m c_m \hat{H} \phi_m. \tag{5.31}$$

After left-multiplying with ϕ_n^* and integrating, the equation of motion for the coefficient c_n is obtained:

$$\dot{c}_n + \dot{\mathbf{R}} \cdot \sum_m c_m \langle \phi_n | \nabla_R | \phi_m \rangle = -\frac{i}{\hbar}(\mathbf{Hc})_n, \tag{5.32}$$

which in matrix notation reads

$$\dot{\mathbf{c}} = -\frac{i}{\hbar}(\mathbf{H} + \dot{\mathbf{R}} \cdot \mathbf{D})\mathbf{c}. \tag{5.33}$$

Here $\mathbf{D}$ is the matrix representation of the *non-adiabatic coupling vector operator* $\hat{D} = -i\hbar \hat{\nabla}_R$:

$$D_{nm} = \frac{\hbar}{i} \langle \phi_n | \nabla_R | \phi_m \rangle. \tag{5.34}$$

Note that $\mathbf{D}$ is purely imaginary. $\mathbf{D}$ is a vector in the multidimensional space of the relevant nuclear coordinates $\mathbf{R}$, and each vector element is an operator (or matrix) in the Hilbert space of the basis functions.

In terms of the density matrix, the equation of motion for ρ for non-stationary basis functions then is

$$\dot{\rho} = \frac{i}{\hbar}[\rho, \mathbf{H} + \dot{\mathbf{R}} \cdot \mathbf{D}]. \tag{5.35}$$

The shape of this equation is the same as in the case of time-independent basis functions (5.29): the Hamiltonian is modified by a (small) term, which can often be treated as a perturbation. The difference is that in (5.29) the perturbation comes from a, usually real, term in the potential energy, while in (5.35) the perturbation is imaginary and proportional to the velocity of the sources of the potential energy. Both perturbations may be present simultaneously. We note that in the literature (see, e.g., Tully, 1990) the real and antisymmetric (but non-Hermitian) matrix element (which is a vector in $\boldsymbol{R}$-space)

$$\boldsymbol{d}_{nm} \stackrel{\text{def}}{=} \langle \phi_n | \boldsymbol{\nabla}_R | \phi_m \rangle = \frac{i}{\hbar} \boldsymbol{D}_{nm} \tag{5.36}$$

is often called the *non-adiabatic coupling vector*; its use leads to the following equation of motion for the density matrix:

$$\dot{\rho} = \frac{i}{\hbar}[\rho, \mathbf{H}] + \dot{\boldsymbol{R}}[\rho, \mathbf{d}]. \tag{5.37}$$

In the proof of (5.35), given below, we make use of the fact that $\mathbf{D}$ is a *Hermitian* matrix (or operator):

$$\mathbf{D}^\dagger = \mathbf{D}. \tag{5.38}$$

This follows from the fact that $\hat{D}$ cannot change the value of $\langle \phi_n | \phi_m \rangle$, which equals δ_{nm} and is thus independent of $\boldsymbol{R}$:

$$\hat{D} \left(\int \phi_n^* \phi_m \, d\boldsymbol{r} \right) = 0. \tag{5.39}$$

Hence it follows that

$$\int (\hat{D}\phi_n^*) \phi_m \, d\boldsymbol{r} + \int \phi_n^* \hat{D}\phi_m \, d\boldsymbol{r} = 0 \tag{5.40}$$

or, using the fact that $\mathbf{D}^* = -\mathbf{D}$,

$$-\left(\int \phi_m^* \hat{D}\phi_n \, d\boldsymbol{r} \right)^* + \int \phi_n^* \hat{D}\phi_m \, d\boldsymbol{r} = 0, \tag{5.41}$$

meaning that

$$\mathbf{D}^\dagger = \mathbf{D}. \tag{5.42}$$

Since $\mathbf{D}$ is imaginary, $D_{nm} = -D_{mn}$ and $D_{nn} = 0$.

Proof We prove (5.35). Realizing that $\rho = \mathbf{c}\mathbf{c}^\dagger$ and hence $\dot{\rho} = \dot{\mathbf{c}}\mathbf{c}^\dagger + \mathbf{c}\dot{\mathbf{c}}^\dagger$, we find that

$$\dot{\rho} = -\frac{i}{\hbar}(\mathbf{H} + \dot{\boldsymbol{R}} \cdot \mathbf{D})\rho + \frac{i}{\hbar}\rho(\mathbf{H}^\dagger + \dot{\boldsymbol{R}} \cdot \mathbf{D}^\dagger)$$

. Using the fact that both $\mathbf{H}$ and $\mathbf{D}$ are Hermitian, (5.35) follows. $\qquad \square$

5.2.4 The two-level system

It is instructive to consider a quantum system with only two levels. The extension to many levels is quite straightforward. Even if our real system has multiple levels, the interesting *non-adiabatic* events that take place under the influence of external perturbations, such as tunneling, switching from one state to another or relaxation, are active between two states that lie close together in energy. The total range of real events can generally be built up from events between two levels. We shall use the density matrix formalism (see Section 14.8),[4] as this leads to concise notation and is very suitable for extension to ensemble averages.

We start with a description based on a diagonal zero-order Hamiltonian $\mathbf{H}^0$ plus a time-dependent perturbation $\mathbf{H}^1(t)$. The perturbation may arise from interactions with the environment, as externally applied fields or fluctuating fields from thermal fluctuations, but may also arise from motions of the nuclei that provide the potential field for electronic states, as described in the previous section. The basis functions are two eigenfunctions of $\mathbf{H}^0$, with energies E_1^0 and E_2^0. The equation of motion for the density matrix is (5.29):

$$\dot{\rho} = \frac{i}{\hbar}[\rho, \mathbf{H}] = \frac{i}{\hbar}[\rho, \mathbf{H}^0 + \mathbf{H}^1(t)]. \tag{5.43}$$

Since $\operatorname{tr}\rho = 1$, it is convenient to define a variable:

$$z = \rho_{11} - \rho_{22}, \tag{5.44}$$

instead of ρ_{11} and ρ_{22}. The variable z indicates the population difference between the two states: $z = 1$ if the system is completely in state 1 and $z = -1$ if it is in state 2. We then have the complex variable ρ_{12} and the real variable z obeying the equations:

$$\dot{\rho}_{12} = \frac{i}{\hbar}[\rho_{12}(H_{22} - H_{11}) + zH_{12}], \tag{5.45}$$

$$\dot{z} = \frac{2i}{\hbar}[\rho_{12}H_{12}^* - \rho_{12}^*H_{12}], \tag{5.46}$$

where we have used the Hermitian property of ρ and H. Equations (5.45) and (5.46) imply that the quantity $4\rho_{12}\rho_{12}^* + z^2$ is a constant of the motion since the time derivative of that quantity vanishes. Thus, if we define a real

---

[4] See Berendsen and Mavri (1993) for density-matrix evolution (DME) in a two-level system; the theory was applied to proton transfer in aqueous hydrogen malonate by Mavri *et al.* (1993). The proton transfer case was further extended to multiple states by Mavri and Berendsen (1995) and summarized by Berendsen and Mavri (1997). Another application is the perturbation of a quantum oscillator by collisions, as in diatomic liquids (Mavri *et al.*, 1994).

three-dimensional vector $\boldsymbol{r}$ with components x, y, z, with:

$$x = \rho_{12} + \rho_{12}^*, \tag{5.47}$$

$$y = -i(\rho_{12} - \rho_{12}^*), \tag{5.48}$$

then the length of that vector is a constant of the motion. The motion of $\boldsymbol{r}$ is restricted to the surface of a sphere with unit radius. The time-dependent perturbation causes this vector to wander over the unit sphere. The equation of motion for $\boldsymbol{r}(t)$ can now be conveniently expressed in the perturbations when we write the latter as three real time-dependent *angular frequencies*:

$$\omega_x(t) \overset{\text{def}}{=} \frac{1}{\hbar}(H_{12} + H_{12}^*), \tag{5.49}$$

$$\omega_y(t) \overset{\text{def}}{=} \frac{1}{i\hbar}(H_{12} - H_{12}^*), \tag{5.50}$$

$$\omega_z(t) \overset{\text{def}}{=} \frac{1}{\hbar}(H_{11} - H_{22}^*), \tag{5.51}$$

yielding

$$\dot{x} = y\omega_z - z\omega_y, \tag{5.52}$$

$$\dot{y} = z\omega_x - x\omega_z, \tag{5.53}$$

$$\dot{z} = x\omega_y - y\omega_x. \tag{5.54}$$

These equations can be summarized as one vector equation:

$$\dot{\boldsymbol{r}} = \boldsymbol{r} \times \boldsymbol{\omega}. \tag{5.55}$$

Equation (5.55) describes a rotating top under the influence of a torque. This equivalence is in fact well-known in the quantum dynamics of a two-spin system (Ernst *et al.*, 1987), where $\boldsymbol{r}$ represents the magnetization, perturbed by fluctuating local magnetic fields. It gives some insight into the relaxation behavior due to fluctuating perturbations.

The off-diagonal perturbations ω_x and ω_y rotate the vector $\boldsymbol{r}$ in a vertical plane, causing an oscillatory motion between the two states when the perturbation is stationary, but a relaxation towards equal populations when the perturbation is stochastic.[5] In other words, off-diagonal perturbations cause *transitions* between states and thus limit the lifetime of each state. In the language of spin dynamics, off-diagonal stochastic perturbations cause *longitudinal* relaxation. Diagonal perturbations, on the other hand, rotate the vector in a horizontal plane and cause *dephasing* of the wave functions; they cause loss of phase coherence or *transverse* relaxation. We see that the

---

[5] In fact, the system will relax towards a Boltzmann equilibrium distribution due to a balance with spontaneous emission, which is not included in the present description.

effect of the non-adiabatic coupling vector (previous section) is off-diagonal: it causes transitions between the two states.

In a macroscopic sense we are often interested in the *rate* of the transition process from state 1 to state 2 (or vice versa). For example, if the two states represent a *reactant* state R and a *product* state P (say a proton in the left and right well, respectively, of a double-well potential), the macroscopic transfer rate from R to P is given by the rate constant k in the "reaction"

$$R \underset{k'}{\overset{k}{\rightleftharpoons}} P, \tag{5.56}$$

fulfilling the rate equation

$$\frac{dc_R}{dt} = -kc_R + k'c_P \tag{5.57}$$

on a *coarse-grained* time scale. In terms of simulation results, the rate constant k can be found by observing the *ensemble-averaged* change in population $\Delta\rho_{11}$ of the R-state, starting at $\rho_{11}(0) = 1$, over a time Δt that is large with respect to detailed fluctuations but small with respect to the inverse of k:

$$k = -\frac{\langle \Delta\rho_{11} \rangle}{\Delta t}. \tag{5.58}$$

In terms of the variable z, starting with $z = 1$, the rate constant is expressed as

$$k = -\frac{\langle \Delta z \rangle}{2\Delta t}. \tag{5.59}$$

For the two-level system there are analytical solutions to the response to stochastic perturbations in certain simplified cases. Such analytical solutions can give insight into the ongoing processes, but in simulations there is no need to approximate the description of the processes in order to allow for analytical solutions. The full wave function evolution or – preferably – the density matrix evolution (5.43) can be followed on the fly during a dynamical simulation. This applies also to the multilevel system (next section), for which analytical solutions do not exist. We now give an example of an analytical solution.

The perturbation $\boldsymbol{w}(t)$ is a stochastic vector, i.e., its components are fluctuating functions of time. Analytical solutions can only be obtained when the fluctuations of the perturbations decay fast with respect to the change of $\boldsymbol{r}$. This is the limit considered by Borgis *et al.* (1989) and by Borgis and Hynes (1991) to arrive at an expression for the proton transfer rate in a double-well potential; it is also the *Redfield limit* in the treatment of relaxation in spin systems (Redfield, 1965). Now consider the case that the off-diagonal perturbation is real, with

$$\omega_x = \frac{2C(t)}{\hbar} \tag{5.60}$$

$$\omega_y = 0. \tag{5.61}$$

We also define the diagonal perturbation in terms of its integral over time, which is a fluctuating *phase*:

$$\phi(t) \stackrel{\text{def}}{=} \int_0^t \omega_z(t') \, dt'. \tag{5.62}$$

We start at $t = 0$ with $x, y = 0$ and $z = 1$. Since we consider a time interval in which z is nearly constant, we obtain from (5.55) the following equations to first order in t by approximating $z = 1$:

$$\dot{x} = y\omega_z, \tag{5.63}$$

$$\dot{y} = \omega_x - x\omega_z, \tag{5.64}$$

$$\dot{z} = -y\omega_x. \tag{5.65}$$

¿From the first two equations a solution for y is obtained by substituting

$$g(t) = (x + iy)e^{i\phi}, \quad g(0) = 0, \tag{5.66}$$

which yields

$$\dot{g} = i\omega_x e^{i\phi}, \tag{5.67}$$

with solution

$$g(t) = i \int_0^t \omega_x(t') e^{i\phi(t')} \, dt'. \tag{5.68}$$

¿From (5.66) and (5.68) $y(t)$ is recovered as

$$y(t) = \int_0^t \omega_x(t') \cos[\phi(t') - \phi(t)] \, dt'. \tag{5.69}$$

Finally, the rate constant is given by

$$k = -\frac{1}{2}\langle \dot{z} \rangle = \frac{1}{2}\langle y\omega_x \rangle, \tag{5.70}$$

which, with $\tau = t - t'$ and extending the integration limit to ∞ because t is much longer than the decay time of the correlation functions, leads to the following expression:

$$k = \frac{1}{2} \int_0^\infty \langle \omega_x(t)\omega_x(t - \tau) \cos[\phi(t) - \phi(t - \tau)] \rangle \, d\tau. \tag{5.71}$$

After inserting (5.60), this expression is equivalent to the one used by Borgis *et al.* (1989):

$$k = \frac{2}{\hbar^2} \int_0^\infty d\tau \langle C(t)C(t-\tau) \cos \int_{t-\tau}^t w_z(t') \, dt' \rangle. \qquad (5.72)$$

This equation teaches us a few basic principles of perturbation theory. First consider what happens when w_z is constant or nearly constant, as is the case when the level splitting is large. Then the cosine term in (5.72) equals $\cos w_z \tau$ and (5.72) represents the Fourier transform or *spectral density* (see Chapter 12, Eq. (12.72) on page 326) of the correlation function of the fluctuating coupling term $C(t)$ at the angular frequency w_z, which is the frequency corresponding to the energy difference of the two levels. An example is the transition rate between ground and excited state resulting from an oscillating external electric field when the system has a nonzero off-diagonal *transition dipole moment* μ_{12} (see (2.92) on page 35). This applies to optical absorption and emission, but also to proton transfer in a double-well potential resulting from electric field fluctuations due to solvent dynamics.

Next consider what happens in the case of level crossing. In that case, at the crossing point, $w_z = 0$ and the transfer rate is determined by the integral of the correlation function of $C(t)$, i.e., the zero-frequency component of its spectral density. However, during the crossing event the diagonal elements are not identically zero and the transfer rate is determined by the time-dependence of both the diagonal and off-diagonal elements of the Hamiltonian, according to (5.72). A simplifying assumption is that the fluctuation of the off-diagonal coupling term is not correlated with the fluctuation of the diagonal splitting term. The transfer rate is then determined by the integral of the product of two correlation functions $f_x(\tau)$ and $f_z(\tau)$:

$$f_x(\tau) = \langle C(t)C(t-\tau) \rangle, \qquad (5.73)$$

$$f_z(\tau) = \langle \cos \phi(\tau) \rangle, \quad \text{with } \phi(\tau) = \int_{t-\tau}^t w_z(t') \, dt'. \qquad (5.74)$$

For stationary stochastic processes ϕ is a function of τ only. When $w_z(t)$ is a memoryless random process, $\phi(\tau)$ is a Wiener process (see page 253), representing a *diffusion* along the ϕ-axis, starting at $\phi = 0$, with diffusion constant D:

$$D = \int_0^\infty \langle w_z(0)w_z(t) \rangle \, dt. \qquad (5.75)$$

This diffusion process leads to a distribution function after a time τ of

$$p(\phi, \tau) = \frac{1}{\sqrt{4\pi D\tau}} \exp\left(-\frac{\phi^2}{4D\tau}\right), \tag{5.76}$$

which implies an exponentially decaying average cosine function:

$$\langle \cos \phi(\tau) \rangle = \int_{-\infty}^{\infty} p(\phi, \tau) \cos \phi \, d\phi = e^{-D\tau}. \tag{5.77}$$

When the random process $w_z(t)$ is not memoryless, the decay of $\langle \cos \phi \rangle$ will deviate from exponential behavior for short times, but will develop into an exponential tail. Its correlation time D^{-1} is given by the inverse of (5.75); it is *inversely* proportional to the correlation time of w_z. The faster w_z fluctuates, the slower $\langle \cos \phi \rangle$ will decay and the smaller its influence on the reaction rate will be.

We end by noting, again, that in simulations the approximations required for analytical solutions need not be made; the reaction rates can be computed by solving the complete density matrix evolution based on time-dependent perturbations obtained from simulations.

5.2.5 The multi-level system

The two-state case is able to treat the dynamics of tunneling processes involving two nearby states, but is unable to include transitions to low-lying excited states. The latter are required for a full *non-adiabatic* treatment of a transfer process. The extension to the multi-level case is straightforward, but the analogy with a three-dimensional rotating top is then lost. The basis functions should be chosen orthogonal, but they need not be solutions of any stationary Schrödinger equation. Nevertheless, it is usually convenient and efficient to consider a stationary *average* potential and construct a set of basis functions as solutions of the Schrödinger equation with that potential. In this way one can be sure that the basis set adequately covers the required space and includes the flexibility to include low-lying excited states. Mavri and Berendsen (1995) found that five basis functions, constructed by diagonalization of five Gaussians, were quite adequate to describe proton transfer over a hydrogen bond in aqueous solution. They also conclude that the use of only two Gaussians is inadequate: it underestimates the transfer rate by a factor of 30! A two-level system can only describe ground-state tunneling and does not allow paths involving excited states; it also easily underestimates the coupling term because the barrier region is inadequately described.

In the multi-level case the transfer rate cannot simply be identified with the course-grained decay rate of the population, such as ρ_{11} in the two-level case (5.58). This is because a "reactant state" or "product state" cannot be identified with a particular quantum level. The best solution is to define a spatial *selection* function $S(\boldsymbol{r})$ with the property that it is equal to one in the spatial domain one wishes to identify with a particular state and zero elsewhere. The probability p_S to be in that state is then given by

$$p_S = \int \Psi^* \Psi S(\boldsymbol{r}) \, d\boldsymbol{r} = \text{tr}\,(\boldsymbol{\rho S}), \qquad (5.78)$$

with

$$S_{nm} = \int \phi_n^* \phi_m S(\boldsymbol{r}) \, d\boldsymbol{r}. \qquad (5.79)$$

5.3 Embedding in a classical environment

Thus far we have considered how a quantum (sub)system develops when it is subjected to time-dependent perturbing influences from classical degrees of freedom (or from external fields). The system invariably develops into a mixed quantum state with a wave function consisting of a superposition of eigenfunctions, even if it started from a pure quantum state. We did not ask the question whether a single quantum system indeed develops into a mixed state, or ends up in one or another pure state with a certain probability governed by the transition rates that we could calculate. In fact that question is academic and unanswerable: we can only observe an ensemble containing all the states that the system can develop into, and we cannot observe the fate of a single system. If a single system is observed, the measurement can only reveal the probability that a given final state has occurred. The common notion among spectroscopists that a quantum system, which absorbs a radiation quantum, suddenly *jumps* to the excited state, is equally right or wrong as the notion that such a quantum system gradually mixes the excited state into its ground state wave function in the process of absorbing a radiation quantum. Again an academic question: we don't need to know, as the outcome of an experiment over an ensemble is the same for both views.

We also have not considered the related question how the quantum system *reacts back* onto the classical degrees of freedom. In cases where the coupling between quantum and classical degrees if freedom is weak (as, e.g., in nuclear spins embedded in classical molecular systems), the back reaction has a negligible effect on the dynamics of the classical system and can be disregarded. The classical system has its autonomous dynamics. This is also true for a reaction (such as a proton transfer) in the very first beginning, when the wave function has hardly changed. However, when the coupling is *not* weak, the back reaction is important and essential to fulfill the conservation laws for energy and momentum. Now it is important whether the single quantum system develops into a mixed state or a pure state, with very different strengths of the back reaction. For example, as already discussed on page 112, after a "crossing event" the system "chooses" one branch or another, but not both, and it reacts back onto the classical degrees of freedom from one branch only. Taking the back reaction from the mixed quantum wave function – which is called the *mean field* back reaction – is obviously incorrect. Another example (Tully, 1990) is a particle colliding with a solid surface, after which it either reflects back or gets adsorbed. One observes 20% reflection and 80% absorption, for example, but not 100% of something in between that would correspond to the mean field reaction.

It now seems that an academic question that cannot be answered and has no bearing on observations, suddenly becomes essential in simulations. Indeed, that is the case, and the reason is that we have been so stupid as to separate a system into quantum and classical degrees of freedom. If a system is treated completely by quantum mechanics, no problems of this kind arise. For example, if the motion along the classical degree of freedom in a level-crossing event is handled by a Gaussian wave packet, the wave packet splits up at the intersection and moves with certain probability and phase in each of the branches (see, for example, Hahn and Stock (2000) for a wave-packet description of the isomerization after excitation in rhodopsin). But the artificial separation in quantum and classical parts calls for some kind of contraction of the quantum system to a single trajectory and an artificial handling of the interaction.[6] The most popular method to achieve a consistent overall dynamics is the surface hopping method of Tully (1990), described in Section 5.3.3.

5.3.1 Mean-field back reaction

We consider how the evolution of classical and quantum degrees of freedom can be solved simultaneously in such a way that total energy and momentum are conserved. Consider a system that can be split up in classical coordinates (degrees of freedom) R and quantum degrees of freedom r, each with its conjugated momenta. The

total Hamiltonian of the system is a function of all coordinates and momenta. Now assume that a proper set of orthonormal basis functions $\phi_n(r; R)$

---

[6] For a review of various methods to handle the dynamics at level crossing (called *conical intersections* in more-dimensional cases), see Stock and Thoss (2003).

has been defined. The Hamiltonian is an operator $\hat{H}$ in $\boldsymbol{r}$-space and is represented by a matrix with elements

$$H_{nm}(q,p) = \langle n|\hat{H}(\boldsymbol{r}, \boldsymbol{R}, \boldsymbol{P})|m\rangle, \qquad (5.80)$$

where $\boldsymbol{P}$ are the momenta conjugated with $\boldsymbol{R}$. These matrix elements can be evaluated for any configuration $(\boldsymbol{R}, \boldsymbol{P})$ in the classical phase space. Using this Hamiltonian matrix, the density matrix ρ evolves according to (5.29) or (5.37), which reduces to (5.35) for basis functions that are independent of the classical coordinates. The classical system is now propagated using the *quantum expectation* of the forces $\boldsymbol{F}$ and velocities:

$$\boldsymbol{F} = \dot{\boldsymbol{P}} = -\operatorname{tr}(\boldsymbol{\rho}\mathbf{F}), \quad \text{with } F_{nm} = \langle n| - \boldsymbol{\nabla}_R\hat{H}|m\rangle, \qquad (5.81)$$
$$\dot{\boldsymbol{R}} = \operatorname{tr}(\boldsymbol{\rho}\boldsymbol{\nabla}_P\mathbf{H}). \qquad (5.82)$$

The latter equation is – for conservative forces – simply equal to

$$\dot{\boldsymbol{R}} = \boldsymbol{\nabla}_P K, \qquad (5.83)$$

or $\boldsymbol{V} = \dot{\boldsymbol{R}}$ for cartesian particle coordinates, because the classical kinetic energy K is a separable term in the total Hamiltonian.

It can be shown (see proof below) that this combined quantum/classical dynamics conserves the total energy of the system:

$$\frac{dE_{\text{tot}}}{dt} = \frac{d}{dt}\operatorname{tr}(\boldsymbol{\rho}\mathbf{H}) = 0. \qquad (5.84)$$

This must be considered a minimum requirement for a proper non-stochastic combined quantum/classical dynamics scheme. Any average energy increase (decrease) in the quantum degrees of freedom is compensated by a decrease (increase) in energy of the classical degrees of freedom.[7] Note that the force on the classical particles (5.81) is the expectation of the force matrix, which is the expectation of the gradient of $\hat{H}$ and *not* the gradient of the expectation of $\hat{H}$. The latter would also contain a contribution due to the gradients of the basis functions in case these are a function of $\boldsymbol{R}$. The force calculated as the expectation of the negative gradient of the Hamiltonian is called the *Hellmann–Feynman force*; the matrix elements F_{nm} in (5.81) are the Hellmann–Feynman force matrix elements. The use of forces averaged over the wave function is in accordance with Ehrenfest's principle (see (3.15) on page 43). Note that the energy conservation is exact, even when the basis

---

[7] The DME method with average back reaction has been applied to a heavy atom colliding with a quantum harmonic oscillator (Berendsen and Mavri, 1993) and to a quantum harmonic oscillator in a dense argon gas by Mavri and Berendsen (1994). There is perfect energy conservation in these cases. In the latter case thermal equilibration occurs between the harmonic oscillator and the gas.

set is incomplete and the wave function evolution is therefore not exact. The forces resulting from gradients of the basis functions are called the *Pulay forces*; as shown above, they should *not* be included in the forces on the classical particles when the perturbation term due to the non-adiabatic coupling is included in the dynamics of the quantum subsystem. In the next section we return to the Pulay forces in the adiabatic limit.

Proof First we split (5.84) into two parts:

$$\frac{dE_{\text{tot}}}{dt} = \text{tr}\,(\dot{\boldsymbol{\rho}}\mathbf{H}) + \text{tr}\,(\boldsymbol{\rho}\dot{\mathbf{H}}), \tag{5.85}$$

and consider the first part, using (5.37):

$$\text{tr}\,(\dot{\boldsymbol{\rho}}\mathbf{H}) = \frac{i}{\hbar}\,\text{tr}\,([\boldsymbol{\rho},\mathbf{H}]\,\mathbf{H}) + \dot{\boldsymbol{R}}\cdot\text{tr}\,([\boldsymbol{\rho},\mathbf{d}]\,\mathbf{H}) = \dot{\boldsymbol{R}}\cdot\text{tr}\,([\boldsymbol{\rho},\mathbf{d}]\,\mathbf{H}), \tag{5.86}$$

because $\text{tr}\,([\boldsymbol{\rho},\mathbf{H}]\,\mathbf{H}) = 0$, since $\text{tr}\,(\boldsymbol{\rho}\mathbf{H}\mathbf{H}) = \text{tr}\,(\mathbf{H}\boldsymbol{\rho}\mathbf{H})$.[8] The second part of (5.85) can be written out as follows:

$$\text{tr}\,(\boldsymbol{\rho}\dot{\mathbf{H}}) = \dot{\boldsymbol{R}}\cdot\text{tr}\,(\boldsymbol{\rho}\nabla_R\mathbf{H}) + \dot{\boldsymbol{P}}\cdot\text{tr}\,(\boldsymbol{\rho}\nabla_P\mathbf{H}). \tag{5.87}$$

The tricky term is $\nabla_R\mathbf{H}$:

$$\nabla_R H_{nm} = \langle\nabla_R\phi_n|\hat{H}|\phi_m\rangle + \langle\phi_n|\nabla_R\hat{H}|\phi_m\rangle + \langle\phi_n|\hat{H}|\nabla_R\phi_m\rangle.$$

The middle term equals $-F_{nm}$, according to (5.81); in (5.87) it produces a term $-\dot{\boldsymbol{R}}\cdot\dot{\boldsymbol{P}}$, which exactly cancels the last term in (5.87). The first term can be rewritten with the use of the Hermitian property of $\hat{H}$:

$$\langle\nabla_R\phi_n|\hat{H}|\phi_m\rangle = \int d\boldsymbol{r}\,\nabla_R\phi_n^*\hat{H}\phi_m = \int d\boldsymbol{r}\,(\hat{H}\nabla_R\phi_n)^*\phi_m$$
$$= (\mathbf{Hd})_{mn}^* = [(\mathbf{Hd})^\dagger]_{nm},$$

while the third term equals $(\mathbf{Hd})_{nm}$. Realizing that

$$(\mathbf{Hd})^\dagger = \mathbf{d}^\dagger\mathbf{H}^\dagger = -\mathbf{dH},$$

the first and third term together are equal to $[\mathbf{H},\mathbf{d}]_{nm}$. Thus (5.87) reduces to

$$\text{tr}\,(\boldsymbol{\rho}\dot{\mathbf{H}}) = \dot{\boldsymbol{R}}\cdot\text{tr}\,(\boldsymbol{\rho}\,[\mathbf{H},\mathbf{d}]) = -\dot{\boldsymbol{R}}\cdot\text{tr}\,([\boldsymbol{\rho},\mathbf{d}]\,\mathbf{H}),$$

which exactly cancels the term (5.86) left over from the first part of (5.85).

□

---

[8] The trace of a matrix product is invariant for cyclic permutation of the matrices in the product.

5.3.2 Forces in the adiabatic limit

The considerations in the previous section allow an extrapolation to the pure Born–Oppenheimer approximation (or *adiabatic limit*) and allow an analysis of the proper forces that act on the classical degrees of freedom in the adiabatic limit, which have been the subject of many discussions. The leading principle, again, is the conservation of total energy.

In the Born–Oppenheimer approximation it is assumed that for any point in classical phase space (R, P) the time-independent Schrödinger equation $\hat{H}\psi = E\psi$ has been solved *exactly* for the ground state. The classical coordinates R are only parameters in this solution. The system is assumed to be *and remain* in the pure ground state, i.e., in terms of a density matrix with the exact solutions of the Schrödinger equation as basis functions, numbered $n = 0, 1, \ldots$, $\rho_{00} = 1$ and $\dot{\rho} = 0$. No transitions to excited states are allowed. It is this assumption that forms the crucial approximation. It also implies that the system always remains in its exact ground state, i.e., that it follows adiabatic dynamics.

The assumption $\dot{\rho} = 0$ implies not only that the Hamiltonian is diagonal, including whatever small perturbations there are, but also that the second term in (5.35): $\dot{R}[\rho, \mathbf{d}]$, is completely neglected.

The ground-state energy $E_0(R)$ functions as the potential energy for the Hamiltonian dynamics of the classical degrees of freedom. Hence the forces must be the negative gradients of the ground-state energy in order to conserve the total energy:

$$F = -\nabla \int d\mathbf{r}\, \psi_0^* \hat{H}\psi_0. \tag{5.88}$$

Since all three elements in the integral depend on R, this force is not equal to the Hellmann-Feynman force:

$$F_{\mathrm{HF}} = -\int d\mathbf{r}\, \psi_0^* \nabla \hat{H}\psi_0. \tag{5.89}$$

The difference is the Pulay force due to the dependence of the wave function on the nuclear coordinates:

$$F_{\mathrm{Pulay}} = -\int d\mathbf{r}\, \nabla\psi_0^* \hat{H}\psi_0 - \int d\mathbf{r}\, \psi_0^* \hat{H}\nabla\psi_0 = [\mathbf{d}, \mathbf{H}]_{00}, \tag{5.90}$$

where the last equality follows from the same reasoning as was followed in the proof of (5.84) on page 132. It is not surprising that the Pulay force reappears, because it only canceled in (5.84) against the now neglected term in the density matrix evolution.

The Pulay force seems to be a nasty complication, but it isn't. When the Hamiltonian is really diagonal, the term $[\mathbf{d}, \mathbf{H}]_{00}$ is equal to zero and the Pulay force vanishes. So, for a pure adiabatic process, the Hellmann-Feynman forces suffice.

5.3.3 Surface hopping dynamics

The method of surface hopping (SH), originating from ideas of Pechukas (1969a, 1969b), was introduced by Tully and Preston (1971) and specified with the minimum number of switches by Tully (1990). The method was designed to incorporate the influence of excited electronic states into the atomic motion. The basic notion is that there is no single *best* atomic trajectory subject to the influence of electronic transitions (which would lead to mean-field behavior), but that trajectories must split into branches. The splitting is accomplished by making sudden switches between electronic states, based on the diagonal elements of the quantum density matrix. The density matrix is evolved as usual by (5.35). The sudden switches are determined in each time step Δt by a random process based on the transition probabilities from and to the state before the time step. More precisely, if the present state is n, then consider the rate of change of the population ρ_{nn} from (5.35):

$$\dot{\rho}_{nn} = \sum_{m \neq n} \left[\frac{i}{\hbar}(\rho_{nm}H^*_{nm}) - \dot{\mathbf{R}} \cdot (\rho_{nm}d_{nm} - \rho^*_{nm}d_{nm}) \right] \qquad (5.91)$$

$$= \sum_{m \neq n} \left[\frac{2}{\hbar}\Im(\rho^*_{nm}H_{nm}) - 2\Re(\dot{\mathbf{R}} \cdot \rho_{nm}d_{nm}) \right], \qquad (5.92)$$

which can be written as

$$\dot{\rho}_{nn} = \sum_{m \neq n} b_{nm}. \qquad (5.93)$$

Now define a switching probability g_{nm} within a time step Δt from the current state n to other states m as

$$g_{nm} = \frac{\Delta t b_{mn}}{\rho_{nn}}. \qquad (5.94)$$

If $g_{nm} < 0$, it is set to zero (i.e., only switches to states with higher probabilities are allowed; this in fact corresponds to the condition of a minimal number of switches). The cumulative probabilities $h_m = \sum_{k=1}^{m} g_{nk}$ are determined. A uniform random number between $0 \leq \zeta < 1$ is generated and a switch to state m occurs when $h_m < \zeta < h_{m+1}$.

In the vast majority of cases no switch will occur and the system remains in the same state n. The classical forces are now calculated as Hellmann-Feynman forces from the nth state, not from the complete density matrix. If a switch occurs, one must take *ad hoc* measures to conserve the total energy: scale the velocity of the classical degrees of freedom (in the direction of the nonadiabatic coupling vector). If the kinetic energy of the particle does not suffice to make up for the increase in energy level after the switch, the switch is not made.

There are many applications of surface hopping. Mavri (2000) has made an interesting comparison between mean-field DME, SH and exact quantum calculation of a particle colliding with a quantum oscillator and found in general that SH is closer to exact behavior than mean-field DME, but also concluded that both have their advantages and disadvantages, depending on the case under study. There is nothing in the Schrödinger equation that compels a treatment one way or another.

5.3.4 Other methods

The situation regarding the consistent treatment of the back reaction from quantum-dynamical subsystems is not satisfactory. Whereas mean-field DME fails to select a trajectory based on quantum probabilities, SH contains too many unsatisfactory *ad hoc* assumptions and fails to keep track of the coherence between various quantum states. Other approaches have similar shortcomings.[9] A possible, but not practical, solution is to describe the system as an *ensemble* of surface-hopping classical trajectory segments, keeping simultaneously track of the trajectories belonging to each of the states that mix into the original state by DME (Ben-Nun and Martinez, 1998; Kapral and Ciccotti, 1999; Nielsen *et al.*, 2000). The effect of quantum decoherence was addressed by Prezhdo and Rossky (1997a, 1997b).

A promising approach is the use of Bohmian particle dynamics (quantum hydrodynamics, see section 3.4 on page 64). The quantum particle is sampled from the initial distribution $\psi^2(r, 0)$ and moves as a classical particle in an effective potential that includes the *quantum potential* Q (see (3.93) on page 68). The latter is determined by the wave function, which can be computed either by DME in the usual way or by evaluating the gradient of the density of trajectories. Thus the quantum particle follows a single well-defined trajectory, different for each initial sample; the branching occurs automatically as a distribution of trajectories. The back reaction now is

---

[9] There is a vast literature on non-adiabatic semiclassical dynamics, which will not be reviewed here. The reader may wish to consult Meyer and Miller(1979), Webster *et al.* (1991), Laria *et al.* (1992), Bala *et al.* (1994), Billing (1994), Hammes-Schiffer (1996), Sun and Miller (1997), Müller and Stock (1998).

dependent on the Bohmian particle position. The method has been applied by Lepreore and Wyatt (1999),[10] Gindensperger *et al.* (2000, 2002, 2004) and Prezhdo and Brooksby (2001). The latter two sets of authors do not seem to agree on whether the quantum potential should be included in the classical back reaction. There are still problems with energy conservation[11] (which is only expected to be obeyed when averaged over a complete ensemble), and it is not quite clear how the Bohmian particle approach should be implemented when the quantum subsystem concerns some *generalized coordinates* rather than particles. Although at present the method cannot be considered quite mature, it is likely to be the best overall solution for the trajectory-based simulation of mixed quantum/classical dynamics.

Mixed quantum-classical problems may often be approximated in a practical way, when the details of the crossing event are irrelevant for the questions asked. Consider, for example, a chromophore in a protein that uses light to change from one conformation to another. Such systems are common in vision (rhodopsin), in energy transformation (bacteriorhodopsin) and in biological signaling processes. After excitation of the chromophore, the system evolves on the excited-state potential surface (generally going downhill along a dihedral angle from a *trans* state towards a *cis* state), until it reaches the *conical intersection* between excited state and ground state.[12] It then crosses over from the excited state to either a *trans* or a *cis* ground state, proceeding down-hill. The uphill continuation of the excited state is unlikely, as its probability in the damped, diffusion-like multidimensional motion is very small; if it happens it will lead to re-entry into the conical intersection and can be disregarded as an irrelevant process. The fate of the system upon leaving the conical intersection is of much more interest than the details during the crossing event. Groenhof *et al.* (2004), in a simulation study of the events after light absorption in the bacterial signaling protein PYP (photoactive yellow protein), used a simple approach with a single surface hop from excited to ground state after the excited state was found to cross the ground state. The potential surface of ground and excited states were determined by CASSCF (complete active space SCF, see page 105). After each time step the configuration-interaction vector is determined and it is seen whether the system crossing has occurred. If it has, the classical forces are switched from the excited to the ground state and the system continues

---

[10] See also Wyatt (2005).

[11] See comment by Salcedo (2003).

[12] A conical intersection is the multidimensional analog of the two-state crossing, as depicted in Fig. 5.2. The main coordinate in retinal-like chromophores is a dihedral angle or a combination of dihedral angles, but there are other motions, called skeletal deformations, that aid in reaching the intersection.

on one of the descending branches. Thus the crossing details are disregarded, and hopping between states before or after they cross are neglected. Still, the proper system evolution is obtained with computed quantum yield (fraction of successful evolutions to the *cis* state) close to the experimentally observed one.

Exercises

5.1 Derive the adiabatic wave functions and energies for two states with diabatic energy difference ΔE_0 and off-diagonal real coupling energies C (see (5.1)).

5.2 How do the results differ from those of the previous exercise when the off-diagonal coupling is purely imaginary?

5.3 Show that (5.45) and (5.46) imply that the quantity $4\rho_{12}\rho_{12}^* + z^2$ is a constant of the motion.

5.4 Prove that (5.77) follows from (5.76).

6

Molecular dynamics

6.1 Introduction

In this chapter we consider the motion of nuclei in the classical limit. The laws of classical mechanics apply, and the nuclei move in a conservative potential, determined by the electrons in the Born–Oppenheimer approximation. The electrons are assumed to be in the ground state, and the energy is the ground-state solution of the time-independent Schrödinger equation, with all nuclear positions as parameters. This is similar to the assumptions of Car–Parrinello dynamics (see Section 6.3.1), but the derivation of the potential *on the fly* by quantum-mechanical methods is far too compute-intensive to be useful in general. In order to be able to treat large systems over reasonable time spans, a *simple* description of the potential energy surface is required to enable the simulation of motion on that surface. This is the first task: design a suitable *force field* from which the forces on each atom, as well as the total energy, can be efficiently computed, given the positions of each atom.[1] Section 6.3 describes the principles behind force fields, and emphasizes the difficulties and insufficiencies of simple force field descriptions. But before considering force fields, we must define the system with its *boundary conditions* (Section 6.2). The way the interactions over the boundary of the simulated systems are treated is in fact part of the total potential energy description.

The force field descriptions take the covalent structure of molecules into account. They are not valid when chemical reactions may take place, changing the covalent structure, or the redox state, or even the protonation state of a molecule. In such cases at least the reactive part of the molecular system should be treated differently, e.g., by quantum-chemical methods. The

---

[1] The name "atom" is used interchangeably with "nucleus;" as the electronic motion is not separately considered, the difference is immaterial.

"*ab initio* molecular dynamics" method (Section 6.3.1) solves the electronic and nuclear equation simultaneously. Methods that use quantum-chemical approaches for a subsystem, embedded in a larger system described by a "standard" force field, are called QM/MM methods, and are the subject of Section 6.3.10.

The methods to solve the equations of motion are described in Section 6.4. We focus on methods that retain cartesian coordinates for the atomic positions, as those are by far the easiest to implement, and generally also the fastest to execute. Internal constraints are then implemented by special algorithms. A review of the principles of classical mechanics and a more complete description of the special techniques for rigid-body and constrained dynamics is given in Chapter 15.

In Section 6.5 coupling of the simulated system to external temperature and pressure baths is described. This includes *extended system* methods that extend the system with extra degrees of freedom allowing the control of certain variables within the system. Such controls can also be used to invoke *non-equilibrium* molecular dynamics, driving the system away from thermal equilibrium in a specified way, and allowing the direct study of transport properties.

Some straightforward applications are discussed in Section 6.7. The computation of macroscopic quantities that depend on the extent of accessible phase space rather than on microscopic averages, *viz.* entropy and free energy, is left for the next chapter.

6.2 Boundary conditions of the system

The first item to consider is the overall shape of the system and the boundary conditions that are applied. Long-range interactions, notably of electrostatic origin, are dependent on such conditions. Only isolated molecules in the dilute gas phase are an exception, but these have a very limited interest in practice. In general we are concerned with systems consisting of molecules in the condensed phase, with a size much larger than the system we can afford to simulate. Thus the simulated system interacts over its boundaries with an environment that is (very) different from vacuum. The important consideration is that the environment must respond to changes in the system; such a response is not only static, involving an average interaction energy, but is also dynamic, involving *time-dependent* forces reacting to changes in the system.

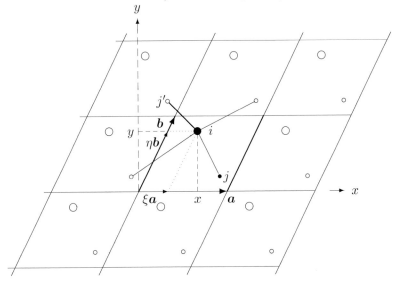

Figure 6.1 Periodic box (2D, for simplicity) determined by base vectors a and b. Particles i and j in the unit cell are given with their images; the nearest image to i is not j, but j' in the NW neighboring cell.

6.2.1 Periodic boundary conditions

The simplest, and most often applied, conditions are *periodic boundary conditions* (Fig. 6.1). The system is exactly replicated in three dimensions, thus providing a periodic lattice consisting of *unit cells*. Each unit cell can have an arbitrary *triclinic* shape, defined by three basis vectors a, b, c, with arbitrary angles α, β, γ (α is the angle between b and c, etc.) between the basis vectors. If there is only one angle different from $90°$, the cell is *monoclinic*; if all angles are $90°$, the cell is *rectangular*, and if in addition all basis vectors have equal length, the cell is *cubic*. Note that the unit cell of a periodic system is not uniquely defined: the origin can be placed arbitrarily, the unit cell can be arbitrarily rotated, and to each base vector a linear combination of integer multiples of other base vectors may be added. The volume of the unit cell does not change with any of these operations. For example, if c is changed into $c + n_a a + n_b b$ ($n_a, n_b = 0, \pm 1, \pm 2, \ldots$), the volume (see below) $V = (a \times b) \cdot c$ does not change since a and b are both perpendicular to $a \times b$. One can always choose a unit cell for which the projection of b on a is smaller than $\frac{1}{2}a$, and make sure that $\gamma \geq 60°$.

Several other cell shapes have been devised, such as the *truncated octahedron* and the *rhombic dodecahedron*, both of which pack in three dimensions and have a smaller volume than a cubic cell for the same minimal distance

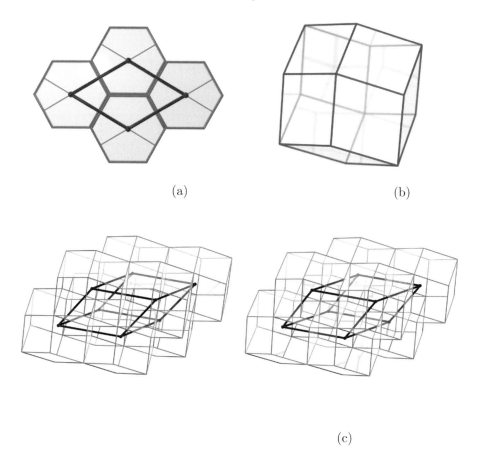

(a)   (b)

(c)

Figure 6.2 (a) Close-packed 2D hexagons, with one of the many possible unit cells describing a corresponding lattice. Each unit cell contains parts of four hexagons and each hexagon is spread over four cells. (b) The 3D rhombic octahedron with 12 faces. (c) A stereo pair depicting a packed array of rhombic dodecahedra and the triclinic unit cell of the corresponding lattice (parallel view: left picture for left eye). Figures reproduced from Wassenaar (2006) by permission of Tsjerk Wassenaar, Groningen.

to a neighboring cell. However, they can all be defined in a triclinic periodic lattice, and there is no need to invoke such special unit cells (Bekker, 1997). Figure 6.2 shows how a periodic, but more near-spherical shape (as the hexagon in two dimensions or the rhombic octahedron in three dimensions) packs in a monoclinic or triclinic box.

Coordinates of particles can always be expressed in a cartesian coordinate system x, y, z, but expression in relative coordinates ξ, η, ζ in the periodic unit cell can be convenient for some purposes, e.g., for Fourier transforms (see page 331). These latter are the *contravariant* components[2] of the position vector in the oblique coordinate system with unit vectors $\boldsymbol{a}, \boldsymbol{b}, \boldsymbol{c}$.

$$\boldsymbol{r} = x\boldsymbol{i} + y\boldsymbol{j} + z\boldsymbol{k} \tag{6.1}$$

$$\boldsymbol{r} = \xi\boldsymbol{a} + \eta\boldsymbol{b} + \zeta\boldsymbol{c}. \tag{6.2}$$

Here $\boldsymbol{i}, \boldsymbol{j}, \boldsymbol{k}$ are cartesian unit vectors, and $\boldsymbol{a}, \boldsymbol{b}, \boldsymbol{c}$ are the base vectors of the triclinic unit cell. The origins of both coordinate systems coincide. For transformation purposes the matrix $\mathbf{T}$ made up of the cartesian components of the unit cell base vectors is useful:[3]

$$\mathbf{T} = \begin{pmatrix} a_x & b_x & c_x \\ a_y & b_y & c_y \\ a_z & b_z & c_z \end{pmatrix}. \tag{6.3}$$

This matrix can be inverted when the three base vectors are linearly independent, i.e., when the volume of the cell is not zero. It is easily verified that

$$\mathbf{r} = \mathbf{T}\rho \tag{6.4}$$

$$\rho = \mathbf{T}^{-1}\mathbf{r}, \tag{6.5}$$

where $\mathbf{r}$ and ρ denote the column matrices $(x, y, z)^{\mathsf{T}}$ and $(\xi, \eta, \zeta)^{\mathsf{T}}$, respectively. These equations represent the transformations between cartesian and oblique contravariant components of a vector (see also page 331). Another characteristic of the oblique coordinate system is the *metric tensor* $\mathbf{g}$, which defines the length of a vector in terms of its contravariant components:

$$(dr)^2 = (dx)^2 + (dy)^2 + (dz)^2 = \sum_{i,j} g_{ij} d\xi_i \, d\xi_j, \tag{6.6}$$

where ξ_i stands for ξ, η, ζ. The metric tensor is given by

$$\mathbf{g} = \mathbf{T}^{\mathsf{T}}\mathbf{T}. \tag{6.7}$$

Finally, the volume V of the triclinic cell is given by the determinant of the transformation vector:

$$V = (\boldsymbol{a} \times \boldsymbol{b}) \cdot \boldsymbol{c} = \det \mathbf{T}. \tag{6.8}$$

This determinant is also the Jacobian of the transformation, signifying the

---

[2] One may also define *covariant* components, which are given by the projections of the vector onto the three unit vectors, but we shall not need those in this context. We also do not follow the convention to write contra(co)variant components as super(sub)scripts. In cartesian coordinates there is no difference between contra- and covariant components.

[3] In many texts this transformation matrix is denoted with the symbol $\mathbf{h}$.

Table 6.1 *Unit cell definitions and volumes for the cubic box, the rhombic dodecahedron and the truncated octahedron (image distance d)*

Box type	Box volume	Box vectors a b c	Box angles α β γ
cubic	d^3	$\begin{pmatrix} d & 0 & 0 \\ 0 & d & 0 \\ 0 & 0 & d \end{pmatrix}$	90° 90° 90°
rhombic dodecahedron	$0.707d^3$	$\begin{pmatrix} d & 0 & d/2 \\ 0 & d & d/2 \\ 0 & 0 & \sqrt{2}d/2 \end{pmatrix}$	60° 60° 90°
truncated octahedron	$0.770d^3$	$\begin{pmatrix} d & d/3 & -d/3 \\ 0 & 2\sqrt{2}d/3 & \sqrt{2}d/3 \\ 0 & 0 & \sqrt{6}d/3 \end{pmatrix}$	70.5° 70.5° 70.5°

modification of the volume element $dx\,dy\,dz$:

$$dx\,dy\,dz = \frac{\partial(x,y,z)}{\partial(\xi,\eta,\zeta)} d\xi\,d\eta\,d\zeta = \det \mathbf{T}\, d\xi\,d\eta\,d\zeta. \tag{6.9}$$

In practice, it is always easier to express forces and energies in cartesian rather than oblique coordinates. Oblique coordinates are useful for manipulation with images.

In many applications, notably proteins in solvent, the optimal unit cell has a minimal volume under the condition that there is a prescribed minimum distance between any atom of the protein and any atom of any neighboring image. This condition assures that the interaction between images (which is an artefact of the periodicity) is small, while the minimal volume minimizes the computational time spent on the less interesting solvent. For approximately spherical molecules the rhombic dodecahedron is the best, and the truncated octahedron the second-best choice (see Table 6.1). The easy, and therefore often used, but suboptimal choice for arbitrary shapes is a properly chosen rectangular box. As Bekker *et al.* (2004) have shown, it is also possible to automatically construct an optimal *molecular-shaped box*, that minimizes the volume while the distances between atoms of images remain larger than a specified value. An example is given in Fig. 6.3. For the particular protein molecule shown, and with the same minimum distance between atoms of images, the volumes of the cube, truncated octahedron, rhombic dodecahedron and molecular-shaped boxes were respectively 817,

629, 578 and 119 nm^3; even more dramatically the numbers of solvent (water) molecules were respectively 26 505, 20 319, 18 610 and 3 474, the latter reducing the computational time for MD simulation from 9h41 for the cubic box to 1h25 for the molecular-shaped box. Using a molecular-shaped box in MD simulations, it is mandatory to prevent overall rotation of the central molecule in the box in order not to destroy the advantages of the box shape when time proceeds. An algorithm to constrain overall rotation and translation due to Amadei *et al.* (2000) can be easily implemented and it has been shown that rotational constraints do not to influence the internal dynamics of the macromolecule (Wassenaar and Mark, 2006).

Artifacts of periodic boundary conditions

Periodic boundary conditions avoid the perturbing influence of an artificial boundary like a vacuum or a reflecting wall, but add the additional artefact of periodicity. Only if one wishes to study a crystal, periodic boundary conditions are natural, but even then they suppress all motions in neighboring unit cells that are different from the motions in the central cell. In fact, periodic boundary conditions suppress all phenomena that – in reciprocal space – concern k-vectors that do not fit multiples of the reciprocal basis vectors in (see Section 12.9 on page 331 for a description of the reciprocal lattice) $2\pi/a^*, 2\pi/b^*, 2\pi/c^*$.

Periodic boundary conditions imply that potential functions are also periodic. For example, the Coulomb energy between two charges $q_1(\mathbf{r}_1), q_2(\mathbf{r}_2)$ contains the interaction with all images, including the self-images of each particle. Algorithms to implement *lattice sums* for Coulomb interactions are described in Chapter 13. This leads to the requirement of overall electrical neutrality of the unit cell to avoid diverging electrostatic energies, to correction terms for the self-energy and to considerable artifacts if the system should correspond to a non-periodic reality. For example, consider two opposite charges at a distance d in the x-direction in a cubic cell with an edge of length a. The Coulomb force is depicted in Fig. 6.4 as a function of d/a. When d/a is not small, the periodicity artefact is considerable, amounting to a complete cancelation of the force at $d = a/2$ and even a sign reversal.

Artifacts of periodicity are avoided if modified interaction potentials are used that vanish for distances larger than half the smallest length of the unit cell. In this way only the *nearest image* interactions occur. Of course, this involves a modification of the potential that causes its own artifacts, and needs careful evaluation. Care should also be taken with the handling of cut-offs and the long-range parts in the potential function, as described on page 159: sudden cut-offs cause additional noise and erroneous behavior; smooth

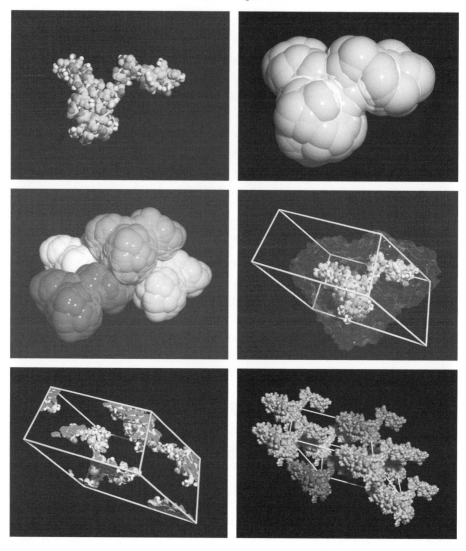

Figure 6.3 Construction of a molecular-shaped triclinic box. The molecule (top left) is expanded with a shell of size equal to half the minimum distance required between atoms of images (top right). Subsequently these shapes are translated into a close-packed arrangement (middle left). Middle right: the unit cell depicted with one molecule including its shel l of solvent. Bottom left: the unit cell as simulated (solvent not shown); bottom right: reconstructed molecules. Figures reproduced by permission of Tsjerk Wassenaar, University of Groningen (Wassenaar, 2006). See also Bekker *et al.*, 2004.

cutoffs strongly modify the interaction. There is no good solution to avoid periodicity artifacts completely. The best strategy is to use consistent forces and potentials by inclusion of complete lattice sums, but combine this with

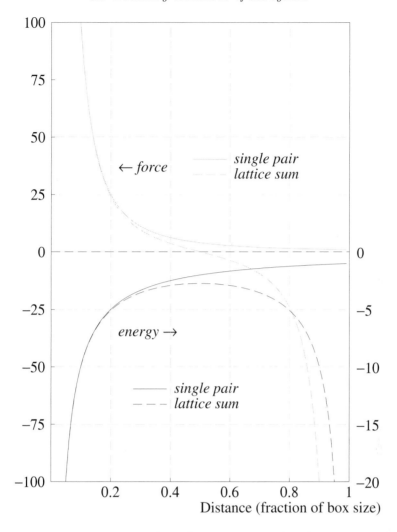

Figure 6.4 The Coulomb energy (black) and the force (grey) between an isolated positive and a negative unit charge at a distance d (solid curves) is compared with the energy and force between the same pair in a cubic periodic box (dashed curves).

studying the behavior of the system as a function of box size. In favorable cases it may also be possible to find analytical (or numerical) corrections to the effects of either periodicity or modifications of the interaction potentials (see the discussion on electrostatic continuum corrections on page 168).

6.2.2 *Continuum boundary conditions*

Other boundary conditions may be used. They will involve some kind of reflecting wall, often taken to be spherical for simplicity. The character of the problem may require other geometries, e.g., a flat wall in the case of molecules adsorbed on a surface. Interactions with the environment outside the "wall" should represent in the simplest case a *potential of mean force* given the configuration of atomic positions in the explicit system. In fact the system has a reduced number of degrees of freedom: all degrees of freedom outside the boundary are not specifically considered. The situation is identical to the reduced system description, treated in Chapter 8. The omitted degrees of freedom give rise to a combination of systematic, frictional and stochastic forces. Most boundary treatments take only care of the systematic forces, which are derivatives of the potential of mean force. If done correctly, the thermodynamic accuracy is maintained, but erroneous dynamic boundary effects may persist.

For the potential of mean force a simple approximation must be found. Since the most important interaction with the environment is of electrostatic nature, the electric field inside the system should be modified with the influence exerted by an environment treated as a continuum dielectric, and – if appropriate – conducting, material. This requires solution of the Poisson equation (see Chapter 13) or – if ions are present – the Poisson–Boltzmann equation. While for general geometries numerical solutions using either finite-difference or boundary-element methods are required, for a spherical geometry the solutions are much simpler. They can be either described by adding a field expressed in spherical harmonics, or by using the method of image charges (see Chapter 13).

Long-range contributions other than Coulomb interactions involve the r^{-6} dispersion interaction. Its potential of mean force can be evaluated from the average composition of the environmental material, assuming a homogeneous distribution outside the boundary. Since the dispersion interaction is always negative, its contribution from outside the boundary is not negligible, despite its fast decay with distance.

Atoms close to the boundary will feel modified interactions and thus deviate in behavior from the atoms that are far removed from the boundary. Thus there are non-negligible boundary effects, and the outer shell of atoms must not be included in the statistical analysis of the system's behavior.

6.2.3 Restrained-shell boundary conditions

A boundary method that has found applications in hydrated proteins is the incorporation of a *shell of restrained molecules*, usually taken to be spherical, between the system and the outer boundary with a continuum. One starts with a final snapshot from a full, equilibrated simulation, preferably in a larger periodic box. One then defines a spherical shell in which the atoms are given an additional restraining potential (such as a harmonic potential with respect to the position in the snapshot), with a force constant depending on the position in the shell, continuously changing from zero at the inner border of the shell to a large value at the outer border. Outside the shell a continuum potential may be added, as described above. This boundary-shell method avoids the insertion of a reflecting wall, gives smooth transitions at the two boundaries, and is easy to implement. One should realize, however, that molecules that would otherwise diffuse are now made rigid, and the response is "frozen in." One should allow as much motion as possible, e.g., restrain only the centers of mass of solvent molecules like water, leaving rotational freedom that allows a proper dielectric response. This and most other boundary methods do not allow elastic response of the environment and could produce adverse building up of local pressure (positive or negative) that cannot relax due to the rigidity of the boundary condition.

Examples of spherical boundary conditions are the SCAAS (surface-constrained all-atom solvent) model of King and Warshel (1989), which imposes harmonic position and orientation restraints in a surface shell and treats the shell by stochastic Brownian dynamics, and the somewhat more complex boundary model of Essex and Jorgensen (1995). In general one may question the efficiency of boundary methods that are sophisticated enough to yield reliable results (implying a rather extensive water shell around the solute, especially for large hydrated (macro)molecules), compared to periodic systems with efficient shapes, as discussed above.

6.3 Force field descriptions

The fact that there are many force fields in use, often developed along different routes, based on different principles, using different data, specialized for different applications and yielding different results, is a warning that the theory behind force fields is not in a good shape. Ideally a force field description should consist of terms that are *transferable* between different molecules, and valid for a wide range of environments and conditions. It is often the non-additivity of constituent terms, and the omission of important contributions, that renders the terms non-transferable. Since most force fields contain parameters adjusted to empirical observations, an error or omission in one term is compensated by changes in other terms, which are then not accurate when used for other configurations, environments or conditions than those for which the parameters were adjusted.

Ideally, *ab initio* quantum calculations should provide a proper potential energy surface for molecules and proper descriptions for the interaction between molecules. Density functional theory (DFT) has – for small systems – advanced to the point that it is feasible to compute the energy and forces by DFT at every time step of the molecular motion and thus evolve the system dynamically. The "*ab initio* molecular dynamics" method of Car and Parrinello (1985), described in Section 6.3.1, employs a clever method to solve the electronic and nuclear equations simultaneously. Other quantum approximations that scale more linearly with the number of particles, such as the *divide-and-conquer* and the *tight-binding* approximations (see Section 4.9) are candidates for on-the-fly quantum calculations of energies and force during dynamic evolution. In general, however, for large systems such direct methods are not efficient enough and simpler descriptions of force fields are required.

There are several reasons that quantum calculations on isolated molecules do not suffice to produce reliable force fields, and empirical adjustments are still necessary:

- For condensed systems, interaction with the (infinite) environment must be properly accounted for,

- The force field description must necessarily be simplified, if possible to additive local terms, and this simplification involves approximation of the full quantum potential energy surface.

- Even high-quality *ab initio* calculations are not accurate enough to produce overall accuracies better than $k_B T$, as required to yield accurate thermodynamic properties. Note that $k_B T = 2.5$ kJ/mol for room temperature, while an error of 6 kJ/mol in the free energy difference of two states corresponds to an error of a factor of 10 in concentrations of components participating in an equilibrium.

- Small effects that are not incorporated in the Born–Oppenheimer quantum mechanics, such as nuclear quantum effects, must be accounted for. The choice must be made whether or not such corrections are applied to the result of calculations, before parameter adjustments are made. If they are not applied afterwards, the quantum effects are mimicked by adjustments in the force field contributions. In fact, this consideration does not only apply to quantum corrections, but to all effects that are not explicitly accounted for in the force field description.

Still, it is through the study of quantum chemistry that insight is obtained in the additivity of constituent terms and the shape of the potential terms can be determined. Final empirical adjustments of parameters can then optimize the force field.

6.3.1 Ab-Initio molecular dynamics

Considering nuclei as classical point particles and electrons as providing a force field for the nuclear motion in the Born–Oppenheimer approximation, one may try to solve the energies and forces for a given nuclear configuration by quantum-chemical methods. The nuclear motion may then be advanced in time steps with one of the standard molecular dynamics algorithms. For efficiency reasons it is mandatory to employ the fact that nuclear configurations at successive time steps are very similar and therefore the solutions for the electronic equations are similar as well.

In a seminal article, Car and Parrinello (1985) described the simultaneous solution of the nuclear equations of motion and the evolution of the wave function in a density-functional description. The electron density $n(\boldsymbol{r})$ is written in terms of occupied single-particle orthonormal *Kohn–Sham orbitals* (K–S orbitals, see Section 4.7 for a more detailed description);

$$n(\boldsymbol{r}) = \sum_i |\psi_i(\boldsymbol{r})|^2, \tag{6.10}$$

where each $\psi_l(\boldsymbol{r})$ is a linear combination of well-chosen basis functions. Car and Parrinello chose as basis functions a set of plane waves $\exp(i\boldsymbol{k} \cdot \boldsymbol{r})$ compatible with the
periodic boundary conditions. Thus every K–S orbital is a vector in reciprocal space. A point of the Born–Oppenheimer potential energy surface is given by the minimum with respect to the K–S orbitals of the energy functional (4.59):

$$E = -\frac{\hbar^2}{2m_e} \sum_i \int d\boldsymbol{r}\, \psi_i^*(\boldsymbol{r})\nabla^2\psi_i(\boldsymbol{r}) + U[n(\boldsymbol{r}); \boldsymbol{R}]. \tag{6.11}$$

Here the first term is the kinetic energy of the electrons and the second term is a density functional, containing both the electron-nuclear and electron-electron interaction. The latter consists of electronic Coulomb interactions, and exchange and correlation contributions. The K–S wave functions ψ_i

(i.e., the plane wave coefficients that describe each wave function) must be varied to minimize E while preserving the orthonormality conditions

$$\int d\boldsymbol{r}\, \psi_i^*(\boldsymbol{r})\psi_j(\boldsymbol{r}) = \delta_{ij}. \tag{6.12}$$

The nuclear coordinates are constant parameters in this procedure. Once the minimum has been obtained, the forces on the nuclei follow from the gradient of E with respect to the nuclear coordinates. With these forces the nuclear dynamics can be advanced to the next time step.

The particular innovation introduced by Car and Parrinello lies in the method they use to solve the minimization problem. They consider an *extended* dynamical system, consisting of the nuclei *and* the K–S wave functions. The wave functions are given a *fictitious mass* μ and a Lagrangian (see (15.2)) is constructed:

$$\mathcal{L} = \sum \frac{1}{2} \int d\boldsymbol{r}\, |\dot{\psi}_i|^2 + \sum_I M_I \dot{\boldsymbol{R}}_I^2 - E(\psi, \boldsymbol{R}). \tag{6.13}$$

This Lagrangian, together with the constraints (6.12), generate the following equations of motion (see Section 15.8):

$$\mu \ddot{\psi}_i(\boldsymbol{r}, t) = -\frac{\partial E}{\partial \psi_i} + \sum_k \Lambda_{ik}\psi_k, \tag{6.14}$$

$$M_I \ddot{\boldsymbol{R}}_I = -\nabla_{\boldsymbol{R}_I} E, \tag{6.15}$$

where I numbers the nuclei, M_I are the nuclear masses and $\boldsymbol{R}_I$ the nuclear coordinates. The Λ_{ik} are Lagrange multipliers introduced in order to satisfy the constraints (6.12). The equations are integrated with a suitable constraint algorithm (e.g., Shake, see Section 15.8).

When the fictitious masses of the wave functions are chosen small enough, the wave function dynamics is much faster than the nuclear dynamics and the two types of motion are virtually uncoupled. Reducing the "velocities" and consequently the kinetic energy or the "temperature" of the wave functions will cause the wave functions to move close to the B–O minimum energy surface (in the limit of zero temperature the exact minimum is reached). In fact, this dynamic cooling, reminiscent of the "simulated annealing" method of Kirkpatrick *et al.* (1983), is an effective multidimensional minimization method.

In practice the temperature of the electronic degrees of freedom can be kept low enough for the system to remain close to the B–O energy surface, even when the temperature of the nuclei is considerably higher. Because both systems are only weakly coupled, heat exchange between them is very weak. Both systems can be coupled to separate thermostats (see Section 6.5) to stabilize their individual temperatures. There is a trade-off between computational efficiency and accuracy: when the "wave function mass" is small, wave functions and nuclei are effectively uncoupled and the system can remain accurately on its B–O surface, but the electronic motions become fast and a small time step must be used. For larger masses the motions of the electronic and nuclear degrees of freedom will start to overlap and the B–O surface is not accurately followed, but a larger time step can be taken. In any case the Car–Parrinello method is time-consuming, both because a large number of (electronic) degrees of freedom are added and because the time step must be taken considerably smaller than in ordinary molecular dynamics.

For algorithmic details and applications of *ab initio* molecular dynamics the reader is referred to Marx and Hutter (2000).

6.3.2 Simple molecular force fields

The simplest force fields, useful for large molecular systems, but not aiming at detailed reproduction of vibrational spectroscopic properties, contain the following elements:

- Atoms are the mass points that move in the force field. In *united-atom* approaches some hydrogen atoms are "incorporated" into the atom to which they are bound. In practice this is used for hydrogen atoms bound to aliphatic carbon atoms; the resulting CH_2 or CH_3 groups are "united atoms," acting as a single moving mass.

- Atoms (or united atoms) are also the source points for the different terms in the force field description. This means that the various contributions to the forces are expressed as functions of the atomic positions.

- There are two types of interactions: *bonded* interactions between dedicated groups of atoms, and *non-bonded* interactions between atoms, based on their (changing) distance. These two types are computationally different: bonded interactions concern atoms that are read from a fixed list, but atoms involved in non-bonded interactions fluctuate and must be updated regularly. *Non-bonded interactions are assumed to be pairwise additive.*

Bonded interactions are of the following types:

(i) A covalent *bond* between two atoms is described by a harmonic potential of the form:[4]

$$V_b(\boldsymbol{r}_i, \boldsymbol{r}_j) = \tfrac{1}{2} k_b (r - b)^2, \tag{6.16}$$

where

$$r = |\boldsymbol{r}_i - \boldsymbol{r}_j|, \tag{6.17}$$

and k_b and b are parameters which differ for each bond type.
The harmonic potential may be replaced by the more realistic Morse
potential:

$$V_{\mathrm{morse}}(\boldsymbol{r}_i, \boldsymbol{r}_j) = D_{ij}[1 - \exp(-\beta_{ij}(r_{ij} - b))]^2. \tag{6.18}$$

Other forms contain harmonic plus cubic terms.

(ii) A covalent *bond angle* is described by a harmonic angular potential
of the form

$$V_a(\boldsymbol{r}_i, \boldsymbol{r}_j, \boldsymbol{r}_k) = \tfrac{1}{2} k_\theta (\theta - \theta_0)^2, \tag{6.19}$$

where

$$\theta = \arccos \frac{\boldsymbol{r}_{ij} \cdot \boldsymbol{r}_{kj}}{r_{ij} r_{jk}}, \tag{6.20}$$

or by the simpler form

$$V_a = \tfrac{1}{2} k'(\cos \theta - \cos \theta_0)^2. \tag{6.21}$$

(iii) *Dihedral angles* ϕ are defined by the positions of four atoms i, j, k, l
as the angle between the normals $\boldsymbol{n}$ and $\boldsymbol{m}$ to the two planes i, j, k
and j, k, l:

$$\phi = \arccos \frac{\boldsymbol{n} \cdot \boldsymbol{m}}{nm} \tag{6.22}$$

where

$$\boldsymbol{n} = \boldsymbol{r}_{ij} \times \boldsymbol{r}_{kj}$$
$$\boldsymbol{m} = \boldsymbol{r}_{jk} \times \boldsymbol{r}_{lk}. \tag{6.23}$$

The dihedral potential is given by a periodic function

$$V_d(\phi) = k_\phi(1 + \cos(n\phi - \phi_0)). \tag{6.24}$$

This makes all minima equal (e.g., the *trans* and the two *gauche*
states for a threefold periodic dihedral, as between two sp^3 carbon
atoms). The actual difference between the minima is caused by the

---

[4] The GROMOS force field uses a quartic potential of the form $V = (k_b b^{-2}/8)(r^2 - b^2)^2$, which
for small deviations is virtually equivalent to (6.16), but computationally much faster because
it avoids computation of a square root.

introduction of an extra interaction between atoms i and l, called the 1-4 interaction.

Instead of using a 1-4 interaction, one may also use a set of periodic functions with different periodicity, or a set of powers of cosine functions, as in the *Ryckaert–Bellemans potential*

$$V_{RB}(\phi) = \sum_{n=0}^{5} C_n \cos^n \phi. \tag{6.25}$$

(iv) In order to keep planar groups (as aromatic rings) planar and prevent molecules from flipping over to their mirror images, *improper dihedrals* are defined, based on four atoms i, j, k, l and given a harmonic restraining potential:

$$V_{improper}(\xi) = \tfrac{1}{2} k_\xi (\xi - \xi_0)^2. \tag{6.26}$$

Bonded interactions, if they are so stiff that they represent high-frequency vibrations with frequency $\nu \gg k_B T/h$, can be replaced by *constraints*. In practice this can only be done for bond length constraints, and in some cases for bond-angle constraints as well (van Gunsteren and Berendsen, 1977). The implementation of constraints is described in Chapter 15, Section 15.8 on page 417.

Non-bonded interactions are pair-additive, and a function of the distance $r_{ij} = r$ between the two particles of each pair. Pairs that are already involved in bonded interactions are excluded from the non-bonded interaction; this concerns 1-2 and 1-3 interactions along a covalently-bonded chain. The 1-4 interactions are either excluded, or used in modified form, depending on the dihedral functions that are used. Non-bonded interactions are usually considered within a given *cut-off radius*, unless they are computed as full lattice sums over a periodic lattice (only in the case of periodic boundary conditions). They are of the following types:

(i) *Lennard–Jones interactions* describe the short-range repulsion and the longer-range dispersion interactions as

$$v_{LJ}(r) = \frac{C_{12}}{r^{12}} - \frac{C_6}{r^6}, \tag{6.27}$$

which can be alternatively expressed as

$$v_{LJ}(r) = 4\varepsilon \left(\left(\frac{\sigma}{r}\right)^{12} - \left(\frac{\sigma}{r}\right)^6 \right). \tag{6.28}$$

The treatment of the long-range part of the dispersion will be separately considered below in Section 6.3.4. The r^{-12} repulsion term is of rather arbitrary shape, and can be replaced by the more realistic exponential form

$$v_{\text{rep}} = A \exp(-Br), \tag{6.29}$$

which, combined with the dispersion term, is usually referred to as the *Buckingham potential*.

(ii) *Coulomb interactions* between charges or partial charges on atoms:

$$V_C(r) = f_{\text{el}} \frac{q_i q_j}{\varepsilon_r r} \tag{6.30}$$

Here $f_{\text{el}} = (4\pi\varepsilon_0)^{-1}$ and ε_r is a relative dielectric constant, usually taken equal to 1, but in some force fields taken to be a function of r itself (e.g., equal to r measured in Å) to mimic the effect of dielectric screening. The latter form must be considered as ad hoc without physical justification. Special care is needed for the treatment of the long-range Coulomb interaction, which is separately described below in Section 6.3.5. The partial charges are often derived from empirical dipole and quadrupole moments of (small) molecules, or from quantum calculations. A simple Mulliken analysis of atomic charges resulting from the occupation of atomic orbitals does not suffice; the best partial charges are *potential-derived* charges: those that reproduce the electric potential in the environment of the molecule, with the potential determined from a high-level *ab initio* quantum calculation. Once the potential has been determined on a grid, and suitable weight factors are chosen for the grid points, such charges can be found by a least squares optimization procedure. This method suffers from some arbitrariness due to the choice of grid points and their weights. Another, more robust, method is to fit the charges to multipoles derived from accurate quantum calculations.[5]

With the use of pair-additive Coulomb interactions, the omission of explicit polarization, and/or the incomplete treatment of long-range interactions, the empirically optimized partial charges do not and should not correspond to the *ab initio*-derived charges. The Coulomb interactions should then include the *average* polarization and the average effects of the omission of polarizing particles in the environment. Modification of partial charges cannot achieve the correct results, however, as it will completely miss the dielectric solva-

---

[5] See Jensen (2006) for a general, and Sigfridsson and Ryde (1998) for a more specific discussion. The latter authors advocate the multipole-fitting method, as do Swart *et al.* (2001), who use density-functional theory to derive the multipole moments, and list partial charges for 31 molecules and all aminoacids.

tion energy of (partial) charges in electronically polarizable environments. If average polarization enhances dipole moments, as in water, the partial charges are enhanced, while reducing the charges may be appropriate to mimic the interactions in a polarizable environment. These deficiencies are further discussed in Section 6.3.6.

6.3.3 More sophisticated force fields

Force fields that go beyond the simple type described above may include the following extra or replacing features:

(i) *Polarizability* This is the single most important improvement, which is further detailed in Section 6.3.6.

(ii) *Virtual interaction sites* Several force fields use interaction sites that do not coincide with atomic positions. For example, one may place a partial charge at the position of a "bond" midway between two atoms 1 and 2: $r = \frac{1}{2}(r_1 + r_2)$. Such sites are always a (vector) function of n atomic positions:

$$r = r(r_1, r_2, \ldots r_n), \tag{6.31}$$

and move with these positions. Virtual sites have no mass and do not participate directly in the equations of motion; they are reconstructed after every dynamic step. The potential energy $V(r, \ldots)$ depends on $r_1 \ldots r_n$ *via* its dependence on r and the force F acting on a virtual site is *distributed* among the atoms on which the site depends (we write the α-component):

$$F_{1\alpha} = -\frac{\partial V}{\partial r} \cdot \frac{\partial r}{\partial x_{1\alpha}} = F \cdot \frac{\partial r}{\partial x_{1\alpha}}. \tag{6.32}$$

For the simple case of the halfway-site $F_1 = F_2 = \frac{1}{2}F$. Other linear combinations are similarly simple; more complex virtual sites as out-of-plane constructions are more complicated but follow from the same equation.

(iii) *Dummy particles* These are sites that carry mass and participate in the equations of motion. They replace real atoms and are meant to simplify rigid-body motions. The replaced atoms are reconstructed as virtual sites. For example, the 12 atoms with 12 constraints of a rigid benzene molecule C_6H_6 can be replaced by three dummy atoms with three constraints, having the same total mass and moments of inertia. All interaction sites (atoms in this case) can be reconstructed by linear combinations from the dummies.[6] Feenstra *et al.* (1999) have used dummy sites to eliminate fast motions of hydrogen atoms in proteins, enabling an increase of time step from the usual 2 fs to as much as 7 fs. Dummy atoms do not really belong in this list because they are not a part of the force-field description.

(iv) *Coupling terms* Force fields that aim at accurate reproduction of vibrational properties include coupling terms between bond, bond-angle and dihedral displacements.

(v) *Flexible constraints* Internal vibrations with frequencies higher than $k_B T/h$ exhibit essential quantum behavior. At very high frequencies the corresponding degrees of freedom are in the ground state and can be considered static. It seems logical, therefore, to treat such degrees of freedom as constraints. However, it is quite tricky to separate the real quantum degree of freedom from the classical degrees of freedom, as the fluctuating force on anharmonic bonds shifts the harmonic oscillator both in position and in energy. Constraining the quantum vibration amounts to constraining the bond length to the position where the net force vanishes. Such "flexible constraints" were first proposed by Zhou *et al.* (2000) and have been implemented in a polarizable *ab initio* water model by Hess *et al.* (2002) in molecular dynamics and by Saint-Martin *et al.* (2005) in Monte Carlo algorithms. The difference with the usual holonomic constraints is not very large.

(vi) *Charge distributions* Descriptions of the electronic charge distributions in terms of point charges is not quite appropriate if accuracy at short distances between sources is required. In fact, the electron distributions in atoms have a substantial width and nearby distributions will interpenetrate to a certain extent. The modified (damped) short-range interactions are better represented by charge distributions than by point charges. Exponential shapes (as from the distribution of Slater-type orbitals) are the most appropriate.

(vii) *Multipoles* To increase the accuracy of the representation of charge distributions while avoiding too many additional virtual sites, dipoles – and sometimes quadrupoles – may be added to the description. The disadvantage is that the equations of motion become more complicated, as even for dipoles the force requires the computation of electric field gradients, due to charges, dipoles and quadrupoles. Dipoles and quadrupoles are subjected to torques, which requires distribution of

---

[6] The term "dummy" is not always reserved for sites as described here, but is often used to indicate virtual sites as well.

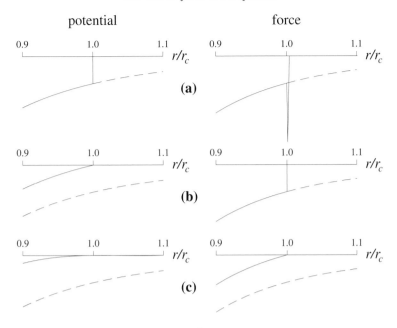

Figure 6.5 Truncation schemes for an r^{-6} dispersion potential. Thick lines: potential (left) and force (right) for (a) truncated potential, (b) truncated force = shifted potential, (c) shifted force. Dashed lines give the correct potential and force.

forces over particles that define the multipole axes. Its use is not recommended.

(viii) *QM/MM methods* combine energies and forces derived from quantumchemical calculations for selected parts of the system with force fields for the remainder (see Section 6.3.10).

(ix) *Ab initio molecular dynamics* applies DFT-derived forces during dynamics evolution: see Section 6.3.1.

6.3.4 Long-range dispersion interactions

The non-bonded potential and force calculations are usually based on lists of atom or group pairs that contain only pairs within a given distance. The reason for this is a computational one: if all pairwise interactions are included, the algorithm has an N^2 complexity, which runs out of hand for large systems. This implies the use of a cut-off distance, beyond which the interaction is either neglected, or treated in a different way that is less computationally demanding than N^2. For the r^{-6} dispersion potential such cut-offs can be applied without gross errors, but for the r^{-1} Coulomb potential simple cut-offs give unacceptable errors.

When an abrupt force and potential cut-off is used, the force is no longer the derivative of the potential, and therefore the potential is no longer conservative. As illustrated in Fig. 6.5a, the derivative of a truncated potential contains a delta function at the cut-off radius. Incorporation of this unphysical delta function into the force leads to intolerable artifacts. The use of a truncated force without delta function (Fig. 6.5b) implies that the effective potential function in the simulation is not the truncated real potential but a *shifted* potential obtained by integration of the truncated force. One may then expect that equilibrium ensembles generated by dynamic trajectories differ from those generated by Monte Carlo simulations based on the truncated potential.

But even truncated forces with shifted potentials generate artifacts. When particles diffuse through the limit of the interaction range they encounter a sudden force change leading to extra *noise*, to heating artifacts and to artifacts in the density distributions. The discontinuity of the force causes errors when higher-order integration algorithms are used that rely on the existence of force derivatives. Such effects can be avoided by shifting the force function to zero at the cut-off distance (Fig. 6.5c), but this has an even more severe influence on the effective potential, which now deviates from the exact potential over a wide range. Several kinds of *switching functions*, switching the force smoothly off at the cut-off radius, are used in practical MD algorithms. The user of such programs, which seemingly run error-free even for short cut-offs, should be aware of the possible inadequacies of the effective potentials.

Of course, the error due to neglect of the long-range interaction beyond the cut-off radius r_c can be reduced by increasing r_c. This goes at the expense of longer pair lists (which scale with r_c^3) and consequently increased computational effort. In addition, r_c should not increase beyond half the smallest box size in order to restrict interactions to the nearest image in periodic systems. One should seek an optimum, weighing computational effort against required precision. But what is the error caused by the use of a modified interaction function?

Let us consider a homogeneous fluid with an interatomic dispersion interaction $v^{\mathrm{disp}} = -C_6 r^{-6}$ and compute the *correction* terms to energy and pressure for three commonly used short-range interactions v^{sr}: truncated potential, truncated force, shifted force. The average number density is ρ and the radial distribution function is $g(r)$: given the presence of a particle at the origin, the probability of finding another particle in a volume element $d\boldsymbol{r}$ equals $\rho\, g(r)\, d\boldsymbol{r}$. Noting that the correction $\Delta v(r)$ involves the full dispersion interaction minus the employed short-range interaction, the correction to the potential energy and therefore to the internal energy u *per particle* is

$$\Delta u = \frac{1}{2}\rho \int_0^\infty \Delta v(r) 4\pi r^2 g(r)\, dr, \tag{6.33}$$

and the pressure correction is

$$\Delta P = -\frac{2\pi}{3}\rho^2 \int_0^\infty r^3 g(r)\frac{d\Delta v(r)}{dr}\, dr, \tag{6.34}$$

as is easily derived from the virial expression for the pressure (see Section 17.7.2 on page 484)

$$PV = Nk_bT + \frac{1}{3}\sum_i\sum_{j>i}\langle \mathbf{r}_{ij} \cdot \mathbf{F}_{ij}\rangle, \tag{6.35}$$

with $\mathbf{r}_{ij} = \mathbf{r}_i - \mathbf{r}_j$ and $\mathbf{F}_{ij}$ is the force exerted by j on i:

$$\mathbf{F}_{ij} = -\left(\frac{dv(r)}{dr}\right)_{r=r_{ij}}\frac{\mathbf{r}_{ij}}{r_{ij}}. \tag{6.36}$$

We obtain the following results, assuming that $g(r) = 1$ for $r >= r_c$ and that the number of particles within $r_c \gg 1$:

(i) *Truncated potential*

$$v^{\mathrm{sr}} = -C_6 r^{-6}, \tag{6.37}$$

$$\Delta u = -\frac{2\pi}{3}\rho C_6 r_c^{-3}, \tag{6.38}$$

$$\Delta P = -2\pi\rho^2 C_6 r_c^{-3}. \tag{6.39}$$

The pressure correction consists for two-thirds of a contribution from the missing tail and for one third of a contribution due to the discontinuity of the potential at the cut-off radius.

(ii) *Truncated force*

$$v^{\mathrm{sr}} = -C_6 r^{-6} + C_6 r_c^{-6}, \tag{6.40}$$

$$\Delta u = -\frac{4\pi}{3}\rho C_6 r_c^{-3}, \tag{6.41}$$

$$\Delta P = -\frac{4\pi}{3}\rho^2 C_6 r_c^{-3}. \tag{6.42}$$

(iii) *Shifted force*

$$v^{\mathrm{sr}} = -C_6 r^{-6} - 6C_6 r r_c^{-7} + 7C_6 r_c^{-6}, \tag{6.43}$$

$$\Delta u = -\frac{7\pi}{3}\rho C_6 r_c^{-3}, \tag{6.44}$$

$$\Delta P = -\frac{7\pi}{3}\rho^2 C_6 r_c^{-3}. \qquad (6.45)$$

In order to avoid difficulties caused by the discontinuity in the potential, even for MC simulations where the pressure is affected, the potential should never be truncated, neither in MD nor in MC. In the derivation of the corrections we have neglected details of the radial distribution function in the integrals, which is justified when r_c extends beyond the region where $g(r)$ differs from 1. Thus Δu depends only – and linearly – on density, with the consequences that the change in Helmholtz free energy equals Δu, that there is no change in entropy, and that $\Delta P = \rho \Delta u$. The effect of the correction is appreciable, especially affecting vapor pressure and the location of the critical point.[7] For example, for the Lennard–Jones liquid *argon*[8] at 85 K, not far from the boiling point (87.3 K), the energy correction for a shifted force with $r_c = 1$ nm is -0.958 kJ/mol, and the pressure correction is -20.32 kJ mol^{-1} nm^{-3} $= -337$ bar! In Table 6.2 the corrections in the thermodynamic properties at this particular density-temperature state point are compared to the thermodynamic values themselves. The effects are non-negligible, even for this rather long cut-off radius of about 3 σ. For *water* with[9] $C_6 = 0.002617$ and at room-temperature liquid density the shifted force correction for 1 nm cut-off is -0.642 kJ/mol for the energy and -357 bar for the pressure. Such corrections are essential. Still, they are usually not applied, and the models are parameterized to fit empirical data using MD with a given cut-off method. It is clear that the model parameters are then *effective* parameters that incorporate the effect of restrictions of the simulations; such parameters must be readjusted when other cut-offs or long-range methods are applied. While such effective potentials are convenient (commonly they do not only imply the effects of long-range treatment but also of lack of polarizability and other contributions that are not pair-additive, neglect of specific interaction terms and neglect of quantum effects), they tend to restrict the generality and transferability of force fields.

At this point it is worthwhile to remark that for dispersion (and other power-law interactions) long-range contributions can be evaluated under pe-

---

[7] See for the Lennard–Jones equation-of-state and a discussion on truncation of LJ interactions: Nicolas *et al.* (1979), Smit (1992), Johnson *et al.* (1993) and Frenkel and Smit (2002). The critical temperature for the truncated force shifts as much as 5% downward when the cut-off is decreased from 3.5 to 3σ.

[8] $\sigma = 0.34$ nm; $\varepsilon = 119.8\, k_B = 0.9961$ kJ/mol; $C_6 = 4\varepsilon\sigma^6 = 0.006155$ kJ mol^{-1} nm^6. Density at 85 K and 1 bar: 35.243 mol/dm^3 (number density 21.224 nm^{-3}).

[9] Value for the SPC (Berendsen *et al.*, 1981) and SPC/E model (Berendsen *et al.*, 1987), which is the value recommended by Zeiss and Meath (1975), based on experimental data.

Table 6.2 *Corrections to thermodynamic properties of liquid argon at 85 K and a density of 35.243 mol/dm^{-3}, if "measured" by isobaric/isochoric MD with shifted force, cut-off at 1 nm. The second column gives the corresponding thermodynamic properties of argon (data from the Handbook of Chemistry and Physics (Lide, 1995))*

Correction on MD value	Thermodynamic value	Unit
$\Delta U = -0.958$	$U = -4.811$	kJ/mol
$\Delta H = -1.915$	$H = -4.808$	kJ/mol
$\Delta S = 0$	$S = 53.6$	$J\,mol^{-1}\,K^{-1}$
$\Delta P = -337$	$P = 1$	bar
$\Delta\mu = -1.915$	$\mu = -9.634$	kJ/mol
$p_{sat}^{corr} = 0.067$	$p_{sat} = 1$	bar at 87.3 K

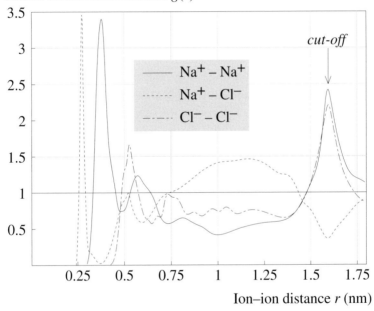

Radial distribution function $g(r)$

Ion–ion distance r (nm)

Figure 6.6 The ion-ion radial distribution functions for an aqueous NaCl solution, simulated with a Coulomb cut-off radius of 1.6 nm by Auffinger and Beveridge (1995).

riodic boundary conditions by Fourier methods similar to the mesh methods that have been worked out for Coulombic interactions (see below).[10]

---

[10] Essmann *et al.* (1995) describe the implementation of long-range dispersion forces.

6.3.5 Long-range Coulomb interactions

The *Coulomb interactions* are considerably longer-ranged than dispersion interactions, but because of overall charge neutrality they tend to cancel at large distances. Important long-range effects due to polarization of the medium beyond the cut-off radius persist and must be accounted for. The Coulomb interactions can be cut off at a given distance, but easily produce severe artifacts when the cut-off concerns full charges instead of dipoles. It has been remarked by several authors[11] that fluids containing ions show an accumulation of *like* ions and a depletion of oppositely charged ions near the cut-off radius. The ion–ion radial distribution function shows severe artifacts near the cut-off radius, as depicted in Fig. 6.6. It is understandable that like ions accumulate at the cut-off: they repel each other until they reach the cut-off distance, after which they will try to diffuse back in. Such effects are in fact intolerable; they do not occur with smooth forces and with electrostatic interactions computed by lattice sums. The use of a straight particle–particle cut-off is also detrimental when dipolar interactions represented by two opposite charges are considered: when the cut-off cuts between the two charges of a dipole, a full charge is in effect created and forces fluctuate wildly with distance (see Fig. 6.7). The effect is minimized when using force functions that taper off smoothly towards the cut-off radius (with vanishing force and derivative), but of course such forces deviate appreciably from the true Coulomb form. Some force fields use cut-offs for *charge groups* that are neutral as a whole, rather than for individual partial charges.

The effect of dielectric response of the medium beyond the cut-off radius can be incorporated by the introduction of a *reaction field*.[12] It is assumed that the medium outside the cut-off radius r_c has a relative dielectric constant ε_{RF} and – if applicable – an ionic strength κ. We first assume that the system contains no explicit ions and truncation is done on a neutral-group basis; ionic reaction fields will be considered later. When a spherical force truncation is used, any charge q_i will fully interact with all other charges q_j within the cut-off range r_c, but misses the interaction with (induced) dipoles and charge densities outside this range. The latter can be added as a potential of mean force, obtained by integrating the forces due to the reaction

---

[11] Brooks *et al.* (1985) consider the effects of truncation by integral equation techniques and by Monte Carlo simulations; Auffinger and Beveridge (1995) apply MD with truncation; Tironi *et al.* (1995) compare truncation with other long-range techniques.

[12] See Chapter 13, (13.82) on page 347 and (13.87) on page 347 for the derivation of reaction fields. For application to simulation see the original paper (Barker and Watts, 1973) and Barker (1994); Tironi *et al.* include ionic strength, and Hünenberger and van Gunsteren (1998) compare different reaction field schemes.

Force on charge pair

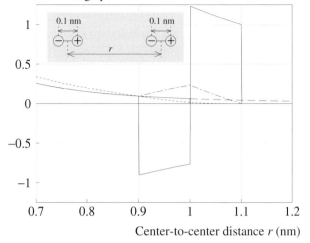

Figure 6.7 The effect of various cut-off methods on the force acting between two pairs of charges. Thick line: truncation on the basis of 1 nm group cut-off; thick line with dashed extension: exact force. Thin line: truncation based on 1 nm atomic cut-off. Dash–dot line: *shifted force* $1/r^2 - 1/r_c^2$ with 1 nm atomic cut-off. Dotted line: force from *cubic spread function,* see page 369 and (13.194), with 1 nm atomic cut-off, meant to be completed with long-range component.

field from the medium outside r_c. We assume that the interactions within range are based on a group cut-off, and denote the inclusion condition by $R_{ij} \leq r_c$; R_{ij} is the distance between the *reporter positions* of the neutral groups to which i and j belong. Pairs that are excluded from non-bonded short-range interactions, because their interactions are already accounted for in other bonded force-field terms, are indicated to belong to an *exclusion list* exclst. We need to define the *total* dipole moment in the sphere M:

$$M = \sum_{j;R_{ij}\leq r_c} q_j(r_j - r_i). \tag{6.46}$$

The reaction field at the center of the sphere, i.e., at position r_i, is determined by M and given by (13.87) on page 347:

$$E_{\mathrm{RF}}(r_i) = \frac{1}{4\pi\varepsilon_0 r_c^3} f(\varepsilon_r, \kappa) \sum_{j;R_{ij}\leq r_c} q_j(r_j - r_i), \tag{6.47}$$

$$f(\varepsilon_r, \kappa) = \frac{2(\varepsilon' - 1)}{(2\varepsilon' + 1)}, \tag{6.48}$$

$$\varepsilon' = \varepsilon_r \left(1 + \frac{\kappa^2 r_c^2}{2(1 + \kappa r_c)} \right), \tag{6.49}$$

where ε_r is the relative dielectric constant of the medium outside the cut-off range, and κ is the inverse Debye length (see (13.55) on page 342). Note that the central charge q_i should be included in the sum. The total dipole moment within the sphere gives a field in the origin; the total quadrupole moment gives a field gradient, etc. When the system contains charges only, and no explicit higher multipoles, we need only the reaction *field* to compute the forces on each particle.[13] Including the direct interactions, the force on i is given by

$$F_i = \frac{q_i}{4\pi\varepsilon_0}\left[\sum_{\substack{j;\,R_{ij}\leq r_c \\ (i,j)\notin\mathrm{exclst}}} q_j\frac{r_{ij}}{r_{ij}^3} - \sum_{\substack{j \\ R_{ij}\leq r_c}} q_j f(\varepsilon_r,\kappa)\frac{r_{ij}}{r_c^3}\right]. \qquad (6.50)$$

Here, $r_{ij} \overset{\text{def}}{=} r_i - r_j$, hence the minus sign in the reaction field term. Note that the inclusion of the reaction field simply modifies the force function; in the *tin-foil* or *conducting* boundary condition ($\varepsilon_r = \infty$ or $\kappa = \infty$: $f = 1$) the modification produces a shifted force, which is continuous at the cut-off radius. This shifted force is well approximated for media with high dielectric constant, such as water with $f(\varepsilon_r,\kappa) = 0.981$. Because of the smooth force the RF modification yields acceptable dynamics, even in cases where a reaction field is not appropriate because of anisotropy and inhomogeneity of the medium.

The potential energy function that generates the forces of (6.50) is obtained by integration:

$$V(r) = \frac{1}{4\pi\varepsilon_0}\sum_i q_i\left[\sum_{\substack{j>i;\,R_{ij}\leq r_c \\ (i,j)\notin\mathrm{exclst}}} q_j(r_{ij}^{-1} - r_c^{-1}) + \sum_{\substack{j\geq i \\ R_{ij}\leq r_c}} q_j\frac{f(\varepsilon_r,\kappa)}{2r_c^3}(r_{ij}^2 - r_c^2)\right].$$

$$(6.51)$$

The forces are not the exact derivatives of this potential because the truncation of F_{ij} is based on distances R_{ij} between reporter positions of groups, which depend not only on r_{ij}, but also on the position of other particles. Since $R_{ij} \approx r_{ij}$, this effect is small and is neglected. In addition, there are discontinuities in the reaction-field energies when dipoles cross the cut-off boundary, which lead to impulsive contributions to the forces. Since these are not incorporated into the forces, the effective (i.e., integrated-force) po-

---

[13] See Tironi *et al.* (1995) and Hünenberg and van Gunsteren (1998) for the full multipole equations. Note that force fields with explicit dipoles need to consider the reaction field gradient as well!

tentials will slightly deviate from the real reaction-field potentials. This situation is similar to the effect of truncated long-range dispersion potentials as discussed above (page 159).

The reaction field addition does not account for the polarization effects in the medium beyond the cut-off due to *charges* rather than dipoles. Naively one may say that the Born reaction potential of a charge (see (13.72) on page 345) is a constant potential with zero gradient, which does not lead to extra forces on the charges and will therefore not influence the equilibrium distributions. Therefore the Born correction can be applied afterwards to the results of a simulation. However, this reasoning disregards the discontinuities that occur in the potential when two charges enter or leave the interaction range. When the impulsive derivative of such potentials are not incorporated into the force, the effective potential – which is the integral of the forces – does not equal the Born correction. For charges these effects are more severe than for dipoles. Let us consider the simple case of two charges. The Born-corrected, r_c-truncated, interaction energy for a set of charges is:

$$U = \frac{1}{4\pi\varepsilon_0} \sum_i q_i \left[\sum_{\substack{j>i \\ r_{ij} \leq r_c}} \frac{q_j}{r_{ij}} - \frac{1}{2} \sum_{\substack{j \\ r_{ij} \leq r_c}} q_j \frac{g(\varepsilon, \kappa)}{r_c} \right], \tag{6.52}$$

$$g(\varepsilon, \kappa) \overset{\text{def}}{=} 1 - \frac{1}{\varepsilon_r(1 + \kappa r_c)}. \tag{6.53}$$

Applying this to two charges q_1 and q_2 at a distance r, we find for the potential energy:

$$r > r_c: \quad V(r) = V_\infty = -\frac{1}{4\pi\varepsilon_0}\frac{1}{2}(q_1^2 + q_2^2)\frac{g(\varepsilon, \kappa)}{r_c},$$

$$r \leq r_c: \quad V(r) = \frac{1}{4\pi\varepsilon_0}\left[\frac{q_1 q_2}{r} - \frac{1}{2}(q_1 + q_2)^2\frac{g(\varepsilon, \kappa)}{r_c}\right]$$

$$= V_\infty + \frac{1}{4\pi\varepsilon_0}q_1 q_2(r^{-1} - g(\varepsilon, \kappa)r_c^{-1}). \tag{6.54}$$

The effective potential obtained from the integrated truncated force would simply yield a shifted potential:

$$V_{\text{eff}}(r) = \frac{1}{4\pi\varepsilon_0}q_1 q_2(r^{-1} - r_c^{-1}). \tag{6.55}$$

It is interesting that in conducting-boundary conditions ($g(\varepsilon, \kappa) = 1$) the potential equals a shifted potential, plus an overall correction V_∞ equal to the sum of the Born energies of the isolated charges.

While reaction fields can be included in the force field with varying degrees of sophistication,[14] they are never satisfactory in inhomogeneous systems and systems with long-range correlation. *The latter include in practice all systems with explicit ions.* The polarization in the medium is not simply additive, as is assumed when reaction fields are included per particle. For example, the field in the medium between a positive and a negative charge at a distance larger than r_c is strong and induces a strong polarization, while the field between two positive charges cancels and produces no polarization. The total polarization energy is then (in absolute value) larger, resp. smaller than predicted by the Born-corrected force field of (6.52), which is indifferent for the sign of the charge. One is faced with a choice between inaccurate results and excessive computational effort.

There are three ways out of this dilemma, neither using a reaction field correction. The first is the application of continuum corrections, treated below. The second is the use of the *fast multipole method* (FMM), which relies on a hierarchical breakdown of the charges in clusters and the evaluation of multipole interactions between clusters. The method is implemented in a few software packages, but is rather complex and not as popular as the lattice summation methods. See the discussion in Section 13.9 on page 362. The third, most recommendable method, is the employment of efficient lattice summation methods. These are, of course, applicable to periodic systems, but even non-periodic clusters can be cast into a periodic form. There are several approaches, of which the accurate and efficient *smooth-particle mesh-Ewald* (SPME) method of Essmann *et al.* (1995) has gained wide popularity. These methods are discussed at length in Chapter 13, Section 13.10 on page 362, to which the reader is referred.

Continuum correction methods are due to Wood (1995).[15] Consider a "model world" in which the Hamiltonian is given by the force field used, with truncated, shifted or otherwise modified long-range Coulomb interaction, and possibly periodic boundary conditions. Compare with a "real world" with the full long-range interactions, possibly of infinite extension without periodic boundary conditions. Now assume that the *difference* in equilibrium properties between the two worlds can be computed by electrostatic continuum theory, since the difference concerns long-range effects on a scale much coarser than atomic detail, and at such distances from real

---

[14] See for advanced reaction fields, e.g., Hummer *et al.* (1996), Bergdorf *et al.* (2003) and Christen *et al.* (2005).

[15] These methods are based on earlier ideas of Neumann (1983), who applied continuum methods to interpret simulation results on dielectric behavior. Several authors have made use of continuum corrections, and most applications have been reviewed by Bergdorf *et al.* (2003), who considered the effects of truncation, reaction field functions and periodic boundary conditions on ionic hydration and on the interaction between two ions in a dielectric medium.

charges that the dielectric response can be assumed to be linear and field-independent. Now correct the simulation results of the model world with the difference obtained by continuum theory. Separate corrections can be obtained for charge–dipole and dipole–dipole cutoffs, and for the effect of periodic boundary conditions. The principle is as simple as that, but the implementation can be quite cumbersome and not applicable to all possible cases. The principle of the method is as follows.

Consider a system of sources, taken for simplicity to be a set of charges q_i at positions r_i, in a dielectric medium with linear (i.e., field-independent) local dielectric constant $\varepsilon = \varepsilon_r \varepsilon_0$ and without electrostriction (i.e., field-dependent density). The local polarization $P(r)$ (dipole density) is given by a product of the (local) *electric susceptibility* χ and the electric field E:[16]

$$P = \varepsilon_0 \chi E, \quad \chi = \varepsilon_r - 1, \tag{6.56}$$

The field is determined by the sum of the direct field of the sources and the dipolar fields of the polarizations elsewhere:

$$E(r) = \sum_i G(r - r_i) + \int T(r - r') P(r') \, dr'. \tag{6.57}$$

Here $G(r)$ is the field produced at r by a unit charge at the origin, and $T(r)$ is the tensor which – multiplied by the dipole vector – yields the field at r due to a dipole at the origin. The vector function G equals minus the gradient of the potential $\Phi(r)$ by a unit charge at the origin. For G and T one can fill in the actual truncated or modified functions as used in the simulation, including periodic images in the case of periodic conditions. For example, for a truncated force, these influence functions are:

$$G(r) = \frac{1}{4\pi\varepsilon_0} \frac{r}{r^3} \text{ for } r \leq r_c,$$
$$= 0 \text{ for } r > r_c, \tag{6.58}$$
$$T_{\alpha\beta} = \frac{1}{4\pi\varepsilon_0} (3x_\alpha x_\beta - r^2 \delta_{\alpha\beta}) r^{-5} \text{ for } r \leq r_c,$$
$$= 0 \text{ for } r > r_c. \tag{6.59}$$

The first term in (6.57) is the vacuum field of the set of charges. The integral equation (6.56) with (6.57) can then be numerically solved for P and E and from there energies can be found. The total electrostatic energy is $\frac{1}{2} \int \varepsilon E^2 \, dr$, which diverges for point charges and must be corrected for the vacuum self-energy (see (13.42) on page 340); the polarization energy is

---

[16] See Chapter 13 for the basic electrostatic equations.

Table 6.3 *Continuum corrections to the simulated hydration free energy of a sodium ion in 512 water molecules in a periodic box, simulated with Coulomb forces truncated at r_c^{i-w} for ion-water interactions and r_c^{w-w} for water–water interactions. Column 3 gives the simulation results (Straatsma and Berendsen, 1988), column 4 gives the Born correction resulting from the ion-water cutoff, column 5 gives the Born-corrected results, column 6 gives the continuum correction resulting from water-water cutoff (Wood, 1995) and column 7 gives the corrected results. Column 8 gives the corrections resulting from periodicity*

1	2	3	4	5	6	7	8
r_c^{i-w}	r_c^{w-w}	ΔG_{sim}	C_{Born}^{i-w}	$3+4$	$C_{\mathrm{w-w}}$	$5+6$	C_{PBC}
nm	nm	kJ/mol	kJ/mol	kJ/mol	kJ/mol	kJ/mol	kJ/mol
0.9	0.9	−424	−76.2	−500	+21.0	−479	−0.1
1.05	0.9	−444	−65.3	−509	+32.5	−477	−0.2
1.2	0.9	−461	−57.2	−518	+42.5	−476	−0.2
0.9	1.2	−404	−76.2	−480	+03.6	−477	−0.1
1.2	1.2	−429	−57.2	−486	+13.9	−472	−1.3

given by the interaction of each charge with the induced dipoles:

$$U_{\mathrm{pol}} = -\frac{1}{2} \sum_i q_i \int \boldsymbol{P}(\boldsymbol{r}) \cdot \boldsymbol{G}(\boldsymbol{r} - \boldsymbol{r}_i)\, d\boldsymbol{r}. \tag{6.60}$$

The factor $\frac{1}{2}$ is implicit in the basic equations (see Chapter 13, Section 13.6 on page 339), but can also be derived from changing the charges from zero to the full charge and integrating the work it costs to increment the charges against the existing polarization field. While the solution of the integral equation (which can be transformed into a matrix equation, to be solved by a suitable iterative method, see Bergdorf *et al.*, 2003) is needed for the "model world," the reference calculation for the "real world" can be cast into a Poisson equation and solved by more elegant methods. Finally, the difference in solvation energy should be used to correct "model-world" simulations. The continuum approximations are rather poor in the immediate neighborhood of an ion, where specific solvent structures and dielectric saturation may occur; such regions must be covered correctly by simulation with atomic detail. These regions, however, do not contribute significantly to the difference between "model world" and "real world" because they fall well within the cutoff range of the model potentials.

Wood (1995) applied continuum corrections to a series of MD simulations by Straatsma and Berendsen (1988) on the free energy of hydration of an ion, with various values of water–water and water–ion cutoff radii. Because these early simulations were limited in size (one ion and 512 water molecules in a cubic periodic box, with cutoffs up to 1.2 nm), they serve well for demonstration purposes because the corrections are substantial. Table 6.3 shows the results of the free energy gained by charging a neon atom to become a Na^+ ion. Different values for the ion–water and the water–water cutoff were used (columns 1 and 2). When the simulation results (column 3) are corrected for the Born energy (column 4), using the ion–water cutoff, the results (column 5) appear to be inconsistent and depend strongly on the water–water cutoff. The reason for this is that the ion polarizes the water with the result that there is a substantial water–water interaction beyond the water–water cutoff radius. When the continuum correction is applied to the water–water cutoff as well (column 6), the result (column 7) is now consistent for all simulation conditions, i.e., within the accuracy of the simulations of about 3 kJ/mol. The correction for the periodic condition (column 8) appears to be rather small and to increase with the cutoff radius. This is consistent with the more detailed calculations of Bergdorf *et al.* (2003), who observe that the corrections for periodicity are quite small when applied to truncated potentials, but much larger when applied to full lattice sums. This can be interpreted to mean that the use of full lattice sums enhances the periodicity artifacts. On the other hand, it prevents the – generally worse – artifacts due to truncation.

6.3.6 Polarizable force fields

The dominant and most relevant omission in the usual force fields is the incorporation of *electronic polarizability*. The electron distribution around a given nuclear configuration depends on the presence of external electric fields; in first order a *dipole moment* μ is induced proportional to the (homogeneous) electric field E:

$$\mu = \alpha E, \tag{6.61}$$

where α is the *polarizability tensor* of the electron distribution.[17] In non-homogeneous fields there will be induced higher multipoles proportional to field gradients; these we shall not invoke as it is much easier to consider small fragments with induced dipoles than large fragments with induced multipoles. For small fields the induced moments are proportional to the

---

[17] Note that α is to be expressed in SI units $F\,m^2$; it is convenient to define a modified polarizability $\alpha' \stackrel{\text{def}}{=} \alpha/(4\pi\varepsilon_0)$, which has the dimension volume and is to be expressed in m^3.

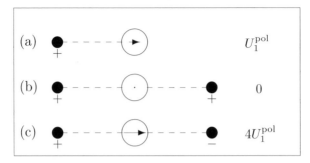

Figure 6.8 The non-additivity of polarization energies. (a) a single charge that polarizes a neutral particle causes an induced dipole and a (negative) polarization energy; (b) The fields of two equal, but oppositely placed, charges compensate each other; there is no polarization energy; (c) When the fields add up, the polarization energy due to two charges is four times the polarization energy due to one charge.

field; for large fields a significant nonlinear *hyperpolarizability* may occur, usually in the form of a reduction of the induced moment.

Polarizability is a *non-additive* electrical interaction. While the polarization energy of a neutral particle with polarizability α at a distance r from a non-polarizable charge q equals $-\alpha' q^2/(8\pi\varepsilon_0 r^4)$ (here $\alpha' = \alpha/(4\pi\varepsilon_0)$), the energy is not additive when multiple sources are present. This is demonstrated in Fig. 6.8 for the simple case of a polarizable atom, situated between two (non-polarizable) point charges, each at a distance r from the atom. When both charges have the same sign, the polarization energy is zero because the field cancels at the origin; two charges of opposite sign double the field, leading to a polarization energy four times the polarization energy caused by a single charge. Thus polarizability cannot be simply incorporated by an extra r^{-4}-type pair interaction.

When polarizable particles interact, as they physically do by electrical interaction between induced dipoles, the field on each particle depends on the dipoles on the other particles. Thus one needs to either solve a matrix equation or iterate to a self-consistent solution. The additional computational effort has thus far retarded the introduction of polarizability into force fields. One may choose to neglect the interaction between induced dipoles, as has been done by a few authors,[18] in which case the extra computational effort

[18] Straatsma and McCammon (1990a, 1990b, 1991) introduced a non-iterative polarizable force field with neglect of mutual interaction between induced dipoles and applied this to free energy calculations. In the case of a xenon atom dissolved in water the polarization free energy equals -1.24 kJ/mol. In this case the applied model is exact and the polarization is due to the fluctuating electric field produced by water molecules. The free energy change between a non-polarizable and a non-mutual polarizable empirical model of water was found to be -0.25 kJ/mol. Linssen (1998, see also Pikkemaat *et al.*, 2002) applied a restricted non-mutual polarizability model to the enzyme *haloalkane dehalogenase* which contains a chloride ion in a

is negligible. The effect of neglecting the mutual induced-dipole interaction is an *exaggeration* of the polarization produced by ions: the field of neighboring induced dipoles generally opposes the direct Coulomb field. Soto and Mark (2002) showed that the polarization energy of an ion in cyclohexane, using non-mutual polarizability, equals 1.5 times the correct Born energy. While non-mutual polarizability does repair the main deficiencies of non-polarizable force fields (such as ion solvation in non-polar solvents), its use in accurate force fields is not recommended. If it is used, empirical *effective* polarizabilities (smaller than the physical values) should be employed that reproduce correct average ionic solvation energies.

One may wonder how important the introduction of polarizability will be. If we put some reasonable numbers in the example of Fig. 6.8, the polarization energy of a CH_2 group ($\alpha' = 1.8 \times 10^{-3}$ nm^3) at a distance of 0.3 nm from an elementary charge (say, a chlorine ion) equals -15.4 kJ/mol. Linssen (1998) estimated the polarization energy of an internal chloride ion in the enzyme *haloalkane dehalogenase* as -112 kJ/mol.[19] The polarization interaction of a single carbonyl oxygen atom ($\alpha' = 0.84 \times 10^{-3}$ nm^3) liganded at 0.3 nm distance to a doubly charged calcium ion (as occurs in several enzymes) is as high as -29 kJ/mol. These are not negligible amounts, and polarization cannot be omitted in cases where full, or even partial, charges occur in non-polar or weakly polar environments. In polar environments, such as water, electronic polarization is less important, as it is dominated by the orientational polarization. The latter is fully accounted for by non-polarizable force fields. Still, the electronic dipole moment induced in a water molecule ($\alpha' = 1.44 \times 10^{-3}$ nm^3) at a distance of 0.3 nm from an elementary charge (say, a potassium ion) amounts to 40% of the intrinsic dipole moment of water itself! In non-polarizable water models the increased intrinsic dipole moment takes partially care of the ionic polarizing effect, but is not properly dependent on the charge and distance of the nearby ion.

Considering these large effects it is quite amazing that one can get away at all with non-polarizable force fields. The reason for the success of simple non-polarizable empirical force fields is that the *average* effects of polarizability are incorporated into phenomenologically adjusted force fields. For example, water models that give correct densities, sublimation energies, dielectric constant and structure for the liquid state, all have considerably

---

hydrophobic environment. Soto and Mark (2002) investigated the effect of polarizability on the stability of peptide folding in a non-npolar environment and used a non-mutual polarization model.

[19] This value may overestimate the full polarization interaction by a factor of 1.5, as discussed above. Without polarization the ion is far too weakly bound and in simulations it is quickly expelled from its binding site into the external solution.

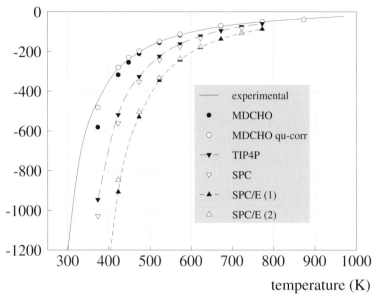

Second virial coefficient (cm³/mol)

temperature (K)

Figure 6.9 The second virial coefficients for water for four models (symbols), compared to experimental values (solid curve). TIP4P, SPC and SPC/E are non-polarizable effective pair potentials; MCDHO is an *ab initio* polarizable model, given with and without quantum correction. The experimental curve is a cubic spline fitted to the average of seven data sets dated after 1980 (Millot *et al.*, 1998, and those quoted by Saint-Martin *et al.*, 2000). Data for TIP4P and SPC/E (1) are from Kusalik *et al.* (1995), for MCDHO from Saint-Martin *et al.* (2000) and for SPC and SPC/E (2) from Guissani and Guillot (1993).

enhanced dipole moments. These enhanced moments are likely to be close to the real average dipole moment (intrinsic plus induced) in the liquid state. The solid state is not very different and may also be reasonably represented, but the liquid-state models are accurate neither for the dilute gas phase nor for non-polar environments. Thus non-polarizable force fields are based on *effective pair potentials*, which can only be valid for environments not too different from the environment in which the model parameters were empirically adjusted. Effective pair potentials do not only incorporate the average induced dipole moments, but also incorporate average quantum effects and average non-additivity of repulsion and dispersion contributions.

In the case of water, effective pair potentials exaggerate the interaction between iso lated pairs. Thus the hydrogen-bonding energy between two molecules in the dilute gas phase is too large: -27.6 kJ/mol for SPC and -30.1 kJ/mol for SPC/E, compared to -22.6 kJ/mol derived from dimer spectroscopy, and -21.0 kJ/mol from accurate quantum calculations. This is also apparent from the second virial coefficient, which deviates from experimental values in the sense that the dimer attraction is too

large (Fig. 6.9).[20] As noted by Berendsen *et al.* (1987), there is an inconsistency in the use of effective pair potentials when these incorporate average induced dipoles: the full electrostatic interaction between two molecules is taken into account, while the polarization energy of induced dipoles equals only half the electrostatic energy. Half of the electrostatic energy is "used" for the formation of the induced dipoles (see below for the correct equations). While polarizable models take this factor of two correctly into account, effective pair potentials do not. The heat of vaporization should be corrected with the self-energy of the induced dipoles. It was found that a water model with such corrections (the extended simple point charge model, SPC/E) gives better values for density, radial distribution function, diffusion constant and dielectric constant than the same model (SPC) without corrections. On the other hand, the heat of vaporization, and also the free energy change from liquid to gas, is too large, as the molecule retains its enhanced dipole moment also in the gas phase. As seen in Fig. 6.9, the discrepancy from gas phase dimer interaction is even larger than for the "classical" effective pair models. The boiling point should therefore be higher than the experimental value and free energies of solvation of water into non-polar environments should also be too large.[21]

There is no remedy to these artifacts other than replacing effective potentials with polarizable ones.

---

[20] It is through the pioneering work of A. Rahman and F. H. Stillinger in the early 1970s (Rahman and Stillinger, 1971; Stillinger and Rahman, 1972, 1974) that the importance of effective potentials became clear. Their first simulation of liquid water used the Ben-Naim-Stillinger (BNS) model that had been derived on the basis of both gas-phase data (second virial coefficient related to the pure pair potentials) and condensed-phase data (ice). This pair potential appeared too weak for the liquid phase and could be improved by a simple scaling of energy. When a modified version, the ST2 potential (Stillinger and Rahman, 1974), was devised, the notion of an effective potential was already developed (Berendsen *et al.*, 1987).

[21] The SPC dipole moment of 2.274 D is enhanced to 2.351 D in SPC/E (compare the gas phase value of 1.85 D). The heat of vaporization at 300 K (41.4 kJ/mol) increases to 46.6 kJ/mol, which decreases the vapor pressure tenfold and increases the boiling point by several tens of degrees. Amazingly, according to Guissani and Guillot (1993), the SPC/E model follows the liquid-vapor coexistence curve quite accurately with critical parameters ($T_c = 640$ K, $\rho_c = 0.29$ g/cm^{-3}, $P_c = 160$ bar) close to those of real water ($T_c = 647.13$ K, $\rho_c = 0.322$ g/cm^{-3}, $P_c = 220.55$ bar). It does better than SPC with $T_c = 587$ K and $\rho_c = 0.27$ g/cm^{-3} (de Pablo *et al.*, 1990).

6.3.7 Choices for polarizability

The first choice to be made is the representation of the induced dipoles in the model. We assume that the fixed sources in the non-polarized model are charges q_i only.[22] One may:

- include induced dipoles (*dipolar model*),
- represent the induced dipoles by a positive and negative charge, connected with a harmonic spring (*shell model*),[23] or
- modify charges at given positions to include induced dipoles (*fluctuating charge model*).[24]

The dipole, or in the second case one of the charges, can be placed on atoms or on virtual points that are defined in terms of atom positions. For the shell model, the other charge is in principle a massless interaction site. In the third case there must be a sufficient number of properly placed charges (four for every polarizability). It is highly recommended to use *isotropic* polarizabilities on each site; in that case the induced dipole points in the direction of the electric field and – if iterated to self-consistency – there is no torque $\boldsymbol{\mu} \times \boldsymbol{E}$ acting on the dipole or the spring. With anisotropic polarizabilities there are torques, also in the shell model where the spring constant would be an anisotropic tensor, which considerably complicates the equations of motion. In the isotropic dipolar model there is a force only, given by the tensor product of dipole moment and field gradient: one needs to evaluate the field gradient on each dipolar site. In the isotropic shell model there are only charges and no field gradients are needed; however, the number of particles doubles per polarizable site and the number of Coulomb interactions quadruples. The fluctuating charge model can accommodate anisotropic polarizabilities, but requires an even larger number of interactions.

The question arises if the choice of isotropic polarizabilities is adequate. After all, many molecules possess an *anisotropic* polarizability, and this can never be generated by simple addition of isotropic atomic polarizabilities. However, the mutual interaction of induced dipoles saves the day: neighboring induced dipoles enhance an external electric field when they are lined up

---

[22] The inclusion of fixed dipoles is a trivial complication, but one should be careful to obtain forces from the correct derivatives of the dipolar fields. Forces on dipoles include torques, which are to be transferred to the rigid molecular frame in which the dipoles are defined.

[23] The shell model idea originates from solid state physics, where an ion is represented by a positive core and a negative shell which carries the interactions of electronic nature. Such models correctly predict phonon spectra. It was first used by Dick and Overhauser (1958) on alkali halide crystals.

[24] Although used by Berendsen and van der Velde (1972), the fluctuating charge model was first introduced into the literature by Zhu *et al.* (1991) and by Rick *et al.* (1994). See also Stern *et al.* (2001).

in the field direction and depress the field when they are situated in lateral directions, thus producing an anisotropic overall polarization. In general it is very well possible to reproduce experimental molecular polarizability tensors based on interacting isotropic atomic polarizabilities.[25] Consider as example a homonuclear diatomic molecule, with two points with isotropic polarizability α at the nuclear positions, separated by a distance d, in an external field $\boldsymbol{E}$. The field at each point includes the dipolar field from the other point. For the induced dipole per point in the parallel and perpendicular directions to the molecular axis we obtain, denoting $\alpha' = \alpha/(4\pi\varepsilon_0)$:

$$\mu_\| = \alpha\left(E + \frac{2\mu_\|}{4\pi\varepsilon_0 d^3}\right) \quad \rightarrow \quad \mu_\| = \frac{\alpha E}{1 - 2\alpha'/d^3}, \tag{6.62}$$

$$\mu_\perp = \alpha\left(E - \frac{\mu_\perp}{4\pi\varepsilon_0 d^3}\right) \quad \rightarrow \quad \mu_\perp = \frac{\alpha E}{1 + \alpha'/d^3}, \tag{6.63}$$

resulting in a molecular anisotropic polarizability:

$$\alpha_\| = \frac{2\alpha}{1 - 2\alpha'/d^3} \tag{6.64}$$

$$\alpha_\perp = \frac{2\alpha}{1 + \alpha'/d^3} \tag{6.65}$$

$$\alpha_{\text{iso}} = (\alpha_\| + 2\alpha_\perp)/3. \tag{6.66}$$

The parallel component diverges for a distance of $(2\alpha')^{1/3}$. This is a completely unphysical behavior, leading to far too large anisotropies for diatomics (see Fig. 6.10). The remedy (Thole, 1981) is to introduce a *damping* at short distances by considering the induced dipole not as a point, but as a distribution. Both an exponential decay and a linear decay yield satisfactory results, with a single polarizability for each atom irrespective of its chemical environment, and with a single screening length, if distances between two atoms are scaled by the inverse sixth power of the product of the two atom polarizabilities (van Duijnen and Swart, 1998).

Another empirical approach to predict isotropic polarizabilities of organic molecules, based on additivity, uses the number of electrons, their effective quantum numbers and effective nuclear shielding (Glen, 1994).

---

[25] Applequist *et al.* (1972) included interactions between point dipoles and lists atomic polarizabilities both for a poor additive model and a more successful interaction model. To amend the unrealistic approach to diverging behavior at short distances, Thole (1981) proposed to insert a damping at small distances, as can be realized by considering the polarizable sites not as points, but as smeared-out distributions. A simple triangular distribution with a width scaled by the polarizabilities appeared to yield good results with a single isotropic polarizability for each atom type (H,C,N,O), irrespective of its chemical nature. Van Duijnen and Swart (1998) extended Thole's model for a wider set of polarizabilities, including *ab initio* data for 70 molecules and including sulphur and the halogens in addition to Thole's atom set. They compared a linear as well as an exponential distribution, with a slight preference for the latter.

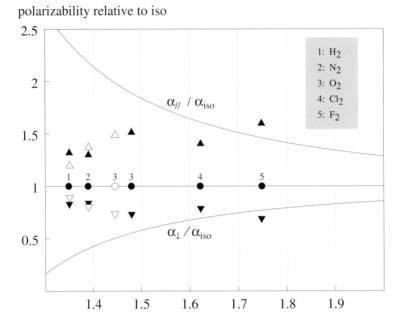

Figure 6.10 Inadequacy of the interacting dipole model for polarizability of diatomics: The parallel and perpendicular components, relative to the isotropic value of the polarizability are drawn as a function of the internuclear distance, the latter expressed relative to the 1/3 power of the atomic polarizability that reproduces the isotropic molecular polarizability. The symbols indicate the values of the parallel (up-triangles), isotropic (circles) and perpendicular (down-triangles) polarizabilities for diatomic molecules. Filled symbols represent theoretical values from recent NIST tables; open symbols are from experimental values as quoted by Thole (1981).

It should be noted that shell models with interacting shell distributions will be equally capable of producing correct polarizabilities. Let us discuss the question of which description should be preferred in polarizable force fields: induced dipoles, harmonic shells or fluctuating charges? Computational considerations were given above, and combining a preference for both simplicity and speed, they favor the shell model. More important are scientific considerations that favor the shell model as well. We follow the argument of Jordan *et al.* (1995). Consider two neon atoms: if described by a repulsion/dispersion interaction and dipolar polarizabilities centered on the nuclei, there are no electric effects when the atoms approach each other. However, we know that collision-induced infrared absorption can be observed, and the (computed) electrical quadrupole moment of a neon pair is not zero. The logical explanation is that when the electron clouds attract each other at larger distances by dispersion, they are pulled in the direction of the other atom and produce a positive quadrupole moment; when they repel each other at shorter distances, they are pushed away and produce a negative quadrupole moment. The quadrupole moment, computed by a high-level *ab initio* method, does indeed follow the interatomic force. A shell model with the interatomic interaction acting between the shells reproduces such results. Pursuing this idea, Jordan *et al.* (1995) succeeded in devising a shell model for the nitrogen molecule which reproduces experimental static and dynamic properties in the solid, liquid and gas phases, including a subtle pressure-induced phase transition between a cubic and a hexagonal solid phase. The model contains three shells, one bound to both nitrogens (a "bond") and the other two bound to one nitrogen ("lone pairs").

Models of this type could be devised for other molecules, but they tend to become complex and computationally intensive. The number of interaction sites would be quite large: one shell per bond and at least one additional shell for each non-hydrogen atom; thus far, complete force fields along these lines have not been constructed. A somewhat similar and quite successful shell model for water (the MCDHO model) was published by Saint-Martin *et al.* (2000).[26] The model includes a single shell with an exponential charge distribution, which is connected to the oxygen nucleus by a spring but also interacts with the hydrogen charges in the molecule. Parameters were fitted to *ab initio* calculations of dimer and oligomers. The fact that this model reproduces experimental properties of the gas, liquid and solid phases quite well holds the promise that general force fields with transferable terms may be derived from *ab initio* calculations on small molecules, possibly with small empirical corrections, if electronic polarization is properly included.

More simple polarizable models have been quite successful as well. Most work has been done on models for water, with the aim to construct models that yield accurate thermodynamic and dynamic properties for a wide range of phases and conditions. While the development of polarizable models is still proceeding, we shall review only the features from which basic principles can be learned. A comprehensive review of models for simulation of water by Guillot (2002) is available.

---

[26] MCDHO: Mobile Charge Densities in Harmonic Oscillators. The article includes Monte Carlo simulations of the liquid state. See also Hess *et al.* (2002), who performed molecular dynamics on this model.

The successful pair-additive simple point charge models[27] have been modified to include polarization, both with induced dipoles and with shell models. The earliest attempt of this type was a modification of the SPC model by Ahlström *et al.* (1989), who added a polarizable point dipole on the oxygen atom while reducing the charges to obtain the gas phase dipole moment, and others of this type followed with fine-tuning of parameters.[28] Van Maaren and van der Spoel (2001) investigated the properties of a shell model, with the moving charge attached with a spring to a virtual position on the symmetry axis of the molecule about 0.014 nm from the oxygen. They retained the gas phase structure, dipole and quadrupole moment and polarizability, while optimizing the Lennard–Jones interaction. Several water models have a similar position for the negative charge for the simple reason that three-charge models cannot satisfy the experimental quadrupole moment without displacing the negative charge. Yu *et al.* (2003) developed a simple model with three atoms plus moving charge bound with a spring to the oxygen, which they named the "charge-on-spring" model. The model was intended for computational efficiency; it uses a large moving charge and is in fact a shell-implementation of an induced point dipole model. Polarizable models of this kind, whether they use a moving charge or a point dipole, are moderately successful. They repair the main deficiencies of effective pair potentials, but do not show the accurate all-phase behavior that one should wish. For example, the interaction in the critical region – and thereby the

[27] SPC (Berendsen *et al.*, 1981) uses the three atoms as interaction site with partial charges on oxygen (-0.82 e) and hydrogens ($+0.41$ e). The geometry is rigid, $r_{OH} = 0.1$ nm; HOH angle $= 109.47°$. There is a Lennard–Jones interaction on the oxygens: $C_6 = 2.6169 \times 10^{-3}$ kJ mol^{-1} nm^6; $C_{12} = 2.6332 \times 10^{-6}$ kJ mol^{-1} nm^{12}. The similar TIP3 model (Jorgensen, 1981) has the rigid experimental geometry: OH distance of 0.09572 nm, HOH angle of 104.52°, with hydrogen charge of 0.40 e; $C_6 = 2.1966 \times 10^{-3}$; $C_{12} = 2.4267 \times 10^{-6}$ (units as above). This model did not show a second-neighbor peak in the radial distribution function and was modified to TIP3P (Jorgensen *et al.*, 1983), with hydrogen charge of 0.417 e; $C_6 = 2.4895 \times 10^{-3}$ and $C_{12} = 2.4351 \times 10^{-6}$. In a subsequent four-site model (TIPS2: Jorgensen, 1982) the negative charge was displaced to a point M on the bisectrix in the direction of the hydrogens at 0.015 nm from the oxygen position, while the Lennard–Jones interaction remained on the oxygen, and the parameters were improved in the TIP4P model (Jorgensen *et al.*, 1983). The hydrogen charge is 0.52 e; $C_6 = 2.5522 \times 10^{-3}$ and $C_{12} = 2.5104 \times 10^{-6}$. Finally, the SPC/E model (Berendsen *et al.*, 1987), which includes an energy correction for average polarization, is like the SPC model but with oxygen charge -0.4238. Van der Spoel *et al.* (1998) evaluated these models and optimized them for use with a reaction field. They conclude that the SPC/E model is superior in predicting properties of liquid water.

[28] These include models by Cieplack *et al.* (1990) with polarizabilities on O and H and extra repulsion/dispersion terms; Caldwell *et al.* (1990) with SPC/E modified by O and H polarizabilities and a three-body repulsion term; Dang (1992) with an improvement on the the the latter model; Wallqvist and Berne (1993) with a polarizable and a non-polarizable model with extra terms; Chialvo and Cummings (1996) with an evaluation of displacement of negative charge and a point polarizability on oxygen; Svishchev *et al.* (1996) with the PPC model which has a displaced (by 0.011 nm) negative charge and polarizability only in the molecular plane caused by displacement of the negative charge (this is a partial shell model with enhanced permanent dipole and reduced polarizability), Dang and Chang (1997) with a revised four-site model with the negative charge and dipolar polarizability displaced by 0.0215 nm from the oxygen;

critical temperature – is underestimated,[29] and the dielectric constant is often too large. Both effects are a result of the relatively (too) large polarization interaction at short distances, which is then compensated by (too weak) attraction at longer distances. A shell model developed by Lamoureux *et al.* (2003) could be made to fit many water properties including the dielectric constant, but only with reduced polarizability (1.04 instead of 1.44 Å^3). A study of ion hydration (Spångberg and Hermansson, 2004) with various water models showed too large solvation enthalpies for polarizable models. Giese and York (2004) find an overpolarizability for chains of water molecules using screened Coulomb interactions; they suggest that exchange overlap reduces short-range polarization.

The answer seems to lie in a proper damping of the short-range Coulombic fields and polarization. This can be accomplished by Thole-type smearing of charges and polarization, as was introduced into a water model by Burnham *et al.* (1999), and also by Paricaud *et al.* (2005), both authors using polarizabilities on all three atoms. However, shell models with charge distributions for the shells are far more natural and efficient. In a revealing study, fitting electrostatic models to *ab initio* potentials, Tanizaki *et al.* (1999) showed that point charge models tend to become counterintuitive (as positive charge on lone-pair positions) unless shielding distributed charges are used. As the MCDHO model of Saint-Martin *et al.* (2000) shows, a single shell for the molecule suffices, neglecting the polarizabilities on hydrogen. This model does not only reproduce liquid behavior, but has excellent second virial coefficients (see Fig. 6.9) and excellent liquid–vapor coexistence and critical behavior (Hernández-Cobos *et al.*, 2005). But also with the MCDHO model the polarization is too strong in the condensed phase, leading to a high dielectric constant,[30] and some refinement seems necessary.

6.3.8 Energies and forces for polarizable models

Consider a collection of fixed source terms, which may be charges q_i only or include higher multipoles as well. The sources may be charge or multipole density distributions. They produce an electric field $E_0(r)$ at a position r, which diverges at a point source itself. In addition there will be induced dipoles μ_k^s or shells with charge q_k^s at sites r_k^s (these may be atoms or virtual sites), with the shell connected by a harmonic spring with spring constant k_k^s to an atom or virtual site r_k. The latter may or may not carry a fixed charge as well. These dipoles or shells (which may be points or distributions)

---

[29] See Jedlovszky and Richardi (1999) for a comparison of water models in the critical region.
[30] Saint-Martin *et al.* (2005).

produce an extra field $\boldsymbol{E}_{\text{ind}}(\boldsymbol{r})$ at a position $\boldsymbol{r}$. The total energy can best be viewed as the sum of the *total* electrostatic interaction plus a *positive* polarization energy V_{pol} needed to create the induced dipoles μ:

$$V_{\text{pol}} = \sum_k \frac{(\mu_k^{\text{s}})^2}{2\alpha_k} \tag{6.67}$$

or to stretch the springs

$$V_{\text{pol}} = \sum_k \frac{1}{2} k_k^{\text{s}} |r_k^{\text{s}} - r_k|^2. \tag{6.68}$$

The total electrostatic energy consists of the source–source interaction V_{qq} and the source–dipole plus dipole–dipole interaction $V_{\text{q}\mu} + V_{\mu\mu}$, or for shells the source–shell plus shell–shell interactions $V_{\text{qs}} + V_{\text{ss}}$. In these sums every pair should be counted only once and, depending on the model used, some neighbor interactions must be excluded (minimally a shell has no Coulomb interaction with the site(s) to which it is bound and dipoles do not interact with charges on the same site). The form of the potential function for a pair interaction depends on the shape of the charge or dipole distribution. The polarizations (dipoles or shell positions) will adjust themselves such that the total potential energy is minimized. For dipoles this means:

$$\frac{\partial V_{\text{tot}}}{\partial \mu_k^{\text{s}}} = 0, \quad V_{\text{tot}} = V_{\text{qq}} + V_{\text{q}\mu} + V_{\mu\mu} + V_{\text{pol}}, \tag{6.69}$$

or, for shells:

$$\frac{\partial V_{\text{tot}}}{\partial r_k^{\text{s}}} = 0, \quad V_{\text{tot}} = V_{\text{qq}} + V_{\text{qs}} + V_{\text{ss}} + V_{\text{pol}}. \tag{6.70}$$

It is easily shown that this minimization leads to

$$\boldsymbol{\mu}_k^{\text{s}} = \alpha_k \boldsymbol{E}_k^{\text{tot}}(\boldsymbol{r}_k^{\text{s}}), \tag{6.71}$$

or, for shells:

$$r_k^{\text{s}} = r_k + \frac{q_k^{\text{s}}}{k_k^{\text{s}}} \boldsymbol{E}_k^{\text{tot}}(\boldsymbol{r}_k^{\text{s}}). \tag{6.72}$$

Here $\boldsymbol{E}^{\text{tot}}$ is minus the gradient of V_{tot}. The shell displacement corresponds to an induced dipole moment equal to q^{s} times the displacement and hence corresponds to a polarizability

$$\alpha_k = \frac{(q_k^{\text{s}})^2}{k_k^{\text{s}}}. \tag{6.73}$$

The polarization energy appears to compensate just half of the electrostatic interaction due to induced charges or shells.

This, in fact, completes the equations one needs. The forces on site i are determined as minus the gradient to the position r_i of the total potential energy.[31] The electrostatic interactions between arbitrary charge *distributions* are not straightforward. The interaction energy between a point charge q at a distance r from the origin of a spherical charge distribution $q^s w(r)$ (with $\int_0^\infty 4\pi r^2 w(r)\, dr = 1$) is given (from Gauss' law) by integrating the work done to bring the charge from infinity to r:

$$V_{qq}^s = \frac{qq^s}{4\pi\varepsilon_0} \int_r^\infty \frac{1}{r'^2} \int_0^{r'} 4\pi r''^2 w(r'')\, dr'' = \frac{qq^s}{4\pi\varepsilon_0} \phi^s(r). \tag{6.74}$$

In general the function $\phi^s(r)$ and its derivative (to compute fields) can be best tabulated, using cubic splines for interpolation (see Chapter 19). For a Slater-type exponential decay of the density:[32]

$$w(r) = \frac{1}{\pi\lambda^3} e^{-2r/\lambda}, \tag{6.75}$$

the potential function is given by

$$\phi(r) = 1 - \left(1 + \frac{r}{\lambda}\right) e^{-2r/\lambda}. \tag{6.76}$$

For two interacting charge distributions the two-center integral needs to be tabulated, although Carrillo-Trip *et al.* (2003) give an approximate expression that gives good accuracy at all relevant distances (note that shells are never very close to each other):

$$V_{ss} \approx \frac{q_1^s q_2^s}{r_{12}}[1 - (1+z)e^{-2z}], \quad z = \frac{\lambda_1\lambda_2 + (\lambda_1 + \lambda_2)^2}{(\lambda_1 + \lambda_2)^3} r_{12}. \tag{6.77}$$

6.3.9 Towards the ideal force field

General force fields for molecular systems are not of sufficient quality for universal use. They are often adapted to specific interactions and the terms are not sufficiently transferable. The latter is clear from the ever increasing number of atom types deemed necessary when force fields are refined. A large amount of work has been done on models for water, being important, small, polar, polarizable, hydrogen-bonding, and endowed with a huge

---

[31] In the dipolar case the induced dipoles are kept constant in taking the derivatives. This is correct because the partial derivatives to the dipole are zero, provided the dipoles have been completely relaxed before the force is determined.

[32] λ is the decay length of the corresponding wave function; 32% of the charge density is within a distance λ from the origin.

knowledge body of experimental properties. The excessive number of published water models indicate a state of confusion and a lack of accepted guidelines for the development of force fields. Still, it seems that an advisable approach to constructing accurate molecular models emerges from the present confusion. Let us attempt to clarify the principles involved and make a series of rational choices.

We list a few *considerations*, not necessarily in order of importance:

- *Simplicity*: as such very simple models as SPC/E are already so successful within their realm of applicability, it seems an overkill to devise very sophisticated models with many multipoles and polarizabilities, and many intermolecular interaction terms. However excellent such models may be, they will not become adopted by the simulation community.
- *Robustness*, meaning validity in varying environments. Correct phase behavior can only be expected if the same model represents all phases well, over a wide range of temperature and pressure. The same model should also be valid in very different molecular environments.
- *Transferability*, i.e., the principle of constructing the model, and as much as possible also the parameters, should be applicable to other molecules with similar atoms.
- *Accuracy*, meaning the precision reached in reproducing experimental properties. The nature of such properties can be thermodynamic (e.g., phase boundaries, free energies), static structural (e.g., radial distribution functions), static response (e.g., dielectric constant, viscosity, diffusion constant) or dynamic response (e.g., dielectric dispersion, spectroscopic relaxation times). The choice of properties and the required accuracy depend on the purpose for which the force field is to be used. The good-for-everything force field will be too complex and computationally intensive to be useful for the simulation of large systems with limited accuracy requirements. Thus there is not one ideal force field, but a hierarchy depending on the application.
- *Ab initio computability*, meaning that the model parameters should in principle be obtained from fitting to high-level quantum chemistry calculations. This opens the way to construct reliable terms for unusual molecules for which sufficient experimental data are not readily available. If *ab initio* parametrization does not yield acceptable results and empirical refinement is necessary, the model designer should first consider the question whether some interaction type or inherent approximation has been overlooked.

Ideally, a very accurate parent force field, derivable from quantum calculations, should be constructed to form the top of the hierarchy. Then, by constraining terms or properties that do fluctuate in the parent model, to their average values in a given set of conditions, simpler child models can be derived with more limited applicability. This process may be repeated to produce even simpler and more restricted grandchildren. In this way simple and efficient force fields for a limited range of applications may be derived without the need for extensive reparametrization of each new force field. This strategy has been advocated by Saint-Martin *et al.* (2005) and shown to be successful for the MCDHO model.[33] The full model has internal flexibility and is polarizable; if the flexibility is constrained at the average value obtained in a given environment and under given conditions, a simpler "child" model emerges that is valid for a range of similar environments and conditions. The same applies to constraining induced dipoles to yield a "grandchild": a simple four-site effective pair potential, valid in a more limited range of conditions. In a similar fashion child models with simpler force truncation schemes could be constructed, with corrections obtained from the average long-range contributions within the parent model.

The parent model should explicitly express separate aspects: any omission will lead to effective incorporation of the omitted aspect into terms of another physical nature at the expense of robustness and transferability of the model. We list a number of these effects:

- *Quantum character of nuclear motions* These can be included in a thermodynamically correct (but dynamically questionable) way by replacing each nucleus by a path integral in imaginary time, approximated by a string of beads as explained in Chapter 5. "Grandparent" models with complete path integral implementations are considerably more complex. More approximately, but sufficient for most applications, these quantum effects may be estimated by quantum corrections to second order in $\hbar$, as detailed in Section 3.5, or they may be incorporated in a refinement of the model that includes Feynman–Hibbs quantum widths of the nuclei. Such second-order corrections are not sufficient for oscillators with frequencies much above $k_B T/h$.
- *Quantum character of high-frequency vibrations* There are no good methods to include quantum vibrational states dynamically into a classical system. The best one can do is to make an adiabatic approximation and include the equilibrium distribution over quantum states as a potential of mean force acting on the nuclei. Consider an single oscillating bond (the theory is readily extended to coupled vibrations) between two masses m_1

---

[33] See also Hernández-Cobos *et al.* (2005).

and m_2, with the bond length as a general coordinate q. Let q_0 be the bond length for which there is no net force (potential force plus centrifugal force); q_0 depends on the environment and the velocities. The *deviation* $\xi = q - q_0$ can be separated as a quantum degree of freedom acting in a quadratic potential with its minimum at $\xi = 0$. It will have an oscillator (angular) frequency $\omega = \sqrt{k/\mu}$, where k is the force constant and μ the reduced mass $m_1 m_2/(m_1 + m_2)$. We now wish to treat the total system as a reduced dynamic system with q constrained to q_0, and omit the quantum degree of freedom ξ. This we call a *flexible constraint* because the position q_0 fluctuates with external forces and angular velocities. In order to preserve the correctness of thermodynamic quantities, the potential of the reduced (i.e., flexibly constrained) system must be replaced by the potential of mean force V^{mf} with respect to the omitted degree of freedom (this is the adiabatic approximation):

$$V^{\mathrm{mf}} = V_{\mathrm{class}} + \frac{1}{2}\hbar\omega + k_{\mathrm{B}}T\ln(1 - e^{-\hbar\omega/k_{\mathrm{B}}T}). \tag{6.78}$$

Since in principle ω is a function of the classical coordinates through its dependence on the force constant, the potential of mean force will lead to forces on the classical system and energy exchange with the quantum oscillator. However, this dependence is generally small enough to be negligible. In that case it suffices to make *posteriori* corrections to the total energy. The important point is to impose flexible constraints on the system. As Hess *et al.* (2002), who give a more rigorous statistical-mechanical treatment, have shown, the iterations necessary to implement flexible constraints do not impose a large computational burden when polarization iterations are needed anyway.

It should be noted that fully flexible classical force fields will add an incorrect $k_{\mathrm{B}}T$ per oscillator to the total energy. Appropriate corrections are then mandatory.

- *Intramolecular structural response to external interactions* Both full flexibility and flexible constraints take care of intramolecular structural response to external interactions. Generally the structural deviations are small, but they represent a sizeable energetic effect, as the restoring intramolecular potentials are generally quite stiff. Structural fluctuations also react back on the environment and cause a coupling between mechanical forces and polarization. Not only the structural response to external forces should be taken into account, but also the structural response to external electric fields.

- *Intramolecular electronic response to external interactions* This, in fact, is the polarizability response, through which the intramolecular charge distribution responds to electrical (and in the case of shell models also mechanical) forces from the environment. Their incorporation in parent models is mandatory.

- *Long-range dispersion forces* With reference to the discussion in Section 6.3.4 on page 159, we may state that accurate parent models should include very long range dispersion interactions, preferably by evaluating the corresponding lattice sum for periodic systems.

- *Long-range electrical interactions* From the discussion in Section 6.3.5 on page 164 it is clear that parent models must evaluate long-range Coulomb interactions as lattice sums for periodic systems, or else (e.g., for clusters) use no truncation at all.

- *Non-additivity of repulsion and attraction* Repulsion between electronic distributions based on exchange is not strictly pair-additive. It is expected that the nonadditivity is at least partly taken care of by repulsions between moving charge distributions, but this remains to be investigated. A parent model should at least have an estimate of the effect and preferably contain an appropriate three-body term. Also the dispersion interaction is non-additive, but this so-called Axilrod–Teller effect[34] is of order r^{-9} in the distance and probably negligible.

- *Effects of periodicity* Parent models should at least evaluate the effects of periodicity and preferably determine the infinite box-size limit for all evaluated properties. Especially for electrolyte solutions effects of periodicity are a matter of concern.

Summarizing all considerations we arrive at a preference for the structure of a suitable parent model. For simplicity, the model should consist of sites (atoms and possibly virtual sites) with partial charges only, without higher multipoles. In order to represent real charge distributions faithfully, smeared charge distributions should be allowed. Polarizability should be realized by moving charge distributions attached to atoms, with special care to correctly represent the induced charge distributions in close proximity of external charges. The reason is not only simplicity, but it is also less likely that induced dipoles or fluctuating charges will eventually prove to be adequate, as they do not include polarization induced by exchange and dispersion forces (see the discussion on page 178). Intermolecular repulsion and dispersion interactions should be largely centered on the moving shells, possibly refined with short-range atom-based corrections. The model should

---

[34] The three-body dispersion is a result of third-order perturbation and is inversely proportional to $r_{12}^2 r_{23}^2 r_{13}^2$ (Axilrod and Teller, 1948).

be parameterized by fitting to high-level *ab initio* calculations, with possible empirical fine tuning.

6.3.10 QM/MM approaches

Force fields of the type described above are not suitable for systems of molecules that undergo chemical transformations. Configurations along reaction paths differ so much from the covalently bonded stable states that they cannot be described by simple modifications of the force fields intended for stable molecules. For such configurations quantum calculations are required. However, in complex systems in which reactions take place, as in enzymes and other catalysts, the part that cannot be described by force fields is generally quite limited and it would be an overkill (if at all possible; however, see the *ab initio MD* treated in Section 6.3.1) to treat the whole system by quantum-chemical methods.

For such systems it is possible to use a hybrid method, combining quantum calculations to obtain energies and forces in a limited fragment, embedded in a larger system for which energies and forces are obtained from the much simpler force field descriptions (see Fig. 6.11). These methods are indicated as QM/MM (quantum mechanics/ molecular mechanics) methods. The principle was pioneered by Warshel and Levitt (1976), and has been widely applied since the mid-1980s.[35] Most simulation packages allow coupling with one or more quantum chemistry programs, which include semi-empirical, as well as density functional and *ab initio* treatments of a fragment. QM/MM can be used for optimizations, for example of transition states and reaction paths, but also for dynamic trajectories. The forces and energies can be generated every time step, producing a very costly dynamics trajectory, or the QM calculations can be done at selected instances from approximate trajectories. Note that the quantum part is always solved in the Born–Oppenheimer approximation: QM/MM methods do not produce quantum dynamics.

The coupling between the QM and the MM part must be carefully modeled. In general, the total energy consists of three contributions, arising from the QM part, the MM part and the interaction between the two. When the

---

[35] For a survey see Gao and Thompson (1998). Chandra Singh and Kollman (1986) introduced the cap atom, which is usually hydrogen and which is kept free to readjust. The coupling between the QM and MM part is described by Field *et al.* (1990). A layered approach, allowing layers of increasing quantum accuracy, has been introduced by Svensson *et al.* (1996). In the "AddRemove" scheme of Swart (2003), the capping atom is first added on the line pointing to the MM link atom, included in the QM, and then its interaction is removed from the sum of QM and MM energies. Zhang and Yang (1999) use a pseudobond to a one-free-valence atom with an effective core potential instead of a cap atom.

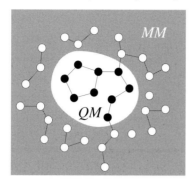

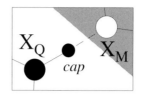

Figure 6.11 A quantum mechanical fraction embedded in a molecular mechanics environment. Covalent bonds that cross the QM-MM boundary are replaced by a cap atom in the QM part.

QM part consists of a separate molecule, the intermolecular interactions couple the two systems. When the QM part consists of a molecular fragment, the covalent bond(s) between the fragment and the MM environment must be replaced by some other construct, most often a "cap" atom or pseudo-atom.

In principle there is a discrepancy between the levels of treatment of the QM part, which includes induction by the electric fields of the environments, and MM part in the case of non-polarizable MM force fields. The reader is referred to Bakowies and Thiel (1996) for a detailed evaluation of the various ways the QM and MM systems can interact.

6.4 Solving the equations of motion

Given a conservative force field, we know the forces acting on atoms and on virtual interaction sites. The forces on virtual sites are first redistributed over the atoms from which the virtual sites are derived, so we end up with (cartesian) forces on mass points. The description may contain constraints of bond lengths and possibly bond angles.

While it is possible to write the equations of motion in *internal coordinates* (see Chapter 15), these tend to become quite complicated and the general recommendation for atomic systems is to stick to cartesian coordinates, even in the presence of constraints. Modern constraint methods are robust and efficient (see Section 15.8 on page 417). However, for completely rigid molecules the use of quaternions may be considered (see page 413).

Consider a system of N particles with mass m_i, coordinates $\boldsymbol{r}_i$ and a defined recipe to compute the total potential energy $E_{\text{pot}} = V(\boldsymbol{r})$ and the forces $F_i(r) = -\nabla_i V(r)$, given the set of all coordinates. Let coordinates and velocities v_i be known at time t. We assume that there are no constraints. The equations of motion are simply Newton's equations (which are Hamiltonian as well):

$$\dot{r}_i = v_i,$$
$$\dot{v}_i = F_i/m_i. \tag{6.79}$$

The total energy $E_{\text{tot}} = K + V$ will be conserved:

$$\frac{dE_{\text{tot}}}{dt} = \frac{d}{dt}\sum_i \frac{1}{2}m_i v_i^2 + \frac{dV(r)}{dt} = \sum_i m_i v_i \cdot \dot{v}_i + \sum_i \frac{\partial V}{\partial r_i} \cdot v_i$$

$$= \sum_i \left(v_i \cdot F_i + \frac{\partial V}{\partial r_i} \cdot v_i \right) = 0. \tag{6.80}$$

Properly solving these equations of motion will produce a microcanonical or N, V, E ensemble. In practice there will be errors that cause deviations from the ideal behavior: the finite time step will cause integration errors and the total energy will not be exactly conserved; errors in forces (e.g., due to truncation) will produce pseudorandom disturbances that cause energies to drift. Since the temperature is determined by the equipartition theorem saying that $K = \frac{3}{2}Nk_{\text{B}}T$, the temperature may drift even when equilibrium has been attained. Therefore there are always modifications to the pure Newtonian equations of motion needed to generate long stable trajectories.

The equations of motion are solved in time steps Δt. Three important considerations influence the choice of algorithm:

(i) Time reversibility, inherent in the Newtonian equations of motion, should be conserved.

(ii) The generated trajectories should conserve volume in phase space, and in fact also wedge products (area) $dq \wedge dp$ in general, i.e., the algorithm should be symplectic (see Chapter 17, page 495). This is important to conserve equilibrium distributions in phase space, because deviation from symplectic behavior will produce time-dependent weight factors in phase space. This importance of the symplectic property has been emphasized in the 1990s (Leimkuhler and Reich, 1994; Leimkuhler and Skeel, 1994) and is now widely recognized.

(iii) Since the computational effort is completely dominated by the force calculation, methods that use only one force evaluation per time step are to be preferred. This rules out the well-known Runge–Kutta methods, which moreover are also not symplectic and lead to erroneous behavior on longer time scales (Leimkuhler, 1999).

In the past practice of MD the *Gear algorithm* has been much used. The Gear algorithm predicts positions and a number of derivatives based on a Taylor expansion of previous values (how many depends on the order of the algorithm); it then evaluates the accelerations from the forces at the predicted position, and corrects the positions and derivatives on the basis of the deviation between predicted and evaluated accelerations. There are several variants and predictor–corrector algorithms of this kind have been applied up to orders as high as eight. They are quite accurate for small time steps but not very stable for larger time steps. When the forces are not very precise, it does not help to use high orders. They are neither time-reversible nor symplectic and have not survived the competition of the simpler, more robust, reversible and symplectic Verlet or leap-frog algorithms.[36] The latter and its variants including multiple time-step versions, are derivable by the Reference System Propagator Algorithms (RESPA) method of Tuckerman *et al.* (1992) that uses simple operator algebra and is outlined below.

The original Verlet algorithm (Verlet, 1967) does not use the velocities, and employs a simple discretization of the second derivative:

$$\ddot{x}(t) \approx \frac{x(t - \Delta t) - 2x(t) + x(t + \Delta t)}{(\Delta t)^2}, \tag{6.81}$$

leading to the predicted position (x stands for every particle coordinate; $f(t) = F_i(x(t))/m_i$ is the corresponding force component, evaluated from the positions at time t and – for convenience – divided by the mass)

$$x(t + \Delta t) = 2x(t) - x(t - \Delta t) + f(t)(\Delta t)^2 + \mathcal{O}((\Delta t)^4). \tag{6.82}$$

The velocity is found in retrospect from

$$v(t) = \frac{v(t + \Delta t) - v(t - \Delta t)}{2\Delta t} + \mathcal{O}((\Delta t)^2), \tag{6.83}$$

but plays no role in the evolution of the trajectory. It can be more accurately estimated from[37]

$$v(t) = \frac{v(t + \Delta t) - v(t - \Delta t)}{2\Delta t} + \frac{f(t - \Delta t) - f(t + \Delta t)}{12} + \mathcal{O}((\Delta t)^3). \tag{6.84}$$

An equivalent scheme is the *leap-frog algorithm*, which uses positions at integer time steps and velocities halfway in between time steps (Hockney

---

[36] Gear algorithms (Gear, 1971) and their variants have been reviewed and evaluated for use in MD including constraints by van Gunsteren and Berendsen (1977) and in relation to the Verlet algorithm by Berendsen and van Gunsteren (1986).
[37] Berendsen and van Gunsteren (1986).

and Eastwood, 1988). Starting from $v(t - \frac{1}{2}\Delta t)$ and $x(t)$ the updates are

$$v\left(t + \frac{1}{2}\Delta t\right) = v\left(t - \frac{1}{2}\Delta t\right) + f(t)\Delta t,$$

$$x(t + \Delta t) = x(t) + v\left(t + \frac{1}{2}\Delta t\right)\Delta t. \tag{6.85}$$

It can be easily shown that this algorithm is equivalent to Verlet's and will generate the same trajectory, if the velocity $v(t - \frac{1}{2}\Delta t)$ is started as $[x(t) - x(t - \Delta t)]/(\Delta t)$. The velocity at integer time steps can be recovered as the average of the two velocities at half time steps earlier and later, but only to the $\mathcal{O}((\Delta t)^2)$ precision of (6.83).

In several applications, for example when velocity-dependent forces are applied, it is desirable to know the velocity at the time the position is predicted, rather than a time step later. There are several algorithms, equivalent to Verlet, that deliver equal-time velocities. One is Beeman's algorithm (Beeman, 1976):

$$x(t + \Delta t) = x(t) + v(t)\Delta t + \left[\frac{2}{3}f(t) - \frac{1}{6}f(t - \Delta t)\right](\Delta t)^2,$$

$$v(t + \Delta t) = v(t) + \left[\frac{1}{3}f(t + \Delta t) + \frac{5}{6}f(t) - \frac{1}{6}f(t - \Delta t)\right]\Delta t, \tag{6.86}$$

but the most popular one is the *velocity-Verlet* algorithm (Swope *et al.*, 1982):

$$x(t + \Delta t) = x(t) + v(t)\Delta t + \frac{1}{2}f(t)(\Delta t)^2,$$

$$v(t + \Delta t) = v(t) + \frac{1}{2}[f(t) + f(t + \Delta t)]\Delta t. \tag{6.87}$$

This algorithm needs the force at the new time step, but there is only one force evaluation per step. Although it is not immediately obvious, all these algorithms are equivalent (Berendsen and van Gunsteren, 1986).

The elegant operator technique considers the exponential Liouville operator to evolve the system in time. We start with (17.152) on page 493:

$$\dot{\mathbf{z}} = i\mathcal{L}\mathbf{z}, \tag{6.88}$$

where $\mathbf{z}$ is the vector of generalized coordinates and conjugate momenta. We apply (6.88) to the cartesian Newtonian equations (6.79) and introduce the time-differential operator $\mathbf{D}$:

$$\begin{pmatrix} \dot{x} \\ \dot{v} \end{pmatrix} = \begin{pmatrix} 0 & 1 \\ \hat{f} & 0 \end{pmatrix} \begin{pmatrix} x \\ v \end{pmatrix} = \mathbf{D} \begin{pmatrix} x \\ v \end{pmatrix}, \tag{6.89}$$

where $\hat{f}$ is an operator acting on x that produces $F_i(x)/m_i$, $(i = 1, \ldots 3N)$. The x, v vector has a length $6N$ for N particles. The solution is

$$\begin{pmatrix} x \\ v \end{pmatrix}(t) = e^{t\mathbf{D}} \begin{pmatrix} x \\ v \end{pmatrix}(0) = \mathbf{U}(t) \begin{pmatrix} x \\ v \end{pmatrix}(0). \tag{6.90}$$

The exponential operator $\mathbf{U} = \exp(t\mathbf{D})$ is time-reversible:

$$\mathbf{U}(-t) = \mathbf{U}^{-1}(t), \tag{6.91}$$

and this solution is exact and symplectic.[38] Unfortunately we cannot solve (6.90) and we have to solve the evolution over small time steps by approximation. First split the operator $\mathbf{D}$ into two simple parts:

$$\begin{pmatrix} 0 & 1 \\ \hat{f} & 0 \end{pmatrix} = \begin{pmatrix} 0 & 1 \\ 0 & 0 \end{pmatrix} + \begin{pmatrix} 0 & 0 \\ \hat{f} & 0 \end{pmatrix}. \tag{6.92}$$

These two parts do not commute and we can use the Trotter–Suzuki expansion (see Chapter 14, page 386)

$$e^{t(A+B)} \approx e^{(t/2)A} e^{tB} e^{(t/2)A}, \tag{6.93}$$

which can be further subdivided into higher products with higher-order accuracy. Substituting the exponential operators A and B by

$$\mathbf{U}_v(t) = t \exp \begin{pmatrix} 0 & 1 \\ 0 & 0 \end{pmatrix} \quad \text{and} \quad \mathbf{U}_f(t) = t \exp \begin{pmatrix} 0 & 0 \\ \hat{f} & 0 \end{pmatrix}, \tag{6.94}$$

and using a first-order expansion of the exponential:

$$\mathbf{U}_v(t) \begin{pmatrix} x \\ v \end{pmatrix} = \begin{pmatrix} x + vt \\ v \end{pmatrix}, \tag{6.95}$$

$$\mathbf{U}_f(t) \begin{pmatrix} x \\ v \end{pmatrix} = \begin{pmatrix} x \\ v + ft \end{pmatrix}, \tag{6.96}$$

we can, for example, split $\mathbf{U}(\Delta t)$ as follows:

$$\mathbf{U}(\Delta t) \approx \mathbf{U}_f(\Delta t/2) \mathbf{U}_v(\Delta t) \mathbf{U}_f(\Delta t/2). \tag{6.97}$$

Writing this out (exercise 1 on page 209) it is easily seen that the velocity-Verlet scheme (6.87) is recovered. Concatenating the force operator for successive steps yields the leap-frog algorithm, (6.85). The method is powerful enough to derive higher-order and multiple time-step algorithms. For

[38] $\mathbf{U}$ is the transformation matrix of the transformation of $(x, v)(0)$ to $(x, v)(t)$. The Jacobian of the transformation, which is the determinant of $\mathbf{U}$, is equal to 1 because of the general rule $\det[\exp(\mathbf{A})] = \exp(\operatorname{tr} \mathbf{A})$; the trace of $\mathbf{D} = 0$.

example, a double time-step algorithm with short- and long-range forces is obtained (Tuckerman *et al.*, 1992) by applying the propagator

$$\mathbf{U}_l(\Delta t/2)[\mathbf{U}_s(\delta t/2)\mathbf{U}_v(\delta t)\mathbf{U}_s(\delta t/2)]^n\mathbf{U}_l(\Delta t/2), \qquad (6.98)$$

where $\mathbf{U}_s$ and $\mathbf{U}_l$ are the propagators for the short- and long-range forces, respectively, and $\Delta t = n\delta t$.

6.4.1 Constraints

The incorporation of constraints is fully treated in Section 15.8 of Chapter 15, to which we refer. The most popular method is coordinate resetting, as in the routine SHAKE and its variants SETTLE and RATTLE. In the Verlet algorithm, the coordinate prediction $x(t + \Delta t)$ is first made as if no constraints exist and subsequently the coordinates are iteratively reset in the direction of $x(t)$ until all constraints are satisfied. The most robust and stable method is the projection method LINCS. The influence of constraints on the statistical mechanics of canonical averages is treated in Chapter 17, Section 17.9.3 on page 499.

6.5 Controlling the system

In almost all cases it is necessary to make modifications to the Newtonian equations of motion in order to avoid undesirable effects due to the inexact solution of the equations of motion and the inexact evaluation of forces. In most applications it is desirable to simulate at constant temperature, preferably generating a canonical NVT ensemble, and in many applications simulation at constant pressure (NpT) is preferred above constant volume. In some applications simulation at constant chemical potential is desirable. Finally, a very important class of applications are *non-equilibrium* simulations, where the system is externally driven out of equilibrium, usually into a steady-state condition, and its response is measured.

Depending on the purpose of the simulation it is more or less important to generate an exact ensemble and to know the nature of the ensemble distribution. When the sole purpose is to equilibrate an initially disturbed system, any robust and smooth method that does not require intervention is acceptable. Generally, when only *average* equilibrium quantities are required, the exact nature of the generated equilibrium ensemble is less important as long as the system remains close to Hamiltonian evolution. One should be aware that there are system-size effects on averages that will depend on the nature of the ensemble (see Chapter 17, Section 17.4.1 on page 462). When one wishes to use the properties of *fluctuations*, e.g., to determine higher derivatives of thermodynamic quantities, knowledge of the exact nature of the generated distribution function is mandatory.

Four classes of methods are available to control the system externally:

(i) *Stochastic methods*, involving the application of stochastic forces together with friction forces. They are particularly useful to control temperature. Such forces mimic the effect of elastic collisions with light particles that form an ideal gas at a given temperature. They produce a canonical ensemble. Other types of stochastic control make use of reassigning certain variables (as velocities to control temperature) to preset distribution functions. Stochastic methods in general enforce the required ensemble distribution, but disturb the dynamics of the system.

(ii) *Strong-coupling methods* apply a constraint to the desired quantity, e.g., for temperature control one may scale the velocities at every time step to set the total kinetic energy exactly at the value prescribed by the desired temperature. This is the *Gauss isokinetic* thermostat. This method follows an old principle by Gauss stating that external constraints should be applied in such a way that they cause the least disturbance. The system dynamics is non-Hamiltonian. The Gauss thermostat produces a canonical distribution in configuration space, but disturbs the dynamical accuracy.

(iii) *Weak-coupling methods* apply a small perturbation aimed at smoothly reducing the property to be controlled to a preset value by a first-order rate equation. Such couplings can be applied to velocity scaling to control the temperature and/or to coordinate and volume scaling to control pressure. As the dynamics of the system is non-Hamiltonian, weak-coupling methods do not generate a well-defined ensemble. Depending on the coupling strength they generate an ensemble in between microcanonical and canonical for temperature scaling and in between isochoric and isobaric for coordinate scaling. It seems warranted to use ensemble averages but not fluctuations in order to determine thermodynamic quantities. Weak-coupling methods are well suited to impose non-equilibrium conditions.

(iv) *Extended system dynamics* extends the system with extra degrees of freedom related to the controlled quantity, with both a "coordinate" and a conjugate "momentum." The dynamics of the extended systems remains fully Hamiltonian, which enables the evaluation of the distribution function, but the dynamics of the molecular system is disturbed. A proper choice of extended variables and Hamiltonian can combine the control of temperature and/or pressure combined with a canonical distribution in configuration space.

The influence of such external modifications on the equilibrium distribution function can be evaluated by the following considerations. First we assume that the underlying (unperturbed) system is Hamiltonian, not only as a differential equation, but also in the algorithmic implementation. This is never true, and the effects of deviations become apparent only on longer time scales as a result of accumulation of errors. For a true Hamiltonian system, the evolution of density in phase space $f(\mathbf{z})$ is given by the Liouville equation (see Chapter 17, Section 17.8 on page 492, and (17.158)). It can easily be seen that *any* distribution $f(H)$ in phase space that depends only on the total energy $H = K + V$ will be stationary:

$$\frac{\partial f(H)}{\partial t} = \frac{df(H)}{dH}\dot{\mathbf{z}} \cdot \nabla H = f'(H) \sum_{i=1}^{3N} \left(\frac{\partial H}{\partial p_i} \frac{\partial H}{\partial q_i} - \frac{\partial H}{\partial q_i} \frac{\partial H}{\partial p_i} \right) = 0. \quad (6.99)$$

This means that in principle any initial distribution $f(H)$ (e.g., the canonical distribution) will not change in time, but there is no restoring force inherent in the dynamics that will correct any deviations that may (slowly) develop. If an external influence drives the system to a given distribution, it provides the necessary restoring force. If the Hamiltonian dynamics is accurate, only a small restoring force is needed.

6.5.1 Stochastic methods

The application of stochastic disturbances to control temperature goes back to Schneider and Stoll (1978) and corresponds to Langevin dynamics (see Section 8.6); we shall call this method *the Langevin thermostat*. The idea is to apply a frictional force and a random force to the momenta:

$$\dot{p}_i = F_i - \gamma p_i + R_i(t), \quad (6.100)$$

where $R_i(t)$ is a zero-average stationary random process without memory:

$$\langle R_i(0) R_i(t) \rangle = 2m_i \gamma_i k_B T \delta(t). \quad (6.101)$$

For convenience we now drop the subscript i; the following must be valid for any particle i and the friction and noise can be chosen differently for every degree of freedom. The random disturbance is realized in time steps Δt and the change in p is a random number drawn from a normal distribution with variance $\langle (\Delta p)^2 \rangle$ given by

$$\langle (\Delta p)^2 \rangle = \left\langle \left(\int_0^{\Delta t} R(t')\, dt' \right)^2 \right\rangle = \int_0^{\Delta t} dt' \int_0^{\Delta t} dt'' \langle R(t')R(t'') \rangle$$
$$= 2m\gamma k_B T (\Delta t). \tag{6.102}$$

In fact, it is not required to draw the change in p from a normal distribution, as long as the distribution has zero mean and finite variance. We now show that this procedure yields extra terms in $\partial f / \partial t$ that force the distribution functions of the momenta to which the noise and friction are applied to a normal distribution. The random force on p causes a *diffusion* of p with a diffusion constant D given by the mean-square displacement of p (one dimension) in the time interval Δt:

$$\langle (\Delta p)^2 \rangle = 2D\Delta t. \tag{6.103}$$

Consequently the diffusion constant is given by

$$D = m\gamma k_B T. \tag{6.104}$$

Diffusion leads to Fick's equation for the distribution function:

$$\frac{\partial f(p,t)}{\partial t} = D\frac{\partial^2 f(p,t)}{\partial p^2} \tag{6.105}$$

and the friction term leads to an extra flux $f\dot{p}$ and therefore to

$$\frac{\partial f}{\partial t} = -\frac{\partial(f\dot{p})}{\partial p} = \gamma f + \gamma p \frac{\partial f}{\partial p}. \tag{6.106}$$

If these two contributions cancel each other, the distribution function will be stationary. The solution of the equation

$$D\frac{\partial^2 f}{\partial p^2} + \gamma f + \gamma p \frac{\partial f}{\partial p} = 0 \tag{6.107}$$

is

$$f(p) \propto \exp\left[-\frac{p^2}{2mk_B T} \right]. \tag{6.108}$$

The total distribution function must be proportional to this normal distribution of p_i, and that applies to all momenta to which random and friction forces are applied. Since H contains $p^2/2m$ as an additive contribution, and the Hamiltonian terms will force the stationary distribution function to be a function of H, it follows that the total distribution will be:

$$f(\mathbf{z}) \propto \exp\left[-\frac{H(\mathbf{z})}{k_B T} \right]. \tag{6.109}$$

Thus the distribution will be canonical and the temperature of the ensemble is set by the relation between applied friction and noise. In a sense the friction and noise driving the canonical distribution compete with accumulating disturbing errors in the numerical solution of the Hamiltonian equations. The applied damping enforces an extra first-order decay on velocity correlation functions and thereby disturb the dynamic behavior on time scales comparable to $1/\gamma$. To minimize dynamic disturbance, the γ's should be taken as small as possible, and need not be applied to all degrees of freedom. But with large errors a strong damping on many particles is required. The Langevin thermostat provides a smooth decay to the desired temperature with first-order kinetics. Note that we have given the equations for cartesian coordinates, for which the mass tensor is diagonal and constant; for generalized coordinates the equations become complicated and difficult to handle because the mass tensor is a function of the configuration.

The velocity rescaling thermostat of Andersen (1980) has a similar effect, but does not distribute the external disturbance as smoothly as the Langevin thermostat. Andersen's method consist of reassigning the velocity of a randomly selected molecule at certain time intervals from the appropriate Maxwellian distribution. As expected, a canonical ensemble is generated.

6.5.2 Strong-coupling methods

The strong-coupling methods artificially constrain a property to the desired value, e.g., the total kinetic energy to a prescribed value determined by the desired temperature. This is accomplished by scaling the velocities with a multiplicative factor that preserves the shape of the distribution function. This amounts to adding an acceleration $\dot{v} = -\alpha v$ t every degree of freedom. This isokinetic or Gauss thermostat was introduced by Hoover *et al.* (1982) and Evans (1983). Following Tuckerman *et al.* (1999), who use the notion of phase-space compressibility (see Chapter 17, Section 17.8, page 495), we shall show that the isokinetic thermostat produces a canonical distribution in configuration space.

Start with the definition of phase-space compressibility κ for non-Hamiltonian flow (17.165):

$$\kappa = \boldsymbol{\nabla} \cdot \dot{\mathbf{z}}. \tag{6.110}$$

Tuckerman *et al.* showed (see page 495) that, if a function $w(\mathbf{z})$ can be defined whose time derivative equals κ, then $\exp[-w(\mathbf{z})]dz_1, \ldots, dz_{2n}$ is an invariant volume element along the trajectory, meaning that $\exp[-w(\mathbf{z})]$ is the equilibrium weight function in phase space. Now κ follows from α:

$$\kappa = \sum_{i=1}^{n} \frac{\partial}{\partial p_i} \dot{p}_i = -n\alpha, \tag{6.111}$$

where n is the number of degrees of freedom $(= 3N - n_c)$. We can express α in phase-space functions by realizing that (in cartesian coordinates)

$$\frac{d}{dt} \sum_{i=1}^{n} \frac{p_i^2}{2m_i} = \sum_{i=1}^{n} \frac{p_i \dot{p}_i}{2m_i} = 0, \tag{6.112}$$

and therefore

$$\sum_{i=1}^{n} \frac{p_i F_i}{m_i} - \alpha \sum_{i=1}^{n} \frac{p_i^2}{m_i} = 0. \tag{6.113}$$

Hence

$$\kappa = -n \frac{\sum_i p_i F_i / m_i}{\sum_i p_i^2 / m_i} = -\beta \sum_i \frac{p_i F_i}{m_i} = \beta \frac{dV(q)}{dt}, \tag{6.114}$$

since

$$\frac{dV(q)}{dt} = \sum_i \frac{\partial V}{\partial q_i} \dot{q}_i = -\sum_i \frac{F_i p_i}{m_i}.$$

Thus the $w = \beta V$ and the weight function equals $\exp[-\beta V]$. We conclude that ensemble averages of a variable $A(\mathbf{z})$ over the isokinetic thermostat are given by

$$\langle A \rangle = \frac{\int A(\mathbf{z}) \exp[-\beta V(q)] \delta(\sum_i p_i^2 / 2m_i - nk_B T) \, d\mathbf{z}}{\int \exp[-\beta V(q)] \delta(\sum_i p_i^2 / 2m_i - nk_B T) \, d\mathbf{z}}. \tag{6.115}$$

Strong coupling can also be applied to constrain the pressure to a preset value (Evans and Morriss, 1983a; 1984). Combined with the isokinetic thermostat Evans and Morriss (1983b) have shown that a NPT ensemble (on the hypersurface of constrained pressure and kinetic energy) is obtained with a weight factor equal to $\exp[-\beta H]$, where H is the enthalpy $U + pV$. We do not pursue this type of pressure control further as it seems not to have become common practice, but refer instead to Allen and Tildesley (1987).

6.5.3 Weak-coupling methods

Weak-coupling methods (Berendsen *et al.*, 1984) are not stochastic in nature, and can be applied both for temperature and pressure control. For temperature control they do have the same effect as a Langevin thermostat on the variance of velocities (i.e., on the temperature). The idea is to rescale velocities per step in such a way that the total temperature T of the system will decay with a first-order process to the desired temperature T_0:

$$\frac{dT}{dt} = \frac{T_0 - T}{\tau}.$$
(6.116)

This rate equation would cause a temperature deviation from T_0 to decay exponentially to zero with a time constant τ. The implementation in terms of a scaling factor λ for the velocities is given by

$$\lambda^2 = 1 + \frac{\Delta t}{\tau}\left(\frac{T_0}{T} - 1\right),$$
(6.117)

where T is given by the kinetic energy found after updating the velocities in a normal dynamic step. For the smallest possible value of time constant $\tau = \Delta t$ the scaling is complete and the temperature is exactly conserved. This corresponds to the Gauss isokinetic thermostat which produces a canonical ensemble in configuration space. For τ much longer than the intrinsic correlation times for internal exchange of energy, the scaling has no effect and a microcanonical ensemble is obtained. This is borne out by the fluctuations in kinetic and potential energy: for small τ the kinetic energy does not fluctuate but the potential energy does; as τ increases, fluctuations in kinetic energy appear at the expense of potential energy fluctuations, to become equal and opposite at large τ. The cross-over occurs roughly from a factor of 10 below to a factor of 10 above the *intrinsic relaxation time* for the exchange between kinetic and potential energy, which is system-dependent.

Morishita (2000), in an attempt to characterize the weak-coupling ensemble, derived the following equation for the compressibility κ:

$$\kappa = \left[\beta - \frac{2}{n}\beta^2\delta K + \mathcal{O}(N^{-1})\right]\frac{dV(q)}{dt},$$
(6.118)

plus some unknown function of p. Here δK is the fluctuation of K, which depends on τ. For small τ, when $\delta K = 0$, this form reduces to $\beta dV/dt$, yielding the canonical configuration space distribution derived above for the isokinetic thermostat. For large τ, when $\delta K = -\delta V$, the configuration space distribution tends to

$$f(q) = \exp\left[-\beta V(q) - \frac{1}{n}\beta^2(\delta V)^2\right],$$
(6.119)

which equals the microcanonical configuration space distribution already derived earlier by Nosé (1984a). For intermediate values of τ Morishita made the assumption that the fluctuation of K is related to that of V by

$\delta K = -\alpha \delta V$, with α depending on τ, and arrives at a configuration space distribution

$$f(q) = \exp\left[-\beta V(q) - \frac{\alpha}{n}\beta^2(\delta V)^2\right]. \tag{6.120}$$

The distribution in momentum space remains unknown, but is less important as the integration over canonical momenta can be carried out separately. Note again that this is valid for cartesian coordinates only. Also note that the weak-coupling algorithm is no longer time-reversible, unless in the two extreme cases $\tau = \Delta t$ and $\tau \to \infty$.

Pressure control by weak coupling is possible by scaling coordinates. In the spirit of weak coupling one attempts to regulate the pressure P according to the first-order equation

$$\frac{dP}{dt} = \frac{1}{\tau_p}(P_0 - P). \tag{6.121}$$

Assume the isothermal compressibility $\beta_T = -(1/V)\partial V/\partial P$ is known, then scaling coordinates and volume,

$$\boldsymbol{r}' = \chi \boldsymbol{r}, \tag{6.122}$$
$$V' = \chi^3 V, \tag{6.123}$$

every time step with a scaling factor χ, given by

$$\chi^3 = 1 - \beta_T \frac{\Delta t}{\tau_p}(P_0 - P), \tag{6.124}$$

will accomplish that task. As the compressibility only enters the algorithm in conjunction with the time constant, its value need not be precisely known. The weak pressure coupling has the advantage of smooth response, but the disadvantages are that it does not generate a known ensemble and fluctuations cannot be used.

6.5.4 Extended system dynamics

The idea to extend the system with an extra degree of freedom that can be used to control a variable in the system, was introduced by Nosé (1984a, 1984b) for temperature control. The method involved a somewhat inconvenient time scaling and was modified by Hoover (1985) into a scheme known as the *Nosé–Hoover thermostat*, which has been widely used since. We treat the thermostat case, but the same principle can be applied to control pressure in a barostat. An extra variable η is introduced, which is a factor scaling the velocities. It has an associated "momentum" $p_\eta = Q\dot{\eta}$, where Q

is the "mass" of the extra degree of freedom. The equations of motion are (p, q, m stand for all p_i, q_i, m_i):

$$\dot{q} = \frac{p}{m},$$ (6.125)

$$\dot{p} = F(q) - p\frac{p_\eta}{Q},$$ (6.126)

$$\dot{p}_\eta = \sum \frac{p_i^2}{2m_i} - nk_\mathrm{B}T.$$ (6.127)

The temperature deviation from the bath temperature drives the *time deriv-ative* of the velocity scaling factor, rather than the scaling factor itself, as is the case in the weak-coupling method. This makes the equations of motion time reversible again, and allows to compute the phase space distribution. On the other hand, it has the practical disadvantage that the temperature control is now a second-order differential equation in time, which leads to oscillatory approaches to equilibrium. Hoover shows that the equilibrium phase space distribution is given by

$$f(q, p, \eta, p_\eta) \propto \exp\left[-\beta\left(V(q) + \sum_i \frac{p_i^2}{2m_i} + \frac{1}{2}Q\eta^2\right)\right],$$ (6.128)

which is canonical. The extra variable is statistically independent of posi-tions and velocities.

The Nosé–Hoover thermostat has been criticized because its behavior is non-ergodic (Toxvaerd and Olsen, 1990), which led Martyna *et al.* (1992) to formulation of the *Nosé–Hoover chain* thermostat. In this thermostat there is a sequence of M additional variables $\eta_1, \eta_2, \ldots, \eta_M$ with their masses and conjugate momenta, each scaling its predecessor in the chain:

$$\dot{q} = \frac{p}{m},$$ (6.129)

$$\dot{p} = F(q) - p\frac{p_{\eta_1}}{Q_1},$$ (6.130)

$$\dot{\eta}_1 = \frac{p_{\eta_1}}{Q_1},$$ (6.131)

$$\dot{p}_{\eta_1} = \sum \frac{p_i^2}{2m_i} - nk_\mathrm{B}T - p_{\eta_1}\frac{p_{\eta_2}}{Q_2},$$ (6.132)

$$\dot{\eta}_j = \frac{p_{\eta_j}}{Q_j} \quad j = 2, \ldots, M,$$ (6.133)

$$\dot{p}_{\eta_j} = \frac{p_{\eta_{j-1}}^2}{Q_{j-1}} - k_\mathrm{B}T - p_{\eta_j}\frac{p_{\eta_{j+1}}}{Q_{j+1}},$$ (6.134)

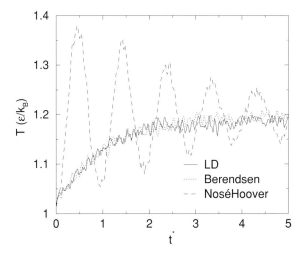

Figure 6.12 The temperature response of a Lennard–Jones fluid under control of three thermostats (solid line: Langevin; dotted line: weak-coupling; dashed line: Nosé–Hoover) after a step change in the reference temperature (Hess, 2002a, and by permission from van der Spoel *et al.*, 2005.)

$$\dot{p}_{\eta_M} = \frac{p_{\eta_{M-1}}^2}{Q_{M-1}} - k_B T. \tag{6.135}$$

6.5.5 Comparison of thermostats

A comparison of the behavior in their approach to equilibrium of the Langevin, weak-coupling and Nosé–Hoover thermostats has been made by Hess (2002a). Figure 6.12 shows that – as expected – the Nosé–Hoover thermostat shows oscillatory behavior, while both the Langevin and weak-coupling thermostats proceed with a smooth exponential decay. The Nosé–Hoover thermostat is therefore much less suitable to approach equilibrium, but it is more reliable to produce a canonical ensemble, once equilibrium has been reached.

D'Alessandro *et al.* (2002) compared the Nosé–Hoover, weak-coupling and Gaussian isokinetic thermostats for a system of butane molecules, covering a wide temperature range. They conclude that at low temperatures the Nosé–Hoover thermostat cannot reproduce expected values of thermodynamic variables as internal energy and specific heat, while the isokinetic thermostat does. The weak-coupling thermostat reproduces averages quite well, but has no predictive power from its fluctuations. These authors also monitored the Lyapunov exponent that is a measure of the rate at which trajectories deviate exponentially from each other; it is therefore an indicator of the tendency for *chaotic behavior*. A high Lyapunov exponent could be interpreted as a more efficient sampling. It turns out that the isokinetic thermostat has the highest exponent, while the Nosé–Hoover thermostat shows very low exponents. The weak coupling is in between. The stochastic thermostat was not studied, but is likely to show a high Lyapunov exponent as well.

6.6 Replica exchange method

Molecular dynamics simulations are usually carried out at a given temperature, using some kind of thermostat, as described in the previous section. A representative initial configuration is chosen and – after an initial equilibration period – one expects the system to reach thermal equilibrium. However, the system of interest may well have two or several potential wells, separated by relatively high barriers that are not effectively crossed during the simulation time. This is the case for proteins and nucleic acids in solution that need a macroscopic time (say, seconds) to fold into a specific conformation, but also for large polymers, e.g., copolymer melts that need long rearrangement times to settle to a structure with minimum free energy. Also glasses below the glass transition temperature will "freeze" into a subset of the possible states. Such systems are not ergodic within the available simulation time; they will be trapped in a limited set of configurations (often called a *conformation*) that is a subset of the complete canonical distribution. How can one be sure that the simulated system is representative for the thermal equilibrium distribution? How can one prepare a proper initial state that samples a conformation with low free energy?

Several methods have been devised in the past to overcome this problem. The most common approach is to use additional *external* information. For example, the dynamics of a folded protein can be studied by using a starting structure derived from experimental X-ray diffraction data; obviously the system is then forced into a predefined conformation without any guarantee that this conformation is the lowest free-energy state compatible with the force field used. Similarly, when crystallization does not occur spontaneously upon lowering the temperature in a simulation of a fluid, one can start from an experimental crystal structure and study its equilibrium with the melt. A more satisfying approach is to start with a high-temperature simulation, which allows the frequent crossing of barriers, and let the system find a low free-energy state by slowly lowering the temperature. This process is called *tempering* as it resembles the tempering of metals by slow cooling. Fast cooling will cause the system to become trapped in a low-energy state, which may not at all be representative for a low free-energy conformation. However, by keeping the system at an elevated temperature just high enough to allow essential barrier crossings, the probability to end up in a low free-energy conformation increases. In metallurgy this *annealing* process leads to the annihilation of lattice defects. A similar computational process, called *simulated annealing*, was proposed in the 1980s in a seminal paper by Kirkpatrick *et al.* (1983) that has stimulated several computational innovations.

A further development that has led to a breakthrough in the efficient generation of a representative ensemble in cases where equilibration is slow, is now known by the name *replica exchange method* (REM). Both Monte Carlo and molecular dynamics versions are possible. In essence, an ensemble is generated that contains not only a number of configurations belonging to a canonical distribution at a given temperature, but also a set of temperatures. By including an exchange mechanism between different temperatures, a total ensemble is generated that encompasses the whole set of temperatures. The effect is a much faster relaxation than a single system would have at a low temperature: each system now rapidly visits a range of temperatures. Replica-exchange methods are ideally suited for parallel computers; each replica runs on a separate processor and there is only communication between processors when exchanges are attempted.

The method is not restricted to a range of temperatures, but may also involve a range of Hamiltonians, e.g., with different interaction parameters. In fact, the method was originally applied to Monte Carlo simulations of spin glasses, involving a range of values for the spin coupling constant (Swendsen and Wang, 1986). Since then there have been several developments, including *multicanonical Monte Carlo* (Berg and Neuhaus, 1991; Hansmann and Okamoto, 1993) and *simulated tempering* (Marinari and Parisi, 1992), but the most generally useful method with application to protein folding is the *replica exchange molecular dynamics* (REMD) method of Sugita and Okamoto (1999), which we now describe in a slightly simplified version.[39]

Consider M replicas $S_1, \ldots, S_M$ of a system, subjected to canonical molecular dynamics (or Monte Carlo) simulations at M different temperatures $T_1, \ldots, T_M$ or inverse temperatures divided by k_B, $\beta_1, \ldots, \beta_M$. Initially, system S_i is at temperature T_i, but we allow exchange between the temperatures of two systems, so that in general system S_i has temperature T_m. We order the temperatures always sequentially: $m = 1, 2, \ldots, M$, but the sequence $\{i\} = i(1), i(2), \ldots, i(M)$ of the systems is a *permuta-*

---

[39] We use the canonical probability in configuration space only, not in the full phase space.

tion of the sequence $1, 2, \ldots, M$. A *state* X in the generalized ensemble consists of M configurations $\boldsymbol{r}_{i(1)}, \boldsymbol{r}_{i(2)}, \ldots, \boldsymbol{r}_{i(M)}$, with potential energies $E_{i(1)}, E_{i(2)}, \ldots, E_{i(M)}$. Because there are no interactions between the systems and each temperature occurs exactly once, the probability of this state is

$$w \propto \exp[-(\beta_1 E_{i(1)} + \beta_2 E_{i(2)} + \cdots + \beta_M E_{i(M)})]. \tag{6.136}$$

Now consider a possible exchange between two systems, e.g., systems S_i at β_m and S_j at β_n. After the exchange, system S_i will be at β_n and S_j will be at β_m. The probabilities before and after the exchange must be, respectively,

$$w_{\text{before}} \propto \exp[-(\beta_m E_i + \beta_n E_j)], \tag{6.137}$$

and

$$w_{\text{after}} \propto \exp[-(\beta_n E_i + \beta_m E_j)]. \tag{6.138}$$

Their ratio is

$$\frac{w_{\text{after}}}{w_{\text{before}}} = e^{-\Delta}, \tag{6.139}$$

where

$$\Delta = (\beta_n - \beta_m)(E_i - E_j). \tag{6.140}$$

The transition probabilities $W_{\rightarrow}$ (meaning from "before" to "after") and $W_{\leftarrow}$ (meaning from "after" to "before") must fulfill the detailed balance condition:

$$w_{\text{before}} W_{\rightarrow} = w_{\text{after}} W_{\leftarrow}. \tag{6.141}$$

Thus it follows that

$$\frac{W_{\rightarrow}}{W_{\leftarrow}} = e^{-\Delta}. \tag{6.142}$$

This is accomplished by the Metropolis acceptance criterion:

$$\text{for} \quad \Delta \leq 0: \quad W_{\rightarrow} = 1 \tag{6.143}$$
$$\text{for} \quad \Delta > 0: \quad W_{\rightarrow} = e^{-\Delta}, \tag{6.144}$$

as is easily seen by considering the backward transition probability:

$$\text{for} \quad \Delta < 0: \quad W_{\leftarrow} = e^{\Delta} \tag{6.145}$$
$$\text{for} \quad \Delta \geq 0: \quad W_{\leftarrow} = 1, \tag{6.146}$$

which fulfills (6.142).

Although exchange between any pair may be attempted, in practice only neighboring temperatures yield non-zero acceptance ratios and the exchange attempt can be limited to neighbors. An acceptance ratio of 20% is considered reasonable. One should choose the set of temperatures such that the acceptance ratio is more or less uniform over the full temperature range. This will depend on the system; Sugita and Okamoto (1999) find for a peptide that an exponential distribution (equal ratios) is satisfactory. They use eight temperatures between 200 and 700 K, but a higher number (10 to 20) is recommended.

The exchange, once accepted, can be accomplished in various ways. In Monte Carlo simulations, one exchanges both the random step sizes and the β's in the acceptance criterion. In dynamics, the least disturbing implementation is to use a weak-coupling or a Langevin thermostat and switch the reference temperatures of the thermostats. The simulation should then extend over several time constants of the thermostat before another exchange is attempted. Using an isokinetic thermostat, the velocities should be scaled proportional to the square root of the temperature ratio upon exchange. In principle, it is also possible to switch not only the thermostats, but also all velocities of the particles between the two systems. This will drastically break up the time correlation of concerted motions; it is not clear whether this is advantageous or disadvantageous for the sampling efficiency.

6.7 Applications of molecular dynamics

Molecular dynamics simulations with atomic detail can be routinely applied to systems containing up to a million particles over several nanoseconds. The time step for stable simulations is determined by the highest frequencies in the system; as a rule of thumb one may assume that at least ten, but preferably 50 time steps should be taken within the shortest period of oscillation.[40] If the system contains mobile hydrogen atoms, bond oscillation periods may be as short as 10 fs; bond vibration involving heavy atoms typically exceed 20 fs. When covalent bonds are constrained, the highest frequencies are rotational and librational modes that involve hydrogen; dihedral angle rotations involving hydroxyl groups have similar periods. For example, in liquid water librational frequencies up to 800 cm^{-1} (40 fs) occur. A usual time step for simulations of systems containing liquid water is 1 to 2 fs when internal bond constraints are imposed; with further restrictions of hydrogen motions, "hydrogen-rich" systems as hydrated proteins remain stable with time steps up to 7 fs (Feenstra *et al.*, 1999). In 2005, simulation of a typical medium-sized protein (lysozyme) in water, totalling some 30 000

---

[40] See Berendsen and van Gunsteren (1981). Mazur (1997) concludes that even less than 10 steps per period suffice for the leap-frog algorithm.

atoms, reached about 1 ns per day on a single state-of-the-art processor (van der Spoel *et al.*, 2005), and this performance is expected to increase by a factor of ten every five years, according to Murphy's law, which has been followed quite closely over the last decades. The availability of massively parallel clusters of processors allows simulation of much larger system sizes and much longer time scales. Proteins can be followed over microseconds, which is not yet sufficient to simulate realistic folding processes and reliably predict protein structures from sequence data. With the near future petaflop computers, the protein folding problem, which has been called the Holy Grail of biophysics (Berendsen, 1998), is likely to be solved. In material science, micron-size solids simulated for microseconds will become a reality.

Figure 6.13 shows several snapshots of a simulation involving more than a billion (10^9) particles by Abraham *et al.* (2002).[41] The simulated system is a crystal of Lennard–Jones particles, modeled to mimic a copper crystal, which is subjected to external tension forces that cause a *crack* to increase in size. The purpose of this simulation is to investigate the formation and propagation of dislocations that characterize the crack and model the process of work-hardening of metals. The system is a slab with 1008 atoms along the three orthogonal sides. Two notches are centered midway along the x-direction, at $y = 0$ and $y = L_y$, with a y-extension of 90 atomic layers which extends through the entire thickness L_z. The exposed notch faces are in the $y - z$ planes with (110) faces, and the notch is pointed in the $(1, -1, 0)$ direction. Periodic boundary conditions are imposed between the $x - y$ faces at $z = 0$ and $z = L_z$. This notched slab geometry has a total of 1 023 103 872 atoms. The total simulation time for this study is 200 000 time-steps or 2 ns. The slab is initialized at zero temperature, and an outward strain of 4% is imposed on the outermost columns of atoms defining the opposing vertical yz faces of the slab. The figures show only atoms with a potential energy less than 97 % of the bulk value magnitude.

In the figures, we see a spaghetti-like network of atomic strings flying from the vertices of the two opposing crack edges. This is simply a large number of mobile dislocations being created at each crack edge, rapidly flowing through the stretched solid in an erratic manner, and eventually colliding with intersecting dislocations from the opposite edge. For the simple face-centered-cubic solid, dislocations are easily created at the apex of the two microcracks where the stress is at a maximum and easily flow through

---

[41] The author is indebted to Dr Farid Abraham for providing the pictures in Fig. 6.13 and the accompanying description. The interested reader is referred to Abraham (2003) for an introductory text on cracks and defects, and ductile and brittle behavior of metals.

the solid giving rise to the ductility of the solid. The simulation supports the prevailing view that even though there may not be enough dislocations originally present in a crystal to account for the extensive slip in a ductile material (in this simulation there are initially no dislocations), their creation in vast amounts can occur at severe stress concentrations, such as at crack tips, enabling a stressed solid to be rapidly filled with dislocations and giving rise to material deformation under a steady load. The figures show snapshots of the propagating dislocations and rigid junctions evolving into a complex topology of the defect-solid landscape.

Colliding dislocations can cause permanent atomic relocation along a line, called a *rigid junction* or *sessile dislocation*. A coarse-grain three-dimensional skeleton of such sessile dislocations becomes apparent from a distant view. They are obstacles to further dislocation mobility. If their density is sufficiently high, dislocation mobility becomes insignificant, and ductility of the solid ceases. The solid no longer can deform through dislocation motion: the ductile solid becomes brittle through this work-hardening process. Thus the simulation, with an impressive billion atoms still representing only a very small solid of 0.3 μm size, gives detailed insight into the mechanisms of work-hardening. The dynamical time span is on the order of a few nanoseconds, enough time for the phenomenon to achieve a final structure state for the small size solid cube.

Exercises

6.1   Write out the operator product $\mathbf{U}(\Delta t) \approx \mathbf{U}_\mathrm{f}(\Delta t/2)\mathbf{U}_\mathrm{v}(\Delta t)\mathbf{U}_\mathrm{f}(\Delta t/2)$ to obtain the velocity-Verlet algorithm.

6.2   Obtain another algorithm by interchanging $\mathbf{U}_\mathrm{v}$ and $\mathbf{U}_\mathrm{f}$.

6.3   Solve (6.107).

6.4   Compute (in reduced units) the period of oscillation of two Lennard–Jones particles at the distance where their interaction energy is minimal. What would be an appropriate time step (in reduced units) for a leap-frog simulation of a Lennard–Jones fluid? How many fs is that time step for argon?

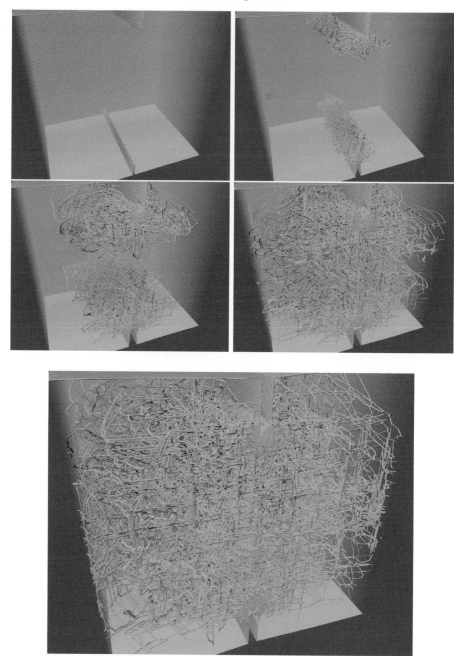

Figure 6.13 Five snapshots from a one-billion particle MD simulation of the propagation of a crack in a copper crystal under tension, showing the massive formation of dislocations. See text for details. Figures were kindly provided by Dr Farid Abraham of IBM Research and Lawrence Livermore National Laboratory, Livermore, CA, USA.

7

Free energy, entropy and potential of mean force

7.1 Introduction

As we know from the applications of thermodynamics, free energy is much more important than energy, since it determines phase equilibria, such as melting and boiling points and the pressure of saturated vapors, and chemical equilibria such as solubilities, binding or dissociation constants and conformational changes. Unfortunately, it is generally much more difficult to derive free energy differences from simulations than it is to derive energy differences. The reason for this is that free energy incorporates an entropic term $-TS$; entropy is given by an integral over phase space, while energy is an ensemble average. Only when the system is well localized in space (as a vibrating solid or a macromolecule with a well-defined average structure) is it possible to approximate the multidimensional integral for a direct determination of entropy. This case will be considered in Section 7.2.

Free energies of substates can be evaluated directly from completely equilibrated trajectories or ensembles that contain all accessible regions of configurational space. In practice it is hard to generate such complete ensembles when there are many low-lying states separated by barriers, but the ideal distribution may be approached by the *replica exchange method* (see Section 6.6). Once the configurational space has been subdivided into substates or *conformations* (possibly based on a cluster analysis of structures), the free energy of each substate is determined by the number of configurations observed in each substate. One may also observe the *density* of configurations along a defined parameter (often called an *order parameter* or a *reaction coordinate*, which is a function of the coordinates) and derive the potential of mean force along that parameter. When a replica-exchange method has been used, the free energies of substates are obtained simultaneously for a range of temperatures, providing energies, entropies and specific heats.

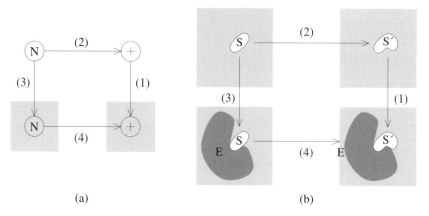

(a)   (b)

Figure 7.1 Two thermodynamic cycles, allowing the replacement of one path by the sum of three other paths (see text). (a) Hydration of an ion (+) through the intermediate of a neutral atom (N), (b) binding of a substrate S′ to an enzyme E (dark gray), compared to the binding of another, similar, substrate S. Light gray squares represent an aqueous environment.

In general we cannot evaluate highly multidimensional integrals and all we have is a trajectory or ensemble in phase space or in configurational space. Then we must apply tricks to derive free energy differences from ensemble averages.

This chapter is mostly about such tricks. In Section 7.3 Widom's *particle insertion method* is treated, which relates the ensemble average of the Boltzmann factor of an inserted particle to its thermodynamic potential. Section 7.4 is about perturbation and integration methods that relate the ensemble average of the Boltzmann factor of a small perturbation to the difference in free energy, or the ensemble average of a Hamiltonian derivative to the derivative of the free energy. In Section 7.5 we first make a clear distinction between two kinds of free energy: free energy of a thermodynamic state, and free energy in a restricted space as a function of one or more *reaction coordinates*. The latter is called a "potential of mean force," or PMF, because its derivative is the ensemble-averaged force. The PMF is indispensable for simulations in reduced dimensionality (Chapter 8), as it provides the systematic forces in such cases. In Section 7.6 the connection between PMF and free energy is made and Section 7.7 lists a number of methods to determine potentials of mean force. Finally, Section 7.8 considers the determination of free energy differences from the work done in non-equilibrium pathways.

Practical questions always concern free energy *differences* between thermodynamic states. So we need to obtain free energy differences from simulations. But, by the use of *thermodynamic cycles*, the computed pathways and

intermediates may differ from the experimental ones as long as the end results match. Thus one may choose even physically unrealistic or impossible
intermediates to arrive at the required free energy differences. For example,
if we wish to compute the free energy of hydration of a sodium ion, the
interest lies in the difference in standard free energy ΔG_1^0 of process (1) below.[1] Here, (g) means ideal gas referred to standard concentration and (aq)
means infinitely dilute aqueous solution referred to standard concentration.
For all processes, constant pressure and temperature are assumed. Now,
process (1) can be built up from processes (2), (3) and (4), with a neutral,
sodium-like atom N as intermediate (see Fig. 7.1a). Here N may be any
convenient intermediate, e.g., a repulsive cavity with a size close to that of
a sodium ion.

$$
\begin{array}{cccc}
(1) & \mathrm{Na^+(g)} & \rightarrow & \mathrm{Na^+(aq)} & \Delta G_1^0 \\
(2) & \mathrm{N(g)} & \rightarrow & \mathrm{Na^+(g)} & \Delta G_2^0 \\
(3) & \mathrm{N(g)} & \rightarrow & \mathrm{N(aq)} & \Delta G_3^0 \\
(4) & \mathrm{N(aq)} & \rightarrow & \mathrm{Na^+(aq)} & \Delta G_4^0
\end{array}
$$

Going clockwise around the thermodynamic cycle, the total free energy
change must be zero:

$$
\Delta G_1^0 - \Delta G_4^0 - \Delta G_3^0 + \Delta G_2^0 = 0. \tag{7.1}
$$

Therefore, ΔG_1^0 can be determined from three different processes, which are
each simpler and more efficient to compute. The intermediate N can be
chosen to optimize computational efficiency, but it can also be chosen to
provide a single intermediate for the hydration of many different ions.

Process (1) in Fig. 7.1b represents the binding of a substrate S' to an enzyme E. ΔG_1^0 of this process yields the equilibrium binding constant $K_{ES'}$.
The direct determination of ΔG_1^0 by simulation is difficult, but the thermodynamic cycle allows to determine the binding constant of S' relative to
another similar substrate S, for which the binding constant K_{ES} of process
(3) is known, by two simple processes (2) and (4).

7.2 Free energy determination by spatial integration

Consider a system of N particles that are on the average situated at positions
$\langle r_i \rangle$, $i = 1, \ldots, N$, and fluctuate randomly about those positions. Such systems are called *non-diffusive* when the mean-squared fluctuations are finite

---

[1] This is a simplified model; even process (1) is not a physically realistic process. It must be
considered as part of a larger cycle involving a negative ion as well, as realistic thermodynamic
experiments require electroneutral species.

and stationary.[2] Without loss of generality we may assume that the system as a whole has no translational and rotational motion and only consider the intra-system free energy. If the system is freely translating and rotating, as a macromolecule in dilute solution, there will be ideal-gas translational and rotational contributions to the total free energy (see Section 17.5.3 on page 468); the partition function (p.f.) is the product of a translational p.f (17.75), a rotational p.f. (17.82) and an internal partition function $Q_{\rm int}$. If the system is a non-translating, non-rotating crystal, the total free energy will consist of the free energy due to lattice vibrations and the internal free energy, assuming that the coupling between the two is negligible. What interests us here is the computation of the internal free energy, based on an ensemble of configurations in a system-fixed cartesian coordinate system, obtained by a proper molecular dynamics or Monte Carlo procedure. More-over, as the interest lies in the relative stability of different conformations (or clusters of configurations), we are not interested in the absolute value of free energies or entropies. Kinetic contributions will cancel out in the difference between conformations.

The determination of entropy directly from simulations was first discussed by Karplus and Kushick (1981). Applications were published – among oth-ers – by Edholm *et al.* (1983) on the entropy of bilayer membranes and by DiNola *et al.*(1984) on the entropy differences between macromolecular conformations. The method has been evaluated with an error analysis by Edholm and Berendsen (1984) and the systematic errors on incomplete equi-libration were discussed by Berendsen (1991a). A major improvement has been proposed by Schlitter (1993). The method has not become a standard: it is not applicable to diffusive systems, it cannot be easily applied to macro-molecules in solution and considerable computational effort is required for slowly relaxing systems.

The essential contribution to the free energy that depends on a multi-dimensional integral is the *configurational entropy*:

$$S_{\rm conf} = {\rm const} - k_{\rm B} \int w(\boldsymbol{q}) \ln w(\boldsymbol{q}) \, d\boldsymbol{q}, \qquad (7.2)$$

where $\boldsymbol{q}$ is the set of generalized coordinates that exclude translational and rotational degrees of freedom (and also constrained degrees of freedom if applicable) and $w(\boldsymbol{q})$ is the joint probability distribution for all the remaining $n = 3N - 6 - n_c$ coordinates. The constant in this equation comes from integration over the kinetic degrees of freedom (the conjugated momenta),

---

[2] This is in contrast to *diffusive systems*, such as liquids, in which the mean-squared fluctuations increase with – usually proportional to – time.

but also contains a term due to the mass tensor (see page 401) that may depend on $\boldsymbol{q}$. The influence of the mass tensor on conformational differences is often negligible and usually neglected or disregarded.

Equation 7.2 is still an unsolvable multidimensional integral. But in non-diffusive systems it is possible to derive the most relevant information on the multidimensional distribution. For example, we can construct the *correlation matrix* of the coordinate fluctuations:

$$C_{ij} = \langle (q_i - \langle q_i \rangle)(q_j - \langle q_j \rangle) \rangle, \tag{7.3}$$

or

$$\mathbf{C} = \langle (\Delta \mathbf{q})(\Delta \mathbf{q})^{\mathsf{T}} \rangle, \tag{7.4}$$

where $i, j = 1, \ldots, n$ run over the generalized coordinates and $\Delta \mathbf{q} = \mathbf{q} - \langle \mathbf{q} \rangle$. It is generally not possible to assess higher correlations with any accuracy. If only the matrix of fluctuations $\mathbf{C}$ is known, one can estimate the *maximum entropy* of any multidimensional distribution with a given $\mathbf{C}$. By maximizing S_{conf} with respect to w under the conditions:

$$\int w(\boldsymbol{q}) \, d\boldsymbol{q} = 1, \tag{7.5}$$

$$\int w(\boldsymbol{q}) \Delta q_i \Delta q_j \, d\boldsymbol{q} = C_{ij}, \tag{7.6}$$

using Lagrange multipliers (see page 456), it appears that $w(\boldsymbol{q})$ must be a multivariate Gaussian distribution in $\boldsymbol{q}$:

$$w(\boldsymbol{q}) = (2\pi)^{-n/2} (\det \mathbf{C})^{-1/2} \exp[-\frac{1}{2}\Delta \mathbf{q}^{\mathsf{T}} \mathbf{C}^{-1} \Delta \mathbf{q}]. \tag{7.7}$$

The entropy of this distribution is

$$S_{\text{max}} = \frac{1}{2}k_{\text{B}}Tn(1 + \ln 2\pi) + \frac{1}{2}k_{\text{B}}T\ln(\det \mathbf{C}). \tag{7.8}$$

Thus, if the distribution really is a multivariate Gaussian, S_{max} is the configurational entropy; for any other distribution S_{max} is an *upper bound* for the entropy of the distribution:

$$S_{\text{conf}} \leq S_{\text{max}}. \tag{7.9}$$

The constants in (7.8) are irrelevant and all we need is the determinant of the correlation matrix of the positional fluctuations. Generally it is possible to determine this entropy accurately; when equilibration is slow, the computed entropy tends to increase with the length of the simulation and approach the limit with a difference that is inversely proportional to the length of the simulation (DiNola *et al.*, 1984; Berendsen, 1991).

It is also possible to derive the entropy from a *principal component anal-ysis* of the positional fluctuations in cartesian coordinates. In that case translational and rotational degrees of freedom must have been constrained, which is most easily done by subjecting all configurations to a standard translational–rotational fit. The principal component analysis, often re-ferred to as "essential dynamics" (Amadei *et al.*, 1993), diagonalizes the correlation matrix of the positional fluctuations, thus producing a new set of collective coordinates with uncorrelated fluctuations. Each eigenvalue is proportional to the contribution of its corresponding collective degree of freedom to the total fluctuation. There are $6 + n_c$ zero (or very small) eigen-values corresponding to the translation, rotation and internal constraints; these should be omitted from the entropy calculation. The determinant of $\mathbf{C}$ in (7.8) is now the product of all remaining eigenvalues.

When there is more knowledge on the fluctuations than the positional correlation matrix, the value of the entropy can be refined. Each refinement on the basis of additional information will *decrease* the computed entropy. For example, the *marginal* distributions $w_i(q_i)$ over single coordinates[3] can usually be evaluated in more detail than just its variance. This is partic-ularly important for dihedral angle distributions that have more than one maximum and that deviate significantly from a Gaussian distribution. Di-hedral angles in alkane chains have three populated ranges corresponding to *trans*, *gauche*[−] and *gauche*[+] configurations. It is then possible to compute the configurational entropy for each degree of freedom from

$$S_{\text{marg}} = -k_B \int w_i(q_i) \ln w_i(q_i) \, dq_i \qquad (7.10)$$

and use that - after subtraction of the entropy of the marginal distribution had the latter been Gaussian with the same variance – as a correction to the entropy computed from the correlation matrix (Edholm and Berendsen, 1984). Another refinement on the basis of extra knowledge is to exploit an observed clustering of configurations in configurational space: each cluster may be considered as a different species and its entropy determined; the total entropy consists of a weighted average with an additional mixing term (see (7.55)).

This determination of the classical entropy on the basis of positional fluc-tuations has an important drawback: it computes in fact the classical en-tropy of a harmonic oscillator (h.o.), which is very wrong for high frequen-

---

[3] The *marginal* distribution of q_i is the full distribution integrated over all coordinates except q_i.

cies. For a one-dimensional classical h.o., the entropy is given by

$$S_{\text{cl}}^{\text{ho}} = k_B + k_B \ln \frac{k_B T}{\hbar \omega}, \qquad (7.11)$$

which has the unfortunate property to become negative and even go to $-\infty$ for large frequencies. Expressed in the variance $\langle x^2 \rangle$ the classical entropy is

$$S_{\text{cl}}^{\text{ho}} = k_B + \frac{1}{2} k_B \ln \left(\frac{k_B T}{\hbar^2} m \langle x^2 \rangle \right), \qquad (7.12)$$

which becomes negative for small fluctuations. This is entirely due to the neglect of quantum behavior. The effect is unphysical, as even a constraint with no freedom to move – which should physically have a zero contribution to the entropy – has a negative, infinite entropy. Equation (7.8) must be considered wrong: if any eigenvalue of the correlation matrix is zero, the determinant vanishes and no entropy calculation is possible.

The correct quantum expression for the h.o. is

$$S_{\text{qu}}^{\text{ho}} = -k_B \ln \left(1 - e^{-\xi} \right) + \frac{k_B \xi}{e^{\xi} - 1}, \qquad (7.13)$$

where $\xi = \hbar \omega / k_B T$. This can be expressed in terms of the *classical* variance, using $m\omega^2 \langle x^2 \rangle = k_B T$,

$$\xi = \frac{\hbar}{\sqrt{k_B T m \langle x^2 \rangle}}. \qquad (7.14)$$

The entropy now behaves as expected; it goes properly to zero when the fluctuation goes to zero. In the multidimensional case one can use (7.13) after diagonalizing the correlation matrix of *mass-weighted* positional fluctuations:

$$C'_{ij} = \langle (\Delta x'_i)(\Delta x'_j) \rangle, \qquad (7.15)$$

where

$$x'_i = x_i \sqrt{m_i}. \qquad (7.16)$$

Let us denote the eigenvalues of the $\mathbf{C}'$ matrix by λ_k. Each eigenvalue corresponds to an independent harmonic mode with

$$\xi_k = \frac{\hbar}{\sqrt{k_B T \lambda_k}} \qquad (7.17)$$

and each mode contributes the the total entropy according to (7.13). Now zero eigenvalues do not contribute to the entropy and do not cause the total entropy to diverge.

For the exact quantum case one can no longer express the entropy in terms of the determinant of the correlation matrix as in (7.8). However, by a clever invention of Schlitter (1993), there is an approximate, but good and efficient, solution. Equation (7.13) is well approximated by S':

$$S_{\text{qu}}^{\text{ho}} \leq S' = 0.5k_{\text{B}} \ln \left(1 + \frac{e^2}{\xi^2} \right),$$ (7.18)

yielding for the multidimensional case

$$S' = 0.5k_{\text{B}} \ln \Pi_k \left(1 + \frac{e^2 k_{\text{B}} T}{\hbar^2} \lambda_k \right).$$ (7.19)

Since the diagonal matrix $\boldsymbol{\lambda}$ is obtained from the mass-weighted correlation matrix $\mathbf{C}'$ by an orthogonal transformation, leaving the determinant invariant, (7.19) can be rewritten as

$$S' = 0.5k_{\text{B}} \ln \det \left(1 + \frac{e^2 k_{\text{B}} T}{\hbar^2} \mathbf{C}' \right).$$ (7.20)

This equation is a convenient and more accurate alternative to (7.8). Note that the mass-weighted positional fluctuations are needed.

In an interesting study, Schäfer *et al.* (2000) have applied the Schlitter version of the entropy calculation to systems where the validity of a maximum-entropy approach based on covariances is doubtful, such as an ideal gas, a Lennard–Jones fluid and a peptide in solution. As expected, the ideal gas results deviate appreciably from the exact value, but for the Lennard–Jones fluid the computed entropy comes close (within $\approx 5\%$) to the real entropy. For a β-heptapeptide in methanol solution, which shows reversible folding in the simulations, the configurational entropy of the peptide itself can be calculated on the basis of the positional fluctuations of the solute atoms. However, the contribution of the solvent that is missing in such calculations appears to be essential for a correct determination of free energy differences between conformational clusters.

7.3 Thermodynamic potentials and particle insertion

Thermodynamic potentials μ_i of molecular *species i* are very important to relate microscopic to macroscopic quantities. Equilibrium constants of reactions, including phase equilibria, partition coefficients of molecules between phases and binding and dissociation constants, are all expressed as changes in standard Gibbs free energies, which are composed of standard thermodynamic potentials of the participating molecules (see Section 16.7). How do thermodynamic potentials relate to free energies and potentials of mean force and how can they be computed from simulations?

The thermodynamic potential of a molecular species is defined as the derivative of the total free energy of a system of particles with respect to the number of moles n_i of the considered species (see (16.8) on page 428):

$$\mu_i = \left(\frac{\partial G}{\partial n_i}\right)_{p,T,n_{j\neq i}} = N_A\{G(p,T,N_i+1,N_j) - G(p,T,N_i,N_j)\}, \quad (7.21)$$

or, equivalently (see (16.27) on page 430),

$$\mu_i = \left(\frac{\partial A}{\partial n_i}\right)_{V,T,n_{j\neq i}} \approx N_A\{A(V,T,N_i+1,N_j) - A(V,T,N_i,N_j)\}, \quad (7.22)$$

where N_i is the number of particles of the i-th species, assumed to be large. The latter equation is the basis of the *particle insertion method* of Widom (1963) to determine the thermodynamic potential from simulations. The method is as ingenious as it is simple: place a *ghost particle* of species i in a random position and orientation in a simulated equilibrium configuration. It is immaterial how the configuration is obtained, but it is assumed that an equilibrium ensemble at a given temperature – and hence β – of configurations is available. Compute the ensemble average of the Boltzmann factor $\exp(-\beta V_{\text{int}})$, where V_{int} is the interaction energy of the ghost particle with the real particles in the system. For a homogeneous system ensemble averaging includes averaging over space. The ghost particle senses the real particles, but does not interact with them and does not influence the ensemble. Now the thermodynamic potential is given by

$$\mu^{\text{exc}} = -RT\ln\langle e^{-\beta V_{\text{int}}}\rangle, \quad (7.23)$$

where μ^{exc} is the *excess* thermodynamic potential, i.e., the difference between the thermodynamic potential of species i in the simulated ensemble and the thermodynamic potential of species i in the ideal gas phase at the same density. This results follows from (7.22) and $A = -k_BT\ln Q$, yielding

$$\mu = -RT\ln\frac{Q(N_i+1)}{Q(N_i)}. \quad (7.24)$$

With (7.48) we can write

$$\frac{Q(N_i+1)}{Q(N_i)} = \left(\frac{2\pi m_i k_B T}{h^2}\right)^3 \frac{1}{N_i+1} \frac{\int d\boldsymbol{r}_{\text{ghost}} \int d\boldsymbol{r}\, \exp[-\beta\{V(\boldsymbol{r}) + V_{\text{int}}\}]}{\int d\boldsymbol{r}\, \exp[-\beta V(\boldsymbol{r})]}, \quad (7.25)$$

where $\boldsymbol{r}$ stands for the coordinates of all real particles and $\boldsymbol{r}_{\text{ghost}}$ for the coordinates of the ghost particle. The ratio of integrals is the ensemble average of $\exp(-\beta V_{\text{int}})$, integrated over the volume. But since – in a homogeneous system – the ensemble average does not depend on the position of the ghost particle, this integration simply yields the volume V. Therefore:

$$\mu = -RT \ln \left[\left(\frac{2\pi m_i k_B T}{h^2} \right)^3 \frac{V}{N_i + 1} \right] - RT \ln \langle e^{-\beta V_{\text{int}}} \rangle. \tag{7.26}$$

The first term is the thermodynamic potential of the i-th species in an ideal gas of non-interacting particles at a density $(N_i + 1)/V$. For large N_i this density is to a good approximation equal to N_i/V. Equation (7.26) is also valid for small N_i; for example, when a water molecule is inserted into a fluid consisting of another species (e.g., hexane), $N_i = 0$ and the excess thermodynamic potential is obtained with respect to the thermodynamic potential of ideal-gas water at a density of one molecule in the entire volume.

For solutions we are interested in the *standard thermodynamic potential* and *activity coefficient* of the solute and – in some cases – the solvent. How are these obtained from particle insertion? On the basis of molar concentration c, the thermodynamic potential (16.55) is expressed as

$$\mu(c) = \mu_c^0 + RT \ln \left(\frac{\gamma_c c}{c^0} \right), \tag{7.27}$$

where c^0 is an agreed standard concentration (e.g., 1 molar) and μ_c^0 is defined by (16.58):

$$\mu_c^0 \stackrel{\text{def}}{=} \lim_{c \to 0} \left[\mu(c) - RT \ln \frac{c}{c^0} \right]. \tag{7.28}$$

This definition guarantees that the activity coefficient γ_c approaches the value 1 for infinitely dilute solutions. A single measurement of $\mu(c)$ at one concentration can never determine both μ_c^0 and γ_c. The standard thermodynamic potential requires a measurement at "infinite" dilution. A single solute molecule in a pure solvent can be considered as infinitely dilute since there is no interaction between solute particles. Thus, for inserting a single solute particle, the activity coefficient will be unity and the thermodynamic potential is given by

$$\mu = \mu_c^0(\text{solution}) + RT \ln \frac{c}{c^0}, \tag{7.29}$$

where

$$c = \frac{1}{N_A V}. \tag{7.30}$$

But since μ is also given by

$$\mu = \mu(\text{id.gas}, c) + \mu^{\text{exc}} \tag{7.31}$$

$$= \mu(\text{id.gas}, c^0) + RT \ln \frac{c}{c^0} + \mu^{\text{exc}}, \tag{7.32}$$

it follows that

$$\mu_c^0(\text{solution}) = \mu_c^0(\text{id.gas}) + \mu^{\text{exc}}. \tag{7.33}$$

Here, μ^{exc} is "measured" by particle insertion according to (7.23). The standard concentration for the solution and the ideal gas must be the same.

Note Thus far we have treated particle insertion into a canonical (N, V, T) ensemble, which yields a relation between the averaged Boltzmann factor and the Helmholtz free energy A, based on (7.22). This equation is not exact and not valid for small numbers N_i. However, (7.21) is exact, as $G = \sum n_i \mu_i$ under conditions of constant pressure and temperature (see (16.12) on page 428); this relation is valid because both p and T are intensive quantities. Such a relation does not exist for A. It is possible to make corrections to A, using the compressibility, but it is more elegant to use a (N, p, T) ensemble. The N, p, T average of the Boltzmann factor yields the thermodynamic potential exactly (see (17.33) and (17.31) on page 461).

The problem with the particle-insertion method is that in realistic dense fluids the insertion in a random position nearly always results in a high, repulsive, interaction energy and hence in a negligible Boltzmann factor. Even with computational tricks that avoid the full computation of all interaction energies it is very difficult to obtain good statistics on the ensemble average. A way out is to insert a *smaller* particle and – in a second step – let the particle grow to its full size and determine the change in free energy by thermodynamic integration (see next section). In cases where the difference in thermodynamic standard potential between two coexisting phases is required – as for the determination of partition coefficients – a suitable method is to determine the potential of mean force over a path that leads from one phase into the other. The difference between the two plateau levels of the PMF in the two phases is also the difference in standard thermodynamic potential.

7.4 Free energy by perturbation and integration

Consider a potential function $V(\boldsymbol{r}, \lambda)$ with a parametric dependence on a *coupling parameter* $0 \leq \lambda \leq 1$ that modifies the interaction. The two extremes, $\lambda = 0$ and $\lambda = 1$, correspond to two different systems, A and B, respectively, with interaction functions $V_A(\boldsymbol{r})$ and $V_B(\boldsymbol{r})$:

$$V_A(\boldsymbol{r}) = V(\boldsymbol{r}, \lambda = 0), \tag{7.34}$$

$$V_B(\boldsymbol{r}) = V(\boldsymbol{r}, \lambda = 1). \tag{7.35}$$

For example, system A may consist of two neon atoms dissolved in 1000 water molecules, while system B consists of one sodium ion and one fluoride ion dissolved in 1000 water molecules. The parameter λ changes the neon–water interaction into an ion–water interaction, essentially switching on the Coulomb interactions. Or, system A may correspond to a protein with a bound ligand L_A in solution, while system B corresponds to the same protein in solution, but with a slightly modified ligand L_B. The end states A and B represent real physical systems, but intermediate states with a λ unequal to either zero or one are artificial constructs. The dependence on λ is not prescribed; it can be a simple linear relation like

$$V(\lambda) = (1 - \lambda)V_A + \lambda V_B, \tag{7.36}$$

or have a complex non-linear form. The essential features are that the potential is a continuous function of λ that satisfies (7.34) and (7.35).

Now consider the Helmholtz free energy of the system at a given value of λ:

$$A(\lambda) = -k_B T \ln \left[c \int e^{-\beta V(\mathbf{r}, \lambda)} \, d\mathbf{r} \right]. \tag{7.37}$$

It is impossible to compute this multidimensional integral from simulations. But it is possible to compute $A(\lambda + \Delta\lambda)$ as a *perturbation* from an *ensemble average*:[4]

$$A(\lambda + \Delta\lambda) - A(\lambda) = -k_B T \ln \frac{\int \exp[-\beta V(\mathbf{r}, \lambda + \Delta\lambda)] \, d\mathbf{r}}{\int \exp[-\beta V(\mathbf{r}, \lambda)] \, d\mathbf{r}} \tag{7.38}$$

$$= -k_B T \ln \left\langle e^{-\beta[V(\lambda + \Delta\lambda) - V(\lambda)]} \right\rangle_\lambda. \tag{7.39}$$

It is also possible to compute $dA/d\lambda$ from an *ensemble average*:

$$\frac{dA}{d\lambda} = \frac{\int \frac{\partial V}{\partial \lambda}(\mathbf{r}, \lambda) \exp[-\beta V(\mathbf{r}, \lambda)] \, d\mathbf{r}}{\int \exp[-\beta V(\mathbf{r}, \lambda)] \, d\mathbf{r}} = \left\langle \frac{\partial V}{\partial \lambda} \right\rangle_\lambda. \tag{7.40}$$

The averages must be taken over an equilibrium ensemble using $V(\lambda)$. Thus, if the λ-path from 0 to 1 is constructed from a number of intermediate points, then the total $\Delta A = A_B - A_A = A(\lambda = 1) - A(\lambda = 0)$ can be reconstructed from the ensemble averages at the intermediate points. In general the most convenient and accurate reconstruction is from the derivatives at intermediate points by integration with an appropriate numerical procedure (Press *et al.*, 1993), e.g., by computing a cubic spline with the given derivatives (see

---

[4] This equation was first given by Torrie and Valleau (1974) in the context of Monte Carlo simulations of
Lennard–Jones fluids. Pearlman and Kollman (1989a) have refined windowing techniques for thermodynamic integration.

Chapter 19).[5] This procedure to find differences in free energies is called *thermodynamic integration*. Integration can also be accomplished from a series of perturbations, using (7.39); in that case there may be systematic deviations if the interval is not very small and it is recommended to apply both positive and negative perturbations from each point and check the closure.

The derivatives or perturbations are only reliable when the ensemble, over which the derivative or perturbation is averaged, is a proper equilibrium ensemble. In slowly relaxing systems there may be remains of history from the previous λ-point and the integration may develop a systematic deviation. It is recommended to perform the thermodynamic integration from both directions, i.e., changing λ from 0 to 1 as well as from 1 to 0. Systematic deviations due to insufficient equilibration are expected to have opposite signs in both cases. So the obtained hysteresis is an indication of the equilibration error.

A limiting case of thermodynamic integration is the *slow-growth* method, in which λ is changed with a small increment $(\Delta\lambda)_i$ at the i-th step in a molecular dynamics simulation, starting at 0 and ending at 1 (or vice versa).[6] This increment may or may not be taken as a constant. Then the total change in free energy is approximated by

$$\Delta A = \sum_i \left(\frac{\partial V(\lambda)}{\partial \lambda} \right)_i (\Delta\lambda)_i. \tag{7.41}$$

Only in the limit of an infinitely slow change of λ a true free energy difference will be obtained; if the growth is too rapid, the ensemble will "lag behind" the proper equilibrium at the actual value of λ. This will lead to an easily detected hysteresis when slow-growth simulations in forward and backward directions are compared. The average between the results of a forward and backward integration is always more accurate than either value. Figure 7.2 gives an early example of hysteresis in a simulation that changes a model neon atom into a sodium ion in aqueous solution by charging the atom proportional to time (Straatsma and Berendsen, 1988). In this case the free energy appears to change quadratically with time and the ensemble appears to relax quickly. Pearlman and Kollman (1989b) and Wood (1991) have

---

[5] The standard error in the integral can be evaluated if the standard error σ_i in each ensemble average A'_i is known. Numerical integration yields an integral that can be expressed as $\Delta A = \sum w_i(A'_i \pm \sigma_i)$, where w_i are weights depending on the interval and the procedure used. The standard error in ΔA equals $\sqrt{\sum w_i^2 \sigma_i^2}$.

[6] The slow-growth method was pioneered by Postma (1985), see Berendsen *et al.* (1985), and applied – among others – by Straatsma and Berendsen (1988) and by Pearlman and Kollman (1989b).

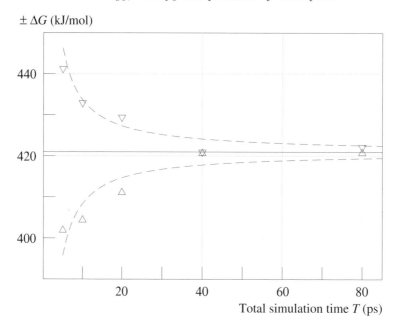

Figure 7.2 Free energy change resulting from transforming a model neon atom into a sodium ion in a bath of 216 water molecules, by linearly increasing the charge on the atom in a total simulation time T. Upward triangles are for growth from Ne to Na^+, yielding a negative ΔG; downward triangles are for the opposite change. Dashed curves are predictions from the theory of Wood (1991) assuming a relaxation time for the ensemble of 0.15 ps. Data are from Straatsma and Berendsen (1988).

analyzed the effects of the rate of slow-growth free energy determinations. At present slow growth is used less frequently than integration based on a number of well-equilibrated intermediate points because the latter allows a better evaluation and optimization of the overall accuracy.

Note A similar remark as was made in connection with particle insertion can be made here as well. In most applications one is interested in constant pressure rather than constant volume conditions. For example, in order to find equilibrium constants, one needs ΔG^0. Using N, V, T ensembles one may need corrections to connect the end points at constant pressure rather than constant volume. It is much more elegant to use N, p, T ensembles, with partition function Δ (see (17.31) on page 461) that relate directly to Gibbs free energies. Ensemble averages of Hamiltonian derivatives now yield derivatives of G rather than A.

There are several practical considerations concerning the method used for the integration of free energy from initial to final state. As computational integration is not limited to physically realistic systems (i.e., as long as the initial and final states *are* realistic), there is almost no bound to the phantasy that can be introduced into the methodology. The word "computational alchemy" is not misplaced, as one may choose to change lead into gold, be it that the gold must – unfortunately – be returned to lead before a realistic result is obtained. We list a few tricks and warnings.

- The free energy as a function of λ may not be well-behaved, so that numerical integration from a limited number of points becomes inaccurate. The density of points in the range $0 \leq \lambda \leq 1$ can be chosen to optimize the integration accuracy, but ideally $A(\lambda)$ should be a smooth function without steep derivatives, being well-represented by a polynomial of low order. One can manipulate the function by changing the functional dependence of the Hamiltonian on λ.

- Ideally the free energy curve should be monotonous; if a large intermediate maximum or minimum occurs, computational effort must be spent to compute compensating free-energy changes. A maximum may easily occur when there are highly repulsive configurations at intermediate values of λ's. Such repulsive intermediates can be avoided by choosing appropriately smoothed potentials.

- Replacing diverging functions as the repulsive r^{-12} or dispersion and Coulomb interactions by *soft-core interactions* for intermediate λ's removes the singularities and allows particles to move *through* each other rather than having to avoid each other.[7] The GROMACS software (van der Spoel *et al.*, 2005) uses a modification of the *distance* between particles of the form

$$V(r) = (1 - \lambda)V_A(r_A) + \lambda V_B(r_B), \tag{7.42}$$
$$r_A = (c\lambda^2 + r^6)^{1/6}, \tag{7.43}$$
$$r_B = [c(1 - \lambda)^2 + r^6]^{1/6}, \tag{7.44}$$

while Tappura *et al.* (2000) switch to a function $ar^6 + b$ below a specified (short) distance, with a and b such that the potential function and its derivative are continuous at the switch distance.

- Another non-physical intervention is to allow particles to move into a *fourth spatial dimension* for intermediate λ's (van Gunsteren *et al.*, 1993). Since there is much more space in four than in three dimensions, particles can easily avoid repulsive configurations. But they also loose their structural coherence and the motion in the fourth dimension must be carefully restrained. The method is more aesthetically appealing than it is practical.

---

[7] Soft-core potentials were originally developed for structure optimization and protein folding (Levitt, 1983; Huber *et al.*, 1997; Tappura *et al.*, 2000).

- There can be problems when particles "vanish" at either end point of the integration path. When the interactions of an atom with its environment are made to vanish, the particle is still there in the simulation as an ideal gas atom. It has mass and velocity, but is uncoupled to other degrees of freedom and therefore does not equilibrate properly. Problems are avoided by constraining the vanishing particle to a fixed position where it has neither a kinetic energy nor a configurational entropy and does not contribute to the free energy. One should take care that the λ-dependence is not diverging near the value of λ where the particle vanishes. Strictly speaking, one should also correct for the vanishing kinetic term $(2\pi m k_{\mathrm{B}} T/h^2)^{-1/2}$ in the free energy, but that term will always be compensated when a complete, physically realistic, cycle is completed.
- When particles are changed into other particles with different mass, the free energy change has a different kinetic term. It is possible to change the masses also with a coupling parameter, but there is no need to do that, as – just as in the case of a vanishing particle – the kinetic effect will always be compensated when a complete, physically realistic, cycle is completed. Real free energy differences always concern the same number and type of particles on both sides of the reaction.
- Be careful when the coupling parameter involves a *constraint*.[8] For example, if one wishes to change a hydrogen atom in benzene into a methyl group (changing benzene into toluene), the carbon–particle distance will change from 0.110 to 0.152 nm. In a simulation with bond constraints, the constraint length is modified as a function of the coupling parameter. Each length modification in the presence of a constraint force involves a change in free energy, as work is done against (or with) the constraint force. So the work done by the constraint force must be monitored. The constraint force F_c follows from the procedure used to reset the constraints (see Section 15.8 on page 417); if the constraint distance r_c is changed by a small increment $\Delta r_c = (dr_c/d\lambda)\,\Delta\lambda$, the energy increases with $F_c\Delta r_c$. Thus there is a contribution from every bond length constraint to the ensemble average of $\partial V/\partial\lambda$:

$$\left\langle \left(\frac{\partial V}{\partial \lambda}\right)_{\mathrm{constr}} \right\rangle = \langle F_c\rangle \frac{dr_c}{d\lambda}. \tag{7.45}$$

In addition, there may be a contribution from the Jacobian of the transformation from cartesian to generalized coordinates, or – equivalently – from the mass-metric tensor (see Section 17.9.3 and specifically (17.199) on page 501). The extra weight factor $|Z|^{-1/2}$ in the constrained ensemble

---

[8] See van Gunsteren *et al.* (1993), pp 335–40.

may well be a function of λ and contribute a term in $dA/d\lambda$:

$$\left(\frac{dA}{d\lambda}\right)_{\text{metric}} = \frac{1}{2}k_BT\left\langle |Z|^{1/2}\frac{\partial |Z|}{\partial \lambda}\right\rangle. \qquad (7.46)$$

The same arguments that are given in Section 17.9.3 to show that the metric effects of constraints are often negligible (see page 502) are also valid for its λ-dependence. Even more so: in closed thermodynamic cycles the effect may cancel.

- A large improvement of the efficiency to compute free energy changes for many different end states (e.g., finding the binding constants to a protein for many compounds) can be obtained by using a *soft* intermediate (Liu *et al.*, 1996; Oostenbrink and van Gunsteren, 2003). Such an intermediate compound does not have to be physically realistic, but should be constructed such that it covers a broad part of configurational space and allows overlap with the many real compounds one is interested in. If well chosen, the change from this intermediate to the real compound may consist of a single perturbation step only.

7.5 Free energy and potentials of mean force

In this section the potential of mean force (PMF) will be defined and a few remarks will be made on the relations between PMF, free energy, and chemical potential. The potential of mean force is a *free energy* with respect to certain defined variables, which are functions of the particle coordinates, and which are in general indicated as *reaction coordinates* because they are often applied to describe reactions or transitions between different potential wells. What does that exactly mean, and what is the difference between a free energy and a potential of mean force? What is the relation of both to the chemical potential?

The potential energy as a function of all coordinates, often referred to as the *energy landscape*, has one global minimum, but can have a very complex structure with multiple local minima, separated by barriers of various heights. If the system has *ergodic* behavior, it visits in an equilibrium state at temperature T all regions of configuration space that have an energy within a range of the order of k_BT with respect to the global minimum. It is generally assumed that realistic systems with non-idealized potentials in principle have ergodic behavior,[9] but whether all relevant regions of config-

---

[9] Idealized systems may well be non-ergodic, e.g., an isolated system of coupled harmonic oscillators will remain forever in the combination of eigenstates that make up its initial configuration and velocities; it will never undergo any transitions to originally unoccupied eigenstates unless there are external disturbances or non-harmonic terms in the interaction function.

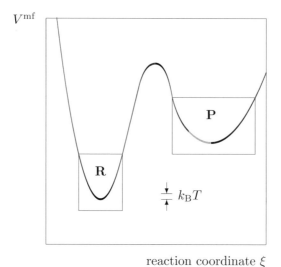

reaction coordinate ξ

Figure 7.3 Potential of mean force in one "reaction coordinate" ξ. There are two regions of configurational space ($\mathbf{R}$ and $\mathbf{P}$) that can be designated as confining a thermodynamic state.

uration space will indeed be accessed in the course of the observation time is another matter. If the barriers between different local minima are large, and the observation time is limited, the system may easily remain trapped in certain regions of configuration space. Notable examples are metastable systems (as a mixture of hydrogen and oxygen), or polymers below the melting temperature or glasses below the glass transition temperature. Thus ergodicity becomes an academic problem, and the *thermodynamic state* of the system is defined by the region of configurational space actually visited in the observation time considered.

Consider a system that can undergo a slow reversible reaction, and in the observation time is either in the *reactant state* R or in the *product state* P. In the complete multidimensional energy landscape there are two local minima, one for the R and one for the P state, separated by a barrier large enough to observe each state as metastable equilibrium. Let the local minima be given by the potentials V_0^R and V_0^P. Then the Helmholtz free energy A (see Chapter 17) is given – for classical systems in cartesian coordinates – by

$$A = -k_B T \ln Q, \tag{7.47}$$

with

$$Q = c \int e^{-\beta V(\boldsymbol{r})} \, d\boldsymbol{r}, \tag{7.48}$$

where the integration is carried out for all particle coordinates over all space. The constant c derives from integration over momenta, which for N particles consisting of species s of N_s indistinguishable particles (with mass m_s) equals

$$c = \left(\frac{2\pi k_{\mathrm{B}} T}{h^2} \right)^{3N/2} \Pi_s \frac{m_s^{3N_s/2}}{N_s!}. \tag{7.49}$$

Expressed in de Broglie wavelengths Λ_s for species s:

$$\Lambda_s = \frac{h}{\sqrt{2\pi m_s k_{\mathrm{B}} T}}, \tag{7.50}$$

and using Stirling's approximation for $N_s!$, the constant c becomes

$$c = \Pi_s \left(\frac{e}{N_s \Lambda_s^3} \right)^{N_s}. \tag{7.51}$$

Note that c has the dimension of an inverse volume in $3N$-dimensional space V^{-N}, and the integral in (7.48) has the dimension of a volume to the power N. Thus, taking logarithms, we cannot split Q in c and an integral without loosing the metric independence of the parts. It is irrelevant what zero point is chosen to express the potential; addition of an arbitrary value V_0 to the potential will result in multiplying Q with a factor $\exp(-\beta V_0)$, and adding V_0 to A.

When each of the states R and P have long life times, and have *local* ergodic behavior, they can be considered as separate thermodynamic states, with Helmholtz free energies

$$A^{\mathrm{R}} = -k_{\mathrm{B}} T \ln Q^{\mathrm{R}} \qquad Q^{\mathrm{R}} = c \int_{\mathrm{R}} e^{-\beta V(\boldsymbol{r})} \, d\boldsymbol{r} \tag{7.52}$$

$$A^{\mathrm{P}} = -k_{\mathrm{B}} T \ln Q^{\mathrm{P}} \qquad Q^{\mathrm{P}} = c \int_{\mathrm{P}} e^{-\beta V(\boldsymbol{r})} \, d\boldsymbol{r}, \tag{7.53}$$

where the integrations are now carried out over the parts of configuration space defined as the R and P regions, respectively. We may assume that these regions encompass all local minima and that the integration over space outside the R and P regions does not contribute significantly to the overall Q.

We immediately see that, although $Q = Q^{\mathrm{R}} + Q^{\mathrm{P}}$, $A \neq A^{\mathrm{R}} + A^{\mathrm{P}}$. Instead, defining the relative probabilities to be in the R and P state, respectively, as w^R and w^P:

$$w^R = \frac{Q^R}{Q} \quad \text{and} \quad w^Q = \frac{Q^P}{Q}, \tag{7.54}$$

it is straightforward to show that

$$A = w^R A^R + w^P A^P + k_B T(w^R \ln w^R + w^P \ln w^P). \tag{7.55}$$

The latter term is due to the mixing entropy resulting from the distribution of the system over two states. Note that the zero point for the energy must be the same for both R and P.

Now define a *reaction coordinate* $\xi(\mathbf{r})$ as a function of particle coordinates, chosen in such a way that it connects the R and P regions of configuration space. There are many choices, and in general ξ may be a complicated nonlinear function of coordinates. For example, a reaction coordinate that will describe the transfer of a proton over a hydrogen bond X-H$\cdots$Y may be defined as $\xi = r_{XH}/r_{XY}$; ξ will encompass the R state around a value of 0.3 and the P state around 0.7. One may also choose several reaction coordinates that make up a reduced configuration space; thus ξ becomes a multidimensional vector. Only in rare cases can we define the relevant degrees of freedom as a subset of cartesian particle coordinates.

We first separate integration over the reaction coordinate from the integral in Q:

$$Q = c \int d\xi \int d\mathbf{r}\, e^{-\beta V(\mathbf{r})} \delta(\xi(\mathbf{r}) - \xi). \tag{7.56}$$

Here $\xi(\mathbf{r})$ is a function of $\mathbf{r}$ defining the reaction coordinate, while ξ is a value of the reaction coordinate (here the integration variable).[10] In the case of multidimensional reaction coordinate spaces, the delta-function should be replaced by a product of delta-functions for each of the reaction coordinates. Now define the potential of mean force $V^{mf}(\xi)$ as

$$V^{mf}(\xi) \overset{\text{def}}{=} -k_B T \ln \left[c \int d\mathbf{r}\, e^{-\beta V(\mathbf{r})} \delta(\xi(\mathbf{r}) - \xi) \right], \tag{7.57}$$

so that

$$Q = \int e^{-\beta V^{mf}(\xi)} \, d\xi, \tag{7.58}$$

and

$$A = -k_B T \ln \left[\int e^{-\beta V^{mf}(\xi)} \, d\xi \right]. \tag{7.59}$$

---

[10] Use of the same notation ξ for both the function and the variable gives no confusion as long as we write the function explicitly with its argument.

Note that the potential of mean force is an integral over multidimensional hyperspace. Note also that the integral in (7.57) is not dimensionless and therefore the PMF depends on the choice of the unit of length. After integration, as in (7.58), this dependency vanishes again. Such inconsistencies can be avoided by scaling both components with respect to a standard multidimensional volume, but we rather omit such complicating factors and always keep in mind that the absolute value of PMFs have no meaning without specifying the underlying metric.

It is generally not possible to evaluate such integrals from simulations. The only tractable cases are homogeneous distributions (ideal gases) and distribution functions that can be approximated by (multivariate) Gaussian distributions (harmonic potentials). As we shall see, however, it will be possible to evaluate *derivatives* of V^{mf} from ensemble averages. Therefore, we shall be able to compute V^{mf} by integration over multiple simulation results, up to an unknown additive constant.

7.6 Reconstruction of free energy from PMF

Once the PMF is known, the Helmholtz free energy of a thermodynamic state can be computed from (7.59) by integration over the relevant part of the reaction coordinate. Thus the PMF is a free energy for the system excluding the reaction coordinates as degrees of freedom. In the following we consider a few practical examples: the harmonic case, both one- and multidimensional and including quantum effects; reconstruction from observed probability densities with dihedral angle distributions as example; the PMF between two particles in a liquid and its relation to the pair distribution function; the relation between the partition coefficient of a solute in two immiscible liquids to the PMF.

7.6.1 Harmonic wells

Consider the simple example of a PMF that is quadratic in the (single) reaction coordinate in the region of interest, e.g in the reactant region R (as sketched in Fig. 7.3):

$$V^{\mathrm{mf}} \approx V_0^{\mathrm{mf}} + \frac{1}{2}k^{\mathrm{R}}\xi^2. \tag{7.60}$$

Then the Helmholtz free energy of the reactant state is given by integration

$$A^{\mathrm{R}} \approx -k_{\mathrm{B}}T \ln\left[\int_{-\infty+\infty} e^{-\beta V^{\mathrm{mf}}(\xi)}\, d\xi\right] = V_0^{\mathrm{mf}} + \frac{1}{2}k_{\mathrm{B}}T \ln \frac{k^{\mathrm{R}}}{2\pi k_{\mathrm{B}}T}. \tag{7.61}$$

Beware that the term under the logarithm is not dimensionless, but that the metric dependence is compensated in V_0^{mf}. We see that A becomes lower when the force constant decreases; the potential well then is broader and the entropy increases.

In the multidimensional harmonic case the PMF is given by a quadratic term involving a symmetric matrix $\mathbf{K}^{\mathrm{R}}$ of force constants, which is equal to the Hessian of the potential well, i.e., the matrix of second derivatives:

$$V^{\mathrm{mf}} \approx V_0^{\mathrm{mf}} + \tfrac{1}{2}\xi^{\mathsf{T}}\mathbf{K}^{\mathrm{R}}\xi. \tag{7.62}$$

Integration according to (7.59) now involves first an orthogonal transformation to diagonalize the matrix, which yields a product of one-dimensional integrals; carrying out the integrations yields a product of eigenvalues of the matrix, which equals the determinant of the diagonalized matrix. But the determinant of a matrix does not change under orthogonal transformations and we obtain

$$A^{\mathrm{R}} \approx V_0^{\mathrm{mf}} + \frac{1}{2}k_{\mathrm{B}}T\ln\frac{\det\mathbf{K}^{\mathrm{R}}}{2\pi k_{\mathrm{B}}T}. \tag{7.63}$$

Thus far we have considered the system to behave classically. However, we know that particles in harmonic wells (especially protons!), as they occur in molecular systems at ordinary temperature, are not at all close to the classical limit and often even reside in the quantum ground state. The classical expressions for the free energy are very wrong in such cases. The PMF well itself is generally determined from simulations or computations with constrained reaction coordinates in which the quantum character of the motion in the reaction coordinate does not appear. It is therefore relevant to ask what quantum effects can be expected in the reconstruction of free energies from harmonic PMF wells.

Quantum corrections to harmonic oscillator free energies can be easily made, if the *frequencies* of the normal modes are known (see Chapter 3, Section 3.5.4 on page 74). The problem with PMFs is that they do not represent pure Hamiltonian potentials in which particles move, and since the reaction coordinates are generalized coordinates which are (in general) non-linear functions of the particle coordinates, the effective masses (or a mass tensor in multidimensional cases) are complicated functions of the coordinates. Instead of computing such effective masses, the frequencies of the normal modes can much more easily be determined from a relatively short MD run with full detail in the potential well. Monitoring and Fourier-transforming the velocities $\dot{\xi}(t)$ of the reaction coordinates will reveal the eigenfrequencies of the motion of the reaction coordinates in the well of the

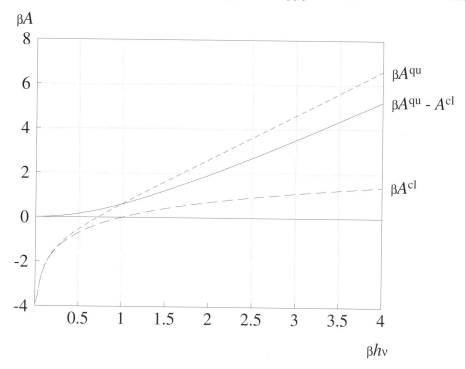

Figure 7.4 Helmholtz free energies divided by k_BT for a single harmonic oscillator, as a function of $h\nu/k_BT$, for both classical and quantum-mechanical statistics. The drawn line gives the quantum correction to a classical free energy.

PMF without bothering about the effective masses of the resulting motion. There are as many eigenfrequencies (but they may be degenerate) as there are independent reaction coordinates. According to quantum statistics (see Chapter 17), each eigenfrequency ν leads to a contribution to the free energy of

$$A_\nu^{\mathrm{qu}} = k_BT \ln\left[\frac{1}{2}\sinh\left(\frac{1}{2}\beta h\nu\right)\right],\tag{7.64}$$

which is to be compared to the classical contribution

$$A_\nu^{\mathrm{cl}} = k_BT\ln(\beta h\nu).\tag{7.65}$$

One may use the difference to correct the classical free energy determination, and – from temperature derivatives – the enthalpy and entropy (see Fig. 7.4).

7.7 Methods to derive the potential of mean force

In general a potential of mean force $V^{mf}(r')$ describes the effective potential that determines the motion of coordinates r' in a *reduced system*, averaged over an equilibrium ensemble of the other coordinates r''. In Chapter 8 the use of potentials of mean force in reduced systems is treated in detail. For simplicity we write r', r'' as a subdivision of cartesian space, but often the reduced system is described by a set of generalized coordinates. In this section we look at methods to derive potentials of mean force, which may then be useful for implementation in the reduced systems dynamics of Chapter 8.

In most cases of interest (containing improbable areas of the primed space) it is impossible to determine V^{mf} directly from an equilibrium simulation of the whole system. If it were, there would not be much point in reducing the number of degrees of freedom in the first place. The following possibilities are open to derive a suitable potential of mean force:

- From a *macroscopic (generally a mean-field) theory*. For example, if we wish to treat a solvent as "irrelevant," its influence on the electrostatic interactions of charges within the "relevant" particles and on the electrostatic contribution to the solvation free energy of (partially) charged particles, can be computed from electrostatic continuum theory (see Section 13.7). This requires solving the Poisson equation (or the Poisson–Boltzmann equation) with a finite-difference or Fourier method on a grid or with a boundary-element method on a triangulated surface. A computationally less demanding approximation is the *generalized Born model* (see Section 13.7.5 on page 351). Since such a treatment cannot be accurate on the atomic scale and misses non-electrostatic contributions, the electrostatic potential of mean force must be augmented by local interaction terms depending on the chemical nature and the surface accessibility of the primed particles. Another example is the treatment of all particles outside a defined boundary as "irrelevant". If the boundary of the primed system is taken to be spherical, the electrostatic terms may be represented by a reaction field that is much simpler to compute than the Poisson equation for an irregular surface (see Section 13.7.4).
- By *thermodynamic integration*. It is possible to obtain the derivative(s) of the potential of mean force at a given configuration of r' by performing a constrained equilibrium simulation of the full system and averaging the constraint forces (which are easily obtained from the simulation) over the double-primed ensemble. By performing a sufficient number of such simulations at strategically chosen configurations of the primed particles,

the potential of mean force can be obtained from numerical integration of the average constraint forces. This method is only feasible for a few (one to three) dimensions of the primed degrees of freedom, because the number of points, and hence full simulations, that is needed to reconstruct a V^{mf} surface in n dimensions increases with the number of points in one dimension (say, 10) to the power n. By taking the gradient of (8.11), we find that

$$
\begin{aligned}
\frac{\partial V^{\mathrm{mf}}(\boldsymbol{r}')}{\partial r_i'} &= \frac{\int \frac{\partial V(\boldsymbol{r}',\boldsymbol{r}'')}{\partial r_i'} e^{-\beta V(\boldsymbol{r}',\boldsymbol{r}'')}\, d\boldsymbol{r}''}{\int e^{-\beta V(\boldsymbol{r}'\boldsymbol{r}'')}\, d\boldsymbol{r}''} \\
&= \left\langle \frac{\partial V(\boldsymbol{r}',\boldsymbol{r}'')}{\partial r_i'} \right\rangle_{''} \\
&= \langle F_i^c \rangle.
\end{aligned}
\tag{7.66}
$$

The second line in the above equation gives – except for the sign – the average over the constrained ensemble of the internal force acting on the i-th primed particle; in a constrained simulation this force is exactly balanced by the constraint force $\boldsymbol{F}_i^c$ on that particle. These equations are modified for generalized coordinates (see den Otter and Briels, 1998; Sprik and Ciccotti, 1998; den Otter, 2000).

- By *thermodynamic perturbation*. Instead of averaging the derivatives of the potential, we may also average the Boltzmann factor of a (small but finite) perturbation:

$$
\begin{aligned}
V^{\mathrm{mf}}(\boldsymbol{r}' + \Delta \boldsymbol{r}') - V^{\mathrm{mf}}(\boldsymbol{r}') &= -k_{\mathrm{B}}T \ln \frac{\int e^{-\beta V(\boldsymbol{r}'+\Delta\boldsymbol{r}',\boldsymbol{r}'')}\, d\boldsymbol{r}''}{\int e^{-\beta V(\boldsymbol{r}',\boldsymbol{r}'')}\, d\boldsymbol{r}''} \\
&= -k_{\mathrm{B}}T \ln \left\langle e^{-\beta[V(\boldsymbol{r}'+\Delta\boldsymbol{r}',\boldsymbol{r}'')-V(\boldsymbol{r}',\boldsymbol{r}'')]} \right\rangle_{''}.
\end{aligned}
\tag{7.67}
$$

This equation is exact, but statistically only accurate for small displacements. By choosing a sufficiently dense net of configurations to generate the ensembles, the potentials of mean force can be reconstructed by fitting perturbations of one point to those of a nearby point.

- By *umbrella sampling*. This method, pioneered by Torrie and Valleau (1977), restrains, rather than constrains, the primed coordinates around a given configuration by adding a restraining potential $V^u(\boldsymbol{r}')$ to the potential $V(\boldsymbol{r}', \boldsymbol{r}'')$. This umbrella potential could, for example, be harmonic in shape. The resulting canonical *umbrella distribution* $w^u(\boldsymbol{r}')$ in the primed coordinates will in equilibrium be given by

$$
w^u(\boldsymbol{r}') \propto \int d\boldsymbol{r}''\, e^{-\beta V(\boldsymbol{r}',\boldsymbol{r}'')-\beta V^U(\boldsymbol{r}')}.
\tag{7.68}
$$

Therefore,

$$V^{\mathrm{mf}}(\mathbf{r}') = \mathrm{constant} - kT \ln \left[w^u(\mathbf{r}') e^{+\beta V^u(\mathbf{r})} \right]$$

$$= \mathrm{constant} - kT \ln[w^u(\mathbf{r}')] - V^u(\mathbf{r}), \qquad (7.69)$$

which says that the potential of mean force can be reconstructed in the neighborhood of the restrained configuration by keeping track of the distribution over the primed coordinates and correcting the bias caused by the umbrella potential. This reconstruction is only accurate in a region where sufficient statistics is obtained to determine $w^u(\mathbf{r}')$ accurately. The full potential of mean force can again be reconstructed by fitting adjacent umbrella distributions to each other.

An alternative to reconstructing the local V^{mf} from the distribution function is averaging of the umbrella force, which is easily monitored in a MD simulation.[11] In the case of a harmonic umbrella, the average force is also equal to the mean displacement of the coordinate(s) on which the umbrella is imposed (with respect to the umbrella center), divided by the harmonic force constant. The average umbrella force is approximately, but not exactly, equal to the derivative of the potential of mean force at the umbrella center. In fact, it is given exactly by a weighted average of the derivative over the umbrella distribution $w^u(\mathbf{r}')$:

$$\langle F^u \rangle_u = \frac{\int d\mathbf{r}' \, w^u(\mathbf{r}')[\partial V^{\mathrm{mf}}(\mathbf{r}')/\partial r_i']}{\int d\mathbf{r}' \, w^u(\mathbf{r}')}, \qquad (7.70)$$

which is accurate to second order (i.e., including the second derivative of the potential) to the derivative at the *average position* of the primed coordinate in the umbrella ensemble. This average can also be used to reconstruct the potential of mean force.

Proof The average umbrella force cancels the average internal force acting on the primed coordinate, and thus is equal to the average derivative of the total potential $V = V(\mathbf{r}', \mathbf{r}'')$ in the umbrella ensemble:

$$\langle F^u \rangle_u = \left\langle \frac{\partial V}{\partial r_i'} \right\rangle_u = \frac{\int d\mathbf{r}' \int d\mathbf{r}''[\partial V/\partial r_i'] \exp[-\beta V - \beta V^u]}{\int d\mathbf{r}' \int d\mathbf{r}'' \exp[-\beta V - \beta V^u]}. \qquad (7.71)$$

In the nominator of (7.71) the term $\exp(\beta V^u)$ can be taken out of the integration over $\mathbf{r}''$ and the remainder can be replaced by the derivative

---

[11] Kästner and Thiel (2005) describe a method for the determination of the derivative of V^{mf}.

of V^{mf} (8.11):

$$\int \frac{\partial V}{\partial r_i'} e^{-\beta V}\, dr'' = \frac{\partial V^{\mathrm{mf}}}{\partial r_i'} \int e^{-\beta V}\, dr''. \qquad (7.72)$$

We now obtain

$$\langle F^u \rangle_u = \frac{\int dr' \exp(-\beta V^u)[\partial V^{\mathrm{mf}}/\partial r_i'] \int dr'' \exp(-\beta V)}{\int dr' \int dr'' \exp(-\beta V - \beta V^u)}, \qquad (7.73)$$

from which (7.70) follows. □

- By *particle insertion* by the method of Widom (1963) in the special case that the potential of mean force is a function of the position of a specific *particle type* and in addition the density of the medium is low enough to allow successful insertions. Let us consider the case that the potential of mean force of a system that is inhomogeneous in the z-direction but homogeneous in the x, y-plane (e.g., containing a planar phase boundary situated at $z = 0$ between two immiscible fluids, or between a fluid and a polymer, or containing a membrane between two liquid phases). Assume an equilibrium simulation is available. One may place a *ghost particle* (not exerting any forces on its environment) at a given z- but randomly chosen x, y-position in a configuration of the generated ensemble, and registers the total interaction energy V between the particle and the surrounding molecules. This insertion is repeated many times, and $\mu_{\mathrm{ex}}(z) = -kT \ln\langle \exp(-\beta V) \rangle_z$ is determined. The standard chemical potential $\mu^0(z)$ of the particle type at position z in the inhomogeneous system is equal to the standard chemical potential μ_{id}^0 of that particle type as an ideal gas, plus the measured excess $\mu_{\mathrm{ex}}(z)$; the potential of mean force as a function of z is, but for an additive constant, equal to $\mu_{\mathrm{ex}}(z)$. One may choose the number of insertions per value of z to satisfy statistical requirements.
- By directly measuring the *particle concentration* $c(z)$ (as a number density per unit volume) in an equilibrium simulation in the special case (as above) that the potential of mean force is a function of the position of a specific particle type, in regions where that concentration is high enough to allow its determination with sufficient accuracy. In those regions the potential of mean force can be constructed from

$$V^{\mathrm{mf}}(z) = \mathrm{const} - kT \ln c(z). \qquad (7.74)$$

- By enforcing the system to move from one part of configurational space to another. In such *pulling* simulations, also called *steering molecular dynamics* (SMD),[12] an extra external force is exerted on the system such that it will move much more quickly over intervening barriers than it would do in an equilibrium simulation. The advantage is that the steering can be executed in a global way without the need for a detailed description of the reaction coordinate; e.g., one may choose to distribute the pulling force over many atoms and let the system decide to find the easiest pathway under such a global force. The disadvantage is that the work exerted by the external force contains a frictional contribution and is therefore not related to the potential of mean force in a straightforward way. Only in the uninteresting limit of zero pulling rate the force acts in a reversible manner and equals the derivative of a potential of mean force.

However, in the case of small, but finite forces or pulling rates, it is possible to derive the potential of mean force V^{mf} from the work W exerted by the external force. It is necessary to use a thermostat that prevents local heating by frictional forces. Although W will always exceed the reversible work ΔV^{mf}, it can be shown (Jarzynsky, 1997a, 1997b) that

$$e^{-\beta \Delta V^{\mathrm{mf}}} = \langle e^{-\beta W} \rangle, \tag{7.75}$$

which is valid if a sufficiently large set of pulling simulations, starting from samples from an initial equilibrium ensemble, is performed. It is quite difficult to obtain sufficient statistics for the evaluation of $\langle \exp(-\beta W) \rangle$; Park *et al.* (2003) found that more reliable results are obtained with a second-order expansion:

$$\ln \langle e^{-\beta W} \rangle = -\beta \langle W \rangle + \frac{\beta^2}{2} \left(\langle W^2 \rangle - \langle W \rangle^2 \right), \tag{7.76}$$

using a time-dependent external potential of the form

$$V^{\mathrm{ext}} = \frac{k}{2} [\xi(\boldsymbol{r}) - \lambda]^2, \qquad \lambda(t) = \lambda_0 + vt, \tag{7.77}$$

with a large force constant k. The method is very similar to the imposition of a slowly changing constraint. Except for near-equilibrium SMD, the method is not preferred above thermodynamic integration using constrained or umbrella-restrained intermediate simulations. In the next sec-

---

[12] Simulations involving pulling were first performed with the explicit purpose to mimic experiments with the atomic force microscope (AFM), e.g., pulling a ligand bound to a protein away from the protein. In such simulations the external force is due to a spring that is slowly displaced, like the lever of an AFM, but for the computation of a potential of mean force the spring is not required. The first simulation of this kind was on the "unbinding" of biotin from the protein streptavidin by Grubmüller *et al.* (1996), soon followed by Izrailev *et al.* (1997). See also Berendsen (1996). The name *Steering Molecular Dynamics* originates from Schulten, see Lu *et al.* (1998). The topic was reviewed by Isralewitz *et al.* (2001).

tion the non-equilibrium methods are more fully treated and a proof of Jarzynski's equation is given.

7.8 Free energy from non-equilibrium processes

In the previous section we have seen how free energies or potentials of mean force can be computed through perturbation and integration techniques. The steered dynamics is reminiscent of the *slow-growth* methods, where an external agent changes the Hamiltonian of the system during a prolonged dynamics run, by adding an additional time-dependent potential or force, or changing the value of a constraint imposed on the system. Thus the system is literally forced from one state to another, possibly over otherwise unsurmountable barriers. If such changes are done very slowly, such that the system remains effectively in equilibrium all the time, the change is a reversible process and in fact the change in free energy from the initial to the final state is measured by the work done to change the Hamiltonian.

In most practical cases the external change cannot be realized in a sufficiently slow fashion, and a partial irreversible process results. The ensemble "lags behind" the change in the Hamiltonian, and the work done on the system by the external agent that changes the Hamiltonian, is partially irreversible and converted to heat. The second law of thermodynamics tells us that the total work W done on the system can only exceed the reversible part ΔA:

$$W \geq \Delta A. \tag{7.78}$$

This is an inequality that enables us to bracket the free energy change between two measured values when the change is made both in the forward and the backward direction, but it does not give any help in quantifying the irreversible part. It would be desirable to have a quantitative relation between work and free energy!

Such a relation indeed exists. Jarzynski (1997a, 1997b) has shown that for an *irreversible process* the Helmholtz free energy change follows from the *work* W done to change Hamiltonian $H(\lambda)$ of the system from $\lambda = 0$ to $\lambda = 1$, if averaged over an equilibrium ensemble of initial points for $\lambda = 0$:

$$A_1 - A_0 = -k_B T \ln \langle e^{-\beta W} \rangle_{\lambda=0}. \tag{7.79}$$

This is the remarkable *Jarzynski equation*, which at first sight is a counterintuitive expression, relating a thermodynamic quantity to a rather ill-defined and very much process-dependent amount of work. Cohen and Mauzerall

(2004) have criticized Jarzynski's derivation on the basis of improper hand-ling of the heat exchange with a heat bath, which induced Jarzynski (2004) to write a careful rebuttal. Still, the validity of this equation has been con-firmed by several others for various cases and processes, including stochastic system evolution (Crooks, 2000; Hummer and Szabo, 2001; Schurr and Fu-jimoto, 2003; Athènes, 2004). Since the variety of proofs in the literature is confusing, we shall give a different proof below, which follows most closely the reasoning of Schurr and Fujimoto (2003). This proof will enable us to specify the requirements for the validity of Jarzynskyi's equation. In this proof we shall pay extra attention to the role of the temperature, clarifying what requirements must be imposed on β.

7.8.1 Proof of Jarzynski's equation

Consider a system of interacting particles with Hamiltonian H_0 that has been allowed to come to equilibrium with an environment at temperature T_0 or Boltzmann parameter $\beta_0 = (k_B T)^{-1}$ and has attained a canonical distribution

$$p_0(z) = \frac{\exp[-\beta_0 H_0(z)]}{\int \exp[-\beta_0 H_0(z')]\, dz'}. \tag{7.80}$$

Here z stands for the coordinates (spatial coordinates and conjugate mo-menta) $q_1, \ldots, p_1, \ldots$ of a point in phase space. At time t_0 we pick a sample from the system with phase space coordinates z_0. When we speak later about averaging over the initial state, we mean averaging over the canonical distribution p_0 of z_0.

Now the system undergoes the following treatment (see Fig. 7.5): at time t_0 the Hamiltonian is abruptly changed from H_0 to H_1 by an external agent; from t_0 to t_1 the system is allowed to evolve from z_0 to z_1 under the (con-stant) Hamiltonian H_1. The evolution is not necessarily a pure Hamiltonian evolution of the isolated system: the system may be coupled to a thermostat and/or barostat or extended with other variables, and the process may be deterministic or stochastic. *The only requirement is that the evolution pro-cess conserves a canonical distribution* $\exp[-\beta_1 H_1(z)]$, where $H_1(z)$ is the total energy of the system at phase point z. Note that we do not require that the temperature during evolution (e.g., given by the thermostat or the friction and noise in a stochastic evolution) equals the temperature before the jump. Now at t_1 the external agent changes the Hamiltonian abruptly from H_1 to H_2, after which the system is allowed to evolve under H_2 from t_1 to t_2, changing from z_1 to z_2. Again, the evolution process is such that it

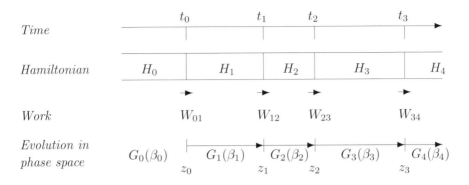

Figure 7.5 Irreversible evolution with changing Hamiltonian. The Hamiltonian is abruptly changed by an external agent at times $t_0, t_1, \ldots$, who exerts *work* $W_{01}, W_{12}, \ldots$ on the system. In the intervening time intervals the system is allowed to evolve with the *propagator* $G_i(\beta_i)$, when the Hamiltonian H_i is valid. The points in phase space, visited by the system at $t_0, t_1, \ldots$, comprising all coordinates and momenta, are indicated by $z_0, z_1, \ldots$.

would conserve a canonical distribution $\exp[-\beta_2 H_2(z)]$. These processes of autonomous evolution followed by a Hamiltonian change may be repeated as often as required to reach the desired end state.

Two definitions before we proceed: the work done by the external agent to change H_i to H_{i+1} we define (following Jarzynski) as $W_{i,i+1}$. But with changing β we also require the change in work relative to the temperature, which we shall denote by $F_{i,i+1}$:

$$W_{i,i+1} \stackrel{\text{def}}{=} H_{i+1}(z_i) - H_i(z_i), \tag{7.81}$$

$$F_{i,i+1} \stackrel{\text{def}}{=} \beta_{i+1} H_{i+1}(z_i) - \beta_i H_i(z_i). \tag{7.82}$$

The sum of these quantities over more than one step is similarly denoted. For example, $W_{0,i}$ is the total work done in all steps up to and including the step to H_i, and $F_{0,i}$ is the total relative work done in all steps up to and including the step to H_i. These quantities have specified values for each realization; useful results require averaging over an initial distribution.

In the following we shall prove that after each Hamiltonian jump to H_i, the free energy A_i is given by

$$\beta_i A_i - \beta_0 A_0 = -\ln\langle e^{-F_{0,i}}\rangle_{p_0}, \tag{7.83}$$

Averaging is done over the initial canonical distribution p_0 of z_0, and also over all possible stochastic evolutions. The latter averaging is automatically fulfilled when the original distribution is sufficiently sampled.

This is the *generalized Jarzynski equation*. It reduces to the original Jarzynski equation when all β_i are equal:

$$A_i - A_0 = -k_B T \ln \langle e^{-\beta W_{0,i}} \rangle_{p_0}. \tag{7.84}$$

H_i can be taken as the required end state; the process may contain any number of steps and the intermediate evolution times may have any value from zero (no evolution) to infinite (evolution to complete equilibrium). So the allowed processes encompass the already well-known single-step (no evolution) free energy perturbation, the few-step perturbation with intermediate equilibrations, and the limiting slow-growth process, the latter taking a large number of steps, each consisting of a small Hamiltonian change followed by a single MD time step.

The proof follows by induction. Consider the free energy change after the first step, before any evolution has taken place:

$$
\begin{aligned}
\beta_1 A_1 - \beta_0 A_0 &= -\ln \frac{\int dz_0 \exp[-\beta_1 H_1(z_0)]}{\int dz_0 \exp[-\beta_0 H_0(z_0)]} \\
&= \langle e^{-[\beta_1 H_1(z_0) - \beta_0 H_0(z_0)]} \rangle_{p_0} = \langle e^{-F_{0,1}} \rangle_{p_0}, \tag{7.85}
\end{aligned}
$$

which is the generalized Jarzynski's equation applied to a single step without evolution.

Now, from t_0 to t_1 the system is left to evolve under the Hamiltonian H_1. Its evolution can be described by a *propagator* $G_1(z, t; z_0, t_0; \beta_1)$ that specifies the probability distribution of phase points z at time t, given that the system is at z_0 at time t_0. For pure Hamiltonian dynamics the path is deterministic and thus G_1 is a delta-function in z; for stochastic evolutions G_1 specifies a probability distribution. In general G_1 describes the evolution of a probability distribution in phase space:

$$p(z, t) = \int G_1(z, t; z_0, t_0; \beta_1) p(z_0, t_0) \, dz_0. \tag{7.86}$$

The requirement that G_1 preserves a canonical distribution $\exp[-\beta_1 H_1(z)]$ can be written as

$$\int G_1(z, t; z_0, t_0) \exp[-\beta_1 H_1(z_0)] \, dz_0 = \exp[-\beta_1 H_1(z)] \tag{7.87}$$

for all t. In fact, G maps the canonical distribution onto itself.

The actual distribution $p_0(z_0)$ at t_0 is not the canonical distribution for H_1, but rather the canonical distribution for H_0. So the property (7.87)

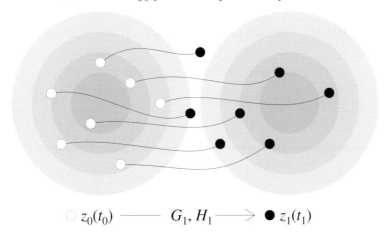

$$z_0(t_0) \;\;\text{———}\;\; G_1, H_1 \;\longrightarrow\; \bullet\, z_1(t_1)$$

Figure 7.6 Paths extending from sampled points z_0 at t_0 to z_1 at t_1. Each of the paths is weighted (indicated by line thickness) such that the distribution of weights becomes proportional to the canonical distribution $\exp[-\beta_1 H_1(z_1)]$. The grey areas indicate the equilibrium distributions for H_0 (left) and H_1 (right).

cannot be applied to the actual distribution at t_1. But we can apply a trick, pictured schematically in Fig. 7.6. Let us give every point z_0 a *weight* such that the *distribution of weights*, rather than of points, becomes the canonical distribution for H_1. This is accomplished if we give the point z_0 a weight $\exp[-\beta_1 H_1(z_0) + \beta_0 H_0(z_0)]$. Note that this weight equals $\exp[-F_{01}]$. Since the distribution of weights, indicated by $p^w(z_0)$, is now proportional to the canonical distribution for H_1:

$$p^w(z_0) = p_0(z_0)e^{-\beta_1 H_1(z_0)+\beta_0 H_0(z_0)} = \frac{\exp[-\beta_1 H_1(z_0)]}{\int dz_0 \exp[-\beta H_0(z_0)]}, \tag{7.88}$$

the distribution of *weights* will remain invariant during the evolution with G_1 to z_1, and hence also

$$p^w(z_1) = p^w(z_0). \tag{7.89}$$

¿From this we can derive the *unweighted* distribution of points z_1 by dividing $p^w(z_1)$ with the weight given to z_0:

$$p(z_1) = p^w(z_1)e^{\beta_1 H_1(z_0)-\beta_0 H_0(z_0)]} = p^w(z_1)e^{F_{01}}$$
$$= \frac{\exp[-\beta_1 H_1(z_1) + F_{01}]}{\int dz_0 \exp[-\beta H_0(z_0)]}. \tag{7.90}$$

Next the external agent changes H_1 to H_2, performing the work $W_{1,2} = H_2(z_1) - H_1(z_1)$ on the system. The relative work is

$$F_{1,2} = \beta_2 H_2(z_1) - \beta_1 H_1(z_1). \tag{7.91}$$

Equation (7.90) can now be rewritten as

$$p(z_1) = \frac{\exp[-\beta_2 H_2(z_1) + F_{0,1} + F_{1,2}]}{\int dz_0 \exp[-\beta H_0(z_0)]}. \tag{7.92}$$

If we now ask what the expectation of $\exp[-F_{0,2}]$ will be, we find

$$\langle e^{-F_{0,2}} \rangle = \langle e^{-(F_{0,1}+F_{1,2})} \rangle = \int dz_1 \, p(z_1) e^{-(F_{0,1}+F_{1,2})} \tag{7.93}$$

$$= \frac{\int dz_1 \exp[-\beta_2 H_2(z_1)]}{\int dz_0 \exp[-\beta H_0(z_0)]} = e^{-(\beta_2 A_2 - \beta_1 A_0)}. \tag{7.94}$$

This is the generalized Jarzynski's equation after the second step has been made. The extension with subsequent steps is straightforward: for the next step we start with $p(z_1)$ and give the points a weight $\exp[-F_{0,2}]$. The weight distribution is now the canonical distribution for H_2, which remains invariant during the evolution G_2. From this we derive $p(z_2)$, and – after having changed the Hamiltonian at t_2 to H_3 – we find that

$$\langle e^{-F_{0,3}} \rangle = e^{-(\beta_3 A_3 - \beta_0 A_0)}. \tag{7.95}$$

This, by induction, completes the proof of (7.83). Note that, in the case of varying β during the process, it is

the total *relative* work, i.e., the change in energy *divided by the temperature*, $F_{0,i}$, that must be exponentially averaged rather than the total work itself.

7.8.2 Evolution in space only

When the external change in Hamiltonian involves the potential energy $V(\mathbf{r})$ only (which usually is the case), and the evolution processes are mappings in configurational space that conserve a canonical distribution (e.g., a sequence of Monte Carlo moves or a Brownian dynamics), the Jarzynski equation is still valid. The evolution operator $G_i(\mathbf{r}_i, t_i; \mathbf{r}_{i-1}, t_{i-1}; \beta_i)$ now evolves $\mathbf{r}_{i-1}$ into $\mathbf{r}_i$; it has the property

$$\int G_i(\mathbf{r}', t'; \mathbf{r}, t; \beta_i) \exp[-\beta_i V_i(\mathbf{r})] \, d\mathbf{r} = \exp[-\beta_1 V_i(\mathbf{r}')]. \tag{7.96}$$

Here $\mathbf{r}$ stands for all cartesian coordinates of all particles, specifying a point in configurational space. When we re-iterate the proof given above, replacing z by $\mathbf{r}$ and H by V, we find the same equation (7.84) for the isothermal case, but a correction due to the kinetic contribution to the free energy if the initial and final temperatures differ. Equation (7.83) is now replaced by

$$\beta_i A_i - \beta_0 A_0 = \frac{3N}{2} \ln \frac{\beta_i}{\beta_0} - \ln \langle e^{-F_{0,i}} \rangle_{p_0}. \tag{7.97}$$

7.8.3 Requirements for validity of Jarzynski's equation

Reviewing the proof given above, we can list the requirements for its validity:

(i) The state of the system at time t is completely determined by the point in phase space $z(t)$. The propagator G determines the future probability distribution, given $z(t)$. This is an expression of the *Markovian* character of the propagator: the future depends on the state at t and not on earlier history. This precludes the use of stochastic propagators with memory, such as the generalized Langevin equation. It is likely (but not further worked out here) that the Markovian property is not a stringent requirement, as one can always define the state at time t to include not only $z(t)$, but also z at previous times. However, this would couple the Hamiltonian step with the future propagation, with as yet unforeseen consequences. Hamiltonian (including extended systems), simple Langevin, Brownian and Monte Carlo propagations are all Markovian.[13]

(ii) The propagator must have the property to conserve a canonical distribution. Microscopic reversibility and detailed balance are not primary requirements.

(iii) The sampling must be sufficient to effectively reconstruct the canonical distribution after each step by the weighting procedure. This requires sufficient overlap between the distribution of end points of each relaxation period and the canonical distribution after the following step in the Hamiltonian. When the steps are large and the relaxations are short, sufficient statistics may not be available. This point is further discussed in the next subsection.

(iv) As is evident from the proof, there is no requirement to keep the inverse temperature β constant during the process. Even if the same β is required for the initial and final states, intermediate values may be chosen differently. This property may be exploited to produce a faster sampling.

---

[13] See Park and Schulten (2004) for a discussion of various ensembles.

7.8.4 Statistical considerations

Since the averaging is done over an exponential function of the work done, the trajectories with the smaller work values will dominate the result. This produces erratic jumps in cumulative averages whenever occasional low values appear. The statistical properties and validity of approximations have been considered by several authors (Park *et al.* 2003, Hummer 2001, Ytreberg and Zuckerman 2004). Let us consider a simple example.

We sample a property x (the "work") from a distribution function $p(x)$, and we wish to compute the quantity A:

$$A = -\frac{1}{\beta} \ln \langle e^{-\beta x} \rangle = -\frac{1}{\beta} \ln \int p(x) e^{-\beta x} \, dx. \tag{7.98}$$

Without loss of generality, we make take the average of x as zero, so that all values refer to the average of the distribution. First consider the cumulant expansion[14] in powers of β, obtained from a simple Taylor expansion:

$$A = -\frac{1}{2!} \beta \langle x^2 \rangle + \frac{1}{3!} \beta^2 \langle x^3 \rangle - \frac{1}{4!} \beta^3 [\langle x^4 \rangle - 3 \langle x^2 \rangle^2 + \cdots]. \tag{7.99}$$

For a Gaussian distribution,

$$p(x) = (\sigma \sqrt{2\pi})^{-1} \exp \left[-\frac{x^2}{2\sigma^2} \right], \tag{7.100}$$

only the first term survives, as can easily be seen by direct integration:

$$A = -\frac{1}{2} \beta \sigma^2. \tag{7.101}$$

Figure 7.7 shows that the cumulative average gives very poor convergence. For this figure 1000 points have been sampled from a normal distribution of zero average and unit variance, and the cumulative exponential averages were calculated with values of β equal to 1, 2 and 4. The theoretical values for A are $-0.5, -1$ and -2, respectively. For $\beta = 1$ convergence is reached after about 600 points; for $\beta = 2$ 1000 points are barely enough, and for $\beta = 4$ 1000 points are clearly insufficient to reach convergence. This means that computing the exponential average is hardly an option if the computation of one path takes a considerable computational effort.

The route via the cumulant expansion (7.99) gives very accurate results if the distribution is known to be Gaussian and (7.101) applies. For n independent samples, the variance (mean-square error) in the estimated average

---

[14] See Zwanzig (1954), who defined the cumulant expansion of $\ln \langle \exp(-\beta x) \rangle$ in a power series in $(-\beta)^n / n!$.

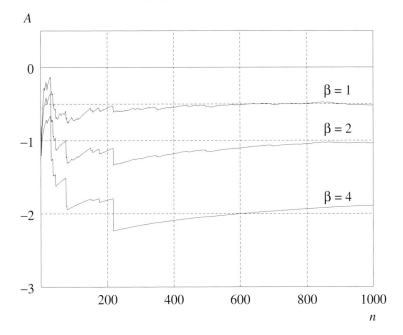

Figure 7.7 Cumulative average of $A = -\beta^{-1}\ln\langle\exp(-\beta x)\rangle$ over n samples drawn from a normal distribution (average 0, variance $\sigma^2 = 1$). The theoretical limits are -0.5β, indicated by dotted lines.

$\langle x \rangle = \sum x_i/n$ is σ^2/n, while the mean-square error in the estimated variance $s^2 = \sum (x - \langle x \rangle)^2/(n-1)$ is $2\sigma^4/(n-1)$ (Hummer, 2001):

$$A = \langle x \rangle - \frac{1}{2}\beta\hat{\sigma}^2 \pm \left(\frac{\hat{\sigma}^2}{n} + \frac{\beta^2\hat{\sigma}^4}{n-1}\right)^{1/2}, \qquad (7.102)$$

where

$$\langle x \rangle = \frac{1}{n}\sum x_i \qquad (7.103)$$

$$\hat{\sigma}^2 = \frac{1}{n-1}\sum (x_i - \langle x \rangle)^2, \qquad (7.104)$$

are best, unbiased, estimates for the average and variance. However, if the distribution is not Gaussian, the higher-order cumulants rapidly add to the inaccuracy.

Ytreberg and Zuckerman (2004) propose, based on an extensive error analysis, to select many random sequences of $m < n$ samples from a set of n data and plot the exponential averages thus obtained versus $m^{-1/2}$. Extrapolation to $n \to \infty$ then corrects for a datasize-dependent systematic bias in the averages, and an error estimate can be obtained.

In practice, Jarzynski's equation can only be used if the irreversible work is small (not exceeding $2k_BT$), i.e., if the process is close to equilibrium. As Oostenbrink and van Gunsteren (2006) have shown, integration from equilibrated intermediate points is generally much more efficient than both irreversible fast growth and near-equilibrium slow growth. It is not clear whether this still holds when optimal corrections are applied to the integration by fast or slow growth.

8

Stochastic dynamics: reducing degrees of freedom

8.1 Distinguishing relevant degrees of freedom

Often the interest in the behavior of large molecular systems concerns global behavior on longer time scales rather than the short-time details of local dynamics. Unfortunately, the interesting time scales and system sizes are often (far) beyond what is attainable by detailed molecular dynamics simulations. In particular, macromolecular structural relaxation (crystallization from the melt, conformational changes, polyelectrolyte condensation, protein folding, microphase separation) easily extends into the seconds range and longer. It would be desirable to simplify dynamical simulations in such a way that the "interesting" behavior is well reproduced, and in a much more efficient manner, even if this goes at the expense of "uninteresting" details. Thus we would like to reduce the number of degrees of freedom that are explicitly treated in the dynamics, but in such a way that the accuracy of global and long-time behavior is retained as much as possible.

All approaches of this type fall under the heading of *coarse graining*, although this term is often used in a more specific sense for models that average over *local* d etails. The relevant degrees of freedom may then either be the cartesian coordinates of special particles that represent a spatial average (the *superatom approach*, treated in Section 8.4), or they may be *densities* on a grid, defined with a certain spatial resolution. The latter type of coarse graining is treated in Chapter 9 and leads to *mesoscopic continuum dynamics*, treated in Chapter 10.

The first choice is to distinguish *relevant degrees of freedom* from *irrelevant degrees of freedom*. With "irrelevant" we do not mean unimportant: these degrees of freedom can have essential influences on the "relevant" degrees of freedom, but we mean that we don't require knowledge of the detailed behavior of those degrees of freedom. This choice is in a sense arbitrary and

249

depends on the system, the properties of interest and the required accuracy. The choice must be judiciously made. It is highly desirable and beneficial for the approximations that will be made, that the "irrelevant" degrees of freedom equilibrate faster (and preferably much faster) than the "relevant" degrees of freedom, as the approximations will unavoidably produce errors on the time scale were these two overlap. However, such a clear distinction is generally not possible, and one must accept the inaccurate prediction of dynamic details of the "relevant" degrees of freedom on short time scales.

Some examples are listed below:

- A rather spherical molecule, as CCl_4. *Relevant*: the center-of-mass motion; *irrelevant*: the rotational and internal degrees of freedom.
- A (macro)molecule in a solvent. *Relevant*: all atoms of the solute; *irrelevant*: all solvent molecules. With this choice there is certainly overlap between the time ranges for the two sets of particles. For example, for a protein in water one may expect incorrect dynamic behavior of charged side chains on a time scale shorter than, or comparable to, the dielectric relaxation time of the solvent.
- A large linear polymer. *Relevant*: the centers of mass of groups of n consecutive atoms; *irrelevant*: all other degrees of freedom. This may work if the polymer shows self-similar behavior, i.e., that macroscopic properties scale in some regular manner with n. These are typical superatom models (see Section 8.4).
- A protein (or other compact non-selfsimilar macromolecule). *Relevant*: a subset of atoms (as C_α atoms, or backbone atoms, or backbone atoms plus a simplified side chain representation); *irrelevant*: all other atoms or degrees of freedom including the surrounding solvent. This is also a superatom approach.
- A protein (or other compact macromolecule). *Relevant*: a set of collective "essential degrees of freedom" generated from an analysis of a detailed simulation, e.g., the first few eigenvectors with largest eigenvalue from a principal component analysis based on atomic fluctuations, or from a quasi-harmonic analysis. *Irrelevant*: all other eigenvectors.
- A chemical reaction or other infrequent process in a complex system. *Relevant*: the reaction coordinate, being a function of internal degrees of freedom of the system that captures the important path between reactants and products in a chemical reaction. This may concern one dimension, or encompass a space of a few dimensions. *Irrelevant*: all other degrees of freedom.
- A colloidal dispersion of relatively large spherical rigid particles in a sol-

vent. *Relevant*: center of mass coordinates of the particles. *Irrelevant*: rotational and internal degrees of freedom of the particles, and solvent degrees of freedom. For non-spherical particles their rotational degrees of freedom may be considered relevant as well.

- A rather homogeneous condensed phase under slowly varying external influences. *Relevant*: Densities of molecular components at grid points on a chosen 3D spatial grid; *irrelevant*: all other degrees of freedom. Instead of a regular 3D grid one may use other, possibly time-dependent, finite element subdivisions of space.

In some cases we can choose *cartesian* degrees of freedom as the relevant ones (e.g., when we divide the particles over both classes), but in most cases we must define the two classes as generalized degrees of freedom. To avoid unnecessary accumulation of complexity, we shall in the following consider cartesian coordinates first, and consider necessary modifications resulting from the use of generalized coordinates later (Section 8.6.1 on page 263). In the case that the relevant coordinate is a distance between atoms or a linear combination of atomic coordinates, the equations are the same as for cartesian coordinates of selected particles, although with a different effective mass.

8.2 The generalized Langevin equation

Assume we have split our system into *explicit* "relevant particles" indicated with a prime and with positions $r'_i(t)$ and velocities $v'_i(t)$, $i = 1, \ldots, N'$, and *implicit* double-primed "irrelevant" particles with positions $r''_j(t)$ and velocities $v''_j(t)$, $j = 1, \ldots, N''$. The force F_i acting on the i-th primed particle comes partly from interactions with other primed particles, and partly from interactions with double-primed particles. The latter are not available in detail. The total force can be split up into:

- *systematic forces* $F^s_i(r')$ which are a function of the primed coordinates; these forces include the mutual interactions with primed particles and the interactions with double-primed particles as far as these are related to the primed *positions*;
- *frictional forces* $F^f_i(v)$ which are a function of the primed velocities (and may parametrically depend on the primed coordinates as well). They include the interactions with double-primed particles as far as these are related to the primed *velocities*;
- *random forces* $F^r_i(t)$. These are a representation of the remainder of the interactions with double-primed particles which are then neither related to

the primed positions nor to the primed coordinates. Such forces are characterized by their statistical distributions and by their time correlation functions. They may parametrically depend on the primed coordinates.

This classification is more intuitive than exact, but suffices (with additional criteria) to derive these forces in practice. A systematic way to derive the time evolution of a selected subsystem in phase space is given by the *projection operator formalism* of Zwanzig (1960, 1961, 1965) and Mori (1965a, 1965b). This formalism uses projection operators in phase space, acting on the Liouville operator, and arrives at essentially the same subdivision of forces as given above. In this chapter we shall not make specific use of it. The problem is that the formalism, although elegant and general, does not make it any easier to solve practical problems.[1]

We make two further assumptions:

(i) the systematic force can be written as the gradient of a potential in the primed coordinate space. This is equivalent to the assumption that the systematic force has no curl. For reasons that will become clear, this potential is called the *potential of mean force*, $V^{\mathrm{mf}}(\boldsymbol{r}')$;

(ii) the frictional forces depend linearly on velocities of the primed particles at earlier times. Linearity means that velocity-dependent forces are truncated to first-order terms in the velocities, and dependence on earlier (and not future) times is simply a result of causality.

Now we can write the equations of motion for the primed particles as

$$
m_i \frac{d\boldsymbol{v}_i}{dt} = -\frac{\partial V^{\mathrm{mf}}}{\partial \boldsymbol{r}_i} - \sum_j \int_0^t \zeta_{ij}(\tau)\boldsymbol{v}_j(t-\tau)\,d\tau + \boldsymbol{\eta}_i(t), \tag{8.1}
$$

where $\zeta_{ij}(\tau)$ (often written as $m_i\gamma_{ij}(\tau)$) is a *friction kernel* that is only defined for $\tau \geq 0$ and decays to zero within a finite time. The integral over past velocities extends to $\tau = t$, as the available history extends back to time 0; when t is much larger than the correlation time of $\zeta_{ij}(\tau)$, the integral can be safely taken from 0 to ∞. This friction term can be viewed as a *linear prediction* of the velocity derivative based on knowledge of the past trajectory. The last term $\boldsymbol{\eta}(t)$ is a random force with properties still to be determined, but surely with

$$
\langle \boldsymbol{\eta}(t) \rangle = 0, \tag{8.2}
$$

$$
\langle \boldsymbol{v}_i(t) \cdot \boldsymbol{\eta}_j(t') \rangle = 0 \ \text{ for any } i,j \text{ and } t' \geq t. \tag{8.3}
$$

---

[1] van Kampen (1981, pg 398) about the resulting projection operator equation: "This equation is exact but misses the point. [...] The distribution cannot be determined without solving the original equation ..."

On the time scale of the system evolution, the random forces are *stationary* stochastic processes, independent of the system history, i.e., their correlation functions do not depend on the time origin, although they may have a weak dependence on system parameters. In principle, the random forces are correlated in time with each other; these correlations are characterized by *correlation functions* $C_{ij}^{\eta}(\tau) = \langle \boldsymbol{\eta}_i(t)\boldsymbol{\eta}_j(t+\tau)\rangle$, which appear (see below) to be related to the friction kernels $\zeta_{ij}(\tau)$.

This is the *generalized Langevin equation* for cartesian coordinates. For generalized coordinates $\{q, p\}$ the mass becomes a tensor; see Section 8.6.1 on page 263.

Note At this point we should make two formal remarks on stochastic equations (like (8.1)) of the form

$$\frac{dy}{dt} = f(y) + c\eta(t), \tag{8.4}$$

where $\eta(t)$ is a random function. The first remark concerns the lack of mathematical rigor in this equation. If the random function represents *white noise*, it can be seen as a sequence of delta functions with random amplitudes. Every delta function causes a jump in y and the resulting function $y(t)$ is not differentiable, so that the notation dy/dt is mathematically incorrect. Instead of a differential equation (8.4), we should write a difference equation for small increments dt, dy, dw:

$$dy = f(y)\, dt + c\, dw(t), \tag{8.5}$$

where $w(t)$ is the *Wiener–Lévy process*, which is in fact the integral of a white noise. The Wiener-Lévy process (often simply called the *Wiener process* is non-stationary, but its increments dw are stationary normal processes. See, for example, Papoulis (1965) for definitions and properties of random processes. While modern mathematical texts avoid the differential equation notation,[2] this notation has been happily used in the literature, and we shall use it as well without being disturbed by the mathematical incorrectness. What we mean by a stochastic differential equation as (8.4) is that the increment of y over a time interval can be obtained by integrating the right-hand side over that interval. The integral r of the random process $\eta(t)$

$$r = \int_t^{t+\Delta t} \eta(t')\, dt' \tag{8.6}$$

is a random number with zero mean and variance given by a double integral over the correlation function of η:

$$\langle r^2 \rangle = \int_t^{t+\Delta t} dt' \int_t^{t+\Delta t} dt''\, \langle \eta(t')\eta(t'')\rangle; \tag{8.7}$$

in the case of a white noise $\langle \eta(t')\eta(t'')\rangle = \delta(t' - t'')$ and therefore $\langle r^2 \rangle = \Delta t$.

The other remark concerns a subtlety of stochastic differential equations with a y-dependent coefficient $c(y)$ in front of the stochastic white-noise term: the widely

---

[2] See, e.g., Gardiner (1990). An early discussion of the inappropriateness of stochastic differential equations has been given by Doob (1942).

debated Itô–Stratonovich "dilemma."[3] Solving the equation in time steps, the variable y will make a jump every step and it is not clear whether the coefficient $c(y)$ should be evaluated before or after the time step. The equation therefore has no meaning unless a recipe is given how to handle this dilemma. Itô's recipe is to evaluate $c(y)$ before the step; Stratonovich's recipe is to take the average of the evaluations before and after the step. The stochastic equation is meant to define a process that will satisfy a desired equation for the distribution function $P(y,t)$ of y. If that equation reads

$$\frac{\partial P}{\partial t} = -\frac{\partial}{\partial y}f(y)P + \frac{1}{2}\frac{\partial^2}{\partial y^2}c(y)^2 P, \tag{8.8}$$

Itô's interpretation appears to be correct. With Stratonovich's interpretation the last term is replaced by

$$\frac{1}{2}\frac{\partial}{\partial y}c(y)\frac{\partial}{\partial y}c(y)P. \tag{8.9}$$

Hence the derivation of the equation will also provide the correct interpretation. Without that interpretation the equation is meaningless. As van Kampen (1981, p. 245) remarks: "no amount of physical acumen suffices to justify a meaningless string of symbols." However, the whole "dilemma" arises only when the noise term is white (i.e., when its time correlation function is a delta function), which is a mathematical construct that never arises in a real physical situation. When the noise has a non-zero correlation time, there is no difference between Itô's and Stratonovich's interpretation for time steps small with respect to the correlation time. So, physically, the "dilemma" is a non-issue after all.

In the following sections we shall first investigate what is required for the potential of mean force in simulations that are meant to preserve long-time accuracy. Then we describe how friction and noise relate to each other and to the stochastic properties of the velocities, both in the full Langevin equation and in the simpler *pure Langevin equation* which does not contain the systematic force. This is followed by the introduction of various approximations. These approximations involve both temporal and spatial correlations in the friction and noise: time correlations can be reduced to instantaneous response involving white noise and spatial correlations can be reduced to local terms, yielding the *simple Langevin equation*. In Section 8.7 we *average* the Langevin dynamics over times long enough to make the inertial term negligible, yielding what we shall call *Brownian dynamics*.[4]

---

[3] See van Kampen (1981) and the original references quoted therein.

[4] The term *Brownian dynamics* in this book is restricted to approximations of particle dynamics that are inertia-free but still contain stochastic forces. There is no universal agreement on this nomenclature; the term Brownian dynamics is sometimes used for any dynamical method that contains stochastic terms.

8.3 The potential of mean force

In a system with reduced dimensionality it is impossible to faithfully retain both thermodynamic and dynamic properties on all time and length scales. Since the interest is in retaining properties on long time scales and with coarse space resolution, we shall in almost all cases be primarily interested in a faithful representation of the thermodynamic properties at equilibrium, and secondarily in the faithful representation of coarse-grained non-equilibrium behavior. If we can maintain these objectives we should be prepared to give up on accurate local and short-time dynamical details.

The criterium of retaining thermodynamic accuracy prescribes that the partition function generated by the reduced dynamics should at least be proportional to the partition function that would have been generated if all degrees of freedom had been considered. Assuming canonical ensembles, this implies that the probability distribution $w(\mathbf{r}')$ in the primed configurational space should be proportional to the integral of the Boltzmann factor over the double-primed space:

$$w(\mathbf{r}') \propto \int e^{-\beta V(\mathbf{r}',\mathbf{r}'')}\, d\mathbf{r}''. \tag{8.10}$$

Now we *define* the *potential of mean force* as

$$V^{\mathrm{mf}}(\mathbf{r}') = -kT \ln \int e^{-\beta V(\mathbf{r}',\mathbf{r}'')}\, d\mathbf{r}'', \tag{8.11}$$

which implies that

$$w(\mathbf{r}')\, d\mathbf{r}' \propto e^{-\beta V^{\mathrm{mf}}(\mathbf{r}')}\, d\mathbf{r}'. \tag{8.12}$$

It follows by differentiation that the forces derived as a gradient of V^{mf} equal the exact force averaged over the ensemble of primed coordinates:

$$\boldsymbol{\nabla}_{\mathbf{r}'} V^{\mathrm{mf}} = \frac{\int (\partial V(\mathbf{r}',\mathbf{r}'')/\partial \mathbf{r}') \exp[-\beta V(\mathbf{r}',\mathbf{r}'')]\, d\mathbf{r}''}{\int \exp[-\beta V(\mathbf{r}',\mathbf{r}'')\, d\mathbf{r}'']}. \tag{8.13}$$

Note that V^{mf} also contains the direct interactions between $\mathbf{r}''$'s, which are separable from the integral in (8.11). It is a *free energy* with respect to the double-primed variables (beware that therefore V^{mf} is temperature-dependent!), but it still is a function of the primed coordinates. It determines in a straightforward manner the probability distribution in the primed space. Note that V^{mf} is *not* a mean potential over an equilibrium double-primed ensemble:

$$V^{\mathrm{mf}} \neq \langle V(\mathbf{r}',\mathbf{r}'')\rangle''. \tag{8.14}$$

Whereas several methods are available to compute potentials of mean force from simulations, as is treated in detail in Chapter 7, Section 7.5 on page 227, empirical validation and generally adjustments are always necessary; the best results are often obtained with completely empirical parametrization because the model can then be fine-tuned to deliver the thermodynamic accuracy required for the application. In the next section we consider the special case of superatom models.

8.4 Superatom approach

A special form of coarse graining is the representation of local groups of atoms by one particle, called a *superatom*. Superatoms are especially useful in chain molecules as polymers and lipids, where they typically represent three to five monomer units. This is a compromise between accuracy and simulation efficiency. Several superatom definitions have been published and most applications do not include additional friction and noise to represent the forces due to the left-out degrees of freedom. This should not influence the equilibrium properties, but is likely to yield a faster dynamics than the real system. The "bead-and-spring" models for polymers, which have a long history in polymer physics, are in fact also superatom models, although they were intended as prototype polymer models rather than as simplified representation of a specific real polymer.[5] The interaction between neighboring beads in a chain is often simply a soft harmonic potential that leads to a Gaussian distribution for the distance between beads; since polymers chains can only be extended to an upper limit r_m, a somewhat more realistic model is the FENE (finitely extendable nonlinear elastic) chain model, with a force between neighboring beads with interbead vector $\boldsymbol{r}$ given by

$$\boldsymbol{F} = -H \frac{\boldsymbol{r}}{1 - (r/r_m)^2}. \tag{8.15}$$

See Fan *et al.* (2003) for a stochastic application of the FENE model.

More recently superatom models have been designed to specifically represent real molecules, e.g., alkanes by the coarse-grained model of Nielsen *et al.* (2003), with Lennard–Jones (LJ) superatoms representing three non-hydrogen atoms. In addition to LJ, this model has soft harmonic bond length and bond angle potentials. The model is parameterized on density and surface tension of liquid alkanes and reproduces end-to-end distribution functions obtained from simulations with atomic details. A more general and very successful coarse-grained force field for lipids and surfactant systems has been defined and tested by Marrink *et al.* (2004). It consists of four types of

---

[5] See Müller-Plathe (2002) for a review on multiscale modelling methods for polymers.

particles (charged, polar, non-polar, apolar) with the charged and non-polar types subdivided into four subtypes depending on their hydrogen-bonding capability. Each particle represents about four non-hydrogen atoms; also four water molecules are represented by one particle. The interactions are of Lennard–Jones and Coulomb type, smoothly switched-off at 1.2 nm, with five possible values for the ε and only one (0.47 nm) for the σ parameters of the LJ interaction. In addition there are fairly soft harmonic bond and bond-angle interaction terms; the total number of parameters is eight. Despite its simplicity, the model reproduces density and isothermal compressibility of water and liquid alkanes within 5% and reproduces mutual solubilities of alkanes in water and water in alkanes to within a free energy of $0.5k_BT$. The melting point of water is 290 K. A time step of 50 fs is possible, and since the dynamics of this model (no friction and noise are added) is about four times faster than reality, an effective time step of 0.2 ps can be realized. One then easily simulates real systems with millions of atoms over microseconds, allowing the study of lipid bilayer formation, micellar formation, vesicle formation and gel/liquid crystalline phase changes with realistic results.[6] See Fig. 8.1 for comparison of a coarse-grained and a detailed simulation of the spontaneous formation of a small lipid vesicle.

8.5 The fluctuation–dissipation theorem

Let us look at the long-time behavior of the kinetic energy $K = \sum_i \frac{1}{2} m_i v_i^2$ in the generalized Langevin equation (8.1). The first term on the r.h.s. (the systematic force) simply exchanges kinetic and potential energy, keeping the total energy constant. The second term (friction or *dissipation*) reduces the kinetic energy and the third stochastic term (noise) increases the kinetic energy. In order for the process to be stationary, the velocity correlation functions $\langle \boldsymbol{v}_i(t)\boldsymbol{v}_j(t+\tau)\rangle$ should become independent of t for large t; in particular the average squared velocities $\langle \boldsymbol{v}_i(t)\boldsymbol{v}_j(t)\rangle$, which are *thermodynamic quantities*, should fulfill the equipartition theorem

$$\langle v_{i\alpha}(t)v_{j\beta}(t)\rangle = \frac{k_B T}{m_i}\delta_{ij}\delta_{\alpha\beta}. \tag{8.16}$$

That is, if the random process is realized many times from the same starting configuration at $t = 0$, then after a sufficiently long time – when the memory of the initial conditions has decayed – the average over all realizations should

---

[6] Spontaneous aggregation of lipid bilayers, gel/liquid crystalline transitions, inverted hexagonal phase formation and formation of Micelles: Marrink *et al.* (2004); hexagonal phase formation: Marrink and Mark (2004); vesicle fusion: Marrink and Mark (2003a); vesicle formation: Marrink and Mark (2003b).

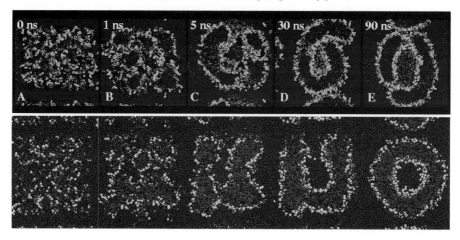

Figure 8.1 Two simulations of the spontaneous formation of a lipid bilayer vesicle. Upper panel: atomic detail molecular dynamics; lower panel: coarse-grained super-atom dynamics simulation (Courtesy of S.-J. Marrink and A. H. de Vries, University of Groningen; reproduced by permission from *J. Comput. Chem.* (van der Spoel *et al.*, 2005)

fulfill the equipartition theorem. This is one – rather restricted – formulation of the *fluctuation–dissipation theorem*.

The general fluctuation–dissipation theorem relates the linear response of some system variable v to the spontaneous fluctuation of v. Kubo (1966) distinguishes a *first* and a *second* fluctuation–dissipation theorem: the first theorem says that the normalized time response of v to a δ-disturbance equals the normalized correlation function of the spontaneously fluctuating v in equilibrium (see Section 18.3 on page 511); the second theorem relates the friction kernel $\zeta(t)$ to the correlation function of the random term $\eta(t)$ in the Langevin equation..

To illustrate these theorems we apply for the sake of simplicity the response to a single *velocity* $v(t)$ that is assumed to follow the *pure* Langevin equation without systematic force:

$$m\dot{v}(t) = -\int_0^t \zeta(\tau)v(t-\tau)\, d\tau + \eta(t) + F^{\text{ext}}(t), \qquad (8.17)$$

with $F^{\text{ext}}(t)$ an external force meant to provide a disturbance to measure the linear response. Now apply an external δ-force at time 0:

$$F^{\text{ext}}(t) = mv_0\delta(t). \qquad (8.18)$$

This produces a jump v_0 in the velocity at time $t = 0$. The velocity $v(t)$ subsequently evolves according to (8.17) (with the external force no longer being present). What we call the δ-response $v_0\Phi(t)$ of the velocity[7] (see (18.1) on page 507) is the *ensemble average* over many realizations of this process, with initial conditions taken randomly from an unperturbed equilibrium distribution and with independent realizations of the noise force $\eta(t)$:

$$v_0\Phi(t) = \langle v(t)\rangle. \tag{8.19}$$

The *first fluctuation–dissipation theorem* states that

$$\Phi(t) = \frac{\langle v(t_0)v(t_0+t)\rangle}{\langle v^2\rangle}, \tag{8.20}$$

where the average is now taken over an unperturbed equilibrium ensemble, for which the velocity correlation function is stationary and hence independent of the time origin t_0.

Proof From (8.19) it follows that

$$\langle \dot{v}(t)\rangle = v_0\frac{d\Phi}{dt}, \tag{8.21}$$

so that on averaging (8.17) immediately gives an equation for Φ, considering that the average over the noise is zero (see (8.2)):

$$m\frac{d\Phi}{dt} = -\int_0^t \zeta(\tau)\Phi(t-\tau)\,d\tau. \tag{8.22}$$

Given the friction kernel $\zeta(\tau)$ and the initial value $\Phi(0) = 1$, this equation determines the response function $\Phi(t)$.

The velocity autocorrelation function $C^v(t) \overset{\text{def}}{=} \langle v(t_0)v(t_0+t)\rangle$ can be found by applying (8.17) to the time t_0+t, multiplying both sides with $v(t_0)$ and taking the ensemble average:

$$m\langle v(t_0)\dot{v}(t_0+t)\rangle = -\int_0^t \zeta(\tau)\langle v(t_0)v(t_0+t-\tau)\rangle\,d\tau + \langle v(t_0)\eta(t_0+t)\rangle, \tag{8.23}$$

which can be written in terms of the velocity correlation function $C^v(t)$, realizing that the last term vanishes because the random force does not correlate with velocities at earlier times (see (8.3)), as

$$m\frac{d}{dt}C^v(t) = -\int_0^t \zeta(\tau)C^v(t-\tau)\,d\tau. \tag{8.24}$$

---

[7] Including v_0 in the response means that Φ is normalized: $\Phi(0) = 1$.

Given the friction kernel $\zeta(\tau)$ and the initial value $C^v(0) = \langle v^2 \rangle$, this equation determines the equilibrium correlation function $C^v(t)$. But this equation is equivalent to the corresponding equation (8.22) for $\Phi(t)$, from which it follows that

$$C^v(t) = \langle v^2 \rangle \Phi(t). \tag{8.25}$$

$\square$

The first fluctuation–dissipation theorem has a solid basis; it applies in general to small deviations from equilibrium, also for systems that include systematic forces (see Section 18.3 on page 511). It is the basis for the derivation of transport properties from equilibrium fluctuations. However, it does not provide the link between friction and noise needed for the implementation of Langevin dynamics.

The *second fluctuation–dissipation theorem* states that

$$\langle \eta(t_0)\eta(t_0 + t) \rangle = m\langle v^2 \rangle \zeta(t) = k_B T \zeta(t). \tag{8.26}$$

This theorem provides the proper connection between friction and noise, but it stands on much weaker grounds than the first theorem. It can be rigorously proven for a pure Langevin equation without systematic force. The proof uses Laplace transforms or one-sided Fourier transforms and rests on the derivation of the stationary velocity autocorrelation function, given the noise correlation function, which must equal the solution of (8.24). We refer the reader for this proof to the literature, where it can be found in several places; a readable discussion is given in the first chapter of Kubo *et al.* (1985). When systematic non-linear forces are present (as is the case in all simulations of real systems), the theorem can no longer be proven to be generally valid. Various special cases involving harmonic forces and heat baths consisting of collections of harmonic oscillators have been considered,[8] and modifications for the general case have been proposed.[9] While the matter appears not to be satisfactorily settled, our recommendation is that *time-dependent friction kernels* should not be used in cases when intrinsic relaxation times, determined by the systematic forces, are of the same order as the characteristic times of the friction kernels.

---

[8] The harmonic-oscillator heat bath was pioneered by Zwanzig (1973) and extended by Cohen (2002); Hernandez (1999) considered the projection of non-equilibrium Hamiltonian systems. Adelman and Doll (1974) simplified Zwanzig's approach for application to atomic collisions with a solid surface.

[9] Ciccotti and Ryckaert (1981) separate the systematic force and obtain a modified friction and noise; Bossis *et al.* (1982) show that the effect of the systematic force is a modification of the second fluctuation–dissipation theorem by the addition of an extra term equal to the correlation of velocity and systematic force. McDowell (2000) considers a chain of heat baths and concludes that an extra bias term should be added to the random force.

However, the memory-free combination of *time-independent friction* and *white noise* does yield consistent dynamics, also in the presence of systematic forces, with proper equilibrium fluctuations. This memoryless approximation is called a *Markovian* process,[10] and we shall call the corresponding equation (which may still be multidimensional) the *Markovian Langevin equation*.

Consider, first for the simple one-dimensional case, the change in kinetic energy due to a Markovian friction force and a white-noise stochastic force. The equation of motion is

$$m\dot{v} = F(t) - \zeta v + \eta(t), \tag{8.27}$$

with

$$\langle \eta(t_0)\eta(t_0 + t)\rangle = A_\eta \delta(t), \tag{8.28}$$

where A_η is the *intensity* of the noise force. Consider the kinetic energy $K(t) = \frac{1}{2}mv^2$. The friction force causes a decrease of K:

$$\left(\frac{dK}{dt}\right)_{\text{friction}} = m v\dot{v} = -\zeta v^2 = -\frac{2\zeta}{m}K. \tag{8.29}$$

The stochastic term causes an increase in K. Consider a small time step Δt, which causes a change in velocity:

$$m\Delta v = \int_t^{t+\Delta t} \eta(t')\, dt'. \tag{8.30}$$

The change in K is

$$(\Delta K)_{\text{noise}} = \frac{1}{2}m[(v + \Delta v)^2 - v^2] = mv\Delta v + \frac{1}{2}m(\Delta v)^2. \tag{8.31}$$

We are interested in the average over the realizations of the stochastic process. The first term on the r.h.s. vanishes as Δv is not correlated with v. The second term yields a double integral:

$$\langle \Delta K \rangle_{\text{noise}} = \frac{1}{2m} \int_t^{t+\Delta t} dt' \int_t^{t+\Delta t} dt'' \langle \eta(t')\eta(t'')\rangle = \frac{1}{2m} A_\eta \Delta t; \tag{8.32}$$

therefore the noise causes on average an increase of K:

$$\left\langle \frac{dK}{dt}\right\rangle_{\text{noise}} = \frac{A_\eta}{2m}. \tag{8.33}$$

Both of these changes are independent of the systematic force. They balance

---

[10] A Markov process is a discrete stochastic process with transition probabilities between successive states that depend only on the properties of the last state, and not of those of previous states.

on average when $\langle K \rangle = A_\eta/(4\zeta)$. Using the equilibrium value at a reference temperature T_0 for one degree of freedom:

$$\langle K \rangle = \frac{1}{2}k_B T_0, \tag{8.34}$$

it follows that a stationary equilibrium kinetic energy is obtained for

$$A_\eta = 2\zeta k_B T_0. \tag{8.35}$$

If the initial system temperature T deviates from $T_0 = A/(2\zeta k_B)$, it will decay exponentially to the reference temperature T_0 set by the noise, with a time constant $m/2\zeta$:

$$\frac{dT}{dt} = -\frac{2\zeta}{m}(T - T_0). \tag{8.36}$$

Thus the added friction and noise stabilize the variance of the velocity fluctuation and contribute to the robustness of the simulation. The flow of kinetic energy into or out of the system due to noise and friction can be considered as heat exchange with a bath at the reference temperature. This exchange is independent of the systematic force and does not depend on the time dependence of the velocity autocorrelation function. It is easy to see that the latter is not the case for a time-dependent (non-Markovian) friction: (8.29) then reads

$$\left\langle \frac{dK}{dt} \right\rangle_{\text{friction}} = m\langle v(t)\dot{v}(t) \rangle = - \int_0^\infty \zeta(\tau)\langle v(t)v(t-\tau)\rangle \, d\tau, \tag{8.37}$$

which clearly depends on the velocity correlation function, which in turn depends on the behavior of the systematic force.

The whole purpose of simplifying detailed dynamics by Langevin dynamics is reducing *fast* degrees of freedom to a combination of friction and noise. When these "irrelevant" degrees of freedom indeed relax fast with respect to the motion of the relevant degrees of freedom, they stay near equilibrium under constrained values of the "relevant" degrees of freedom and in fact realize a good approximation of the constrained canonical ensemble that is assumed in the derivation of the systematic force (8.13) and that allows the simplification without loosing thermodynamic accuracy. "Fast" means that the correlation time of the force due to the "irrelevant" degrees of freedom (the frictional and random force) is short with respect to the relaxation time within the "relevant" system, due to the action of the systematic force. The latter is characterized by the velocity correlation function in the absence of friction and noise. If the forces from the "irrelevant" degrees of freedom are fast in this sense, a Markovian friction and noise friction will be a good approximation that even preserves the slow dynamics of the system; if they are not fast, a Markovian Langevin simulation will perturb the dynamics, but still preserve the thermodynamics.

8.6 Langevin dynamics

In this section we start with the generalized Langevin equation (8.1), which we first formulate in general coordinates. Then, in view of the discussion in the previous section, we immediately reduce the equation to the memory-free Markovian limit, while keeping the multidimensional formulation, and check the fluctuation–dissipation balance. Subsequently we reduce also the spatial complexity to obtain the simple Langevin equation.

8.6.1 Langevin dynamics in generalized coordinates

Consider a full Hamiltonian dynamical system with n degrees of freedom, expressed in $2n$ generalized coordinates and momenta $z = \{q, p\}$. The momenta are connected to the coordinates by the $n \times n$ mass tensor $\mathbf{M}$ (see (15.16) on page 401):

$$\mathbf{p} = \mathbf{M}\dot{\mathbf{q}}, \tag{8.38}$$

with inverse

$$\dot{\mathbf{q}} = \mathbf{M}^{-1}\mathbf{p}. \tag{8.39}$$

The coordinates are distinguished in *relevant* coordinates $\mathbf{q}'$ and *irrelevant* coordinates $\mathbf{q}''$. We partition the inverse mass tensor (as is done in the discussion on constraints, Section 17.9.3 on page 501) as

$$\mathbf{M}^{-1} = \begin{pmatrix} \mathbf{X} & \mathbf{Y} \\ \mathbf{Y}^{\mathsf{T}} & \mathbf{Z} \end{pmatrix}, \tag{8.40}$$

so that

$$(\dot{\mathbf{q}}' \quad \dot{\mathbf{q}}'') = \begin{pmatrix} \mathbf{X} & \mathbf{Y} \\ \mathbf{Y}^{\mathsf{T}} & \mathbf{Z} \end{pmatrix} \begin{pmatrix} \mathbf{p}' \\ \mathbf{p}'' \end{pmatrix}. \tag{8.41}$$

The next step is to find a Langevin equation of motion for $\mathbf{q}'$ by averaging over a canonical distribution for the double-primed subsystem:

$$\dot{\mathbf{q}}' = \langle \mathbf{X} \rangle_{''} \mathbf{p}' + \langle \mathbf{Y}\mathbf{p}'' \rangle_{''}, \tag{8.42}$$

$$\dot{\mathbf{p}}' = -\langle \frac{\partial V}{\partial \mathbf{q}'} \rangle_{''} + \text{ friction } + \text{ noise}. \tag{8.43}$$

The canonical averaging is defined as

$$\langle A \rangle_{''} \stackrel{\text{def}}{=} \frac{\int A(z) \exp[-\beta H(z)]\, dz''}{\int \exp[-\beta H(z)]\, dz''}. \tag{8.44}$$

We recognize the r.h.s. of (8.43) as the Langevin force, similar to the cartesian Langevin force of (8.1), but the l.h.s. is not equal to the simple $m_i \dot{v}_i$ of the cartesian case. Instead we have, in matrix notation, and denoting $\dot{\mathbf{q}}''$ by $\mathbf{v}$,

$$\frac{d}{dt} \langle \mathbf{X} \rangle_{''}^{-1} \mathbf{v}(t) = -\frac{\partial V^{\text{mf}}}{\partial \mathbf{q}} - \int_0^t \boldsymbol{\zeta}(\tau) \mathbf{v}(t - \tau)\, d\tau + \boldsymbol{\eta}(t). \tag{8.45}$$

Here we have omitted the second term in (8.42) because it is an odd function of $\mathbf{p}''$ that vanishes on averaging. In principle, $\mathbf{X}$ can be a function of primed coordinates, in which case the equations become difficult to solve. But practice is often permissive, as we shall see.

Let us have a closer look at the matrix $\mathbf{X}$. Its elements are (Fixman, 1979)

$$X_{kl} = \sum_i \frac{1}{m_i} \frac{\partial q_k''}{\partial \mathbf{r}_i} \cdot \frac{\partial q_l''}{\partial \mathbf{r}_i}. \tag{8.46}$$

Typical "relevant" degrees of freedom are cartesian coordinates of "superatoms" that represents a cluster of real atoms: the radius vector of the center of mass of a cluster of atoms, some linear combination of cartesian coordinates that represent collective motions (principal modes or principal components of a fluctuation matrix), etc. Other cases (e.g., reaction coordinates) may involve distances between two particles or between two groups of particles. The inverse mass matrix $\mathbf{X}$ is particularly simple in these cases. For example, if the relevant coordinates are components of vectors $\boldsymbol{R}_k = \sum_i \alpha_{ki} \boldsymbol{r}_i$, the inverse mass tensor is diagonal with constant terms

$$X_{kl} = \delta_{kl} \sum_i \frac{1}{m_i} \alpha_{ki}^2. \tag{8.47}$$

In the case that the relevant degree of freedom is a distance r_{12} between two particles with mass m_1 and m_2, the inverse mass tensor has one element equal to $(1/m_1) + (1/m_2)$, which is the inverse of the reduced mass of the two particles. The evaluation is equivalent to the evaluation in the case of constraints, treated in Section 17.9.3; see (17.201) on page 502. In all these cases the inverse mass tensor is constant and does not depend on time-dependent coordinates. We shall from hereon restrict ourselves to such cases and write $\mathbf{M}$ for the inverse of $\mathbf{X}$, yielding the general Langevin equation:[11]

$$\mathbf{M}\dot{\mathbf{v}} = \mathbf{F}^{\mathrm{s}}(\mathbf{q}') - \int_0^t \boldsymbol{\zeta}(\tau)\mathbf{v}(t-\tau)\,d\tau + \boldsymbol{\eta}(t), \tag{8.48}$$

$$\mathbf{v} = \dot{\mathbf{q}}, \tag{8.49}$$

$$\mathbf{M} = \mathbf{X}^{-1}. \tag{8.50}$$

For $\mathbf{X}$ see (8.46). Of course, before applying this equation the user should check that the inverse mass tensor is indeed time-independent. For degrees of freedom involving angles this may not always be the case.

We note that the formulation given in (8.48) includes the simple case that the relevant degrees of freedom are the cartesian coordinates of selected particles; the matrix $\mathbf{M}$ is then simply the diagonal matrix of particle masses.

8.6.2 Markovian Langevin dynamics

We shall now consider the dissipation–fluctuation balance for the case of generalized coordinates including a mass tensor. But since we cannot guarantee the validity of the second dissipation–fluctuation theorem for the time-generalized equation (8.48), we shall restrict ourselves to the Markovian multidimensional Langevin equation

$$\mathbf{M}\dot{\mathbf{v}} = \mathbf{F}^{\mathrm{s}}(\mathbf{q}') - \boldsymbol{\zeta}\mathbf{v}(t) + \boldsymbol{\eta}(t). \tag{8.51}$$

Here $\boldsymbol{\zeta}$ is the friction tensor.

Dissipation–fluctuation balance

Consider the generalized equipartition theorem, treated in Section 17.10 and especially the velocity correlation expressed in (8.16) on page 503:

$$\langle \mathbf{v}\mathbf{v}^{\mathsf{T}} \rangle = \mathbf{M}^{-1}k_{\mathrm{B}}T. \tag{8.52}$$

This equation is valid for the primed subsystem, where $\mathbf{M}$ is the inverse of $\mathbf{X}$ as defined above. In order to establish the relation between friction and noise we follow the arguments of the previous section, leading to (8.35) on page 262, but now for the multidimensional case.

The change due to friction is given by

$$\frac{d}{dt}\langle \mathbf{v}\mathbf{v}^{\mathsf{T}} \rangle = \langle \dot{\mathbf{v}}\mathbf{v}^{\mathsf{T}} \rangle + \langle \mathbf{v}\dot{\mathbf{v}}^{\mathsf{T}} \rangle = 2\langle \dot{\mathbf{v}}\mathbf{v}^{\mathsf{T}} \rangle$$

$$= -2\mathbf{M}^{-1}\boldsymbol{\zeta}\langle \mathbf{v}\mathbf{v}^{\mathsf{T}} \rangle. \tag{8.53}$$

---

[11] The notation $\mathbf{M}$ should not cause confusion with the same notation used for the mass tensor of the full system, which would be obviously meaningless in this equation. Note that $\mathbf{M}$ used here in the Langevin equation is *not* a submatrix of the mass tensor of the full system! It should be computed as the inverse of $\mathbf{X}$.

In the first line we have used the symmetry of $\langle \mathbf{vv}^\mathsf{T} \rangle$. The change in a step Δt due to noise is given by

$$\Delta \langle \mathbf{vv}^\mathsf{T} \rangle = \left(\mathbf{M}^{-1} \int_t^{t+\Delta t} \eta(t')\, dt' \right) \left(\int_t^{t+\Delta t} \eta^\mathsf{T}(t')\, dt'\, \mathbf{M}^{-1} \right)$$

$$= \mathbf{M}^{-1}\mathbf{A}\mathbf{M}^{-1}\Delta t. \tag{8.54}$$

The matrix $\mathbf{A}$ is the noise correlation matrix:

$$\langle \eta_i(t_0)\eta_j(t_0 + t) \rangle = A_{ij}\delta(t). \tag{8.55}$$

We used the symmetry of $\mathbf{M}^{-1}$. Balancing the changes due to friction and noise, it is seen that friction and noise are related by

$$\mathbf{A} = 2k_\mathrm{B}T\zeta. \tag{8.56}$$

This is the multidimensional analog of (8.35). It appears that the noise terms for the different degrees of freedom are not independent of each other when the friction tensor is not diagonal, i.e., when the velocity of one degree of freedom influences the friction that another degree of freedom undergoes. We see from (8.56) that the friction tensor must be symmetric, as the Markovian noise correlation is symmetric by construction.

It is also possible, and for practical simulations more convenient, to express the noise forces as linear combinations of *independent normalized* white noise functions $\eta_k^0(t)$ with the properties

$$\langle \eta_k^0(t) \rangle = 0, \tag{8.57}$$

$$\langle \eta_k^0(t_0)\eta_l^0(t_0 + t) \rangle = \delta_{kl}\delta(t), \tag{8.58}$$

$$\eta_i(t) = \sum_k B_{ik}\eta_k^0(t) \quad \text{or} \quad \boldsymbol{\eta} = \mathbf{B}\boldsymbol{\eta}^0. \tag{8.59}$$

It now follows that

$$\mathbf{B}\mathbf{B}^\mathsf{T} = \mathbf{A} = 2\zeta k_\mathrm{B}T. \tag{8.60}$$

In order to construct the noise realizations in a simulation, the matrix $\mathbf{B}$ must be solved from this equation, knowing the friction matrix. The solution of this square-root operation is not unique; a lower trangular matrix is obtained by Choleski decomposition (see Engeln-Müllges and Uhlig, 1996, for algorithms).

Simple Langevin dynamics

The generalized or the Markovian Langevin equation can be further approximated if the assumption can be made that the friction acts locally on each degree of freedom without mutual influence. In that case the friction tensor is diagonal and the *simple Langevin equation* is obtained:

$$(\mathbf{M}\dot{v})_i = F_i^s - \zeta_i v_i(t) + \eta(t), \tag{8.61}$$

with

$$\langle \eta_i(t_0)\eta_j(t_0 + t)\rangle = 2k_B T \zeta_i \delta_{ij}\delta(t). \tag{8.62}$$

Although there is no frictional coupling, these equations are still coupled if the mass tensor $\mathbf{M}$ is not diagonal. In the common diagonal case the l.h.s. is replaced by

$$(\mathbf{M}\dot{v})_i = m_i v_i. \tag{8.63}$$

In the ultimate simplification with negligible systematic force, as applies to a solute particle in a dilute solution, the simple *pure* Langevin equation is obtained:

$$m\dot{v} = -\zeta v + \eta(t). \tag{8.64}$$

As this equation can be exactly integrated, the properties of v can be calculated; they serve as illustration how friction and noise influence the velocity, but are not strictly valid when there are systematic forces as well. The solution is

$$v(t) = v(0)e^{-\zeta t/m} + \frac{1}{m}\int_0^t e^{-\zeta\tau/m}\eta(t-\tau)\,d\tau, \tag{8.65}$$

which, after a sufficiently long time, when the influence of the initial velocity has died out, reduces to

$$v(t) = \frac{1}{m}\int_0^\infty e^{-\zeta\tau/m}\eta(t-\tau)\,d\tau. \tag{8.66}$$

We see from (8.65) that in the absence of noise the velocity decays exponentially with time constant m/ζ, and we expect from the first dissipation–fluctuation theorem (page 259) that the velocity autocorrelation function will have the same exponential decay. This can be shown directly from (8.66) in the case that $\eta(t)$ is a white noise with intensity $2\zeta k_B T$. It follows that, if $\eta(t)$ is stationary, $v(t)$ is also stationary; when the noise intensity is $2\zeta k_B T$, the variance of the velocity is $k_B T/m$. Note that it is not necessary to specify the distribution function of the random variable.

We can also compute the probability distribution $\rho(v)$ for the velocity when equilibrium has been reached. To do this we need an equation for $\rho(v,t)$ as it is generated by the stochastic process defined by (8.64). Such equations are *Fokker–Planck equations*,[12] of which we shall see more examples in the following section. In this one-dimensional case the Fokker–Planck equation is

$$\frac{\partial \rho}{\partial t} = \frac{\zeta}{m}\frac{\partial}{\partial v}(\rho v) + \frac{\zeta k_B T}{m^2}\frac{\partial^2 \rho}{\partial t^2}. \tag{8.67}$$

The equation is an expression of the conservation of total probability, leading to a continuum equation $\partial \rho/\partial t = -\nabla_v(J)$, where J is the probability flux consisting of a drift term due to friction and a diffusional term due to noise. The diffusional term follows from the fact that

$$\frac{\partial \rho}{\partial t} = \frac{1}{2}B\frac{\partial \rho}{\partial t} \tag{8.68}$$

implies that (see exercise 8.3)

$$\langle (\Delta v)^2 \rangle = B\Delta t \tag{8.69}$$

with the variance of the velocity fluctuation given by $(2\zeta k_B T/m^2)\Delta t$ (see (8.31), (8.32) and (8.35) on page 262).

The equilibrium case $(\partial \rho/\partial t = 0)$ has the solution

$$\rho(v) = \sqrt{\frac{m}{2\pi k_B T}}\exp\left(-\frac{mv^2}{2k_B T}\right), \tag{8.70}$$

which is the Maxwell distribution.

8.7 Brownian dynamics

If systematic forces are slow, i.e., when they do not change much on the time scale $\tau_c = m/\zeta$ of the velocity correlation function, we can average the Langevin equation over a time $\Delta t > \tau_c$. The average over the inertial term $M\dot{v}$ becomes small and can be neglected; as a result the acceleration no longer figures in the equation. We obtain *non-inertial* dynamical equations:

$$0 \approx F_i[\mathbf{q}(t)] - \sum_j \zeta_{ij}v_j(t) + \eta_i(t), \tag{8.71}$$

or, in matrix notation:

$$\zeta\mathbf{v} = \mathbf{F} + \boldsymbol{\eta}(t), \tag{8.72}$$

yielding the Brownian equation for the velocities:

$$\mathbf{v} = \dot{\mathbf{q}} = \zeta^{-1}\mathbf{F} + \zeta^{-1}\mathbf{B}\boldsymbol{\eta}^0(t), \tag{8.73}$$

---

[12] See van Kampen (1981) for an extensive treatment of the relation between stochastic equations and the corresponding Fokker–Planck equations.

$$\mathbf{BB}^\mathsf{T} = 2\zeta k_\mathrm{B}T, \tag{8.74}$$

$$\langle \boldsymbol{\eta}^0(t) \rangle = 0, \tag{8.75}$$

$$\langle \boldsymbol{\eta}^0(t_0)(\boldsymbol{\eta}^0)^\mathsf{T}(t_0 + t) \rangle = \mathbf{1}\,\delta(t). \tag{8.76}$$

In simulations the velocity can be eliminated and the positions can be up-dated by a simple Euler step:

$$\mathbf{q}(t + \Delta t) = \mathbf{q}(t) + \boldsymbol{\zeta}^{-1}\mathbf{F}(t)\Delta t + \boldsymbol{\zeta}^{-1}\mathbf{Br}\sqrt{\Delta t}, \tag{8.77}$$

where $\mathbf{r}$ is a vector of random numbers, each drawn independently from a probability distribution (conveniently, but not necessarily, Gaussian) with

$$\langle r \rangle = 0, \quad \langle r^2 \rangle = 1. \tag{8.78}$$

Note that – as the dynamics is non-inertial – the mass does not enter in the dynamics of the system anymore. Apart from coupling through the forces, the coupling between degrees of freedom enters only through mutual friction coefficients.

For the simple Brownian dynamics with diagonal friction matrix, and using the *diffusion constant* $D_i = k_\mathrm{B}T/\zeta_{ii}$, this equation reduces to

$$q_i(t + \Delta t) = q_i(t) + \frac{D}{k_\mathrm{B}T} F_i(t)\Delta t + \xi, \tag{8.79}$$

where ξ is a random number, drawn from a probability distribution with

$$\langle \xi \rangle = 0, \quad \langle \xi^2 \rangle = 2D\Delta t. \tag{8.80}$$

One can devise more sophisticated forms that use the forces at half steps in order to integrate the drift part of the displacement to a higher order, but the noise term tends to destroy any higher-order accuracy.

¿From (8.77) it is seen that friction scales the time: decreasing the friction (or increasing the diffusion constant) has the same effect as increasing the time step. It is also seen that the displacement due to the force is propor-tional to the time step, but the displacement due to noise is proportional to the square root of the time step. This means that slow processes that allow longer time steps are subjected to smaller noise intensities. For macroscopic averages the noise will eventually become negligible.

8.8 Probability distributions and Fokker–Planck equations

In Section 8.6 we used a Fokker–Planck equation to derive the probability distribution for the velocity in the case of the simple pure Langevin equation (see page 267). This led to the satisfactory conclusion that the simple pure

Langevin equation leads to a Maxwellian distribution. In this section we formulate Fokker–Planck equations for the more general Markovian Langevin
equation and for the Brownian dynamics equation. What we wish to gain
from the corresponding Fokker–Planck equations is insight into the steady-
state and equilibrium behavior in order to judge their compatibility with
statistical mechanics, and possibly also to obtain differential equations that
can be solved analytically.

Stochastic equations generate random processes whose distribution functions behave in time according to certain second-order partial differential
equations, called *Fokker–Planck equations*. They follow from the *master
equation* that describes the transition probabilities of the stochastic process. The Fokker–Planck equation is similar to the Liouville equation in
statistical mechanics that describes the evolution of density in phase space
resulting from a set of equations of motion; the essential difference is the
stochastic nature of the underlying process in the case of Fokker–Planck
equations.

8.8.1 General Fokker–Planck equations

We first give the general equations, and apply these to our special cases.
Consider a vector of variables $\mathbf{x}$ generated by a stochastic equation:[13]

$$\dot{\mathbf{x}}(t) = \mathbf{a}(\mathbf{x}(t)) + \mathbf{B}\eta^0(t), \tag{8.81}$$

with $\eta^0(t)$ independent normalized white noise processes, as specified by
(8.58). The variables may be any observable, as coordinates or velocities
or both. The first term is a *drift term* and the second a *diffusion term*.
The corresponding Fokker–Planck equation in the Itô interpretation for the
distribution function $\rho(\mathbf{x}, t)$ (van Kampen, 1981; Risken, 1989) is in matrix
notation

$$\frac{\partial \rho}{\partial t} = -\boldsymbol{\nabla}_x^{\mathsf{T}}(\mathbf{a}\rho) + \frac{1}{2}\operatorname{tr}\left(\boldsymbol{\nabla}_x\boldsymbol{\nabla}_x^{\mathsf{T}}\mathbf{B}\mathbf{B}^{\mathsf{T}}\rho\right), \tag{8.82}$$

or for clarity written in components:

$$\frac{\partial \rho}{\partial t} = -\sum_i \frac{\partial}{\partial x_i}(a_i\rho) + \frac{1}{2}\sum_{ij} \frac{\partial^2}{\partial x_i \partial x_j}\sum_k B_{ik}B_{jk}\rho. \tag{8.83}$$

---

[13] We made a remark on this mathematically incorrect form of a stochastic differential equation
on page 253 in relation to (8.4). The proper equation is $d\mathbf{x} = \mathbf{a}\,dt + \mathbf{B}\,d\mathbf{w}$, where $\mathbf{w}$ is a vector
of Wiener processes.

8.8.2 Application to generalized Langevin dynamics

Let us now apply this general equation to the general Markovian Langevin equation (8.51):

$$\dot{\mathbf{q}} = \mathbf{v}, \tag{8.84}$$

$$\dot{\mathbf{v}} = \mathbf{M}^{-1}\mathbf{F}(\mathbf{q}) - \mathbf{M}^{-1}\boldsymbol{\zeta}\mathbf{v} + \mathbf{M}^{-1}\mathbf{B}\boldsymbol{\eta}^0. \tag{8.85}$$

The single-column matrix $\mathbf{x}$ consists of a concatenation of $\mathbf{q}$ and $\mathbf{v}$. The single-column matrix $\mathbf{a}$ then consists of a concatenation of $\mathbf{v}$ and $\mathbf{M}^{-1}\mathbf{F}(\mathbf{q}) - \mathbf{M}^{-1}\boldsymbol{\zeta}\mathbf{v}$. Carefully applying (8.82) to this $\mathbf{x}$ (and assuming $\mathbf{B}$ to be constant) yields

$$\frac{\partial\rho}{\partial t} = -\mathbf{v}^{\mathsf{T}}\boldsymbol{\nabla}_q\rho - \mathbf{F}^{\mathsf{T}}\mathbf{M}^{-1}\boldsymbol{\nabla}_v\rho + \operatorname{tr}(\mathbf{M}^{-1}\boldsymbol{\zeta})\rho$$

$$+ \mathbf{v}^{\mathsf{T}}\boldsymbol{\zeta}\mathbf{M}^{-1}\boldsymbol{\nabla}_v\rho + \frac{1}{2}\operatorname{tr}(\mathbf{M}^{-1}\mathbf{B}\mathbf{B}^{\mathsf{T}}\mathbf{M}^{-1}\boldsymbol{\nabla}_q\boldsymbol{\nabla}_q\rho). \tag{8.86}$$

Note that the noise coefficient is related to the friction tensor by (8.60) on page 266:

$$\mathbf{B}\mathbf{B}^{\mathsf{T}} = 2\boldsymbol{\zeta}k_{\mathrm{B}}T. \tag{8.87}$$

This rather awesome multidimensional equation can of course be solved numerically and will give the same results as a simulation of the original stochastic equation. More insight is obtained when we reduce this equation to one dimension and obtain the rather famous *Kramers equation* (Kramers, 1940):[14]

$$\frac{\partial\rho}{\partial t} = -v\frac{\partial\rho}{\partial q} - \frac{F}{m}\frac{\partial\rho}{\partial v} + \frac{\rho v}{m}\frac{\partial\rho}{\partial v} + \frac{\zeta}{m}\rho + \frac{\zeta k_{\mathrm{B}}T}{m^2}\frac{\partial^2\rho}{\partial v^2}. \tag{8.88}$$

Even this much simpler equation cannot be solved analytically, but it can be well approximated to obtain classical rates for barrier-crossing processes. Kramer's theory has been used extensively to find damping corrections to the reaction rates derived from Eyring's transition state theory. It is easy to find the equilibrium distribution $\rho_{\mathrm{eq}}(q, v)$ by setting $\partial\rho/\partial t = 0$ (see Exercise 8.5):

$$\rho_{\mathrm{eq}}(q, v) \propto \exp\left[-\frac{mv^2}{2k_{\mathrm{B}}T}\right]\exp\left[-\frac{V(q)}{k_{\mathrm{B}}T}\right], \tag{8.89}$$

where V is the potential defined by $F = -dV/dq$. Again, this is a satisfactory result compatible with the canonical distribution.

---

[14] A generalization to colored noise and friction and with external noise has been given by Banik *et al.* (2000).

8.8.3 Application to Brownian dynamics

For Brownian dynamics the stochastic equations (8.73) and (8.74) are a function of $\mathbf{q}$ only. The corresponding Fokker–Planck equation is

$$\frac{\partial \rho}{\partial t} = -\sum_i \frac{\partial}{\partial q_i}[(\boldsymbol{\zeta}^{-1}\mathbf{F})_i \rho] + k_{\mathrm{B}}T \sum_{ij}(\boldsymbol{\zeta}^{-1})_{ij}\frac{\partial^2 \rho}{\partial q_i \partial q_j}. \tag{8.90}$$

For the case of diagonal friction which reduces to a set of one-dimensional equations (only coupled through the forces), the stochastic equation and the corresponding Fokker–Planck equation read

$$\dot{q} = \frac{D}{k_{\mathrm{B}}T}(F + B\eta^0), \tag{8.91}$$

$$\frac{\partial \rho}{\partial t} = -\frac{D}{k_{\mathrm{B}}T}\frac{\partial}{\partial q}(\rho F) + D\frac{\partial^2 \rho}{\partial q^2}. \tag{8.92}$$

Setting $\partial \rho/\partial t = 0$ and writing $F = -dV/dq$, we find the equilibrium solution

$$\rho(q) \propto \exp\left[-\frac{V(q)}{k_{\mathrm{B}}T}\right], \tag{8.93}$$

which again is the canonical distribution. In order to obtain the canonical distribution by simulation using the stochastic Brownian equation, it is necessary to take the time step small enough for $F\Delta t$ to be a good approximation for the step made in the potential V. If that is not the case, integration errors will produce deviating distributions. However, by applying an acceptance/rejection criterion to a Brownian step, a canonical distribution can be enforced. This is the subject of the following section.

8.9 Smart Monte Carlo methods

The original Metropolis Monte Carlo procedure (Metropolis *et al.*, 1953) consists of a random step in configuration space, followed by an acceptance criterion ensuring that the accepted configurations sample a prescribed distribution function. For example, assume we wish to generate an ensemble with canonical probabilities:

$$w(\mathbf{r}) \propto e^{-\beta V(\mathbf{r})}. \tag{8.94}$$

Consider a random configurational step from $\mathbf{r}$ to $\mathbf{r}' = \mathbf{r} + \Delta \mathbf{r}$ and let the potential energies be given by

$$E = V(\mathbf{r}), \tag{8.95}$$

$$E' = V(\mathbf{r}'). \tag{8.96}$$

The random step may concern just one coordinate or one particle at a time, or involve all particles at once. The sampling must be homogeneous over space. The *transition* probabilities $W_\rightarrow$ from r to r' and $W_\leftarrow$ from r' to r should fulfill the *detailed balance condition*:

$$w(r)W_\rightarrow = w(r')W_\leftarrow, \tag{8.97}$$

leading to the ratio

$$\frac{W_\rightarrow}{W_\leftarrow} = \frac{w(r')}{w(r)} = e^{-\beta(E'-E)}. \tag{8.98}$$

This is accomplished by accepting the step with a probability $p_\rightarrow^{acc}$:

$$\text{for} \quad E' - E \leq 0: \quad W_\rightarrow = p_\rightarrow^{acc} = 1, \tag{8.99}$$
$$\text{for} \quad E' - E > 0: \quad W_\rightarrow = p_\rightarrow^{acc} = e^{-\beta(E'-E)}, \tag{8.100}$$

as is easily seen by considering the backward transition probability:

$$\text{for} \quad E' - E < 0: \quad W_\leftarrow = e^{\beta(E'-E)}, \tag{8.101}$$
$$\text{for} \quad E' - E \geq 0: \quad W_\leftarrow = 1, \tag{8.102}$$

which fulfills (8.98). The acceptance with a given probability $p_\rightarrow^{acc} < 1$ is realized by drawing a uniform random number $0 \leq \eta < 1$ and accepting the step when $\eta < p_\rightarrow^{acc}$. When a step is not accepted, the previous step should be counted again.

In the "smart Monte Carlo" procedure, proposed by Rossky *et al.* (1978), a Brownian dynamic step is attempted according to (8.79) and (8.80), sampling ξ from a Gaussian distribution. We denote the configuration, force and potential energy before the attempted step by r, F and E and after the attempted step by r', F' and E':

$$r' = r + \beta D \Delta t F + \xi. \tag{8.103}$$

The transition probability is not uniform in this case, because of the bias introduced by the force:

$$W_\rightarrow \propto \exp\left(-\frac{(r' - r - \beta D \Delta t F)^2}{4D\Delta t}\right) p_\rightarrow^{acc}, \tag{8.104}$$

because this is the probability that the random variable ξ is chosen such that this particular step results. Now imposing the detailed balance condition (8.98):

$$\frac{W_\rightarrow}{W_\leftarrow} = e^{-\beta(E'-E)}, \tag{8.105}$$

we find for the forward/backward acceptance ratio:

$$\frac{p_{\rightarrow}^{\text{acc}}}{p_{\leftarrow}^{\text{acc}}} = \exp\left[-\beta(E' - E) + \frac{(r' - r - \beta D\Delta t F)^2 - (r - r' - \beta D\Delta t F')^2}{4D\Delta t}\right]$$

$$= e^{-\beta\Delta}, \tag{8.106}$$

with

$$\Delta = E' - E + \frac{1}{2}(r' - r)\cdot(F + F') - \frac{1}{4}\beta D\Delta t(F^2 - F'^2). \tag{8.107}$$

Note that Δ for the forward and backward step are equal in magnitude and opposite in sign. The acceptance is realized, similar to (8.98) and (8.100), by choosing:

$$\text{for}\qquad \Delta \leq 0: \quad p_{\rightarrow}^{\text{acc}} = 1, \tag{8.108}$$

$$\text{for}\qquad \Delta > 0: \quad p_{\rightarrow}^{\text{acc}} = e^{-\beta\Delta}. \tag{8.109}$$

The latter acceptance is implemented by accepting the step when a homogeneous random number $0 \leq \eta < 1$ is smaller than $\exp(-\beta\Delta)$. When a step is not accepted, the previous step should be counted again.

The rejection of a step does destroy the dynamical continuity of the Brownian simulation, but ensures that the proper canonical distribution will be obtained. In practice, the time step – or rather the product $D\Delta t$ – can be chosen such that almost all steps are accepted and the dynamics remains valid, at least within the approximations that have led to the Brownian stochastic equation.

8.10 How to obtain the friction tensor

How can the friction tensor – or, equivalently, the noise correlation matrix – be obtained for use in Langevin or Brownian simulations?

There are essentially three different routes to obtain the friction tensor:

(i) from theoretical considerations,
(ii) from empirical data,
(iii) from detailed MD simulations.

The route to be chosen depends on the system and on the choice of "irrelevant" degrees of freedom over which averaging should take place. In general the accuracy required for the friction tensor is not very high: it only influences the dynamical behavior of the system but not the thermodynamic equilibria. This is seen from the Fokker–Planck equation that appears to yield a canonical distribution in configuration space, even for the rather inaccurate Brownian dynamics, which is independent of the applied friction coefficients as long as the fluctuation–dissipation balance is maintained. It is likely that slow processes on a time scale much longer than the characteristic time scale of the friction, which is around m/ζ, will also be handled with reasonable accuracy. In many applications one is more interested in obtaining a fast sampling of configuration phase than in accurately reproducing the real dynamics; in such cases one may choose a rather low friction in order to obtain a faster dynamical behavior. Ultimately one may choose not to add any friction or noise at all and obtain a fast dynamic sampling by just simulating Hamiltonian molecular dynamics of a reduced system with a proper potential of mean force. This is a quite common procedure in simulations based on "superatoms."

In the following we consider a few examples of friction tensors.

8.10.1 Solute molecules in a solvent

The most straightforward application of stochastic dynamics is the simulation of solute molecules in a solvent. In a dilute solution the friction is determined solely by the difference between the velocity $\boldsymbol{v}$ of the solute particle and the bulk velocity $\boldsymbol{u}$ of the solvent:

$$\boldsymbol{F}^{\mathrm{fr}} = -\zeta(\boldsymbol{v} - \boldsymbol{u}), \tag{8.110}$$

and the friction tensor can at most be a 3×3 matrix for a non-spherical particle. For a spherical particle the friction tensor must be isotropic and equal to $\zeta\mathbf{1}$. We have not introduced terms like "bulk velocities" of the bath particles before, implying that such velocities are assumed to be zero. Langevin and Brownian dynamics do not conserve momentum (and conserve energy only as an average) and should not be applied in the formulation given here when the application requires momentum and/or energy conservation.

The friction coefficient ζ follows from the diffusion coefficient D of the particle and the temperature by the Einstein relation

$$\zeta = \frac{k_{\mathrm{B}}T}{D}. \tag{8.111}$$

D can be obtained from experiment or from a simulation that includes the full solvent. The friction coefficient can also be obtained from hydrodynamics if the solvent can be approximated by a continuum with viscosity η, yielding Stokes' law for a spherical particle with radius a:

$$\zeta = 6\pi\eta a. \tag{8.112}$$

When the solution is not dilute, the most important addition is an interaction term in the systematic force; this can be obtained by thermodynamic integration from detailed simulations with pairs of particles at a series of constrained distances. But the friction force on solute particle i will also be influenced by the velocity of nearby solute particles j. This influence is exerted through the intervening fluid and is called the *hydrodynamic interaction*. It can be evaluated from the Navier–Stokes equations for fluid dynamics. The hydrodynamic interaction is a long-range effect that decays with the inverse distance between the particles. The $1/r$ term in the interaction, averaged over orientations, is expressed as a *mobility matrix*, which forms the interaction part of the inverse of the friction matrix; this is known as the *Oseen tensor*. The equations are

$$\zeta^{-1} = \mathbf{H}, \tag{8.113}$$

$$\mathbf{H}_{ii} = \frac{1}{6\pi\eta a}, \tag{8.114}$$

$$\mathbf{H}_{ij} = \frac{1}{8\pi\eta r}\left(1 + \frac{\mathbf{r}\mathbf{r}^\mathsf{T}}{r^2}\right), \tag{8.115}$$

where $\mathbf{r} = \mathbf{r}_i - \mathbf{r}_j$ and $r = |\mathbf{r}|$. Each element of $\mathbf{H}$, defined above, is a 3×3 cartesian matrix; i, j number the solute particles. Hydrodynamic interactions are often included in stochastic modelling of polymers in solution, where the polymer is modelled as a string of beads and the solution is not modelled explicitly. Meiners and Quake (1999) have compared diffusion measurements on colloidal particles with Brownian simulations using the Oseen tensor and found excellent agreement for the positional correlation functions.

8.10.2 Friction from simulation

In cases where theoretical models and empirical data are unavailable the friction parameter can be obtained from analysis of the "observed" forces in constrained simulations with atomic detail. If detailed simulations are done with the "relevant" degrees of freedom q' constrained, the forces acting on the constrained degrees of freedom are the forces from the double-primed subsystem and – if carried to equilibrium – will approximate the sum of the systematic force and the random force that appear in the Langevin equation. The friction force itself will not appear as there are no velocities in the primed coordinates. The average of the constraint force $\langle F^c \rangle$ will be the systematic force, which on integration will produce the potential of mean force. The fluctuation $\Delta F^c(t)$ will be a realization of the random force. If the second fluctuation–dissipation theorem (8.26) holds, then

$$\langle \Delta F^c(t_0) \Delta F^c(t_0 + t) \rangle = k_{\mathrm{B}} T \zeta(t). \tag{8.116}$$

However, we have simplified the noise correlation function to a δ-function and the friction to a constant, which implies that

$$\zeta = \int_0^\infty \zeta(t)\, dt = \frac{1}{k_{\mathrm{B}} T} \int_0^\infty \langle \Delta F^c(t_0) \Delta F^c(t_0 + t) \rangle \, dt. \tag{8.117}$$

One may also define the friction in terms of the diffusion constant $D = k_{\mathrm{B}} T / \zeta$, so that

$$D = \frac{(k_{\mathrm{B}} T)^2}{\int_0^\infty \langle \Delta F^c(t_0) \Delta F^c(t_0 + t) \rangle \, dt}. \tag{8.118}$$

In the multidimensional case, the cross correlation matrix of the constraint forces will similarly lead to the friction tensor.

Exercises

8.1 Solve $m\dot{v} = -\zeta v + \eta(t)$ for the velocity v, given the velocity at $t = 0$, to yield (8.65).

8.2 Compute $\langle v^2(t) \rangle$ when friction and noise are switched on at $t = 0$ by taking the square of (8.65).

8.3 Show that (8.69) follows from (8.69). Do this by showing that the time derivative of $\langle (\Delta v)^2 \rangle$ equals B.

8.4 Write (8.86) out in components.

8.5 Find the equilibrium solution for the Kramers equation (8.88) by separating variables, considering ρ as a product of $f(q)$ and $g(v)$. This splits the equation; first solve for the $g(v)$ part and insert the result into the $f(q)$ part.

9

Coarse graining from particles to fluid dynamics

9.1 Introduction

In this chapter we shall set out to average a system of particles over space and obtain equations for the variables averaged over space. We consider a Hamiltonian system (although we shall allow for the presence of an external force, such as a gravitational force, that has its source outside the system), and – for simplicity – consider a single-component fluid with isotropic behavior. The latter condition is not essential, but allows us to simplify notations by saving on extra indexes and higher-order tensors that would cause unnecessary distraction from the main topic. The restriction to a single component is for simplicity also, and we shall later look at multicomponent systems.

By averaging over space we expect to arrive at the equations of fluid dynamics. These equations describe the motion of fluid elements and are based on the conservation of mass, momentum and energy. They do not describe any atomic details and assume that the fluid is in local equilibrium, so that an equation of state can be applied to relate local thermodynamic quantities as density, pressure and temperature. This presupposes that such thermodynamic quantities can be locally defined to begin with.

For systems that are locally homogeneous and have only very small gradients of thermodynamic parameters, averaging can be done over very large numbers of particles. For the limit of averaging over an infinite number of particles, thermodynamic quantities can be meaningfully defined and we expect the macroscopic equation to become exact. However, if the spatial averaging procedure concerns a limited number of particles, thermodynamic quantities need to be defined also in terms of spatial averages and we expect the macroscopic equations to be only approximately valid and contain unpredictable noise terms.

The situation is quite comparable to the averaging over "unimportant degrees of freedom" as was discussed in Chapter 8. The "important" degrees of freedom are now the density $\rho(\boldsymbol{r})$ as a function of space, which is described with a limited precision depending on the way the spatial averaging is carried out. All other degrees of freedom, i.e., the particle coordinates within the restriction of a given density distribution, form the "unimportant" degrees of freedom, over which proper ensemble-averaging must be done. The forces that determine the evolution of density with time consist of three types:

(i) *systematic* forces, depending on the coarse-grained density distribution (and temperature) itself;

(ii) *frictional* forces, depending on the coarse-grained velocities;

(iii) *random* forces that make up the unpredictable difference between the exact forces and the systematic plus frictional forces.

In analogy with the behavior of a system with a reduced number of degrees of freedom (Chapter 8), we expect the random force to become of relatively less importance when the spatial averaging concerns a larger number of particles, and, in fact, a decrease in standard deviation with the square root of that number. If the spatial averaging is characterized by a smoothing distance a, then the relative standard deviation of the noise in mechanical properties is expected to be proportional to $a^{-3/2}$. As an example of a specific type of coarse graining, we can consider to simplify the description of particle positions by a density on a cubic spatial grid with spacing a. Instead of na^3 particles (where n is the number density of the particles) we now have one density value per grid cell. So we must sum mechanical properties over roughly na^3 particles: correlated quantities will become proportional to a^3 and the noise will be proportional to the square root of that value. In Section 9.3 more precise definitions will be given.

There are three reasons for obtaining the macroscopic equations for the behavior of fluids by a process of coarse graining:

(i) The assumptions on which the macroscopic equations rest (as validity of local density, bulk fluid velocity, and pressure) are made explicit.

(ii) The limits of application of the macroscopic equations become clear and correction terms can be derived.

(iii) The macroscopic equations valid as approximation for a system of real particles are also an approximation for a system of different and larger particles if their interactions are appropriately chosen. Thus the macroscopic problem can be solved by dynamic simulation of a many-particle system with a much smaller number of particles, be it at the expense of increased noise. This is the basis of *dissipative particle dynamics* described in Chapter 11.

In Section 9.2 an overview is given of the macroscopic equations of fluid dynamics. This is done both as a reminder and to set the stage and notation for the systematic derivation of the macroscopic equations from microscopic equations of motion of the constituent particles, given in Section 9.3. Note that in Section 9.3 the macroscopic quantities are properly defined on the basis of particle properties; in the macroscopic theory these quantities (density, fluid velocity, pressure, etc.) are not really defined, and their existence and validity as spatially-dependent thermodynamic quantities is in most textbooks assumed without further discussion.

9.2 The macroscopic equations of fluid dynamics

Note on notation We shall use vector notation as usual, but in some cases (like the derivatives of tensors) confusion may arise on the exact meaning of compound quantities, and a notation using vector or tensor components gives more clarity. Where appropriate, we shall give either or both notations and indicate cartesian components by greek indexes $\alpha, \beta, \ldots$, with the understanding that summation is assumed over repeated indexes. Thus $\partial v_\beta / \partial x_\alpha$ is the $\alpha\beta$ component of the tensor ∇v, but $\partial v_\alpha / \partial x_\alpha$ is the divergence of v: $\nabla \cdot v$.

The principles of single-component fluid dynamics are really simple. The macroscopic equations that describe fluid behavior express the conservation of mass, momentum and energy. The force acting on a fluid element is – in addition to an external force, if present – given by a thermodynamic force and a frictional force. The thermodynamic force is minus the gradient of the pressure, which is related to density and temperature by a locally valid equation of state, and the frictional force depends on velocity gradients. In addition there is heat conduction if temperature gradients exist. Since we assume perfect homogeneity, there is no noise.

Our starting point is the assumption that at every position in space the *bulk velocity* $u(r)$ of the fluid is defined. Time derivatives of local fluid properties can be defined in two ways:

(i) as the partial derivative in a space-fixed coordinate frame, written as $\partial / \partial t$ and often referred to as the *Eulerian derivative*;
(ii) as the partial derivative in a coordinate frame that moves with the bulk fluid velocity u, written as D/Dt and often referred to as the *Lagrangian derivative* or the *material* or *substantive derivative*.

The latter is related to the former by

$$\frac{D}{Dt} = \frac{\partial}{\partial t} + \boldsymbol{u} \cdot \nabla, \qquad \frac{D}{Dt} = \frac{\partial}{\partial t} + u_\alpha \frac{\partial}{\partial x_\alpha}. \tag{9.1}$$

Some equations (as Newton's equation of motion) are simpler when material derivatives are used.

The next most basic local quantity is the *mass density* $\rho(\boldsymbol{r})$ indicating the mass per unit volume. It is only a precise quantity for locally homogeneous fluids, i.e., fluids with small gradients on the molecular scale, on which no real fluid can be homogeneous). We now *define* the *mass flux density* $\boldsymbol{J}(\boldsymbol{r})$ as the mass transported per unit time and per unit area (perpendicular to the flow direction):

$$\boldsymbol{J} = \rho \boldsymbol{u}. \tag{9.2}$$

9.2.1 Conservation of mass

The *continuity equation* expresses the conservation of mass: when there is a net flow of mass out of a volume element, expressed (per unit of volume) as the divergence of the mass flux density, the total amount of mass in the volume element decreases with the same amount:

$$\frac{\partial \rho}{\partial t} + \nabla \cdot \boldsymbol{J} = 0, \qquad \frac{\partial \rho}{\partial t} + \frac{\partial J_\alpha}{\partial x_\alpha} = 0. \tag{9.3}$$

The continuity equation can also be expressed in terms of the material derivative (using the definition of $\boldsymbol{J}$):

$$\frac{D\rho}{Dt} + \rho \nabla \cdot \boldsymbol{u} = 0. \tag{9.4}$$

¿From this formulation we see immediately that for an *incompressible fluid*, for which ρ must be constant if we follow the flow of the liquid, $D\rho/Dt = 0$ and hence the divergence of the fluid velocity must vanish:

$$\nabla \cdot \boldsymbol{u} = 0 \quad \text{(incompressible fluid)}. \tag{9.5}$$

9.2.2 The equation of motion

Next we apply *Newton's law* to the acceleration of a fluid element:

$$\rho \frac{D\boldsymbol{u}}{Dt} = \boldsymbol{f}(\boldsymbol{r}) = \boldsymbol{f}^{\text{int}} + \boldsymbol{f}^{\text{ext}}, \tag{9.6}$$

where $\boldsymbol{f}(\boldsymbol{r})$ is the total force acting per unit volume on the fluid at position $\boldsymbol{r}$. The total force is composed of *internal* forces arising from interactions within the system and *external* forces, arising from sources outside the system. Internal forces are the result of a pressure gradient, but can also represent friction forces due to the presence of gradients in the fluid velocity (or shear rate). Both kinds of forces can be expressed as the divergence of a *stress tensor* σ:[1]

$$\boldsymbol{f}^{\text{int}} = \nabla \cdot \sigma, \qquad f_\alpha^{\text{int}} = \frac{\partial \sigma_{\alpha\beta}}{\partial x_\beta}. \qquad (9.7)$$

Thus Newton's law reads

$$\rho \frac{D\boldsymbol{u}}{Dt} = \rho \frac{\partial \boldsymbol{u}}{\partial t} + \rho(\boldsymbol{u} \cdot \nabla)\boldsymbol{u} = \nabla \cdot \sigma + \boldsymbol{f}^{\text{ext}},$$

$$\rho \frac{Du_\alpha}{Dt} = \rho \frac{\partial u_\alpha}{\partial t} + \rho u_\beta \frac{\partial u_\alpha}{\partial x_\beta} = \frac{\partial \sigma_{\alpha\beta}}{\partial x_\beta} + f_\alpha^{\text{ext}}. \qquad (9.8)$$

Before elaborating on the stress tensor, we will formulate the equations for momentum conservation.

9.2.3 Conservation of linear momentum

The *momentum density*, or the amount of linear momentum per unit volume, defined with respect to a fixed coordinate system, is given by $\rho\boldsymbol{u}$. This is the same as the mass flux density $\boldsymbol{J}$ (see (9.2)). Conservation of momentum means that – in the absence of external forces – the amount of linear momentum increases with time as a result of the net influx of momentum, or – in other words – that the time derivative of the momentum density equals minus the divergence of the *momentum flux density*. Since momentum density is a vector, the momentum flux density must be a tensor. We call it $\boldsymbol{\Pi}$.

The momentum conservation is expressed by

$$\frac{\partial}{\partial t}(\rho\boldsymbol{u}) = -\nabla \cdot \boldsymbol{\Pi}, \qquad \frac{\partial}{\partial t}(\rho u_\alpha) = -\frac{\partial}{\partial x_\beta}\Pi_{\alpha\beta}. \qquad (9.9)$$

This expression can be proved to be valid (see below) when the following *definition* of the momentum flux density tensor is adopted:

$$\Pi_{\alpha\beta} = -\sigma_{\alpha\beta} + \rho u_\alpha u_\beta. \qquad (9.10)$$

This definition makes sense. There are two contributions: momentum can change either because a force gives an acceleration, or because particles flow in or out of a region. The momentum flux density tensor element $\Pi_{\alpha\beta}$ is

---

[1] For a more detailed discussion of the stress tensor and its relation to pressure, see Chapter 17, Section 17.7

the α component of the outward flow of momentum through a unit area perpendicular to the x_β axis.

Proof

$$\frac{\partial \rho u_\alpha}{\partial t} = \rho \frac{\partial u_\alpha}{\partial t} + u_\alpha \frac{\partial \rho}{\partial t}$$

$$= \rho \frac{D u_\alpha}{Dt} - \rho u_\beta \frac{\partial u_\alpha}{\partial x_\beta} - u_\alpha \frac{\partial \rho u_\beta}{\partial x_\beta}$$

$$= \frac{\partial \sigma_{\alpha\beta}}{\partial x_\beta} - \frac{\partial}{\partial x_\beta}(\rho u_\alpha u_\beta) = -\frac{\partial}{\partial x_\beta}\Pi_{\alpha\beta}.$$

In the first line we have used (9.1) and (9.3) and in the second line (9.8).

$\square$

9.2.4 *The stress tensor and the Navier–Stokes equation*

The stress tensor σ is (in an isotropic fluid) composed of a diagonal pressure tensor and a symmetric *viscous stress tensor τ*:

$$\boldsymbol{\sigma} = -p\mathbf{1} + \boldsymbol{\tau}, \qquad \sigma_{\alpha\beta} = -p\,\delta_{\alpha\beta} + \tau_{\alpha\beta}. \qquad (9.11)$$

In an isotropic Newtonian fluid where viscous forces are assumed to be proportional to velocity gradients, the only possible form[2] of the viscous stress tensor is

$$\tau_{\alpha\beta} = \eta \left(\frac{\partial u_\alpha}{\partial x_\beta} + \frac{\partial u_\beta}{\partial x_\alpha} \right) + \left(\zeta - \frac{2}{3}\eta \right) \delta_{\alpha\beta} \nabla \cdot \boldsymbol{u}. \qquad (9.12)$$

The tensor must be symmetric with $\partial u_\alpha/\partial x_\beta + \partial u_\beta/\partial u_\alpha$ as off-diagonal elements, because these vanish for a uniform rotational motion without internal friction, for which $\boldsymbol{u} = \omega \times \boldsymbol{r}$ (ω being the angular velocity). We can split the viscous stress tensor into a *traceless, symmetric* part and an *isotropic* part:

$$\tau = \eta \begin{pmatrix} 2\frac{\partial u_x}{\partial x} - \frac{2}{3}\nabla \cdot \boldsymbol{u} & \frac{\partial u_x}{\partial y} + \frac{\partial u_y}{\partial x} & \frac{\partial u_x}{\partial z} + \frac{\partial u_z}{\partial x} \\ \frac{\partial u_y}{\partial x} + \frac{\partial u_x}{\partial y} & 2\frac{\partial u_y}{\partial y} - \frac{2}{3}\nabla \cdot \boldsymbol{u} & \frac{\partial u_y}{\partial z} + \frac{\partial u_z}{\partial y} \\ \frac{\partial u_z}{\partial x} + \frac{\partial u_x}{\partial z} & \frac{\partial u_z}{\partial y} + \frac{\partial u_y}{\partial z} & 2\frac{\partial u_z}{\partial z} - \frac{2}{3}\nabla \cdot \boldsymbol{u} \end{pmatrix}$$

$$+\zeta \nabla \cdot \boldsymbol{u} \begin{pmatrix} 1 & 0 & 0 \\ 0 & 1 & 0 \\ 0 & 0 & 1 \end{pmatrix}. \qquad (9.13)$$

---

[2] For a detailed derivation see, e.g., Landau and Lifschitz (1987).

There can be only two parameters: the *shear viscosity coefficient* η related to shear stress and the *bulk viscosity coefficient* ζ related to isotropic (compression) stress.

For *incompressible* fluids, with $\nabla \cdot \boldsymbol{u} = 0$, the viscous stress tensor simplifies to the following traceless tensor:

$$\tau_{\alpha\beta} = \eta \left(\frac{\partial u_\alpha}{\partial x_\beta} + \frac{\partial u_\beta}{\partial x_\alpha} \right) \quad \text{(incompressible)}. \tag{9.14}$$

For incompressible fluids there is only one viscosity coefficient.

The divergence of the viscous stress tensor yields the viscous force. For *space-independent coefficients*, the derivatives simplify considerably, and the viscous force is then given by

$$\boldsymbol{f}^{\text{visc}} = \nabla \cdot \tau = \eta \nabla^2 \boldsymbol{u} + \left(\zeta + \frac{1}{3}\eta \right) \nabla(\nabla \cdot \boldsymbol{u}),$$

$$f_\alpha^{\text{visc}} = \frac{\partial \tau_{\alpha\beta}}{\partial x_\beta} = \eta \nabla^2 u_\alpha + \left(\zeta + \frac{1}{3}\eta \right) \frac{\partial^2 u_\beta}{\partial x_\alpha \partial x_\beta}. \tag{9.15}$$

Combining (9.8) and (9.15) we obtain the *Navier–Stokes* equation (which is therefore only valid for locally homogeneous Newtonian fluids with constant viscosity coefficients):

$$\rho \frac{D\boldsymbol{u}}{Dt} = \rho \frac{\partial \boldsymbol{u}}{\partial t} + \rho(\boldsymbol{u} \cdot \nabla)\boldsymbol{u} = \nabla \cdot \sigma + \boldsymbol{f}^{\text{ext}}$$

$$= -\nabla p + \eta \nabla^2 \boldsymbol{u} + \left(\zeta + \frac{1}{3}\eta \right) \nabla(\nabla \cdot \boldsymbol{u}) + \boldsymbol{f}^{\text{ext}}. \tag{9.16}$$

Note that for incompressible fluids the equation simplifies to

$$\frac{\partial \boldsymbol{u}}{\partial t} + (\boldsymbol{u} \cdot \nabla)\boldsymbol{u} = -\frac{1}{\rho}\nabla p + \frac{\eta}{\rho}\nabla^2 \boldsymbol{u} + \boldsymbol{f}^{\text{ext}} \quad \text{(incompressible)}. \tag{9.17}$$

The viscosity occurs in this equation only as the quotient η/ρ, which is called the *kinematic viscosity* and usually indicated by the symbol ν.

9.2.5 The equation of state

The Navier–Stokes equation (9.16) and the continuity equation (9.3) are not sufficient to solve, for example, the time dependence of the density and velocity fields for given boundary and initial conditions. What we need in addition is the relation between pressure and density, or, rather, the pressure changes that result from changes in density. Under the assumption of local thermodynamic equilibrium, the *equation of state* (EOS) relates pressure, density and temperature:

$$f(\rho, p, T) = 0. \tag{9.18}$$

We note that pressure does not depend on the fluid velocity or its gradient: in the equation of motion (see (9.8) and (9.11)) the *systematic* pressure force has already been separated from the velocity-dependent friction forces, which are gradients of the viscous stress tensor τ.

The equation of state expresses a relation between three thermodynamic variables, and not just pressure and density, and is therefore – without further restrictions – not sufficient to derive the pressure response to density changes. The further restriction we need is the *assumption* that the thermodynamic change is *adiabatic*, i.e., that the change does not involve simultaneous heat exchange with a thermal bath. In real physical systems contact with thermal baths can only be realized at boundaries and is thus incorporated in boundary conditions. There is one exception: in an environment with given temperature the system is in interaction with a radiation field with a black-body distribution typical for that temperature and absorbs and emits radiation, finally leading to thermal equilibration with the radiation field. We may, however, for most practical purposes safely assume that the rate of equilibration with the radiation field is negligibly slow compared to thermal conduction within the system and over its boundaries. The adiabaticity assumption is therefore valid in most practical cases. In simulations, where unphysical heat baths may be invoked, the adiabaticity assumption may be artificially violated.

For small changes, the adiabatic relation between pressure and density change is given by the *adiabatic compressibility* κ_S:

$$\kappa_S = \frac{1}{\rho} \left(\frac{d\rho}{dp} \right)_S, \tag{9.19}$$

or

$$\left(\frac{dp}{d\rho} \right)_S = \frac{1}{\kappa_S \rho}. \tag{9.20}$$

A special case is an ideal gas for which pV^{c_p/c_V} remains constant under an adiabatic change. This implies that

$$\left(\frac{dp}{d\rho} \right)_S = \frac{c_p}{c_V} \frac{p}{\rho}. \tag{9.21}$$

For dense liquids the compressibility is so small that for many applications the fluid can be considered as *incompressible*, and ρ taken as constant in a coordinate system that moves with the fluid. This means that the divergence of the fluid velocity vanishes (see (9.5)) and the Navier–Stokes equation (9.16) simplifies to (9.17).

9.2.6 Heat conduction and the conservation of energy

Note on notation In this section we need thermodynamic quantities per unit mass of material. We use overlined symbols as notation for quantities per unit mass in order not to cause confusion with the same thermodynamic quantities used elsewhere (without overline) per mole of component. These are *intensive* thermodynamic properties; the corresponding *extensive* properties are denoted by capitals (temperature is an exception, being intensive and denoted by T). For internal energy we use $\bar{u}$, not to be confused with fluid velocity $\boldsymbol{u}$. We define the following quantities:

(i) *Internal energy per unit mass* $\bar{u}$. This is the sum of the kinetic energy due to random (thermal) velocities and the potential energy due to interactions within the system.[3] It does not include the kinetic energy per unit mass $\frac{1}{2}u^2$ as a result of the fluid velocity u. It has SI units J/kg or $\mathrm{m^2\,s^{-2}}$.
(ii) *Enthalpy per unit mass* $\bar{h} = \bar{u} + p/\rho$ [J/kg].
(iii) *Entropy per unit mass* $\bar{s}$ $[\mathrm{J\,kg^{-1}\,K^{-1}}]$,
(iv) *Thermodynamic potential per unit mass* $\bar{\mu} = \bar{h} - T\bar{s}$ [J/kg].

Due to adiabatic changes, and to dissipation caused by frictional forces, heat will be locally produced or absorbed, and the

temperature will not be homogeneous throughout the system. Temperature gradients will cause heat flow by conduction, and this heat flow must be incorporated into the total energy conservation. If it is assumed that the heat flux $\boldsymbol{J}_{\mathrm{q}}$ (energy per unit of time flowing through a unit area) is proportional to minus the temperature gradient, then

$$\boldsymbol{J}_{\mathrm{q}} = -\lambda \nabla T, \tag{9.22}$$

where λ is the *heat conduction coefficient*.

The energy per unit mass is given by $\bar{u} + \frac{1}{2}u^2 + \Phi^{\mathrm{ext}}$, and hence the energy per unit volume is $\rho\bar{u} + \frac{1}{2}\rho u^2 + \rho\Phi^{\mathrm{ext}}$. Here Φ^{ext} is the potential energy per unit mass in an external field (such as gz for a constant gravitational field in the $-z$-direction), which causes the external force per unit volume $\boldsymbol{f}^{\mathrm{ext}} = -\rho\nabla\Phi(\boldsymbol{r})$. Note that the external force per unit volume is not equal to minus the gradient of $\rho\Phi$. The energy of a volume element (per unit volume) changes with time for several reasons:

(i) Reversible work is done on the volume element (by the force due to pressure) when the density changes: $(p/\rho)(\partial\rho/\partial t)$.

---

[3] For a discussion on the locality of energy, see Section 17.7.

(ii) Reversible work is done by external forces; however, this work goes at the expense of the potential energy that is included in the definition of the energy per unit volume, so that the energy per unit volume does not change.

(iii) Energy is transported with the fluid (kinetic energy due to fluid velocity plus internal energy), when material flows into the volume element: $-(\bar{u} + \frac{1}{2}u^2 + \Phi^{\text{ext}})\nabla \cdot (\rho\boldsymbol{u})$.

(iv) Heat is produced by irreversible transformation of kinetic energy into heat due to friction: $-\boldsymbol{u} \cdot [\eta\nabla^2\boldsymbol{u} + (\zeta + \frac{1}{3}\eta)\nabla(\nabla \cdot \boldsymbol{u})]$.

(v) Heat flows into the volume element due to conduction: $\nabla \cdot (\lambda\nabla T)$.

Summing up, this leads to the *energy balance equation*

$$\frac{\partial}{\partial t}\left(\rho\bar{u} + \frac{1}{2}\rho u^2 + \rho\Phi^{\text{ext}}\right) = \frac{p}{\rho}\frac{\partial\rho}{\partial t} - \left(\bar{u} + \frac{1}{2}u^2 + \Phi^{\text{ext}}\right)\nabla \cdot (\rho\boldsymbol{u})$$

$$-\boldsymbol{u} \cdot \left[\eta\nabla^2\boldsymbol{u} + \left(\zeta + \frac{1}{3}\eta\right)\nabla(\nabla \cdot \boldsymbol{u})\right] + \nabla \cdot (\lambda\nabla T). \qquad (9.23)$$

This concludes the derivation of the fluid dynamics equations based on the *assumptions* that local density and local fluid velocity can be defined, and local thermodynamical equilibrium is defined and attained. In the next secion we return to a more realistic molecular basis.

9.3 Coarse graining in space

In this section we consider a classical Hamiltonian system of N particles with masses m_i, positions $\boldsymbol{r}_i$, and velocities $\boldsymbol{v}_i$, $i = 1, \ldots, N$. The particles move under the influence of a conservative interaction potential $V(\boldsymbol{r}_1, \ldots, \boldsymbol{r}_N)$ and may be subject to an external force $\boldsymbol{F}_i^{\text{ext}}$, which is minus the gradient of a potential $\Phi(\boldsymbol{r})$ at the position $\boldsymbol{r}_i$.

Instead of considering the individual particle trajectories, we wish to derive equations for quantities that are defined as "local" averages over space of particle attributes. We seek to define the local averages in such a way that the averaged quantities fulfill equations that approximate as closely as possible the equations of continuum fluid dynamics, as described in the previous section. Exact correspondence can only be expected when the averaging concerns an infinite number of particles. For finite-size averaging we hope to obtain modifications of the fluid dynamics equations that contain meaningful corrections and give insight into the effects of finite particle size.

The spatial averaging can be carried out in various ways, but the simplest is a linear *convolution* in space. As stated in the introduction of this chapter, we consider for simplicity an isotropic fluid consisting of particles of one type only. Consider the *number density* of particles $n(r)$. If the particles are point masses at positions r_i, the number density consists of a number of δ-functions in space:

$$n^0(r) = \sum_{i=1}^{N} \delta(r - r_i). \tag{9.24}$$

The coarse-grained number density is now defined as

$$n(r) = \sum_{i=1}^{N} w(r - r_i), \tag{9.25}$$

where $w(r)$ is a *weight function*, with dimension of one over volume. We shall take the weight function to be isotropic: $w(r)$, with the property that it decays fast enough with r for the integral over 3D space to exist. The function is normalized

$$\int_0^\infty w(r)\, 4\pi r^2 \, dr = 1. \tag{9.26}$$

This condition implies that the integral of the number density over a large volume approximates the number of particles within that volume. The weight function is not prescribed in detail, but it should present a *smoothing* over space, and preferably (but not necessarily) be positive and monotonically decreasing with r. A useful and practical example is the 3D Gaussian function

$$w(r) = (\sigma\sqrt{2\pi})^{-3} \exp\left(-\frac{r^2}{2\sigma^2}\right). \tag{9.27}$$

Note on the symmetry of weight functions We have made the weight function $w(r)$ a function of the distance only, and therefore the weight function is perfectly symmetric in space, and invariant for rotation. This is not a necessary condition, and we could take a weight function $w(r)$ that is not rotationally invariant, but still of high, e.g., cubic, symmetry, such as a product function

$$w(r) = w_1(x)w_1(y)w_1(z), \tag{9.28}$$

where $w_1(x)$ is a symmetric function in x. Product functions have the advantage that their Fourier transforms are a product of the Fourier transforms of the one-dimensional weight functions. Normalization according to (9.26) is not valid for product functions in general, but must be replaced by the normalization of each of the 1D functions:

$$\int_{-\infty}^{+\infty} w_1(x)\, dx = 1. \tag{9.29}$$

Simple one-dimensional weight functions are listed below:

(i) *Constant weight*

$$w_1(x) = 1/(2a) \quad \text{for } |x| \le a$$
$$= 0 \quad \text{for } |x| > a. \tag{9.30}$$

The Fourier transform of this function is a sinc function, $\sin ka/(ka)$.

(ii) *Triangular weight*

$$w_1(x) = a^{-2}(a - |x|) \quad \text{for } |x| \le a$$
$$= 0 \quad \text{for } |x| > a. \tag{9.31}$$

This function is in fact a convolution of the previous function with itself, and therefore its Fourier transform is the square of a sinc function $[2\sin(\frac{1}{2}ka)/(ka)]^2$.

(iii) *Sinc function*

$$w_1(x)\frac{1}{a}\frac{\sin(\pi x/a)}{\pi x/a}. \tag{9.32}$$

This function has a *band-limited* Fourier transform that is constant up to $|k| = \pi/a$ and zero for larger $|k|$.

(iv) *Normal distribution*

$$w_1(x) = \frac{1}{\sigma\sqrt{2\pi}} \exp\left(-\frac{x^2}{2\sigma^2}\right). \tag{9.33}$$

The Fourier transform of this function is a Gaussian function of k, proportional to $\exp(-\frac{1}{2}\sigma^2 k^2)$. The 3D product function is a Gaussian function of the distance r. In fact, the Gaussian function is the only 1D function that yields a fully isotropic 3D product function, and is therefore a preferred weight function.

9.3.1 Definitions

We now *define* the following averaged quantities:

(i) *Number density*

$$n(\mathbf{r}) \stackrel{\text{def}}{=} \sum_i w(\mathbf{r} - \mathbf{r}_i). \tag{9.34}$$

(ii) *Mass density*

$$\rho(\mathbf{r}) \stackrel{\text{def}}{=} \sum_i m_i w(\mathbf{r} - \mathbf{r}_i). \tag{9.35}$$

(iii) *Mass flux density or momentum density*

$$\mathbf{J}(\mathbf{r}) \stackrel{\text{def}}{=} \sum_i m_i \mathbf{v}_i w(\mathbf{r} - \mathbf{r}_i). \tag{9.36}$$

(iv) *Fluid velocity*

$$u(r) \stackrel{\text{def}}{=} \frac{J(r)}{\rho(r)}. \tag{9.37}$$

This definition is only valid if ρ differs from zero. The fluid velocity is undetermined for regions of space where both the mass density and the mass flux density are zero, e.g., outside the region to which the particles are confined.

(v) *Force per unit volume*

$$f(r) \stackrel{\text{def}}{=} \sum_i F_i w(r - r_i), \tag{9.38}$$

where F_i is the force acting on particle i. This force consists of an internal contribution due to interactions between the particles of the system, and an external contribution due to external sources.

(vi) *Stress tensor and pressure* The definitions of the stress tensor σ, the pressure, and the viscous stress tensor, are discussed below.

(vii) *Momentum flux density tensor*

$$\Pi_{\alpha\beta}(r) \stackrel{\text{def}}{=} -\sigma_{\alpha\beta}(r) + \sum_i m_i v_{i\alpha} v_{i\beta} w(r - r_i). \tag{9.39}$$

Note that the definition of Π uses the weighted particle velocities and not the fluid velocities as in (9.10). With the present definition linear momentum is conserved, but Newton's equation for the acceleration has extra terms (see below).

(viii) *Temperature*

$$T(r) \stackrel{\text{def}}{=} \frac{\sum_i m_i (v_i - u(r))^2 w(r - r_i)}{3 k_{\mathrm{B}} n(r)}. \tag{9.40}$$

Temperature is only defined for regions where the number density differs from zero. It is assumed that all degrees of freedom behave classically so that the classical equipartition theorem applies. For hard quantum degrees of freedom or for holonomic constraints corrections must be made.

9.3.2 *Stress tensor and pressure*

The coarse-grained stress tensor should be defined such that its divergence equals the internal force per unit volume (see (9.7)). As is elaborated in

Chapter 17 in connection with locality of the virial, there is no unique solution, because any divergence-free tensor can be added to the stress tensor without changing the force derived from it.

For forces between point particles, the stress tensor is localized on *force lines* that begin and end on the particles, but are further arbitrary in shape. Schofield and Henderson (1982) have suggested the following realization of the stress tensor:

$$\sigma_{\alpha\beta} = -\sum_i F_{i\alpha}^{\text{int}} \int_{C_{0i}} \delta(\boldsymbol{r} - \boldsymbol{r}_c)\, dx_{c\beta}, \tag{9.41}$$

where the integral is taken over a path C_{0i} starting at an arbitrary reference point $\boldsymbol{r}_0$ and ending at $\boldsymbol{r}_i$.

The generalization to a coarse-grained quantity is straightforward: the δ-function in (9.41) is replaced by the weight function w and the reference point is chosen at the position $\boldsymbol{r}$. Thus we *define* the averaged stress tensor as

$$\sigma_{\alpha\beta}(\boldsymbol{r}) \stackrel{\text{def}}{=} -\sum_i F_{i\alpha}^{\text{int}} \int_{C_i} w(\boldsymbol{r} - \boldsymbol{r}_c)\, dx_{c\beta}, \tag{9.42}$$

where the integral is taken over a path C_i starting at $\boldsymbol{r}$ and ending at $\boldsymbol{r}_i$. It is logical to choose straight lines for the paths. The divergence of this stress tensor now yields the averaged internal force per unit volume, as defined in (9.38):

$$(\nabla \cdot \sigma)(\boldsymbol{r}) = \boldsymbol{f}^{\text{int}}(\boldsymbol{r}). \tag{9.43}$$

Proof

$$
\begin{aligned}
(\nabla \cdot \sigma)_\alpha &= -\sum_i F_{i\alpha}^{\text{int}} \frac{\partial}{\partial x_\beta} \int_{C_i} w(\boldsymbol{r} - \boldsymbol{r}_c) dx_{c\beta} \\
&= \sum_i F_{i\alpha}^{\text{int}} \int_{\boldsymbol{r}}^{\boldsymbol{r}_i} \frac{\partial}{\partial x_{c\beta}} w(\boldsymbol{r} - \boldsymbol{r}_c) dx_{c\beta} \\
&= \sum_i F_{i\alpha}^{\text{int}} w(\boldsymbol{r} - \boldsymbol{r}_i).
\end{aligned}
\tag{9.44}
$$

□

9.3.3 Conservation of mass

The mass conservation law of continuum mechanics (9.3):

$$\frac{\partial \rho}{\partial t} + \nabla \cdot \boldsymbol{J} = 0, \tag{9.45}$$

is valid and exact for the averaged quantities.

Proof Note that ρ (see (9.35)) is time dependent through the time dependence of $\boldsymbol{r}_i$, and that the gradient of w with respect to $\boldsymbol{r}_i$ equals minus the gradient of w with respect to $\boldsymbol{r}$:

$$\frac{\partial \rho}{\partial t} = -\sum_i m_i (\nabla w(\boldsymbol{r} - \boldsymbol{r}_i)) \cdot \boldsymbol{v}_i$$

$$= -\nabla \cdot \sum_i m_i \boldsymbol{v}_i w(\boldsymbol{r} - \boldsymbol{r}_i)$$

$$= -\nabla \cdot \boldsymbol{J}$$

$\square$

9.3.4 Conservation of momentum

The momentum conservation law of continuum mechanics (9.9):

$$\frac{\partial}{\partial t}(\rho u_\alpha) = -\frac{\partial}{\partial x_\beta} \Pi_{\alpha\beta} \tag{9.46}$$

(valid in the absence of external forces) is valid and exact for the averaged quantities.

Proof After applying (9.37) and (9.39) we must prove that, in the absence of external forces,

$$\frac{\partial J_\alpha}{\partial t} = \frac{\partial \sigma_{\alpha\beta}}{\partial x_\beta} - \frac{\partial}{\partial x_\beta} \sum_i m_i v_{i\alpha} v_{i\beta} w(\boldsymbol{r} - \boldsymbol{r}_i). \tag{9.47}$$

Filling in (9.36) on the l.h.s., we see that there are two time-dependent terms, $v_{i\alpha}$ and $\boldsymbol{r}_i$, that need to be differentiated:

$$\frac{\partial J_\alpha}{\partial t} = \sum_i m_i \dot{v}_{i\alpha} w(\boldsymbol{r} - \boldsymbol{r}_i) - \sum_i m_i v_{i\alpha} \frac{\partial w(\boldsymbol{r} - \boldsymbol{r}_i)}{\partial x_\beta} v_{i\beta}$$

$$= f_\alpha^{\mathrm{int}} - \frac{\partial}{\partial x_\beta} \sum_i m_i v_{i\alpha} v_{i\beta} w(\boldsymbol{r} - \boldsymbol{r}_i).$$

Since the divergence of $\boldsymbol{\sigma}$ equals $\boldsymbol{f}^{\mathrm{int}}(\boldsymbol{r})$ (see (9.43)), we recover the r.h.s. of (9.47). $\square$

9.3.5 The equation of motion

The equation of motion of continuum mechanics (9.6):

$$\rho \frac{D\boldsymbol{u}}{Dt} = \boldsymbol{f}(\boldsymbol{r}) \tag{9.48}$$

now has a slightly different form and contains an additional term. Working out the l.h.s. we obtain

$$\rho \frac{Du_\alpha}{Dt} = \frac{\partial J_\alpha}{\partial t} + u_\beta \frac{\partial J_\alpha}{\partial x_\beta} - u_\alpha \left(\frac{\partial \rho}{\partial t} + u_\beta \frac{\partial \rho}{\partial x_\beta} \right), \tag{9.49}$$

and carrying through the differentiations, using (9.35) and (9.36), we find

$$\rho \frac{Du_\alpha}{Dt} = f_\alpha(\boldsymbol{r}) - \sum_i m_i (v_{i\alpha} - u_\alpha)(v_{i\beta} - u_\beta) \frac{\partial w(\boldsymbol{r} - \boldsymbol{r}_i)}{\partial x_\beta}$$

$$= \frac{\partial}{\partial x_\beta} \left(\sigma_{\alpha\beta} - \sum_i m_i (v_{i\alpha} - u_\alpha)(v_{i\beta} - u_\beta) w(\boldsymbol{r} - \boldsymbol{r}_i) \right). \tag{9.50}$$

The step to the last equation follows since the terms with the partial derivatives $\partial u_\alpha / \partial x_\beta$ and $\partial u_\beta / \partial x_\beta$ vanish. For example:

$$\sum_i m_i \frac{\partial u_\alpha}{\partial x_\beta} (v_{i\beta} - u_\beta) w(\boldsymbol{r} - \boldsymbol{r}_i) = \frac{\partial u_\alpha}{\partial x_\beta} (J_\beta - \rho u_\beta) = 0,$$

because $\boldsymbol{J} = \rho \boldsymbol{u}$.

It thus turns out that there is an extra term in the fluid force that is not present in the equation of motion of continuum mechanics. It has the form of minus the divergence of a tensor that represents the weighted particle velocity *deviation* from the fluid velocity. This term is also exactly the difference between the particle-averaged momentum flux density (9.39) and the momentum flux density (9.10) as defined in fluid mechanics. Let us call this term the *excess momentum flux density* Π^{exc}:

$$\Pi_{\alpha\beta}^{\mathrm{exc}}(\boldsymbol{r}) = \sum_i m_i [v_{i\alpha} - u_\alpha(\boldsymbol{r})][v_{i\beta} - u_\beta(\boldsymbol{r})] w(\boldsymbol{r} - \boldsymbol{r}_i). \tag{9.51}$$

Its divergence gives an extra force per unit volume. Inspection of this term shows that it represents the thermal kinetic energy density, with an equilibrium average determined by equipartition:

$$\langle \Pi_{\alpha\beta}^{\mathrm{exc}}(\boldsymbol{r}) \rangle = n(\boldsymbol{r}) k_{\mathrm{B}} T(\boldsymbol{r}) \delta_{\alpha\beta}. \tag{9.52}$$

This term is indeed the missing term if we compare Π^{exc} to the pressure computed from virial and kinetic energy in statistical mechanics (Chapter 17, (17.127) on page 485). It has no influence on the force unless there is a gradient of number density or a gradient of temperature. In addition to the average contribution, Π^{exc} has a fluctuating component that adds noise to the pressure and to the force.

9.4 Conclusion

As we have seen, coarse graining of a Hamiltonian fluid by spatial averaging with a weight function, yields the conservation laws, if the macroscopic quantities are properly defined. However, the equation of motion has an extra term that can be written as the divergence of an extra pressure term (9.52). It is related to the local thermal kinetic energy and equals the kinetic term required to describe pressure in statistical mechanics. With this term included, and including the local stress tensor derived from the virial of the local force (9.42), the pressure is a property of the system, determined by the density of particles and by the interactions between the particles. This is a manifestation of the local EOS. In fluid dynamics, where the description in terms of interacting particles is lost, the EOS is an *additional* "property" of the system that enables the determination of local pressure based on density and temperature (or energy density or entropy density). Note that local forces between particle pairs, which contribute to the local momentum flux density and therefore to the local pressure, cancel in the coarse-grained force density and do not play a direct role in fluid forces.

Another important difference between the dynamics of a system of interacting particles and a fluid continuum is that the coarse-grained dynamical properties are averages over a *finite* number of particles and are therefore fluctuating quantities with limited precision. This introduces "noise" and will have an influence on chaotic features of fluid dynamics, as turbulence, but only when the length scale of such features approach molecular size ranges. For macroscopic length scales the averaging can be done over such a large number of particles that the fluctuations become negligible. In the intermediate range, where details on an atomic scale are not needed but fluctuations are not negligible, the term *mesoscopic dynamics* is used. Mesoscopic dynamics can be realized either with particles (as Langevin or Brownian dynamics with superatomic system description) or with continuum equations, for example on a grid.

Exercises

9.1   Derive (9.15) from (9.12).

9.2   Derive the second line of (9.50) from the first line. Note that also the fluid velocity is a function of spatial coordinates.

10

Mesoscopic continuum dynamics

10.1 Introduction

The term "mesoscopic" is used for any method that treats nanoscale system details (say, 10 to 1000 nm) but averages over atomic details. Systems treated by mesoscopic methods are typically mixtures (e.g., of polymers or colloidal particles) that show self-organization on the nanometer scale. Mesoscopic behavior related to composition and interaction between constituents comes on top of dynamic behavior described by the macroscopic equations of fluid dynamics; it is on a level between atoms and continuum fluids. In mesoscopic dynamics the inherent noise is not negligible, as it is in macroscopic fluid dynamics.

Mesoscopic simulations can be realized both with particles and with continuum equations solved on a grid. In the latter case the continuum variables are densities of the species occurring in the system. Particle simulations with "superatoms" using Langevin or Brownian dynamics, as treated in Chapter 8, are already mesoscopic in nature but will not be considered in this chapter. Also the use of particles to describe continuum equations, as in dissipative particle dynamics described in Chapter 11, can be categorized as mesoscopic, but will not be treated in this chapter. Here we consider the continuum equations for multicomponent mesoscopic systems in the linear response approximation. The latter means that fluxes are *assumed* to be linearly related to their driving forces. This, in fact, is equivalent to *Brownian dynamics* in which accelerations are averaged-out and *average* velocities are proportional to average, i.e., *thermodynamic*, forces. The starting point for mesoscopic dynamics will therefore be the *irreversible thermodynamics in the linear regime*, as treated in Chapter 16, Section 16.10.

10.2 Connection to irreversible thermodynamics

We start with the irreversible entropy production per unit volume σ of (16.98) on page 446. Replacing the "volume flux" $\boldsymbol{J}_v$ by the bulk velocity $\boldsymbol{u}$ we may write

$$\sigma = \boldsymbol{J}_q \cdot \boldsymbol{\nabla} \frac{1}{T} - \frac{1}{T} \boldsymbol{u} \cdot \boldsymbol{\nabla} p + \frac{1}{T} \boldsymbol{I} \cdot \boldsymbol{E} - \frac{1}{T} \sum_i \boldsymbol{J}_i \cdot (\boldsymbol{\nabla} \mu_i)_{p,T}. \tag{10.1}$$

Here we recognize heat flux $\boldsymbol{J}_q$ and electric current density $\boldsymbol{I}$, driven by a temperature gradient and an electric field, respectively. The second term relates to the irreversible process of bulk flow caused by a force density, which is the gradient of the (generalized) pressure tensor including the viscous stress tensor (see Section 9.2.4 on page 284). The last term is of interest for the *relative* diffusional flux of particle species, driven by the gradient of the thermodynamic potential of that species. Any bulk flow $\boldsymbol{J}_i = c_i \boldsymbol{u}$, with all species flowing with the same average speed, does not contribute to this term since

$$\sum_i c_i (\boldsymbol{\nabla} \mu_i)_{p,T} = 0, \tag{10.2}$$

as a result of the Gibbs–Duhem relation. The term can be written as

$$\sigma_{\text{diff}} = -\frac{1}{T} \sum_i \left(\frac{\boldsymbol{J}_i}{c_i} - \boldsymbol{u} \right) \cdot [c_i (\boldsymbol{\nabla} \mu_i)_{p,T}]. \tag{10.3}$$

The term $\boldsymbol{J}_i/c_i - \boldsymbol{u} = \boldsymbol{u}_i^d$ denotes the average relative velocity of species i with respect to the bulk flow velocity, and we may define the *difference flux* $\boldsymbol{J}_i^d$ as

$$\boldsymbol{J}_i^d \stackrel{\text{def}}{=} c_i \boldsymbol{u}_i^d = \boldsymbol{J}_i - c_i \boldsymbol{u}. \tag{10.4}$$

It is clear that there are only $n-1$ independent difference fluxes for n species, and the sum may be restricted[1] – eliminating species 0 (the "solvent") – to species 1 to $n - 1$, which yields the equivalent form (see also Chapter 16, Eq. (16.104)):

$$\sigma_{\text{diff}} = -\frac{1}{T} \sum_i' \left(\frac{\boldsymbol{J}_i}{c_i} - \frac{\boldsymbol{J}_0}{c_0} \right) \cdot [c_i (\boldsymbol{\nabla} \mu_i)_{p,T}]. \tag{10.5}$$

Simplifying to a two-component system, with components numbered 0 and 1, the diffusional entropy production can be written as

$$\sigma_{\text{diff}} = -\frac{1}{T} [\boldsymbol{u}_1^d c_1 (\boldsymbol{\nabla}) \mu_1)_{p,T} + \boldsymbol{u}_1^d c_1 (\boldsymbol{\nabla} \mu_1)_{p,T}], \tag{10.6}$$

---

[1] This is indicated by the prime in the sum.

with the Gibbs–Duhem relation $c_1(\nabla\mu_1)_{p,T} + c_2(\nabla)\mu_1)_{p,T} = 0$, or alternatively as

$$\sigma_{\text{diff}} = -\frac{1}{T}(\boldsymbol{u}_1 - \boldsymbol{u}_0)c_1(\nabla\mu_1)_{p,T}. \tag{10.7}$$

The *linear response assumption* is that a system that is not in overall equilibrium will develop flows $\boldsymbol{J}_i$ proportional to driving forces $\boldsymbol{X}_j$ (defined such that $\sigma = \sum_i \boldsymbol{J}_i \cdot \boldsymbol{X}_i$) according to the Onsager phenomenological relations (16.109) and (16.111):

$$\boldsymbol{J}_i = \sum_j L_{ij}\boldsymbol{X}_j; \quad L_{ij} = L_{ji}. \tag{10.8}$$

On the mesoscopic level of theory the transport coefficients L_{ij} are input parameters for mesoscopic simulations; they can be derived from experiment or from non-equilibrium simulations at the atomic level, but do not follow from mesoscopic system simulation. One may adopt the simplifying but poor assumption that there are only diagonal transport coefficients.

For a two-component system there is only one coefficient connecting the relative particle flux (i.e., diffusional flux) to the chemical potential gradients. This coefficient is related to the diffusion constant in the following way. For a dilute or ideal solution of component 1 in solvent 0 (i.e., small c_1), the thermodynamic potential (see Chapter 16, Section 16.6 on page 435) is given by

$$\mu_1 = \mu_1^0 + RT\ln(c_1/c^0), \tag{10.9}$$

and hence

$$\nabla\mu_1 = \frac{RT}{c_1}\nabla c_1, \tag{10.10}$$

while the diffusional flux equals the diffusion constant D times the concentration gradient:

$$\boldsymbol{J}_1^d = c_1(\boldsymbol{u}_1 - \boldsymbol{u}_0) = -D\nabla c_1. \tag{10.11}$$

Combined this implies that

$$\boldsymbol{u}_1 - \boldsymbol{u}_0 = -\frac{D}{RT}\nabla\mu_1. \tag{10.12}$$

The negative gradient of μ_1 is the thermodynamic force that tries to move component 1 with respect to component 0; in the steady state the thermodynamic force is counterbalanced by an average frictional force $\zeta(\boldsymbol{u}_1 - \boldsymbol{u}_0)$, where ζ is the friction coefficient. The friction coefficient is therefore related to the diffusion coefficient by

$$\zeta = \frac{RT}{D}. \tag{10.13}$$

For n-component mixtures there are $n-1$ independent concentrations and $\frac{1}{2}n(n-1)$ diffusion coefficients.[2] In the *local coupling approximation* (LCA) it is assumed that the transport coefficient is proportional to the local density and the gradient of the thermodynamic potential.

Now consider the time evolution of the *concentration* c_i of species i. In the mesoscopic literature it is costumary to indicate this quantity by the *density* ρ_i, expressed either in number of particles or in moles per unit volume, and we shall adopt this convention. We shall focus on the structural rearrangements in mixtures following material transport and therefore simplify the system considerably by considering an isothermal/isobaric system, in which there is no heat flux, electric current, or bulk flow. The continuity equation for species i reads

$$\frac{\partial \rho_i}{\partial t} = -\boldsymbol{\nabla} \boldsymbol{J}_i \tag{10.14}$$

with the flux in the local coupling approximation and including a random term $\boldsymbol{J}_i^{\mathrm{rand}}$ due to thermal fluctuation:

$$\boldsymbol{J}_i = -M\rho_i\boldsymbol{\nabla}\mu_i + \boldsymbol{J}_i^{\mathrm{rand}}, \tag{10.15}$$

where we take for simplicity a single transport coefficient

$$M = \frac{D}{RT} = \zeta^{-1} \tag{10.16}$$

and where $\boldsymbol{J}_i^{\mathrm{rand}}$ is the random residual of the flux which cannot be neglected when the coarse-graining averages over a finite number of particles. This "noise" must satisfy the fluctuation–dissipation theorem and is intimately linked with the friction term; it is considered in the next section.

Note The friction can be treated with considerably more detail, e.g., one may distinguish the frictional contribution of different species (if there are more than two species), in which case the flux equation becomes a matrix equation. One may also generalize the local coupling approximation inherent in (10.15) and use a spread function for the local friction. So the general form is

$$\boldsymbol{J}_i(\boldsymbol{r}) = -\sum_j \int_V \Lambda_{ij}(\boldsymbol{r};\boldsymbol{r}')\boldsymbol{\nabla}\mu_j(\boldsymbol{r}')\,d\boldsymbol{r}' + \boldsymbol{J}_i^{\mathrm{rand}}, \tag{10.17}$$

---

[2] The mutual diffusion constants are complicated functions of the concentrations, but the dependencies become much simpler in the *Maxwell–Stefan* description in terms of *inverse* diffusion constants or friction coefficients, because the frictional forces with respect to other components add up to compensate the thermodynamic force. See Wesselingh and Krishna (1990) for an educational introduction to the Maxwell–Stefan approach, as applied to chemical engineering.

with

$$\Lambda_{ij}(\boldsymbol{r};\boldsymbol{r}') = M\rho_i\delta_{ij}\delta(\boldsymbol{r}-\boldsymbol{r}') \tag{10.18}$$

in the local coupling approximation.

The equation for the evolution of the density of species i is given by the continuity equation for each species, provided there are no chemical reactions between species:

$$\frac{\partial\rho_1}{\partial t} = -\boldsymbol{\nabla}\cdot\boldsymbol{J}_i = M\boldsymbol{\nabla}\cdot(\rho_i\boldsymbol{\nabla}\mu_i) - \boldsymbol{\nabla}J_i^{\mathrm{rand}}. \tag{10.19}$$

10.3 The mean field approach to the chemical potential

What we are still missing is a description of the position-dependent chemical potential given the density distribution. When we have such a relation the gradients of the thermodynamic potentials are known and with a proper choice of the mobility matrix the time evolution of a given density distribution can be simulated. Thus we can see how an arbitrary, for example homogeneous, density distribution of, e.g., the components of a block copolymer, develops in time into an ordered structural arrangement.

The thermodynamic potential is in fact a *functional* of the density distribution, and vice versa. In order to find the chemical potential, one needs the total free energy A of the system, which follows in the usual way from the partition function. The Hamiltonian can be approximated as the sum of a local contribution, independent of the density distribution, based on a local description of the unperturbed polymer, and a non-local contribution resulting from the density distribution. Simple models like the Gaussian chain model suffice for the local contribution. The non-local contribution to the chemical potential due to the density distribution is in mesoscopic continuum theory evaluated in the *mean-field approximation*, essentially following Landau–Ginzburg theory.

If the free energy A, which is a functional of the density distribution, is known, the position-dependent chemical potential is its functional derivative to the density:

$$\mu(\boldsymbol{r}) = \frac{\delta A}{\delta\rho(\boldsymbol{r})}. \tag{10.20}$$

When the system is in equilibrium, the density distribution is such that A is a global minimum, and the chemical potential is a constant. By adding an energy term $U(\boldsymbol{r})$, which we call the "external field", to the Hamiltonian, the equilibrium density distribution will change; there is a bijective relation between the density distribution and the external field U. The evaluation of the functionals is quite intricate and the reader is referred to the original literature: Fraaije (1993) and Fraaije *et al.* (1997).

The theory has been applied to several di- and triblock-copolymer melts, such as the industrially important triblock polymer "pluronic" that consists of three consecutive blocks ethylene oxide – propylene oxide – ethylene oxide, e.g., EO_{13}-PO_{30}-EO_{13}. Spontaneous formation of lamellar, hexagonal, bicubic and other structures has been observed, where the order remains local and only very slowly extends to larger distances. When shear is applied, ordering over longer distances is induced. See Fig. 10.1 for an example.[3]

---

[3] Some of the relevant articles are Zvelindovski (1998a, 1998b), van Vlimmeren *et al.* (1999), Maurits *et al.* (1998a, 1998b, 1999), Sevink *et al.* (1999) and Morozov *et al.* (2000).

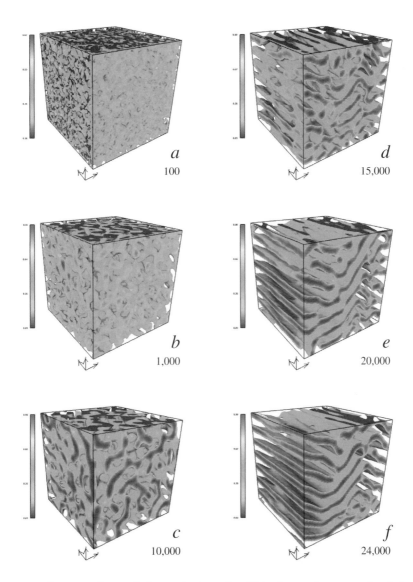

Figure 10.1 Six snapshots of the evolution of a diblock-copolymer melt of the type $A_{10}B_{10}$ in a mesoscopic continuum simulation at $T = 300$ K. At time 0 a homogeneous melt is subjected to a repulsive A–B interaction ($\chi = 0.8$); it develops a lamellar structure (a–c). After 10,000 time steps (c) a shear rate of 0.001 box lengths per time step is imposed; the lamellar structure orients itself along the direction of shear into a co-called perpendicular orientation (d–f). The dimensionless density of A is shown as shades of gray only for values larger than its volume averaged value ($= 0.5$). Figure courtesy of Dr Agur Sevink, Leiden University. See also Zvelindovsky et al. (1998a, 1998b).

11

Dissipative particle dynamics

In this chapter we consider how continuum dynamics, described by continuum equations that are themselves generalizations of systems of particles, can be described by particles again. The particle description in this case is not meant to be more precise than the continuum description and to represent the system in more detail, but is meant to provide an easier and more physically appealing way to solve the continuum equations. There is the additional advantage that multicomponent systems can be modeled, and by varying the relative repulsion between different kinds of particles, phenomena like mixing and spinodal decomposition can be simulated as well. The particles represent lumps of fluid, rather than specified clusters of real molecules, and their size depends primarily on the detail of the boundary conditions in the fluid dynamics problem at hand. The size may vary from superatomic or nanometer size, e.g., for colloidal systems, to macroscopic size. Since usually many (millions of) particles are needed to fill the required volume with sufficient detail, is it for efficiency reasons necessary that the interactions are described in a simple way and act over short distances only to keep the number of interactions low. Yet, the interactions should be sufficiently versatile to allow independent parametrization of the main properties of the fluid as density, compressibility and viscosity. Although *dissipative particle dynamics* (DPD), which is meant to represent continuum mechanics, differs fundamentally from coarse-grained *superatom* models, which are meant to represent realistic molecular systems in a simplified way, the distinction in practice is rather vague and the term DPD is often also used for models of polymers that are closer to a superatom approach.

The origin of DPD can be identified as a paper by Hoogerbrugge and Koelman (1992),[1] who described a rather intuitive way of treating fluid dynamics

---

[1] See also Koelman and Hoogerbrugge (1993), who applied their method to the study of hardsphere suspensions under shear.

problems with particles. Essentially their model consists of particles with very simple, short-ranged conservative interactions with additional friction and noise terms that act pairwise and conserve momentum and average energy. The addition of friction and noise functions as a thermostat and allows an extra parameter to influence the viscosity of the model. But there are predecessors: notably the *scaled particle hydrodynamics* (SPH), reviewed by Monaghan (1988) with the aim to solve the equations of fluid dynamics by the time evolution of a set of points. SPH was originally developed to solve problems in astrophysics (Lucy, 1977). It is largely through the efforts of P. Español[2] that DPD was placed on a firm theoretical footing, and resulted in a formulation where the equation of state (i.e., pressure and temperature as functions of density and entropy or energy) and properties such as viscosity and thermal conductivity can be used as *input values* in the model, rather than being determined by the choice of interparticle interactions (Español and Revenga, 2003). In Español's formulation each particle has four attributes: position, momentum, mass and entropy, for which appropriate stochastic equations of motion are defined.[3] Another model, originated by Flekkøy and Coveney (1999),[4] uses fluid "particles" based on Voronoi tesselation that divides space systematically in polyhedral bodies attributed to moving points in space.

We shall not describe these more complicated DPD implementations, but rather give a short description of a popular and simple implementation of DPD given by Groot and Warren (1997). This implementation is close to the original model of Hoogerbrugge and Koelman (1992). One should be aware that simplicity comes at a price: models of this kind have intrinsic properties determined by the interaction functions and their parameters and simulations are generally needed to set such properties to the desired values. The model contains stochastic noise and friction, and represents therefore a Langevin thermostat (see Chapter 6, page 196). Such a distributed thermostat causes isothermal behavior rather than the adiabatic response that is usually required in realistic fluid dynamics.

---

[2] Español (1995) derived hydrodynamic equations from DPD and evaluated the probability density from the Fokker–Planck equation corresponding to the stochastic equations of motions.

[3] The formal derivation of thermodynamically consistent fluid particle models is based on the GENERIC (General Equation for Non-Equilibrium Reversible-Irreversible Coupling) formalism of Öttinger (Grmela and Öttinger, 1997; Öttinger and Grmela, 1997; Öttinger, 1998). In this formalism the change in a set of variables that characterize the state of a system is expressed in terms of the dependence of energy and entropy on the state variables; this is done in such a way that energy is conserved and entropy cannot decrease, while the fluctuation–dissipation theorem is satisfied. See Español *et al.* (1999) for the application to hydrodynamic generalization.

[4] See also Flekkøy *et al.* (2000) and Español (1998). Serrano and Español (2001) elaborated on this model and the two approaches were compared by Serrano *et al.* (2002).

11.1 Representing continuum equations by particles

The system consists of particles with mass m_i, position $\boldsymbol{r}_i$ and velocity $\boldsymbol{v}_i$. Each particle represents a fluid element that moves coherently. The particles interact pairwise though two types of forces: a potential-derived conservative force and a dissipative friction force that depends on the velocity *difference* between two interacting particles. The energy dissipation due to the dissipative force is balanced by a random force, so that the total average kinetic energy from motion with respect to the local center of mass, excluding the collective kinetic energy (the "temperature"), remains constant. Since all forces act pairwise in the interparticle direction and are short-ranged, the sum of forces is zero and both linear and angular momentum is conserved, even on a local basis. Since mass, energy and momentum conservation are the basis of the continuum equations of fluid dynamics, DPD dynamics will follow these equations on length scales larger than the average particle separation and on time scales larger than the time step used for integration the equations of motion.

The equations of motion are Newtonian:

$$\dot{\boldsymbol{r}}_i = \boldsymbol{v}_i \tag{11.1}$$

$$\dot{\boldsymbol{v}}_i = \boldsymbol{F}_i = \sum_{j \neq i} \boldsymbol{F}_{ij}, \tag{11.2}$$

where

$$\boldsymbol{F}_{ij} = \boldsymbol{F}_{ij}^{\mathrm{C}} + \boldsymbol{F}_{ij}^{\mathrm{D}} + \boldsymbol{F}_{ij}^{\mathrm{R}}. \tag{11.3}$$

The conservative force on particle i due to j is repulsive with a range 1 given by:

$$\boldsymbol{F}_{ij}^{\mathrm{C}} = a_{ij}(1 - r_{ij})\frac{\boldsymbol{r}_{ij}}{r_{ij}} \quad r_{ij} < 1, \tag{11.4}$$

$$= 0 \quad r_{ij} \geq 1, \tag{11.5}$$

with $\boldsymbol{r}_{ij} = \boldsymbol{r}_i - \boldsymbol{r}_j$. This corresponds to a quadratic repulsive potential with one parameter a_{ij}. Note that the distance is scaled such that the maximum interaction range equals 1. The dissipative force is given by

$$\boldsymbol{F}_{ij} = -\gamma w^{\mathrm{D}}(r_{ij})(\boldsymbol{v}_{ij} \cdot \boldsymbol{r}_{ij})\frac{\boldsymbol{r}_{ij}}{r_{ij}^2}. \tag{11.6}$$

It acts in the direction of $\boldsymbol{r}_{ij}$ and is proportional to the component of the velocity difference in the interparticle direction, being repulsive when particles move towards each other and attractive when they move away. Thus it damps the relative motion of the two particles. The parameter γ measures the strength of the damping; $w^D(r_{ij})$ is a weight function vanishing for $r_{ij} > 1$.

The random force also acts in the interparticle direction:

$$\boldsymbol{F}_{ij}^{\text{R}} = \sigma w^{\text{R}}(r_{ij})\frac{\boldsymbol{r}_{ij}}{r_{ij}}\theta_{ij}, \tag{11.7}$$

where σ is the strength, $w^{\text{R}}(r_{ij})$ a weight function vanishing for $r_{ij} > 1$, and θ_{ij} a random function with average zero and with no memory: $\theta(0)\theta(t) = \delta(t)$, uncorrelated with the random function on any other particle pair. The distribution function of θ can be chosen to be normal, but that is not a requirement. Español and Warren (1995) showed that the fluctuation–dissipation theorem, ensuring that the energy changes from dissipation and random force cancel, requires that

$$w^{\text{D}}(r_{ij}) = [w^{\text{R}}(r_{ij})]^2, \tag{11.8}$$

and that the noise intensity must be related to the friction coefficient γ and the temperature T, just as is the case for Langevin dynamics:

$$\sigma^2 = 2\gamma k_{\text{B}}T. \tag{11.9}$$

The form of one of the weight function is arbitrary; Groot and Warren (1997) chose for w^{R} the same functional form as for the conservative force:

$$w^{\text{R}}(r_{ij})^{\text{C}} = (1 - r_{ij})\frac{\boldsymbol{r}_{ij}}{r_{ij}} \quad r_{ij} < 1, \tag{11.10}$$

$$= 0 \quad r_{ij} \geq 1. \tag{11.11}$$

11.2 Prescribing fluid parameters

The unit of length has been set by the choice of the maximum interaction range. The number density n (per cubic length unit) can be chosen, and, together with a choice of a, the strength of the conservative force, the pressure is fixed. The pressure is found from the usual virial equation:

$$p = \frac{N}{V}k_{\text{B}}T + \frac{1}{3V}\sum_{i,j>i}\langle\boldsymbol{r}_{ij}\cdot\boldsymbol{F}_{ij}^{\text{C}}\rangle. \tag{11.12}$$

In the virial part also the total force may be used, but in equilibrium the contributions of friction and random force cancel. One may wish to match the *isothermal compressibility* with a desired value for a given liquid. This is best expressed as the dimensionless value

$$\kappa^{-1} = \frac{1}{k_B T} \left(\frac{\partial p}{\partial n} \right), \tag{11.13}$$

which has a value of 16 for water at 300 K and 30 for 1-propanol at 300 K. From a series of simulations Groot and Warren (1997) found that

$$p = n k_B T + \alpha \frac{an}{k_B T}, \quad \alpha = 0.101 \pm 0.001. \tag{11.14}$$

This determines $an/k_B T$, which is equal to 75 for water.

11.3 Numerical solutions

Since the force F(t) depends on the equal-time velocity v(t), the normal Verlet-type algorithms cannot be used, because they have the equal-time velocities $v(t)$ available after a step that has already used the force F(t). This applies also to the velocity-Verlet version. If earlier velocities are used, the order of the algorithm degrades and the performance becomes unacceptable. Possibilities are to predict the velocity for the force calculation and correct it afterwards, or solve the velocity iteratively, requiring more than one force evaluation per step. Lowe (1999) has devised an alternative algorithm which adds a thermostat much like Andersen's velocity rescaling. The equations of motion are integrated with the velocity-Verlet scheme, but in addition randomly selected pairs of particles exchange their relative velocity for a sample drawn from a Maxwellian distribution, in such a way that momentum (and angular momentum) is conserved. This solves problems with temperature drift that otherwise occur.

11.4 Applications

Applications have been published in many different fields, such as polymer rheology (Schlijper *et al.*, 1995), rheology of colloidal suspensions (Koelman and Hoogerbrugge, 1993; Boek *et al.*, 1997), flow of DNA molecules in microchannels (Fan *et al.*, 2003). The method can be applied to mixtures and to microphase separation (Groot and Warren, 1997). Figure 11.1 shows the impact and subsequent coalescence of two drops of liquid moving towards each other in a liquid environment in which the drops don't mix. The simulation comprises 3.75 million DPD particles and was carried out by Florin O. Iancu, University of Delft, the Netherlands (Iancu, 2005). The collision is characterized by the dimensionless *Weber number* $We = \rho D U_r^2 / \Gamma$, where ρ is the density, D the diameter of the drops, U_r the relative velocity of the drops just before impact and Γ the interfacial surface tension. At low Weber numbers the drops bounce off each other without mixing, and at higher Weber numbers they coalesce after impact. The DPD parameters in this case were a density of 10 particles per r_c^3 (r_c being the cut-off range of the repulsive potential) and a repulsion parameter a (see (11.5)) of 14.5 mutually between drop particles or between environment particles, but 41.5 between drop particles and environment particles. This choice leads to an interfacial tension $\Gamma = 28$ (units of $k_B T / r_c^2$). When the collision is off-center, the drops elongate before they coalesce and may even split up afterwards with formation of small satellite droplets.

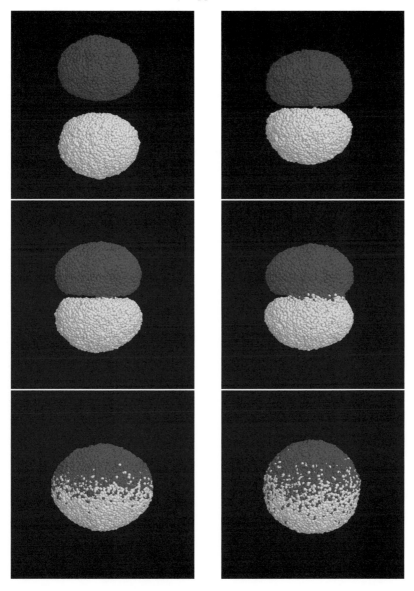

Figure 11.1 Impact and coalescence of two liquid drops moving towards each other in an inmiscible liquid environment at Weber's number of about 4, simulated with 3.75 million DPD particles (courtesy of Dr F.O. Iancu, University of Delft, the Netherlands)

Part II

Physical and Theoretical Concepts

12

Fourier transforms

In this chapter we review the definitions and some properties of Fourier transforms. We first treat one-dimensional non-periodic functions $f(x)$ with Fourier transform $F(k)$, the domain of both coordinates x and k being the set of real numbers, while the function values may be complex. The functions f and F are piecewise continuous and $\int_{-\infty}^{\infty} |f(x)|\, dx$ exists. The domain of x is usually called the *real space* while the domain of k is called *reciprocal space*. Such transforms are applicable to wave functions in quantum mechanics. In Section 12.6 we consider Fourier transforms for one-dimensional *periodic* functions, leading to *discrete* transforms, i.e., Fourier *series* instead of integrals. If the values in real space are also discrete, the computationally efficient *fast Fourier transform* (FFT) results (Section 12.7). In Section 12.9 we consider the multidimensional periodic case, with special attention to triclinic periodic 3D unit cells in real space, for which Fourier transforms are useful when long-range forces are evaluated.

12.1 Definitions and properties

The relations between $f(x)$ and its Fourier transform (FT) $F(k)$ are

$$f(x) = \frac{1}{\sqrt{2\pi}} \int_{-\infty}^{\infty} F(k) \exp(ikx)\, dk, \tag{12.1}$$

$$F(k) = \frac{1}{\sqrt{2\pi}} \int_{-\infty}^{\infty} f(x) \exp(-ikx)\, dx. \tag{12.2}$$

The factors $1/\sqrt{2\pi}$ are introduced for convenience in order to make the transforms symmetric; one could use any arbitrary factors with product 2π. The choice of sign in the exponentials is arbitrary and a matter of convention. Note that the second equation follows from the first by using the definition of the δ-function:

$$\int_{-\infty}^{\infty} \exp[\pm ikx]\, dk = 2\pi\delta(x), \tag{12.3}$$

and realizing that

$$f(x) = \int_{-\infty}^{\infty} \delta(x' - x)\, f(x')\, dx'. \tag{12.4}$$

The following relations are valid:

(i) if $f(x)$ is real then $F(-k) = F^*(k)$

(ii) if $F(k)$ is real then $f(-x) = f^*(x)$;

(iii) if $f(x)$ is real and $f(-x) = f(x)$ then $F(k)$ is real and $F(-k) = F(k)$ (cosine transform);

(iv) if $f(x)$ is real and $f(-x) = -f(x)$ then $F(k)$ is imaginary and $F(-k) = -F(k)$ (sine transform);

(v) the FT of $g(x) \stackrel{\mathrm{def}}{=} f(x + x_0)$ is $G(k) = F(k)\exp(ikx_0)$;

(vi) the FT of $g(x) \stackrel{\mathrm{def}}{=} f(x)\exp(ik_0 x)$ is $G(k) = F(k - k_0)$;

(vii) the FT of $g(x) \stackrel{\mathrm{def}}{=} df(x)/dx$ is $G(k) = ikF(k)$;

(viii) the FT of $g(x) \stackrel{\mathrm{def}}{=} xf(x)$ is $G(k) = -i\, dF(k)/dk$.

12.2 Convolution and autocorrelation

The *convolution* $h(x)$ of two functions $f(x)$ and $g(x)$ is defined as

$$h(x) \stackrel{\mathrm{def}}{=} \int_{-\infty}^{\infty} f^*(\xi - x)g(\xi)\, d\xi \tag{12.5}$$

$$= \int_{-\infty}^{\infty} f^*(\xi)g(x + \xi)\, d\xi, \tag{12.6}$$

with short notation $h = f * g$. Its Fourier transform is

$$H(k) = \sqrt{2\pi}\, F^*(k)G(k), \tag{12.7}$$

and hence

$$h(x) = \int_{-\infty}^{\infty} F^*(k)G(k)\exp(ikx)\, dk. \tag{12.8}$$

If $h(x) \stackrel{\mathrm{def}}{=} \int_{-\infty}^{\infty} f(\xi - x)g(\xi)\, d\xi$ then $H(k) = \sqrt{2\pi}F(k)G(k)$.
A special case is

$$h(0) = \int_{-\infty}^{\infty} f^*(x)g(x)\, dx = \int_{-\infty}^{\infty} F^*(k)G(k)\, dk. \tag{12.9}$$

The *autocorrelation function* is a self-convolution:

$$h(x) \stackrel{\text{def}}{=} \int_{-\infty}^{\infty} f^*(\xi - x) f(\xi) \, d\xi, \tag{12.10}$$

$$H(k) = \sqrt{2\pi} F^*(k) F(k), \tag{12.11}$$

$$h(x) = \int_{-\infty}^{\infty} F^*(k) F(k) \exp(ikx) \, dk, \tag{12.12}$$

$$h(0) = \int_{-\infty}^{\infty} f^*(x) f(x) \, dx = \int_{-\infty}^{\infty} F^*(k) F(k) \, dk. \tag{12.13}$$

Equation (12.13) is known as *Parseval's theorem*. It implies that, if the function $f(x)$ is *normalized* in the sense that $\int_{-\infty}^{\infty} f^* f \, dx = 1$, then its Fourier transform is, in the same sense in k-space, also normalized.

We note that the definitions given here for square-integrable functions differ from the autocorrelation and spectral density functions for infinite time series discussed in Section 12.8 (page 325).

12.3 Operators

When the function $f^* f(x)$ is interpreted as a *probability density*, the *expectation* of some function of x (indicated by triangular brackets) is the average of that function over the probability density:

$$\langle h(x) \rangle \stackrel{\text{def}}{=} \int_{-\infty}^{\infty} h(x) f^* f(x) \, dx. \tag{12.14}$$

Functions of k are similarly defined by averages over the probability density $F^* F(k)$ in k-space:

$$\langle h(k) \rangle \stackrel{\text{def}}{=} \int_{-\infty}^{\infty} h(k) F^* F(k) \, dk. \tag{12.15}$$

It can be shown that for polynomials of k the average can also be obtained in x-space by

$$\langle h(k) \rangle = \int_{-\infty}^{\infty} f^*(x) \hat{h} f(x) \, dx, \tag{12.16}$$

where $\hat{h}$ is an *operator* acting on $f(x)$ with the property that

$$\hat{h} \exp(ikx) = h(k) \exp(ikx). \tag{12.17}$$

Examples are

$$h(k) = k, \qquad \hat{h} = -i \frac{\partial}{\partial x}, \tag{12.18}$$

$$h(k) = k^2, \qquad \hat{h} = -\frac{\partial^2}{\partial x^2}, \qquad (12.19)$$

$$h(k) = k^n, \qquad \hat{h} = i^{-n}\frac{\partial^n}{\partial x^n}. \qquad (12.20)$$

Proof We prove (12.16). Insert the Fourier transforms into (12.16), using (12.3) and (12.17):

$$\int_{-\infty}^{\infty} f^*\hat{h}f\,dx = \frac{1}{2\pi}\int_{-\infty}^{\infty} dx \int_{-\infty}^{\infty} dk \int_{-\infty}^{\infty} dk'\, F^*(k')e^{-ik'x}F(k)\hat{h}e^{ikx}$$

$$= \frac{1}{2\pi}\int_{-\infty}^{\infty} dk \int_{-\infty}^{\infty} dk'\, F^*(k')F(k)h(k)\int_{-\infty}^{\infty} dx\, e^{i(k-k')x}$$

$$= \int_{-\infty}^{\infty} dk\, F^*(k)F(k)h(k) = \langle h(k)\rangle.$$

$\square$

In general, an operator $\hat{A}$ may be associated with a function A of x and/or k, and the *expectation of A* defined as

$$\langle A \rangle \stackrel{\text{def}}{=} \int_{-\infty}^{\infty} f^*(x)\hat{A}f(x)\,dx. \qquad (12.21)$$

An operator $\hat{A}$ is *hermitian* if for any two quadratically integrable functions $f(x)$ and $g(x)$

$$\int_{-\infty}^{\infty} f^*\hat{A}g\,dx = \left(\int_{-\infty}^{\infty} g^*\hat{A}f\,dx\right)^* = \int_{-\infty}^{\infty} g\hat{A}^*f^*\,dx. \qquad (12.22)$$

In particular this means that the expectation of a hermitian operator is real, as is immediately seen if we apply the hermitian condition to $g = f$. Operators that represent physical observables, meaning that expectations must be real physical quantities, are therefore required to be hermitian. It also follows that the eigenvalues of hermitian operators are real because the eigenvalue is the expectation of the operator over the corresponding eigenfunction (the reader should check this).

12.4 Uncertainty relations

If we define the *variances* in x- and k-space as

$$\sigma_x^2 \stackrel{\text{def}}{=} \langle(x - \langle x\rangle)^2\rangle, \qquad (12.23)$$

$$\sigma_k^2 \stackrel{\text{def}}{=} \langle(k - \langle k\rangle)^2\rangle, \qquad (12.24)$$

we can prove that for *any* normalized function $f(x)$ the product of the square root of these two variances (their *standard deviations*) is not less than one half:

$$\sigma_x \sigma_k \geq \tfrac{1}{2}. \tag{12.25}$$

This is the basis of the Heisenberg uncertainty relations for conjugate variables.

Proof The proof[1] starts with the Schwarz inequality for the scalar products of any two vectors u and v:

$$(u, u)(v, v) \geq (u, v)(v, u) = |(u, v)|^2 \tag{12.26}$$

which is valid with the definition of a scalar product of functions.[2]

$$(u, v) \overset{\text{def}}{=} \int_{-\infty}^{\infty} u^* v \, dx. \tag{12.27}$$

This the reader can prove by observing that $(u - cv, u - cv) \geq 0$ for any choice of the complex constant c, and then inserting $c = (v, u)/(v, v)$. We make the following choices for u and v:

$$u = (x - \langle x \rangle) f(x) \exp(ik_0 x) \tag{12.28}$$

$$v = \frac{d}{dx} [f(x) \exp(ik_0 x)], \tag{12.29}$$

where k_0 is an arbitrary constant, to be determined later. The two terms on the left-hand side of (12.26) can be worked out as follows:

$$(u, u) = \int_{-\infty}^{\infty} (x - \langle x \rangle)^2 f^* f \, dx = \sigma_x^2 \tag{12.30}$$

and

$$
\begin{aligned}
(v, v) &= \int_{-\infty}^{\infty} \left[\frac{d}{dx} f^* \exp(-ik_0 x) \right] \left[\frac{d}{dx} f \exp(ik_0 x) \right] dx \\
&= -\int_{-\infty}^{\infty} f^* \exp(-ik_0 x) \left[\frac{d^2}{dx^2} f \exp(ik_0 x) \right] dx \\
&= -\int_{-\infty}^{\infty} f^* f'' \, dx - 2ik_0 \int_{-\infty}^{\infty} f^* f' \, dx + k_0^2 \\
&= \langle k^2 \rangle - 2k_0 \langle k \rangle + k_0^2, \tag{12.31}
\end{aligned}
$$

[1] See, e.g., Kyrala (1967). Gasiorowicz (2003) derives a more general form of the uncertainty relations, relating the product of the standard deviations of two observables A and B to their commutator: $\sigma_A \sigma_b \geq \tfrac{1}{2} |\langle i[\hat{A}, \hat{B}] \rangle|$.

[2] This definition applies to vectors in Hilbert space. See Chapter 14.

where the second line follows from the first one by partial integration. Choosing for k_0 the value for which the last form is a minimum: $k_0 = \langle k \rangle$, we obtain

$$(v, v) = \sigma_k^2. \tag{12.32}$$

Thus, (12.26) becomes

$$
\begin{aligned}
\sigma_x^2 \sigma_k^2 &\geq \left| \int_{-\infty}^{\infty} u^* v \, dx \right|^2 \\
&\geq \left(\mathrm{Re} \int_{-\infty}^{\infty} u^* v \, dx \right)^2 \\
&= \frac{1}{4} \left(\int_{-\infty}^{\infty} (x - \langle x \rangle)(f^* f' + f'^* f) \, dx \right)^2 \\
&= \frac{1}{4} \left(\int_{-\infty}^{\infty} (x - \langle x \rangle) \frac{d}{dx}(f^* f) \, dx \right)^2 \\
&= \frac{1}{4} \left(\int_{-\infty}^{\infty} f^* f \, dx \right)^2 = \frac{1}{4}. \tag{12.33}
\end{aligned}
$$

Hence $\sigma_x \sigma_k \geq \frac{1}{2}$. ☐

12.5 Examples of functions and transforms

In the following we choose three examples of real one-dimensional symmetric functions $f(x)$ that represent a confinement in real space with different shape functions. All functions have expectations zero and are quadratically normalized, meaning that

$$\int_{-\infty}^{\infty} f^2(x) \, dx = 1. \tag{12.34}$$

This implies that their Fourier transforms are also normalized and that the expectation of k is also zero. We shall look at the width of the functions in real and reciprocal space.

12.5.1 Square pulse

The square pulse and its Fourier transform are given in Fig. 12.1. The equations are:

$$
\begin{aligned}
f(x) &= \frac{1}{\sqrt{a}}, \quad &|x| < \frac{a}{2} \\
&= 0, \quad &|x| \geq \frac{a}{2}
\end{aligned} \tag{12.35}
$$

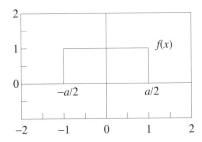
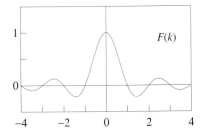

Figure 12.1 Square pulse $f(x)$ with width a (x in units $a/2$, f in units $a^{-1/2}$) and its transform $F(k)$ (k in units $2\pi/a$, F in units $(a/2\pi)^{1/2}$).

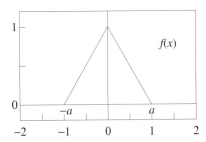
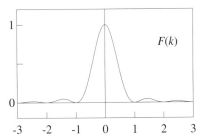

Figure 12.2 Triangular pulse $f(x)$ with width $2a$ (x in units a, f in units $(2a/3)^{-1/2}$) and its transform $F(k)$ (k in units $2\pi/a$, F in units $(3a/4\pi)^{1/2}$).

$$\langle x \rangle = 0 \tag{12.36}$$

$$\sigma_x^2 = \langle x^2 \rangle = \frac{1}{12}a^2 \tag{12.37}$$

$$F(k) = \sqrt{\frac{a}{2\pi}}\frac{\sin\phi}{\phi}, \qquad \phi = \tfrac{1}{2}ka \tag{12.38}$$

$$\langle k \rangle = 0 \tag{12.39}$$

$$\sigma_k^2 = \langle k^2 \rangle = \infty \tag{12.40}$$

$$\sigma_x\sigma_k = \infty \tag{12.41}$$

12.5.2 Triangular pulse

The triangular pulse is a convolution of the square pulse with itself. See Fig. 12.2. The equations are:

$$f(x) = \frac{1}{a}\sqrt{\frac{3}{2a}}(a - |x|), \qquad |x| < a \tag{12.42}$$

$$= 0, \qquad |x| \geq a$$

$$\langle x \rangle = 0 \tag{12.43}$$

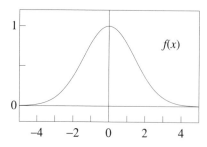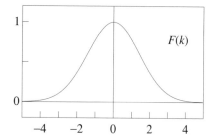

Figure 12.3 Gaussian pulse $f(x)$ with variance σ (x in units σ, f in units $(\sigma\sqrt{2\pi})^{-1/2}$ and its transform $F(k)$ (k in units $1/2\sigma$, F in units $(2\sigma/\sqrt{2\pi})^{-1/2}$).

$$\sigma_x^2 = \langle x^2 \rangle = \frac{1}{10}a^2 \tag{12.44}$$

$$F(k) = \sqrt{\frac{3a}{4\pi}\frac{\sin^2\phi}{\phi^2}} \qquad \phi = \tfrac{1}{2}ka \tag{12.45}$$

$$\langle k \rangle = 0 \tag{12.46}$$

$$\sigma_k^2 = \langle k^2 \rangle = \frac{3}{a^2} \tag{12.47}$$

$$\sigma_x\sigma_k = \sqrt{\frac{3}{10}} = 0.5477 \tag{12.48}$$

12.5.3 Gaussian function

The Gaussian function (Fig. 12.3) is the Fourier transform of itself. the equations are:

$$f(x) = (\sigma\sqrt{2\pi})^{-\frac{1}{2}}\exp\left(-\frac{x^2}{4\sigma^2}\right) \tag{12.49}$$

$$\langle x \rangle = 0 \tag{12.50}$$

$$\sigma_x^2 = \langle x^2 \rangle = \sigma^2 \tag{12.51}$$

$$F(k) = (\sigma_k\sqrt{2\pi})^{-\frac{1}{2}}\exp\left(-\frac{k^2}{4\sigma_k^2}\right); \qquad \sigma_k = \frac{1}{2\sigma} \tag{12.52}$$

$$\langle k \rangle = 0 \tag{12.53}$$

$$\sigma_k^2 = \langle k^2 \rangle = \frac{1}{4\sigma^2} \tag{12.54}$$

$$\sigma_x\sigma_k = \tfrac{1}{2} \tag{12.55}$$

So we see that – of the functions given here – the Gaussian wave function attains the smallest product of variances in real and reciprocal space. In fact, the Gaussian function has the smallest possible product of variances of all (well-behaved) functions.

12.6 Discrete Fourier transforms

Now consider *periodic* functions $f(x)$ with periodicity a:

$$f(x + na) = f(x), \quad n \in \mathbb{Z}. \tag{12.56}$$

The function is assumed to be piecewise continuous and absolutely integrable over the domain $(0, a)$:

$$\int_0^a |f(x)| \, dx \quad \text{exists.} \tag{12.57}$$

The function $f(x)$ can now be expressed in an infinite series of exponential functions that are each periodic functions of x on $(0, a)$:

$$f(x) = \sum_{n \in \mathbb{Z}} F_n \exp\left(\frac{2\pi i n x}{a}\right), \tag{12.58}$$

with

$$F_n = \frac{1}{a} \int_0^a f(x) \exp\left(\frac{-2\pi i n x}{a}\right) dx. \tag{12.59}$$

This can also be written as

$$f(x) = \sum_k F_k e^{ikx}, \quad k = \frac{2\pi n}{a}, \quad n \in \mathbb{Z}, \tag{12.60}$$

$$F_k = \frac{1}{a} \int_0^a f(x) e^{-ikx} \, dx. \tag{12.61}$$

The validity can be checked by computing $\int_0^a [\sum_{k'} \exp(ik'x)] \exp(-ikx) \, dx$. All terms with $k \neq k'$ vanish, and the surviving term yields $a F_k$.

It is clear that $f(x)$ is real when $F_{-k} = F_k^*$. In that case the transform can also be expressed in sine and cosine transforms:

$$f(x) = \tfrac{1}{2}a_0 + \sum_k^\infty (a_k \cos kx + b_k \sin kx), \quad k = \tfrac{2\pi n}{a}, \ n = 1, 2, \ldots \tag{12.62}$$

$$a_k = \frac{1}{2a} \int_0^a f(x) \cos kx \, dx, \tag{12.63}$$

$$b_k = \frac{1}{2a} \int_0^a f(x) \sin kx \, dx. \tag{12.64}$$

12.7 Fast Fourier transforms

A special form of discrete Fourier transforms is the application to periodic *discrete* functions known at regular intervals $\Delta x = a/N$, where N is the number of intervals within the period a. The function values are $f_n = f(x_n)$, with $x_n = na/N$, $n = 0, \ldots, N-1$. The Fourier relations are now

$$f_n = \sum_{m=0}^{N-1} F_m e^{2\pi i n m/N}, \tag{12.65}$$

$$F_m = \frac{1}{N} \sum_{n=0}^{N-1} f_n e^{-2\pi i n m/N}. \tag{12.66}$$

These relations are easily verified by inserting (12.65) into (12.66) and realizing that $\sum_n \exp[2\pi i(m' - m)n/N]$ equals zero unless $m' = m$. In the general case the arrays f and F are complex. One should view both arrays as periodic: they can be shifted to another origin if required.

Consider the values f_n as a variable in time t, defined on a periodic interval $[0, T)$ and discretized in N small time intervals Δt. Thus $T = N\Delta t$. The data are transformed to a discretized *frequency* axis ν_m, with *resolution* $\Delta\nu = 1/T$ and maximum frequency determined by $\nu_{\max} = 1/2\Delta t$. The latter follows from Nyquist's theorem stating that two data points per period of the highest frequency are sufficient and necessary to reproduce a function of time that is band-limited to $\nu_{\max}$ (i.e., which has no Fourier components beyond $\nu_{\max}$).[3] The transform contains N frequency components between $-\nu_{\max}$ and $\nu_{\max}$ in steps of $\Delta\nu$. Because of the periodicity of F_m in $[0, n)$, the negative frequencies are to be found for $N/2 \leq m < N$: $F_{-m} = F_{N-m}$. In the special but common case that f_n is real, the transform is described by a complex array of length $N/2$ because $F_{-m} = F_{N-m} = F_m^*$. If a continuous function that contains frequencies $\nu > \nu_{\max}$ is sampled at intervals $\Delta t = 1/2\nu_{\max}$, these higher frequencies are *aliased* or *folded back* into the frequency domain $[0, \nu_{\max})$ and appear as frequencies $2\nu_{\max} - \nu$, i.e., the signal is mirrored with respect to $\nu_{\max}$. Such a "false" signal is called an *alias*. When noisy signals are sampled at intervals Δt, noise components with frequencies above $1/2\Delta t$ will appear as low-frequency noise and possible mask interesting events. In order to prevent aliasing, one should either apply a low-pass filter to the signal before sampling, or use oversampling with subsequent removal of the high-frequency components.

The assumed periodicity of the series f_n, $n \in [0, N)$, may also cause

---

[3] If a series f_n is interpolated with the sinc function: $f(t) = \sum_n f_n \sin z/z$, with $z = \pi(t - t_i)/\Delta t$, then the Fourier transform of the resulting function vanishes for $\nu > 1/2\Delta t$.

artefacts when f_n are samples over a period T of a truncated, but in principle infinite series of data. In order to avoid unrealistic correlations between the selected time series and periodic images thereof, it is adviza ble to extend the data with a number of zeroes. This is called *zero-padding* and should ideally double the length of the array. The double length of the data series refines resolution of the frequency scale by a factor of two; of course the factual resolution of the frequency distribution is not changed just by adding zeroes, but the distributions look nicer. Distributions can be made smoother and wiggles in spectral lines, caused by sudden truncation of the data, can be avoided by multiplying the time data by a *window function* that goes smoothly to zero near the end of the data series. An example is given below.

Since very fast algorithms (FFT, fast Fourier transform, invented by Cooley and Tukey, 1965)[4] exist for this representation of Fourier transforms, the periodic-discrete form is most often used in numerical problems. A well-implemented FFT scales as $N \log N$ with the array size N. The most efficient implementations are for values of N that decompose into products of small primes, such as powers of 2.

12.8 Autocorrelation and spectral density from FFT

Consider an infinite series of real data f_n, $n \in \mathbb{Z}$. Let the series be stationary in the sense that statistical properties (average, variance, correlation function) evaluated over a specific interval $[i, \ldots i + N)$ are within statistical accuracy independent of the origin i. The *autocorrelation function* C_k, which is discrete in this case, is defined as

$$C_k = \lim_{N \to \infty} \frac{1}{N} \sum_{n=i}^{i+N-1} f_n f_{n+k}. \tag{12.67}$$

The value of C_k does not depend on the origin i; C_0 equals the mean square (also called the mean *power per sample*) of f. The function is symmetric: $C_{-k} = C_k$. Generally the autocorrelation is understood to apply to a function with zero mean: when f_n has zero mean, C_k tends to 0 for large k and C_0 is the variance of f. If the mean of f_n is not zero, the square of the mean is added to each C_k. The discrete autocorrelation function, as defined above, can be viewed as a discretization of the continuous autocorrelation

---

[4] See for a description Press *et al.* (1992) or Pang (1997). Python *numarray* includes an FFT module. A versatile and fast, highly recommended public-domain C subroutine FFT library is available from
http://www.fftw.org/ ("Fastest Fourier Transform in the West").

function $C(\tau)$ of a continuous stationary function of time $f(t)$:[5]

$$C(\tau) \stackrel{\text{def}}{=} \lim_{T \to \infty} \frac{1}{T} \int_{t_0}^{t_0+T} f(t) f(t + \tau) \, dt, \qquad (12.68)$$

where t_0 is an arbitrary origin of the time axis. Note that $C(0)$ is the *mean power per unit time* of $f(t)$ and that $C(-\tau) = C(\tau)$.

The autocorrelation function can be estimated from a truncated series of data f_n, $n \in [0, N)$ as

$$C_k \approx C_k^{\text{trunc}}$$

$$C_k^{\text{trunc}} = C_{-k}^{\text{trunc}} = \frac{1}{N-k} \sum_{n=0}^{N-k-1} f_n f_{n+k} \quad (k \geq 0), \qquad (12.69)$$

or from a periodic series of data f_n, $n \in \mathbb{Z}$, $f_{n+N} = f_n$, as

$$C_k \approx C_k^{\text{per}} = \frac{1}{N-k} \sum_{n=0}^{N-1} f_n f_{n+k}. \qquad (12.70)$$

The factor $1/(N-k)$ instead of $1/N$ in (12.70) corrects for the fact that k terms in the periodic sum have the value zero, provided the data series has been properly zero-padded. Without zero-padding it is better to use the factor $1/N$; now C_k^{per} does not approximate C_k but a mixture of C_k and C_{N-k}:

$$\frac{N-k}{N} C_k + \frac{k}{N} C_{N-k} \approx C_k^{\text{per}} = \frac{1}{N} \sum_{n=0}^{N-1} f_n f_{n+k}. \qquad (12.71)$$

This makes the correlation function symmetric about $N/2$. The estimation is in all cases exact in the limit $N \to \infty$, provided that the correlation dies out to negligible values above a given n.

We now consider the Fourier transform of the continuous autocorrelation function. Because the autocorrelation function is real and even, its Fourier transform can be expressed as a cosine transform over positive time:

$$S(\nu) = 4 \int_0^{\infty} C(\tau) \cos(2\pi\nu\tau) \, d\tau. \qquad (12.72)$$

$S(\nu)$ is a real even function of τ. The factor 4 is chosen such that the inverse

---

[5] Of course, one may read any other variable, such as a spatial coordinate, for t.

transform[6] has a simple form:

$$C(\tau) = \int_0^\infty S(\nu)\cos(2\pi\nu\tau)\,d\nu. \qquad (12.73)$$

We see that the power per unit time $C(0)$ equals the integral of $S(\nu)$ over all (positive) frequencies. Therefore we may call $S(\nu)$ the *spectral density*, as $S(\nu)d\nu$ represents the power density in the frequency interval $d\nu$.

The spectral density can also be determined from the direct Fourier transform of the time function. Consider a time slice $f_T(t), t \in [0,T)$, extended with zero for all other t, with its Fourier transform

$$F_T(\nu) = \int_0^T f(t)e^{2\pi i\nu t}\,dt, \qquad (12.74)$$

$$f(t) = \int_{-\infty}^\infty F(\nu)e^{-2\pi i\nu t}\,dt. \qquad (12.75)$$

We can now see that the spectral density is also given by

$$S(\nu) = 2\lim_{T\to\infty}\frac{1}{T}F_T^*(\nu)F_T(\nu). \qquad (12.76)$$

The right-hand side is the mean *power* (square of absolute value, irrespective of the phase) per unit of time expressed as density in the frequency domain. The factor 2 arises from the fact that $S(\nu)$ is defined for positive ν while the right-hand side applies to the frequency domain $(-\infty,\infty)$. The equality of (12.72) and (12.76) is the *Wiener–Khinchin theorem*.

Proof First we assume that the limit exists. It is intuitively clear that this is true for stationary time series that contain no constant or periodic terms.[7] We skip further discussion on this point and refer the interested reader to the classic paper by Rice (1954) for details. Substituting $C(\tau)$ from (12.68) into the definition of $S(\nu)$ given in (12.72), we find

$$S(\nu) = \lim_{T\to\infty}\frac{2}{T}\int_0^T dt\int_{-\infty}^\infty d\tau\, f_T(t)f_T(t+\tau)e^{2\pi i\nu\tau}. \qquad (12.77)$$

On the other hand, (12.76) can be written as

$$S(\nu) = \lim_{T\to\infty}\frac{2}{T}\int_0^T dt\int_0^T dt'\, f_T(t)e^{-2\pi i\nu t}f_T(t')e^{2\pi i\nu t'}$$

---

[6] The definition of the transform differs slightly from the definitions given in Section 12.1 on page 315, where $2\pi\nu$ is taken as reciprocal variable rather than ν. Because of this the factors $\sqrt{2\pi}$ disappear.

[7] In fact, we do require that the integral from 0 to ∞ of the correlation function exists; this is not the case when constant or periodic components are present.

$$= \lim_{T \to \infty} \frac{2}{T} \int_0^T dt \int_{-t}^{T-t} d\tau \, f_T(t) f_T(t+\tau) e^{2\pi i \nu \tau}. \qquad (12.78)$$

Now, for large T, t "almost always" exceeds the time over which the correlation function is non-zero. Therefore the integral over τ "almost always" includes the full correlation function, so that the limits can be taken as $\pm\infty$. In the limit of $T \to \infty$ this is exact. $\square$

Combination of (12.73) and (12.76) shows that the autocorrelation function can be obtained from the inverse FT of the squared frequency amplitudes:

$$C(\tau) = 2 \lim_{T \to \infty} \frac{1}{T} \int_0^\infty F_T^*(\nu) F_T(\nu) \cos(2\pi\nu\tau) \, d\nu$$

$$= \lim_{T \to \infty} \frac{1}{T} \int_{-\infty}^\infty F_T^*(\nu) F_T(\nu) e^{-2\pi i \nu \tau} \, d\nu. \qquad (12.79)$$

Note that the sign in the exponent is irrelevant because of the symmetry of F^*F.

The discrete periodic case is similar. Given a time series f_n, periodic on $[0, N)$, we obtain the following exact relations:

(i) Autocorrelation function C_k^{per} defined in (12.70).
(ii) Fourier transform F_m of time series is defined by (12.66).
(iii) Spectral density from F_m:

$$S_m^{\mathrm{per}} = \frac{2}{N} F_m^* F_m. \qquad (12.80)$$

(iv) Autocorrelation function from inverse FFT of spectral density:

$$C_k^{\mathrm{per}} = \sum_{m=0}^{N/2-1} S_m^{\mathrm{per}} \cos(2\pi mk/N) = \frac{1}{2N} \sum_{m=0}^{N-1} S_m^{\mathrm{per}} e^{-2\pi i mk/N}. \qquad (12.81)$$

(v) Spectral density from FFT of autocorrelation function:

$$S_m^{\mathrm{per}} = 4 \sum_{k=0}^{N/2-1} C_k^{\mathrm{per}} \cos(2\pi mk/N) = 2 \sum_{k=0}^{N-1} C_k^{\mathrm{per}} e^{2\pi i mk/N}. \qquad (12.82)$$

When the data are samples of a continuous time series taken at intervals Δt during a total time span $T = N\Delta t$, the autocorrelation coefficients C_k^{per} provide an *estimate* for the continuous autocorrelation function $C(k\Delta t)$ of (3.37). If zero-padding has been applied, the estimate is improved by scaling $C(k\Delta t)$ with $N/(N-k)$ (see (12.70) on page 326). Without zero-padding, C_k^{per} provides an estimate for a mixture of $C(k\Delta t)$ and $C(T - k\Delta t)$, as given by (12.71). The spectral resolution is $\Delta\nu = 1/T$ and the coefficients S_m represent the "power" in a frequency range $\Delta\nu$. The continuous power density per unit of frequency $S(\nu)$, given by (12.72) and (12.76), equals $S_m/\Delta\nu = TS_m$.

If the time series is longer than can be handled with FFT, one may break up the series into a number of shorter sections, determine the spectral density of each and average the spectral densities over all sections. One may then proceed with step (iv). This procedure has the advantage that a good error estimate can be made based on the variance of the set of spectra obtained from the sections. Spectra obtained from one data series will be noisy and – if the data length N greatly exceeds the correlation length of the data (as it should) – too finely grained. One may then apply a *smoothing procedure* to the spectral data. Beware that this causes a subjective alteration of the spectrum! This can be done by *convolution* with a local spread function, e.g., a Gaussian function,[8] but the simplest way is to multiply the autocorrelation function with a *window function* which reduces the noisy tail of the correlation function. The smoothed spectral density is then recovered by FFT of the windowed autocorrelation function. The effect is a convolution with the FT of the window function. If a Gaussian window is used, the smoothing is Gaussian as well. See the following example.

In Fig. 12.4 an example is given of the determination of a smoothed spectral density from a given time series. The example concerns an MD simulation at 300 K of the copper-containing protein *azurin* in water and the question was asked which vibration frequencies are contained in the fluctuation of the distance between the copper atom and the sulphur atom of a cysteine, one of the copper ligands. Such frequencies can be compared to experimental resonance-Raman spectra. A time slice of 20 ps with a resolution of 2 fs (10 000 data points) was considered (Fig. 12.4a). It was Fourier-transformed ($\Delta\nu = 10^{12}/20$ Hz = 50 GHz) and its power spectrum, which had no significant components above a range of 400 points (20 THz), computed and plotted in Fig. 12.4b. The complete power spectrum was subsequently inversely Fourier transformed to the autocorrelation function (not shown), which was multiplied by a Gaussian window function with

---

[8] See Press *et al.* (1992) for a discussion of optimal smoothing.

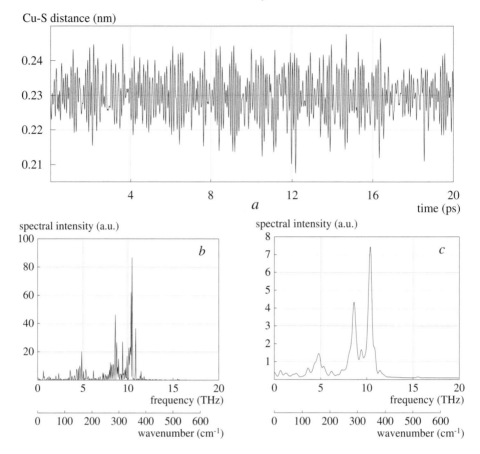

Figure 12.4 Fluctuating distance between Cu and S atoms in the copper protein azurin, from an MD simulation, and its spectral density. (a) Time series of 10 000 points, time step 2 fs, duration 20 ps. (b) Spectral intensity (square of absolute value of FFT) by direct FFT of time series. (c) The same after low-pass filtering by applying a Gaussian window to the autocorrelation function (data from Marieke van de Bosch, Leiden).

s.d. of 300 points (600 fs). Fourier transformation then gave the smoothed spectrum of Fig. 12.4c. Here are the few Python lines that do the trick.

PYTHON PROGRAM 12.1 **Spectrum from time series**
Computes the smoothed spectral density from a simulated time series.

```
01 from fftpack import fft,ifft
02 # load data array f
03 fdev=f-f.mean()
04 N = len(fdev)
05 F = fft(fdev)
```

```
06 FF = F.real**2 + F.imag**2
07 spec1 = FF[:400]
08 acf = ifft(FF).real
09 sigma = 300.
10 window = exp(-0.5*(arange(N)/sigma)**2)
11 acf2 = acf*window
12 spec2 = fft(acf2).real[:400]
```

Comments
Line 2: fill array f with data read from file. Subtract average in line 4. Line 6 computes F^*F and line 7 produces the relevant part of the raw spectrum. Line 12 produces the relevant part of the smoothed spectrum. The window function is a Gaussian function with s.d. (σ) of 300 points. Only some 1000 points (3 σ) are relevant, but the full length of N (10 000) is retained to provide a dense grid for plotting.

12.9 Multidimensional Fourier transforms

Fourier transforms are easily generalized to multidimensional periodic funct-ions. For example, if $f(x, y)$ is periodic in both x and y:

$$f(x + n_1 a, y + n_2 b) = f(x, y), \quad n_1, n_2 \in \mathbb{Z}^2, \tag{12.83}$$

then (12.60) and (12.61) generalize to

$$f(x, y) = \sum_{k_1} \sum_{k_2} F_{k_1 k_2} e^{i(k_1 x + k_2 y)}, \tag{12.84}$$

$$F_{k_1 k_2} = \frac{1}{ab} \int_0^a dx \int_0^b dy \, e^{-i(k_1 x + k_2 y)}, \tag{12.85}$$

where $k_1 = 2\pi n_1/a$ and $k_2 = 2\pi n_2/b$; $n_1, n_2 \in \mathbb{Z}^2$. In vector notation:

$$f(\boldsymbol{r}) = \sum_{\boldsymbol{k}} F_{\boldsymbol{k}} e^{i\boldsymbol{k}\cdot\boldsymbol{r}}, \tag{12.86}$$

$$F_{\boldsymbol{k}} = \frac{1}{V} \int_V d\boldsymbol{r} \, f(\boldsymbol{r}) e^{-i\boldsymbol{k}\cdot\boldsymbol{r}}, \tag{12.87}$$

where V is the volume ab.... Thus 3D FTs for periodic spatial functions *with a rectangular unit cell*, which fulfill the periodicity rule of (12.83), are simple products of three 1D FTs. This simple product decomposition does *not* apply to periodic spaces with monoclinic or triclinic unit cells.[9] If the unit cell is spanned by (cartesian) base vectors $\boldsymbol{a}, \boldsymbol{b}, \boldsymbol{c}$, the periodicity is expressed as

$$f(\boldsymbol{r} + n_1\boldsymbol{a} + n_2\boldsymbol{b} + n_3\boldsymbol{c}) = f(\boldsymbol{r} + \mathbf{Tn}) = f(\boldsymbol{r}), \quad \mathbf{n} \in \mathbb{Z}^3, \tag{12.88}$$

---

[9] See page 142 for the description of general periodicity in 3D space.

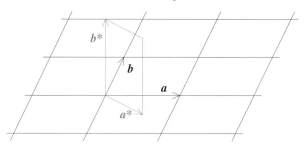

Figure 12.5 A two-dimensional real lattice with base vectors $\boldsymbol{a} = (a, 0)$ and $\boldsymbol{b} = (0.25a, 0.5a)$. The reciprocal vectors are $\boldsymbol{a}^* = (1/a, -0.5/a)$ and $\boldsymbol{b}^* = (0, 2/a)$. For a the value $\sqrt{2}$ is taken. Note that a large spacing in real space means a small spacing in "reciprocal space."

where $\mathbf{T}$ is the transformation matrix

$$\mathbf{T} = \begin{pmatrix} a_x & b_x & c_x \\ a_y & b_y & c_y \\ a_z & b_z & c_z \end{pmatrix}. \tag{12.89}$$

This is not of the form of (12.83), but functions expressed in *relative coordinates* $\boldsymbol{\rho} = (\xi, \eta, \zeta)$:

$$\boldsymbol{r} = \xi\boldsymbol{a} + \eta\boldsymbol{b} + \zeta\boldsymbol{c}, \tag{12.90}$$

$$\mathbf{r} = \mathbf{T}\boldsymbol{\rho}, \tag{12.91}$$

are periodic in the sense of (12.83):

$$f(\boldsymbol{\rho}) = f(\boldsymbol{\rho} + \mathbf{n}), \quad \mathbf{n} \in \mathbb{Z}^3. \tag{12.92}$$

Fourier transforms now involve $\exp(\pm i\boldsymbol{\kappa} \cdot \boldsymbol{\rho})$, with $\boldsymbol{\kappa} = 2\pi\mathbf{m}$; $\mathbf{m} \in \mathbb{Z}^3$. These exponentials can be rewritten as $\exp(\pm i\boldsymbol{k} \cdot \boldsymbol{r})$ as follows (in matrix notation):

$$\boldsymbol{\kappa} \cdot \boldsymbol{\rho} = \boldsymbol{\kappa}^\mathsf{T}\boldsymbol{\rho} = 2\pi\mathbf{m}^\mathsf{T}\mathbf{T}^{-1}\mathbf{r} = \mathbf{k}^\mathsf{T}\mathbf{r} = \boldsymbol{k} \cdot \boldsymbol{r}, \tag{12.93}$$

if $\mathbf{k}$ is defined as

$$\mathbf{k} \stackrel{\text{def}}{=} 2\pi(\mathbf{T}^{-1})^\mathsf{T}\mathbf{m}. \tag{12.94}$$

Defining the (cartesian) *reciprocal lattice vectors*[10] $\boldsymbol{a}^*, \boldsymbol{b}^*, \boldsymbol{c}^*$ by the rows of the inverse transformation matrix $\mathbf{T}^{-1}$:

$$\mathbf{T}^{-1} = \begin{pmatrix} a_x^* & a_y^* & a_z^* \\ b_x^* & b_y^* & b_z^* \\ c_x^* & c_y^* & c_z^* \end{pmatrix}, \tag{12.95}$$

---

[10] Note that the asterisk does not represent a complex conjugate here.

we see that

$$k = 2\pi(m_1 a^* + m_2 b^* + m_3 c^*).$$ (12.96)

With this definition of k in terms of reciprocal lattice vectors, the Fourier pair (12.86) and (12.87) remain fully valid. In crystallography, where $f(r)$ represents an electron density, the quantities F_k are usually called *structure factors* and the indices m_1, m_2, m_3 are often indicated by h, k, l. The volume V of the unit cell equals the determinant of $\mathbf{T}$. The reciprocal lattice vectors have a scalar product of 1 with their corresponding base vectors and are perpendicular to the other base vectors:

$$a \cdot a^* = 1,$$ (12.97)

$$a \cdot b^* = 0,$$ (12.98)

and similar for other products, as follows immediately from the definition of the reciprocal vectors. Fig. 12.5 shows a two-dimensional real lattice and the corresponding reciprocal lattice vectors.

Exercises

12.1   Show that hermitian operators have real eigenvalues.

12.2   Show that the operator for k is hermitian. Use partial integration of $\int f^*(\partial g/\partial x)\, dx$; the product f^*g vanishes at the integration boundaries.

12.3   Show that the expectation of k must vanish when the wave function is real.

12.4   Show that the expectation value of k equals k_0 if the wave function is a real function, multiplied by $\exp(ik_0)$.

12.5   Derive (12.62) to (12.64) from (12.60) and (12.61) and express a_k and b_k in terms of F_k.

12.6   Show that $\xi = a^* \cdot r, \eta = b^* \cdot r, \zeta = c^* \cdot r$.

12.7   Prove (12.97) and (12.98).

12.8   Derive the Fourier pair (12.72) and (12.73) from the original definitions in Section 12.1.

13

Electromagnetism

13.1 Maxwell's equation for vacuum

For convenience of the reader and for unity of notation we shall review the basic elements of electromagnetism, based on Maxwell's equations. We shall use SI units throughout. Two unit-related constants figure in the equations: the electric and magnetic *permittivities* of vacuum ε_0 and μ_0:

$$\varepsilon_0 = \frac{1}{\mu_0\, c^2} = 8.854\,187\,817\ldots \times 10^{-12} \text{ F/m}, \tag{13.1}$$

$$\mu_0 = 4\pi \times 10^{-7} \text{ N/A}^2, \tag{13.2}$$

$$\varepsilon_0\mu_0 = \frac{1}{c^2}. \tag{13.3}$$

The basic law describes the *Lorentz force* $\boldsymbol{F}$ on a particle with charge q and velocity $\boldsymbol{v}$ in an electromagnetic field:

$$F = q(\boldsymbol{E} + \boldsymbol{v} \times \boldsymbol{B}). \tag{13.4}$$

Here, $\boldsymbol{E}$ is the *electric field* and $\boldsymbol{B}$ the *magnetic field* acting on the particle. The fields obey the four *Maxwell equations* which are continuum equations in vacuum space that describe the relations between the fields and their source terms ρ (charge density) and $\boldsymbol{j}$ (current density):

$$\text{div } \boldsymbol{E} = \rho/\varepsilon_0, \tag{13.5}$$

$$\text{div } \boldsymbol{B} = 0, \tag{13.6}$$

$$\text{curl } \boldsymbol{E} + \frac{\partial \boldsymbol{B}}{\partial t} = 0, \tag{13.7}$$

$$\text{curl } \boldsymbol{B} - \frac{1}{c^2}\frac{\partial \boldsymbol{E}}{\partial t} = \mu_0 \boldsymbol{j}. \tag{13.8}$$

Moving charges (with velocity $\boldsymbol{v}$) produce currents:

$$\boldsymbol{j} = \rho\boldsymbol{v}. \tag{13.9}$$

335

The charge density and current obey a *conservation law*, expressed as

$$\operatorname{div} \boldsymbol{j} + \frac{\partial \rho}{\partial t} = 0, \tag{13.10}$$

which results from the fact that charge flowing out of a region goes at the expense of the charge density in that region.

13.2 Maxwell's equation for polarizable matter

In the presence of linearly polarizable matter with electric and magnetic *susceptibilities* χ_e and χ_m, an *electric dipole density* $\boldsymbol{P}$ and a *magnetic dipole density* (M) are locally induced according to

$$\boldsymbol{P} = \varepsilon_0 \chi_e \boldsymbol{E}, \tag{13.11}$$
$$\boldsymbol{M} = \chi_m \boldsymbol{H}. \tag{13.12}$$

The charge and current densities now contain terms due to the polarization:

$$\rho = \rho_0 - \operatorname{div} \boldsymbol{P}, \tag{13.13}$$
$$\boldsymbol{j} = \boldsymbol{j}_0 + \frac{\partial \boldsymbol{P}}{\partial t} + \operatorname{curl} \boldsymbol{M}, \tag{13.14}$$

where ρ_0 and $\boldsymbol{j}_0$ are the *free* or unbound sources. With the definitions of the *dielectric displacement* $\boldsymbol{D}$ and *magnetic intensity*[1] $\boldsymbol{H}$ and the material electric and magnetic permittivities ε and μ,

$$\boldsymbol{D} = \varepsilon_0 \boldsymbol{E} + \boldsymbol{P} = \varepsilon \boldsymbol{E}, \tag{13.15}$$
$$\boldsymbol{H} = \frac{1}{\mu_0} \boldsymbol{B} - \boldsymbol{M} = \frac{1}{\mu} \boldsymbol{B}, \tag{13.16}$$

the Maxwell equations for linearly polarizable matter are obtained:

$$\operatorname{div} \boldsymbol{D} = \rho_0, \tag{13.17}$$
$$\operatorname{div} \boldsymbol{B} = 0, \tag{13.18}$$
$$\operatorname{curl} \boldsymbol{E} + \frac{\partial \boldsymbol{B}}{\partial t} = 0, \tag{13.19}$$
$$\operatorname{curl} \boldsymbol{H} - \frac{\partial \boldsymbol{D}}{\partial t} = \boldsymbol{j}_0. \tag{13.20}$$

---

[1] In older literature $\boldsymbol{H}$ is called the magnetic field strength and $\boldsymbol{B}$ the magnetic induction.

The time derivative of D acts as a current density and is called the *displacement current density*. The permittivities are related to the susceptibilities:

$$\varepsilon = (1 + \chi_e)\varepsilon_0, \tag{13.21}$$

$$\mu = (1 + \chi_m)\mu_0. \tag{13.22}$$

ε is often called the dielectric constant, although it is advisable to reserve that term for the *relative* dielectric permittivity

$$\varepsilon_r = \frac{\varepsilon}{\varepsilon_0}. \tag{13.23}$$

13.3 Integrated form of Maxwell's equations

The Maxwell relations may be integrated for practical use. We then obtain:

- the *Gauss equation*, relating the integral of the normal component of D over a closed surface to the total charge inside the enclosed volume:

$$\iint D \cdot dS = \sum q, \tag{13.24}$$

which leads immediately to the Coulomb field of a point charge at the origin:

$$D(r) = \frac{q}{4\pi} \frac{r}{r^3}; \tag{13.25}$$

- *Faraday's induction law*, equating the voltage along a closed path with the time derivative of the total magnetic flux through a surface bounded by the path:

$$V_{\text{ind}} = \oint E \cdot dl = -\frac{\partial}{\partial t} \iint B \cdot dS; \tag{13.26}$$

- *Ampere's law*, relating the magnetic field along a closed path to the total current i through a surface bounded by the path:

$$\oint H \cdot dl = \sum i + \frac{\partial}{\partial t} \iint D \cdot dS. \tag{13.27}$$

13.4 Potentials

It is convenient to describe electromagnetic fields as spatial derivatives of a *potential field*. This can only be done if four quantities are used to describe the potential, a *scalar potential* ϕ and a *vector potential* A:

$$E = -\operatorname{grad} \phi - \frac{\partial A}{\partial t}, \tag{13.28}$$

$$B = \operatorname{curl} A. \tag{13.29}$$

The definitions are not unique: the physics does not change if we replace ϕ by $\phi - \partial f / \partial t$ and simultaneously A by $A - \operatorname{grad} f$, where f is any differentiable function of space and time. This is called the *gauge invariance*. Therefore the divergence of A can be chosen at will. The *Lorentz convention* is

$$\operatorname{div} A + \frac{1}{c^2} \frac{\partial \phi}{\partial t} = 0, \tag{13.30}$$

implying (in a vacuum) that

$$\left(\nabla^2 - \frac{1}{c^2} \frac{\partial^2}{\partial t^2} \right) \phi = -\rho_0 / \varepsilon_0, \tag{13.31}$$

$$\left(\nabla^2 - \frac{1}{c^2} \frac{\partial^2}{\partial t^2} \right) A = -\mu_0 j_0. \tag{13.32}$$

13.5 Waves

The Maxwell equations support waves with the velocity of light in vacuum, as can be seen immediately from (13.31) and (13.32). For example, a linearly polarized electromagnetic plane wave in the direction of a vector k, with wave length $2\pi/k$ and frequency $\omega/2\pi$, has an electric field

$$E(r, t) = E_0 \exp[i(k \cdot r - \omega t)], \tag{13.33}$$

where E_0 (in the polarization direction) must be perpendicular to k, and a magnetic field

$$B(r, t) = \frac{1}{\omega} k \times E(r, t). \tag{13.34}$$

The wave velocity is

$$\frac{\omega}{k} = c. \tag{13.35}$$

The wave can also be represented by a vector potential:

$$A(r, t) = \frac{i}{\omega} E(r, t). \tag{13.36}$$

The scalar potential ϕ is identically zero. Waves with $A = 0$ cannot exist.
The vector

$$\Sigma = E \times H \tag{13.37}$$

is called the *Poynting vector*. It is directed along k, the direction in which the wave propagates, and its magnitude equals the energy flux density, i.e., the energy transported by the wave per unit of area and per unit of time.

13.6 Energies

Electromagnetic fields "contain" and "transport" energy. The *electromagnetic energy density* W of a field is given by

$$W = \tfrac{1}{2} \boldsymbol{D} \cdot \boldsymbol{E} + \tfrac{1}{2} \boldsymbol{B} \cdot \boldsymbol{H}. \tag{13.38}$$

In vacuum or in a linearly polarizable medium with time-independent permittivities, for which $\boldsymbol{D} = \varepsilon \boldsymbol{E}$ and $\boldsymbol{B} = \mu \boldsymbol{H}$, the time dependence of W is

$$\frac{dW}{dt} = -\operatorname{div} \boldsymbol{\Sigma} - \boldsymbol{j} \cdot \boldsymbol{E}. \tag{13.39}$$

Proof Using the time-independence of ε and μ, we can write

$$\dot{W} = \boldsymbol{E} \cdot \dot{\boldsymbol{D}} + \boldsymbol{H} \cdot \dot{\boldsymbol{B}}$$

(to prove this in the case of tensorial permittivities, use must be made of the fact that ε and μ are symmetric tensors). Now we can replace $\dot{\boldsymbol{D}}$ by $\operatorname{\mathbf{curl}} \boldsymbol{H} - \boldsymbol{j}$ (see (13.20)) and $\dot{\boldsymbol{B}}$ by $-\operatorname{\mathbf{curl}} \boldsymbol{E}$ (see (13.19)) and use the general vector equality

$$\operatorname{div} (\boldsymbol{A} \times \boldsymbol{B}) = \boldsymbol{B} \cdot \operatorname{\mathbf{curl}} \boldsymbol{A} - \boldsymbol{A} \cdot \operatorname{\mathbf{curl}} \boldsymbol{B}. \tag{13.40}$$

Equation (13.39) follows. $\qquad\square$

Equation (13.39) is an energy-balance equation: $-\operatorname{div} \boldsymbol{\Sigma}$ is the energy flowing out per unit volume due to electromagnetic radiation; $-\boldsymbol{j} \cdot \boldsymbol{E}$ is the energy taken out of the field by the friction of moving charges and dissipated into heat (the *Joule heat*). Note that this equation is not valid if permittivities are time-dependent.

For quasi-stationary fields, where radiation does not occur, the total field energy of a system of interacting charges and currents can also be expressed in terms of the source densities and potentials as follows:

$$U_{\text{field}} = \int W(\boldsymbol{r}) \, d\boldsymbol{r} = \int \tfrac{1}{2} (\rho \phi + \boldsymbol{j} \cdot \boldsymbol{A}) \, d\boldsymbol{r}, \tag{13.41}$$

where integration of W is over all space, while the sources ρ and $\boldsymbol{j}$ are confined to a bounded volume in space.

Proof Quasi-stationarity means that the following equations are valid:

$$\boldsymbol{E} = -\operatorname{\mathbf{grad}} \phi,$$
$$\operatorname{div} \boldsymbol{D} = \rho,$$
$$\boldsymbol{B} = \operatorname{\mathbf{curl}} \boldsymbol{A},$$

$$\mathbf{curl}\,\boldsymbol{H} = \boldsymbol{j}.$$

Using the first two equations we see that

$$\int \boldsymbol{D} \cdot \boldsymbol{E}\,d\boldsymbol{r} = -\int \boldsymbol{D} \cdot \mathbf{grad}\,\phi\,d\boldsymbol{r} = \int \phi\,\mathrm{div}\,\boldsymbol{D}\,d\boldsymbol{r} = \int \rho\phi\,d\boldsymbol{r}.$$

The third integral follows from the second by partial integration, whereby the integral over the (infinite) boundary vanishes if the sources are confined to a bounded volume. Similarly

$$\int \boldsymbol{B} \cdot \boldsymbol{H}\,d\boldsymbol{r} = -\int (\mathbf{curl}\,\boldsymbol{A}) \cdot \boldsymbol{H}\,d\boldsymbol{r} = \int \boldsymbol{A} \cdot \mathbf{curl}\,\boldsymbol{H}\,d\boldsymbol{r} = \int \boldsymbol{j} \cdot \boldsymbol{A}\,d\boldsymbol{r}.$$

The reader is invited to check the partial integration result by writing the integrand ount in all coordinates. ☐

This equation is often more convenient for computing energies than the field expression (13.38).

Both expressions contain a *self-energy* for isolated charges that becomes singular for delta-function point charges. This self-energy is not taken into account if the Coulomb interaction energy between a system of point charges is considered. Thus for a set of charges q_i at positions $\boldsymbol{r}_i$, the total interaction energy is

$$U_{\mathrm{int}} = \tfrac{1}{2} \sum_i q_i(\phi_i - \phi_i^{\mathrm{self}}), \tag{13.42}$$

where

$$\phi_i - \phi_i^{\mathrm{self}} = \frac{1}{4\pi\varepsilon} \sum_{j \neq i} \frac{q_j}{|\boldsymbol{r}_j - \boldsymbol{r}_i|}. \tag{13.43}$$

This is indeed the usual sum of Coulomb interactions over all pairs:

$$U_{\mathrm{int}} = \frac{1}{4\pi\varepsilon} \sum_{i<j} \frac{q_i q_j}{|\boldsymbol{r}_j - \boldsymbol{r}_i|}. \tag{13.44}$$

The factor $\tfrac{1}{2}$ in (13.42) compensates for the double counting of all pairs.

13.7 Quasi-stationary electrostatics

In the vast majority of molecular systems of interest, motions are slow enough that the time dependence in the Maxwell equations can be neglected, and the magnetic effects of net currents are also negligible. The relations now simplify considerably. There are no magnetic fields and waves are no longer supported. The result is called *electrostatics*, although slow time dependence is not excluded.

13.7.1 The Poisson and Poisson–Boltzmann equations

Since the curl of the electric field $\boldsymbol{E}$ is now zero, $\boldsymbol{E}$ can be written as a pure gradient of a potential

$$\boldsymbol{E} = -\operatorname{\mathbf{grad}}\phi. \tag{13.45}$$

In a linear dielectric medium, where the dielectric constant may still depend on position, we obtain from (13.15) and (13.17) the *Poisson equation*

$$\operatorname{div}(\varepsilon\operatorname{\mathbf{grad}}\phi) = -\rho, \tag{13.46}$$

which simplifies in a homogeneous dielectric medium to

$$\nabla^2\phi = -\frac{\rho}{\varepsilon}. \tag{13.47}$$

Here, ρ is the density of *free* charges, not including the "bound" charges due to divergence of the polarization. We drop the index 0 in ρ_0, as was used previously.

In ionic solutions, there is a relation between the charge density and the potential. Assume that at a large distance from any source terms, where the potential is zero, the electrolyte contains bulk concentrations c_i^0 of ionic species i, which have a charge (including sign) of $z_i e$ per ion.[2] Electroneutrality prescribes that

$$\sum_i c_i^0 z_i = 0. \tag{13.48}$$

In a *mean-field* approach we may assume that the concentration $c_i(\boldsymbol{r})$ of each species is given by its Boltzmann factor in the potential $\phi(\boldsymbol{r})$:

$$c_i(\boldsymbol{r}) = c_i^0 \exp\left(-\frac{z_i e\phi}{kT_{\mathrm{B}}}\right), \tag{13.49}$$

so that, with

$$\rho = F\sum_i c_i z_i, \tag{13.50}$$

where F is the Faraday constant (96 485.338 C), we obtain the *Poisson–Boltzmann equation* for the potential:

$$\operatorname{div}(\varepsilon\operatorname{\mathbf{grad}}\phi) = -F\sum_i c_i^0 z_i \exp\left(-\frac{z_i F\phi}{RT}\right). \tag{13.51}$$

In the *Debye–Hückel approximation* the exponential is expanded and only

---

[2] Note that concentrations must be expressed in mol/m^3, if SI units are used.

the first two terms are kept.[3] Since the first term cancels as a result of the electroneutrality condition (13.48), the resulting *linearized Poisson–Boltzmann equation* is obtained:

$$\text{div}\,(\varepsilon\,\mathbf{grad}\,\phi) = \frac{F^2 \sum_i c_i^0 z_i^2}{RT}\phi. \tag{13.52}$$

The behavior depends only on the *ionic strength* of the solution, which is defined as

$$I = \tfrac{1}{2}\sum_i c_i^0 z_i^2. \tag{13.53}$$

Note that for a 1:1 electrolyte the ionic strength equals the concentration. In dielectrically homogeneous media the linearized Poisson–Boltzmann equation gets the simple form

$$\nabla^2\phi = \kappa^2\phi, \tag{13.54}$$

where κ is defined by

$$\kappa^2 = \frac{2IF^2}{\varepsilon RT}. \tag{13.55}$$

The inverse of κ is the *Debye length*, which is the characteristic distance over which a potential decays in an electrolyte solution.

The Poisson equation (13.46) or linearized Poisson–Boltzmann equation (13.52) can be numerically solved on a 3D grid by iteration.[4] In periodic systems Fourier methods can be applied and for arbitrarily shaped systems embedded in a homogeneous environment, *boundary element* methods are available.[5] The latter replace the influence of the environment by a boundary layer of charges or dipoles that produce exactly the reaction field due to the environment. For simple geometries analytical solutions are often possible, and in Sections 13.7.2, 13.7.3 and 13.7.4 we give as examples the reaction potentials and fields for a charge, a dipole and a charge distribution in a sphere, embedded in a polarizable medium. Section 13.7.5 describes the generalized Born approximation for a system of embedded charges.

The following *boundary conditions* apply at boundaries where the dielectric properties show a stepwise change. Consider a planar boundary in the x, y-plane at $z = 0$, with $\varepsilon = \varepsilon_1$ for $z < 0$ and $\varepsilon = \varepsilon_2$ for $z > 0$, without free

---

[3] For dilute solutions the Debye–Hückel approximation is appropriate. In cases where it is not appropriate, the full Poisson–Boltzmann equation is also not adequate, since the mean-field approximation also breaks down. Interactions with individual ions and solvent molecules are then required. In general, when κ^{-1} (see (13.55)) approaches the size of atoms, the mean-field approximation will break down.

[4] A popular program to solve the PB equation on a grid is DELPHI, see Nicholls and Honig (1991).

[5] Juffer *et al.* (1991).

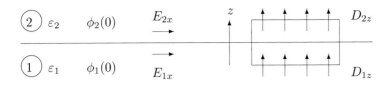

Figure 13.1 Boundary conditions on a planar discontinuity: potential ϕ, tangential electric field E_x and perpendicular displacement D_z are continuous.

charge density at the surface (Fig. 13.1). Since the potential is continuous over the surface, its derivative along the surface, i.e., the component of the electric field along the surface, is also continuous:

$$E_x(z \uparrow 0) = E_x(z \downarrow 0),$$
$$E_y(z \uparrow 0) = E_y(z \downarrow 0). \tag{13.56}$$

Since there is no free charge at the boundary, there is no source for D, and hence div $D = 0$. This implies that the ingoing flux of D through one plane of a cylindrical box equals the outgoing flux through the other plane at the other side of the boundary. Hence

$$D_z(z \uparrow 0) = D_z(z \downarrow 0). \tag{13.57}$$

13.7.2 Charge in a medium

Our first example is a charge q in a cavity (ε_0) of radius a in a homogeneous dielectric environment with $\varepsilon = \varepsilon_r \varepsilon_0$. According to (13.25), the dielectric displacement at a distance $r > a$ equals

$$D(r) = \frac{q}{4\pi r^2} \frac{r}{r}. \tag{13.58}$$

The field energy outside the cavity is

$$U_{\text{field}} = \frac{1}{2\varepsilon} \int_a^\infty D^2(r) 4\pi r^2 \, dr = \frac{q^2}{8\pi\varepsilon a}. \tag{13.59}$$

The self-energy outside the cavity is the same with $\varepsilon = \varepsilon_0$ and the excess field energy due to the polarization of the environment, which is negative, is

$$U_{\text{Born}} = -\frac{q^2}{8\pi\varepsilon_0 a}\left(1 - \frac{1}{\varepsilon_r}\right). \tag{13.60}$$

This polarization energy is often called the *dielectric solvation energy* or *Born energy* after Born (1920), who first described this term. It is the (free) energy change when a charge q is moved from vacuum into a cavity with radius a in a dielectric continuum.

Another description of the Born energy is by using the alternative field energy description of (13.41) in terms of charge-potential products. The polarizable environment produces a potential ϕ_{RP} at the position of the charge, called the *reaction potential*, and the energy in that potential is

$$U_{\text{Born}} = \tfrac{1}{2}q\phi_{RP}. \tag{13.61}$$

The reaction potential is the summed potential of all induced dipoles in the medium:

$$\phi_{RP} = -\frac{1}{4\pi\varepsilon_0} \int_a^\infty \frac{P(r)}{r^2} 4\pi r^2 \, dr, \tag{13.62}$$

where

$$P = \left(1 - \frac{1}{\varepsilon_r}\right) D = \frac{1}{4\pi}\left(1 - \frac{1}{\varepsilon_r}\right)\frac{q}{r^2}. \tag{13.63}$$

Integration yields

$$\phi_{RP} = -\frac{q}{4\pi\varepsilon_0 a}\left(1 - \frac{1}{\varepsilon_r}\right). \tag{13.64}$$

This, of course, yields the same Born energy as was derived directly from integrating the field energy in (13.60).

How does the solvation energy of a charge in a cavity behave in the case that the medium is an electrolyte solution? Let us compute the reaction potential, i.e., the excess potential at $r = 0$ due to the presence of the medium beyond $r = a$. As before, the dielectric displacement is continuous at $r = a$, and therefore

$$D(a) = \frac{q}{4\pi a^2} \tag{13.65}$$

is also valid at the boundary in the medium, so that the boundary conditions for ϕ are

$$\left(\frac{d\phi}{dr}\right)_a = -\frac{q}{4\pi\varepsilon a^2}, \quad \phi(r) \to 0 \text{ for } r \to \infty. \tag{13.66}$$

The differential equation for ϕ in the range $r \geq a$ is

$$\nabla^2\phi = -\frac{F}{\varepsilon}\sum_i c_i^0 z_i \exp\left(-\frac{z_i F\phi}{RT}\right) \tag{13.67}$$

This equation can be solved numerically, given the ionic composition of the medium. In the Debye–Hückel approximation, and using the radial expression for the ∇^2 operator, the equation simplifies to

$$\frac{1}{r^2}\frac{d}{dr}\left(r^2\frac{d\phi}{dr}\right) = \kappa^2\phi, \tag{13.68}$$

with the boundary conditions given above. The solution, valid for $r \geq a$, is

$$\phi(r) = \frac{q\exp[-\kappa(r-a)]}{4\pi\varepsilon(1+\kappa a)r}. \tag{13.69}$$

The potential in the cavity ($r \leq a$) is given by

$$D(r) = -\varepsilon_0\frac{d\phi}{dr} = \frac{q}{4\pi r^2}, \tag{13.70}$$

and hence

$$\phi(r) = \frac{q}{4\pi\varepsilon_0 r} + \phi_{RP}, \tag{13.71}$$

where ϕ_{RP} is a constant given by the boundary condition that ϕ is continuous at $r = a$. This constant is also the excess potential (the reaction potential) due to the mean-field response of the medium. Applying that boundary condition we find

$$\phi_{RP} = -\frac{q}{4\pi\varepsilon_0 a}\left(1 - \frac{1}{\varepsilon_r(1+\kappa a)}\right). \tag{13.72}$$

The excess energy of the charge – due to interaction with the medium outside the cavity in excess to the vacuum self-energy – is, as in the previous case, given by

$$U_{exc} = \tfrac{1}{2}q\phi_{RP}. \tag{13.73}$$

We see that for zero ionic strength the Born reaction potential (13.60) is recovered, but that for large ionic strength the screening is more effective, as if the dielectric constant of the medium increases. The excess energy due to κ:

$$U_{exc}(\kappa) - U_{exc}(\kappa = 0) = -\frac{1}{2}\frac{q^2}{4\pi\varepsilon_0\varepsilon_r}\frac{\kappa}{(1+\kappa a)}, \tag{13.74}$$

is the (free) energy resulting from transferring the cavity with charge from an infinitely dilute solution to the electrolyte (assuming the dielectric constant does not change). This term is due to the mean-field distribution of ions around a charge: the counterions are closer and produce a negative energy. This term is responsible for the reduction of ionic chemical potentials in electrolyte solutions, proportional to the square root of the ionic concentration.

13.7.3 Dipole in a medium

The next example is the dielectric solvation energy for a *dipole* in the center of a spherical cavity with radius a in a medium with homogeneous dielectric constant ε. The problem is similar as the previous case of a charge, except that we cannot make use of spherical symmetry. Let us choose the z-axis in the direction of the dipole moment μ situated at $r = 0$, and use spherical coordinates r, θ (polar angle with respect to the z-axis), and ϕ (azimuthal angle of rotation around z, which must drop out of the problem for reasons of symmetry).

For regions of space where there are no sources, i.e., for our problem everywhere except for $r = 0$, the Poisson equation (13.47) reduces to the *Laplace equation*

$$\nabla^2 \phi = 0, \tag{13.75}$$

which has the general solution in spherical coordinates

$$\phi(r, \theta, \phi) = \sum_{l=0}^{\infty} \sum_{m=-l}^{l} (A_l^m r^l + B_l^m r^{-l-1}) P_l^m(\cos\theta) \exp(im\phi), \tag{13.76}$$

where P_l^m are Legendre functions and A and B are constants that must follow from boundary conditions. In a bounded region of space (e.g., $r \leq a$), the r^l solutions are acceptable, but in an unbounded region with the requirement that $\phi \to 0$ for $r \to \infty$, only the r^{-l-1} solutions are acceptable. The r^{-l-1} solutions are only acceptable for $r \to 0$ if there is a singularity due to a source at $r = 0$. The singularity determines which angular term is acceptable. In the case of a dipole source, the singularity has a cosine dependence on θ ($l = 1$), and only the $l = 1$ terms need be retained.

From these considerations, we can write the potentials in the cavity and in the medium as

$$\phi^{\mathrm{cav}}(\mathbf{r}) = \frac{\mu \cos\theta}{4\pi\varepsilon_0 r^2} + br\cos\theta \quad (r \leq a), \tag{13.77}$$

$$\phi^{\mathrm{med}}(\mathbf{r}) = \frac{c\cos\theta}{r^2} \quad (r \geq a), \tag{13.78}$$

where b and c are constants to be determined from the boundary conditions at $r = a$:

(i) ϕ is continuous:

$$\phi^{\mathrm{cav}}(a) = \phi^{\mathrm{med}}(a); \tag{13.79}$$

(ii) $\boldsymbol{D}$ is continuous in the radial direction:

$$\varepsilon_0 \left(\frac{d\phi^{\mathrm{cav}}}{dr}\right)_a = \varepsilon \left(\frac{d\phi^{\mathrm{med}}}{dr}\right)_a . \tag{13.80}$$

Applying these conditions gives straightforwardly

$$b = -\frac{\mu}{4\pi\varepsilon_0 a^3} \frac{2(\varepsilon_r - 1)}{2\varepsilon_r + 1} . \tag{13.81}$$

The term $br\cos\theta = bz$ in the cavity potential is simply the potential of a homogeneous electric field in the z-direction. This is the *reaction field* resulting from the polarization in the medium, in the direction of the dipole, and with magnitude $-b$:

$$E_{\mathrm{RF}} = \frac{\mu}{4\pi\varepsilon_0 a^3} \frac{2(\varepsilon_r - 1)}{2\varepsilon_r + 1} \tag{13.82}$$

The energy of the dipole in the reaction field is

$$U_{\mathrm{RF}} = -\frac{1}{2}\mu E_{\mathrm{RF}} = -\frac{\mu^2}{8\pi\varepsilon_0 a^3} \frac{2(\varepsilon_r - 1)}{2\varepsilon_r + 1} . \tag{13.83}$$

We can now specify the potential anywhere in space:

$$\phi^{\mathrm{cav}}(\boldsymbol{r}) = \frac{\mu\cos\theta}{4\pi\varepsilon_0}\left(\frac{1}{r^2} - \frac{r}{a^3}\frac{2(\varepsilon_r - 1)}{2\varepsilon_r + 1}\right) \quad (r \leq a), \tag{13.84}$$

$$\phi^{\mathrm{med}}(\boldsymbol{r}) = \frac{\mu\cos\theta}{r^2}\frac{3}{2\varepsilon_r + 1} \quad (r \geq a). \tag{13.85}$$

In the presence of ions in the medium, a similar reasoning can be followed as we used in deriving (13.72) for a single charge. The result is

$$E_{\mathrm{RF}} = \frac{\mu}{4\pi\varepsilon_0 a^3} \frac{2(\varepsilon' - 1)}{2\varepsilon' + 1} , \tag{13.86}$$

with

$$\varepsilon' = \varepsilon_r \left(1 + \frac{\kappa^2 a^2}{2(1 + \kappa a)}\right) . \tag{13.87}$$

The effect of the ionic strength is to increase the effective dielectric constant and therefore the screening of the dipolar field, just as was the case for the reaction potential of a charge. But the extra screening has a different dependence on κ.

Let us finally consider – both for a charge and for a dipole – the special case that either $\varepsilon = \infty$ or $\kappa = \infty$ (these apply to a *conducting medium*). In this case

$$\phi_{RP} = -\frac{q}{4\pi\varepsilon_0 a}, \tag{13.88}$$

$$\boldsymbol{E}_{RF} = \frac{\boldsymbol{\mu}}{4\pi\varepsilon_0 a^3}. \tag{13.89}$$

The potential and the field of the source now vanish at the cavity boundary and are localized in the cavity itself.

13.7.4 Charge distribution in a medium

Finally, we consider a charge *distribution* in a spherical cavity (ε_0) of radius a, centered at the origin of the coordinate system, embedded in a homogeneous dielectric environment with $\varepsilon = \varepsilon_r \varepsilon_0$ and possibly with an inverse Debye length κ. There are charges q_i at positions $\boldsymbol{r}_i$ within the cavity ($r_i < a$). There are two questions to be answered:

(i) what is the reaction potential, field, field gradient, etc., in the center of the cavity? This question arises if Coulomb interactions are truncated beyond a cut-off radius r_c and one wishes to correct for the influence of the environment beyond r_c (see Section 6.3.5 on page 164);

(ii) what is the reaction field due to the environment at any position within the sphere? This question arises if we wish to find energies and forces of a system of particles located within a sphere but embedded in a dielectric environment, such as in molecular dynamics with continuum boundary conditions (see Section 6.2.2 on page 148).

The first question is simple to answer. As is explained in the next section (Section 13.8), the potential outside a localized system of charges can be described by the sum of the potentials of multipoles, each localized at the center of the system of charges, i.e., at the center of the coordinate system. The simplest multipole is the *monopole* $Q = \sum_i q_i$; it produces a reaction *potential* Φ_{RP} at the center given by (13.72), replacing the charge in the center by the monopole charge. The reaction potential of the monopole is homogeneous and there are no reaction fields. The next multipole term is the *dipole* $\boldsymbol{\mu} = \sum_i q_i \boldsymbol{r}_i$, which leads to a reaction *field* $\boldsymbol{E}_{RF}$ at the center given by (13.86) and (13.87). The reaction field is homogeneous and there is no reaction field gradient. Similarly, the quadrupole moment of the distribution will produce a field gradient at the center, etc. When the system is described by charges only, one needs the fields to compute forces, but field gradients are not needed and higher multipoles than dipoles are not

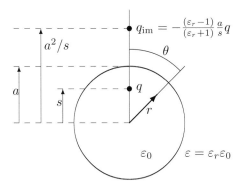

Figure 13.2 A source charge q in a vacuum sphere, embedded in a dielectric medium, produces a reaction potential that is well approximated by the potential of an image charge q_{im} situated outside the sphere.

required. However, when the system description involves dipoles, one needs the field gradient to compute forces on the dipole and the reaction field gradient of the quadrupole moment of the distribution would be required.

The second question is considerably more complicated. In fact, if the system geometry is not spherical, full numerical Poisson (or linearized Poisson-Boltzmann) solutions must be obtained, either with finite-difference methods on a grid or with boundary-element methods on a triangulated surface. For a charge distribution in a sphere (ε_0), embedded in a homogeneous dielectric environment ($\varepsilon = \varepsilon_r \varepsilon_0$), Friedman (1975) has shown that the reaction field is quite accurately approximated by the Coulomb field of *image charges* outside the sphere. The approximation is good for $\varepsilon_r \gg 1$. We shall not repeat the complete derivation (which is based on the expansion of the potential of an excentric charge in Legendre polynomials) and only give the results.

Consider a charge q, positioned on the z-axis at a distance s from the center of a sphere with radius a (see Fig. 13.2). The potential *inside* the sphere at a point with spherical coordinates (r, θ) is given by the direct Coulomb potential of the charge plus the *reaction potential* $\Phi_R(r, \theta)$:

$$\Phi_R(r, \theta) = \frac{q}{4\pi\varepsilon_0 a}(1 - \varepsilon_r) \sum_{n=0}^{\infty} \frac{n+1}{n + \varepsilon_r(n+1)} \left(\frac{rs}{a^2}\right)^n P_n(\cos\theta). \qquad (13.90)$$

Here $P_n(\cos\theta)$ are the Legendre polynomials (see page 361). Because of the axial symmetry of the problem, the result does not depend on the azimuthal angle ϕ of the observation point. This equation is exact, but is hard to evaluate. Now, by expanding the first term in the sum:

$$\frac{n+1}{n+\varepsilon_r(n+1)} = \frac{1}{\varepsilon_r+1}(1+c+c^2+\cdots), \tag{13.91}$$

where

$$c = \frac{1}{(\varepsilon_r+1)(n+1)}, \tag{13.92}$$

one obtains an expansion of $\Phi_R(r,\theta)$:

$$\Phi_R = \Phi_R^{(0)} + \Phi_R^{(1)} + \Phi_R^{(2)} + \cdots. \tag{13.93}$$

The exact expression for the terms in this expansion is

$$\Phi_R^{(k)}(r,\theta) = -\frac{(\varepsilon_r-1)}{(\varepsilon_r+1)^{k+1}}\frac{a}{s}\frac{q}{4\pi\varepsilon_0(a^2/s)}\sum_{n=0}^{\infty}(n+1)^k\left(\frac{r}{a^2/s}\right)^n P_n(\cos\theta). \tag{13.94}$$

The zero-order term $\Phi_R^{(0)}$ is just the Legendre-polynomial expansion (for $r < a^2/s$) of the potential at (r,θ) due to a charge q_{im} at a position a^2/s on the z-axis:

$$\Phi_R^{(0)}(r,\cos\theta) = \frac{q_{im}}{4\pi\varepsilon_0}\left[\left(\frac{a^2}{s}\right)^2 + r^2 - 2r\frac{a^2}{s}\cos\theta\right]^{-1/2}$$

$$= \frac{q_{im}}{4\pi\varepsilon_0|\boldsymbol{r}-\boldsymbol{r}_{im}|}, \tag{13.95}$$

with

$$q_{im} = -\frac{(\varepsilon_r-1)}{(\varepsilon_r+1)}\frac{a}{s}q. \tag{13.96}$$

The first-order term $\Phi_R^{(1)}$ is considerably smaller than $\Phi_R^{(0)}/(\varepsilon_r+1)$, which is also true for the ratio of subsequent terms, so that $\Phi_R^{(0)}$ is a good approximation to the exact reaction potential when $\varepsilon_r \gg 1$.

So, in conclusion: the reaction potential of a source charge q at position $(s,0)$ within a sphere (vacuum, ε_0) of radius a is well approximated by the Coulomb potential (in vacuum) of an image charge q_{im} (13.96) located at position $\boldsymbol{r}_{im} = (a^2/s, 0)$ outside the sphere on the same axis as the source.

The interaction free energy of the source q at distance s from the center $(s < a)$ with its own reaction potential in the image approximation is

$$U_R(s) = \frac{1}{2}q\Phi_R^{(0)}(s) = -\frac{1}{2}\left(\frac{\varepsilon_r-1}{\varepsilon_r+1}\right)\frac{q^2}{4\pi\varepsilon_0 a}\left(\frac{a^2}{a^2-s^2}\right). \tag{13.97}$$

We can compare this result with the exact Born free energy of solvation. For a charge in the center $(s = 0)$, we obtain

$$U_R(0) = -\frac{1}{2}\left(\frac{\varepsilon_r - 1}{\varepsilon_r + 1}\right)\frac{q^2}{4\pi\varepsilon_0 a}, \quad (13.98)$$

while the Born energy, according to (13.60), equals

$$U_{\text{Born}} = -\frac{1}{2}\left(\frac{\varepsilon_r - 1}{\varepsilon_r}\right)\frac{q^2}{4\pi\varepsilon_0 a}. \quad (13.99)$$

The difference is due to the neglect of higher order contributions to the reaction potential and is negligible for large ε_r. We can also compare (13.97) with the exact result in the other limit: $s \to a$ $(d \ll a)$ approaches the case of a charge q at a distance $d = a - s$ from a *planar surface*. The source is on the vacuum side; on the other side is a medium (ε_r). The exact result for the reaction potential is the potential of an image charge $q_{\text{im}} = -q(\varepsilon_r - 1)/(\varepsilon_r + 1)$ at the mirror position at a distance d from the plane in the medium (see Exercise 13.3). In this limit, (13.97) simplifies to

$$U_R(d) = -\frac{1}{2}\left(\frac{\varepsilon_r - 1}{\varepsilon_r + 1}\right)\frac{1}{4\pi\varepsilon_0}\frac{q^2}{2d}, \quad (13.100)$$

which is exactly the free energy of the charge in the reaction potential of its mirror image.

When the sphere contains many charges, the reaction potentials of all charges add up in a linear fashion. Thus all charges interact not only with all other charges, but also with the reaction potential of itself (i.e., with its own image) and with the reaction potentials (i.e., the images) of all other charges.

13.7.5 The generalized Born solvation model

When charges are embedded in an irregular environment, e.g. in a macro-molecule that itself is solvated in a polar solvent, the electrical contribution to the free energy of solvation (i.e., the interaction energy of the charges with the reaction potential due to the polar environment) is hard to compute. The "standard" approach requires a time-consuming numerical solution of the Poisson (or linearized Poisson–Boltzmann) equation. In simulations with implicit solvent, the extra computational effort destroys the advantage of omitting the solvent and explicit solvent representation is often preferred. Therefore, there is a pressing need for approximate solutions that can be rapidly evaluated, for simulations of (macro)molecules in an implicit solvent (see Section 7.7 on page 234). The *generalized Born solvation model* was invented to do just that. The original introduction by Still *et al.* (1990) was followed by several refinements and adaptations,[6] especially for application to macromolecules and proteins (Onufriev *et al.*, 2004).

The general problem is how to compute forces and energies for simulations of explicit (macro)molecules in an implicit polar solvent. The direct interactions between the explicit particles in the macromolecule (described by the "vacuum energy" V_{vac}) must be augmented by the solvation free energy ΔG_{solv} between these particles and the solvent. The total solvation free energy consists of an electrostatic term ΔG_{el} and a surface term ΔG_{surf}. Consider the following sequence of processes:

(i) start with the system in vacuum;
(ii) remove all charges, i.e., remove the direct Coulomb interactions;
(iii) solvate the uncharged system, i.e., add the surface free energy;
(iv) add all charges back in the presence of the solvent.

The total potential of mean force is

$$V_{tot}^{mf} = V_{vac} + \Delta G_{surf} + \Delta G_{el}. \tag{13.101}$$

The surface free energy is usually taken to be proportional to the *solvent accessible surface area*, with a proportionality constant derived from the experimental free energy of solvation of small molecules. Onufriev *et al.* (2004) quote a value of 0.005 kcal/mol par $Å^2$ = 2.1 kJ/mol per nm^2, but this value may be differentiated depending on the particular atom type. The free energy of adding all charges back in the presence of the solvent is the total electrostatic interaction in the polarizable medium. From this the direct Coulomb interaction (in vacuum) should be subtracted in order to obtain ΔG_{el}, because the direct Coulomb interaction has been removed in step (ii) above.

Now consider the total electrostatic interaction for a dilute system of charges q_i, each in a (small) spherical vacuum *cavity* with radius R_i centered at position r_i, embedded in a medium with relative dielectric constant ε_r ("dilute liquid of charges-in-cavities"). When all distances r_{ij} are large compared to all radii R_i, the total electrostatic energy is

$$4\pi\varepsilon_0 U_{el} = \sum_{i<j} \frac{q_i q_j}{\varepsilon_r r_{ij}} - \frac{1}{2}\left(1 - \frac{1}{\varepsilon_r}\right)\sum_i \frac{q_i^2}{R_i}. \tag{13.102}$$

---

[6] To mention a few modifications of the expression (13.105) to compute the effective "distance" f_{ij}^{GB}: Hawkins *et al.* (1996) add another parameter to the term in square brackets; Onufriev *et al.* (2004) modify the $1/\varepsilon_r$ to $\exp(-\kappa f_{ij}^{GB})/\varepsilon_r)$ to take care of an ionic strength in the solvent; Schaefer and Karplus (1996) retain a dielectric constant in the macromolecule.

Here, the first term is the electrostatic energy of a distribution of charges in a medium (see (13.44) on page 340) and the second term is the sum of the Born energies of the charges (see (13.60) on page 343). After subtracting the direct vacuum Coulomb energy we obtain

$$4\pi\varepsilon_0\Delta G_{\mathrm{el}} = -\frac{1}{2}\left(1 - \frac{1}{\varepsilon_r}\right)\sum_i\sum_j\frac{q_iq_j}{f_{ij}^{\mathrm{GB}}}, \qquad (13.103)$$

with

$$\begin{aligned} f_{ij}^{\mathrm{GB}} &= r_{ij} \quad \text{for } i \neq j, \\ &= R_i \quad \text{for } i = j. \end{aligned} \qquad (13.104)$$

Still *et al.* (1990) propose the following form for f_{ij}^{GB} that includes the case of the dilute liquid of charges-in-cavities, but has a much larger range of validity:

$$f_{ij}^{\mathrm{GB}} = \left[r_{ij}^2 + R_iR_j\exp\left(-\frac{r_{ij}^2}{4R_iR_j}\right)\right]^{1/2}. \qquad (13.105)$$

The "effective Born radii" R_i are to be treated as parameters that depend on the shape of the explicitly treated system and the positions of the charges therein. They are determined by comparison with Poisson–Boltzmann calculations on a grid, by free energy perturbation (or integration) calculations of charging a molecule in explicit solvent or from experimental solvation free energies. The effective Born radius is related to the *distance to the surface* as we can see from the following example.

Consider a charge q situated in a spherical cavity at a distance d from the surface. This is equivalent to the situation treated in Section 13.7.4, see Fig. 13.2, with $s = a - d$. The solvation free energy in the image approximation (13.97) then becomes

$$4\pi\varepsilon_0 U_{\mathrm{R}} = -\frac{1}{2}\frac{(\varepsilon_r - 1)}{(\varepsilon_r + 1)}\frac{q^2}{2d(1 - d/2a)}, \qquad (13.106)$$

which is nearly equivalent to the GB equation with an effective Born radius of $2d(1-d/2a)$. This equals twice the distance to the surface of the sphere for small distances, reducing to once the distance when the charge approaches the center.

13.8 Multipole expansion

Consider two *groups of charges*, A and B, with q_i at $\boldsymbol{r}_i, i \in$ A and q_j at $\boldsymbol{r}_j, j \in$ B (see Fig. 13.3). Each group has a defined *central coordinate* $\boldsymbol{r}_{\mathrm{A}}$

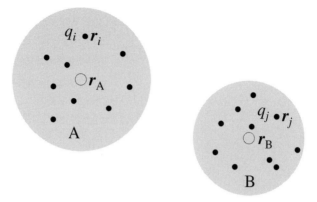

Figure 13.3 Two interacting non-overlapping groups of charges

and r_B, and the groups are non-overlapping. The groups may, for example, represent different atoms or molecules. For the time being the medium is taken as vacuum ($\varepsilon = \varepsilon_0$) and the charges are fixed; i.e., there is no polarizability. It is our purpose to treat the interaction energy of the two distributions in terms of a *multipole expansion*. But first we must clearly define what the interaction energy is.

The total energy of the two groups of charges is, according to (13.42), and omitting the self-energy terms in the potential:

$$U = \tfrac{1}{2} \sum_{i \in A} q_i \phi(r_i) + \tfrac{1}{2} \sum_{j \in B} q_i \phi(r_j). \tag{13.107}$$

Here $\phi(r_i)$ is the sum of $\phi^A(r_i)$ due to all other charges in A, and $\phi^B(r_i)$ due to all charges in B. Furthermore, $U_A = \tfrac{1}{2} \sum_{i \in A} q_i \phi^A(r_i)$ is the internal electrostatic energy of group A (and, similarly, U_B). The total electrostatic energy equals

$$U = U_A + U_B + U_{AB}, \tag{13.108}$$

with U_{AB} being the interaction energy between group A and group B:

$$U_{AB} = \tfrac{1}{2} \sum_{i \in A} q_i \phi^B(r_i) + \tfrac{1}{2} \sum_{j \in B} q_j \phi^A(r_j). \tag{13.109}$$

Both terms in U_{AB} are equal, which can be easily verified by inserting

$$\phi^B(r_i) = \frac{1}{4\pi\varepsilon_0} \sum_{j \in B} \frac{q_j}{r_{ij}}, \quad i \in A$$

and

$$\phi^A(r_j) = \frac{1}{4\pi\varepsilon_0} \sum_{i \in A} \frac{q_i}{r_{ij}}, \quad i \in B$$

into (13.109). Therefore the interaction energy U_{AB} can also be written as

$$U_{AB} = \sum_{i \in A} q_i \phi^B(r_i), \tag{13.110}$$

with

$$\phi^B(r) = \frac{1}{4\pi\varepsilon_0} \sum_{j \in B} \frac{q_j}{|r - r_j|}. \tag{13.111}$$

Thus we consider B as the *source* for the potential acting on A. By omitting the factor $1/2$, (13.110) represents the *total* interaction energy. It would be incorrect to add any interaction of charges in B with potentials produced by A.

We now proceed to write the interaction U_{AB} in terms of a multipole expansion. This can be done in two ways:

(i) The potential $\phi^B(r)$ in (13.110) is expanded in a Taylor series around the center r_A, involving derivatives of the potential at r_A.
(ii) The source terms $q_j(r_j)$ in (13.111) are expanded in a Taylor series around the center r_B.

Both methods lead to nearly equivalent multipole definitions, with subtle differences that we subsequently discuss. Combined they result in a description of the interaction between two charge clouds as a sum of interactions between multipoles. The expansions are only convergent when $r_{iA} < r_{AB}$, resp. $r_{jB} < r_{AB}$, which is fulfilled when the charge distributions do not overlap.

13.8.1 Expansion of the potential

In the following we shall use a notation with greek subscripts $\alpha, \beta, \ldots$ for cartesian components of vectors and tensors. The components of a radiusvector r are indicated by x_α, etc. We use the *Einstein summation convention*: summation over repeated indices is assumed. Thus $\mu_\alpha E_\alpha$ means $\sum_{\alpha=1}^{3} \mu_\alpha E_\alpha$, which is equivalent to the inner vector product $\mu \cdot E$.

We concentrate on the distribution A. The sources of the potential are external to A and we drop the superscript B for the potential to simplify the notation. Also, we take the center of the coordinate system in r_A. Now we expand $\phi(r)$ in a three-dimensional Taylor series around the coordinate center, assuming that all derivatives of ϕ exist:

$$\phi(r) = \phi(0) + x_\alpha \left(\frac{\partial \phi}{\partial x_\alpha}\right)(0) + \frac{1}{2!} x_\alpha x_\beta \left(\frac{\partial^2 \phi}{\partial x_\alpha \partial x_\beta}\right)(0)$$

$$+ \frac{1}{3!} x_\alpha x_\beta x_\gamma \left(\frac{\partial^3 \phi}{\partial x_\alpha \partial x_\beta \partial x_\gamma} \right) (0) + \cdots . \tag{13.112}$$

Inserting this expansion into (13.110), we find

$$U_{AB} = \overline{M}^{(0)} \phi(0) + \overline{M}^{(1)}_\alpha \left(\frac{\partial \phi}{\partial x_\alpha} \right) (0) + \frac{1}{2!} \overline{M}^{(2)}_{\alpha\beta} \left(\frac{\partial^2 \phi}{\partial x_\alpha \partial x_\beta} \right) (0)$$

$$+ \frac{1}{3!} \overline{M}^{(3)}_{\alpha\beta\gamma} \left(\frac{\partial^3 \phi}{\partial x_\alpha \partial x_\beta \partial x_\gamma} \right) (0) + \cdots , \tag{13.113}$$

where $\overline{M}^{(n)}$ is a form of the n-th multipole of the distribution, which is a symmetric tensor of rank n:[7]

$$\overline{M}^{(0)} = \sum_i q_i \quad \text{(monopole)}, \tag{13.114}$$

$$\overline{M}^{(1)}_\alpha = \sum_i q_i x_{i\alpha} = \boldsymbol{\mu} \quad \text{(dipole)}, \tag{13.115}$$

$$\overline{M}^{(2)}_{\alpha\beta} = \sum_i q_i x_{i\alpha} x_{i\beta} = \overline{\boldsymbol{Q}} \quad \text{(quadrupole)}, \tag{13.116}$$

$$\overline{M}^{(3)}_{\alpha\beta\gamma} = \sum_i q_i x_{i\alpha} x_{i\beta} x_{i\gamma} = \overline{\boldsymbol{O}} \quad \text{(octupole)}, \tag{13.117}$$

etc. (hexadecapole,[8] ...). We use an overline to denote this form of the multipole moments, as they are not the definitions we shall finally adopt.

The quadrupole moments and higher multipoles, as defined above, contain parts that do not transform as a tensor of rank n. They are therefore *reducible*. From the quadrupole we can separate the trace $\operatorname{tr} \overline{\boldsymbol{Q}} = \overline{Q}_{\alpha\alpha} = \overline{Q}_{xx} + \overline{Q}_{yy} + \overline{Q}_{zz}$; this part is a scalar as it transforms as a tensor of rank 0. Defining $\boldsymbol{Q}$ as the traceless tensor

$$\boldsymbol{Q} = 3\overline{\boldsymbol{Q}} - (\operatorname{tr} \overline{\boldsymbol{Q}}) \mathbf{1}, \tag{13.118}$$

$$Q_{\alpha\beta} = \sum_j q_j (3x_{j\alpha} x_{j\beta} - r_j^2 \delta_{\alpha\beta}), \tag{13.119}$$

the quadrupolar term in the energy expression (13.113) becomes

$$\frac{1}{2} \overline{Q}_{\alpha\beta} \frac{\partial^2 \phi}{\partial x_\alpha \partial x_\beta} = \frac{1}{6} Q_{\alpha\beta} \frac{\partial^2 \phi}{\partial x_\alpha \partial x_\beta} + (\operatorname{tr} \overline{\boldsymbol{Q}}) \nabla^2 \phi. \tag{13.120}$$

---

[7] A real tensor in 3D space is defined by its transformation property under a rotation $\mathbf{R}$ of the (cartesian) coordinate system, such that tensorial relations are invariant for rotation. For a rank-0 tensor t (a scalar) the transformation is $t' = t$; for a rank-1 tensor $\boldsymbol{v}$ (a vector) the transformation is $\boldsymbol{v}' = \mathbf{R}\boldsymbol{v}'$ or $v'_\alpha = R_{\alpha\beta} v_\beta$. For a rank-2 tensor $\boldsymbol{T}$ the transformation is $\boldsymbol{T}' = \mathbf{R}\boldsymbol{T}\mathbf{R}^\mathsf{T}$ or $T_{\alpha\beta} = R_{\alpha\gamma} R_{\beta\delta} T_{\gamma\delta}$, etc.

[8] The names di-, quadru-, octu- and hexadecapole stem from the minimum number of charges needed to represent the pure general n-th multipole.

Since the potential has no sources in domain A, the Laplacian of ϕ is zero, and the second term in this equation vanishes. Thus the energy can just as well be expressed in terms of the traceless quadrupole moment Q. The latter is a pure rank-2 tensor and is defined by only five elements since it is symmetric and traceless. It can always be transformed to a diagonal tensor with two elements by rotation (which itself is defined by three independent elements as Eulerian angles).

The octupole case is similar, but more complex. The tensor has 27 elements, but symmetry requires that any permutation of indices yields the same tensor, leaving ten different elements. Partial sums that can be eliminated because they multiply with the vanishing Laplacian of the potential are of the form

$$O_{xxx} + O_{yyx} + O_{zzx} = 0. \tag{13.121}$$

There are thee such equations that are not related by symmetry, thus leaving only seven independent elements. These relations are fulfilled if the octupole as defined in (13.135) is corrected as follows:

$$O_{\alpha\beta\gamma} = \sum_j q_j [5x_{j\alpha}x_{j\beta}x_{j\gamma} - r_j^2(x_{j\alpha}\delta_{\beta\gamma} + x_{j\beta}\delta_{\gamma\alpha} + x_{j\gamma}\delta_{\alpha\beta})]. \tag{13.122}$$

This is the rank-3 equivalent of the traceless rank-2 tensor.

The energy expression (13.113) can also be written in tensor notation as

$$U_{\mathrm{AB}} = q\phi(0) - \boldsymbol{\mu} \cdot \boldsymbol{E} - \tfrac{1}{6}Q{:}\boldsymbol{\nabla}\boldsymbol{E} - \tfrac{1}{30}O{:}\boldsymbol{\nabla}\boldsymbol{\nabla}\boldsymbol{E} + \cdots, \tag{13.123}$$

where the semicolon denotes the *scalar product* defined by summation over all corresponding indices. $\boldsymbol{E}$ is the electric field vector, $\boldsymbol{\nabla}\boldsymbol{E}$ its gradient and $\boldsymbol{\nabla}\boldsymbol{\nabla}\boldsymbol{E}$ the gradient of its gradient.[9]

13.8.2 Expansion of the source terms

Now consider the source charge distribution $q_j(\boldsymbol{r}_j)$. The potential at an arbitrary point $\boldsymbol{r}$ (such as $\boldsymbol{r}_A$) is given by

$$\phi(\boldsymbol{r}) = \frac{1}{4\pi\varepsilon_0} \sum_j \frac{q_j}{|\boldsymbol{r} - \boldsymbol{r}_j|}. \tag{13.124}$$

When the point $\boldsymbol{r}$ is outside the charge distribution, i.e., $|\boldsymbol{r}| > |\boldsymbol{r}_j|$ for any j, the right-hand side can be expanded with the Taylor expansion of $|\boldsymbol{r} - \boldsymbol{r}_j|^{-1}$

---

[9] Note that we do not write a dot, as $\boldsymbol{\nabla} \cdot \boldsymbol{E}$ means its divergence or scalar product; we write $\boldsymbol{\nabla}\boldsymbol{\nabla}$ and not $\boldsymbol{\nabla}^2$, as the latter would indicate the Laplacian, which is again a scalar product.

in terms of powers of r_j/r. The general Taylor expansion in three dimensions is

$$f(\boldsymbol{a}+\boldsymbol{x}) = f(\boldsymbol{a}) + x_\alpha \left(\frac{\partial f}{\partial x_\alpha}\right)(\boldsymbol{a}) + \frac{1}{2!}x_\alpha x_\beta \left(\frac{\partial^2}{\partial x_\alpha \partial x_\beta}\right)(\boldsymbol{a}) + \cdots. \quad (13.125)$$

So we shall need derivatives of $1/r$. These are not only needed for expansions of this type, but also come in handy for fields, field gradients, etc. It is therefore useful to list a few of these derivatives:

$$\frac{\partial}{\partial x_\alpha}\frac{1}{r} = -\frac{x_\alpha}{r^3}, \quad (13.126)$$

$$\frac{\partial^2}{\partial x_\alpha \partial x_\beta}\frac{1}{r} = 3\frac{x_\alpha x_\beta}{r^5} - \frac{1}{r^3}\delta_{\alpha\beta}, \quad (13.127)$$

$$\frac{\partial^3}{\partial x_\alpha \partial x_\beta \partial x_\gamma}\frac{1}{r} = -15\frac{x_\alpha x_\beta x_\gamma}{r^7} + \frac{3}{r^5}(x_\alpha \delta_{\beta\gamma} + x_\beta \delta_{\gamma\alpha} + x_\gamma \delta_{\alpha\beta}), (13.128)$$

$$\frac{\partial^4}{\partial x_\alpha \cdots \partial x_\delta}\frac{1}{r} = 105\frac{x_\alpha x_\beta x_\gamma x_\delta}{r^9} - \frac{15}{r^7}(x_\alpha x_\beta \delta_{\gamma\delta} + x_\beta x_\gamma \delta_{\delta\alpha}$$
$$+ x_\gamma x_\delta \delta_{\alpha\beta} + x_\delta x_\alpha \delta_{\beta\gamma} + x_\alpha x_\gamma \delta_{\beta\delta} + x_\beta x_\delta \delta_{\gamma\alpha})$$
$$+ \frac{3}{r^5}(\delta_{\alpha\beta}\delta_{\gamma\delta} + \delta_{\beta\gamma}\delta_{\delta\alpha} + \delta_{\alpha\gamma}\delta_{\beta\delta}). \quad (13.129)$$

These derivatives are in vector notation: $\nabla\frac{1}{r}, \nabla\nabla\frac{1}{r}, \nabla\nabla\nabla\frac{1}{r}, \nabla\nabla\nabla\nabla\frac{1}{r}$. Expanding $\phi(\boldsymbol{r})$ in inverse powers of r is now a matter of substituting these derivatives into the Taylor expansion of $|\boldsymbol{r}-\boldsymbol{r}_j|^{-1}$. Using the same definitions for the multipoles as in (13.114) to (13.117), we obtain:

$$4\pi\varepsilon_0\phi(\boldsymbol{r}) = \overline{M}^{(0)}\frac{1}{r} + \overline{M}^{(1)}_\alpha\frac{x_\alpha}{r^3} + \frac{1}{2!}\overline{M}^{(2)}_{\alpha\beta}\left(\frac{3x_\alpha x_\beta}{r^5} - \frac{\delta_{\alpha\beta}}{r^3}\right)$$
$$+ \frac{1}{3!}\overline{M}^{(3)}_{\alpha\beta\gamma}\left(15\frac{x_\alpha x_\beta x_\gamma}{r^7} - \frac{3}{r^5}(x_\alpha \delta_{\beta\gamma} + x_\beta \delta_{\gamma\alpha} + x_\gamma \delta_{\alpha\beta})\right)$$
$$+ \cdots. \quad (13.130)$$

The terms in this sum are of increasing order in r^{-n}, and represent the potentials of monopole, dipole, quadrupole and octupole, respectively.

Instead of $\overline{M}^{(l)}$ we can also use the traceless definitions, because the trace do not contribute to the potential. For example, the trace part of the quadrupole (which is a constant times the unit matrix) leads to a contribution $\sum_{\alpha=1}^{3} r^{-3}(3x_\alpha^2 - r^2) = 0$. Therefore instead of (13.130) we can write:

$$4\pi\varepsilon_0\phi(\boldsymbol{r}) = \frac{1}{r}M^{(0)} + \frac{1}{r^3}M^{(1)}_\alpha x_\alpha + \frac{1}{6}M^{(2)}_{\alpha\beta}\left(\frac{3x_\alpha x_\beta}{r^5} - \frac{\delta_{\alpha\beta}}{r^3}\right)$$

$$+\frac{1}{10}M^{(3)}_{\alpha\beta\gamma}\left(\frac{5x_\alpha x_\beta x_\gamma}{r^7}-\frac{1}{r^5}(x_\alpha\delta_{\beta\gamma}+x_\beta\delta_{\gamma\alpha}+x_\gamma\delta_{\alpha\beta})\right)$$

$$+\ldots, \tag{13.131}$$

with

$$M^{(0)}=\sum_j q_j=q \quad\text{(monopole)}, \tag{13.132}$$

$$M^{(1)}_\alpha=\sum_j q_j x_{j\alpha}=\boldsymbol{\mu} \quad\text{(dipole)}, \tag{13.133}$$

$$M^{(2)}_{\alpha\beta}=\sum_j q_j(3x_{j\alpha}x_{j\beta}-r_j^2\delta_{\alpha\beta})=\boldsymbol{Q} \quad\text{(quadrupole)}, \tag{13.134}$$

$$M^{(3)}_{\alpha\beta\gamma}=\sum_j q_j[5x_{j\alpha}x_{j\beta}x_{j\gamma}-r_j^2(x_{j\alpha}\delta_{\beta\gamma}+x_{j\beta}\delta_{\gamma\alpha}+x_{j\gamma}\delta_{\alpha\beta})]$$

$$=\boldsymbol{O} \quad\text{(octupole)}. \tag{13.135}$$

Expressed in terms of the derivative tensors, the potential reads:

$$4\pi\varepsilon_0\phi(\boldsymbol{r})=\frac{q}{r}-\boldsymbol{\mu}\cdot\boldsymbol{\nabla}\frac{1}{r}+\frac{1}{6}\boldsymbol{Q}{:}\boldsymbol{\nabla}\boldsymbol{\nabla}\frac{1}{r}-\frac{1}{30}\boldsymbol{O}{:}\boldsymbol{\nabla}\boldsymbol{\nabla}\boldsymbol{\nabla}\frac{1}{r}+\cdots. \tag{13.136}$$

The multipole definitions obviously also apply to continuous charge distributions, when the summations are replaced by integration over space and the charges q_j are replaced by a charge density. *These are the definitions (in cartesian coordinates) that we shall adopt for the multipole moments.* The reader should be aware that there is no consensus on the proper definition of multipole moments and different definitions are used in the literature.[10] Not only the definitions may differ, but also the choice of center is important for all multipoles beyond the lowest non-zero multipole. If the total charge (monopole) is non-zero, the dipole moment depends on the choice of origin; the dipole moment will vanish if the *center of charge* $\sum_i q_i\boldsymbol{r}_i/\sum_i q_i$ is chosen as the center of the expansion. Likewise the quadrupole moment depends on the choice of origin for dipolar molecules, etc.

Another elegant and popular expansion of the source term is in terms of *spherical harmonics* $Y_l^m(\theta,\phi)$. These are functions expressed in polar and azimuthal angles; for use in simulations they are often less suitable than their cartesian equivalents. For higher multipoles they have the advantage of being restricted to the minimum number of elements while the cartesian

---

[10] Our definition corresponds to the one used by Hirschfelder *et al.* (1954) and to the one in general use for the definition of nuclear electric quadrupole moments in NMR spectroscopy (see, e.g., Schlichter, 1963). In molecular physics the quadrupole is often defined with an extra factor $1/2$, corresponding to the Legendre polynomials with $l=2$, as in the reported quadrupole moment of the water molecule by Verhoeven and Dymanus (1970). The definition is not always properly reported and the reader should carefully check the context.

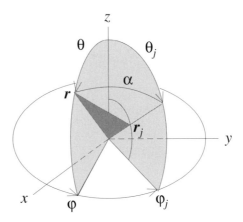

Figure 13.4 The source is at $\boldsymbol{r}_j = (r_j, \theta_j, \phi_j)$ and the potential is determined at $\boldsymbol{r} = (r, \theta, \phi)$. The angle between these two vectors is α.

tensors contain superfluous elements (as 27 cartesian tensor components against the minimum of seven irreducible components for the octupole). On the other hand, for numerical computations it is generally advisable not to use higher multipoles on a small number of centers at all, but rather use lower multipoles (even only monopoles) on a larger number of centers, in order to avoid complex expressions. For example, the computation of a force resulting from dipole–dipole interaction requires the gradient of a dipole field, which involves a rank-3 tensor already; this is generally as far as one is prepared to go. Instead of including quadrupoles, one may choose a larger number of centers instead, without loss of accuracy.

The expansion in spherical harmonics is based on the fact that the inverse distance $1/|\boldsymbol{r} - \boldsymbol{r}_j|$ is a generating function for Legendre polynomials $P_l(\cos \alpha)$, where α is the angle between $\boldsymbol{r}$ and $\boldsymbol{r}_j$ (see Fig. 13.4):

$$(1 - 2\frac{r_j}{r} \cos \alpha + \left(\frac{r_j}{r}\right)^2)^{-1/2} = \sum_{l=0}^{\infty} \left(\frac{r_j}{r}\right)^l P_l(\cos \alpha), \qquad (13.137)$$

where the first four Legendre polynomials are given by P_l^0 in (13.140) to (13.149) below.

These Legendre polynomials of the cosine of an angle α between two directions characterized by polar and azimuthal angles (θ, ϕ) and (θ_j, ϕ_j) can subsequently be expanded by the *spherical harmonics addition theorem:*[11]

$$P_l(\cos \alpha) = \sum_{m=-l}^{l} \frac{(l - |m|)!}{(l + |m|)!} Y_l^m(\theta, \phi) Y_l^{-m}(\theta_j \phi_j), \qquad (13.138)$$

where

$$Y_l^m(\theta, \phi) = P_l^{|m|}(\cos \theta) e^{im\phi} \qquad (13.139)$$

are the spherical harmonic functions and $P_l^{|m|}$ are the associated Legendre functions.[12] For $l \leq 3$ these functions are:

$$l = 0 \quad : \quad P_0^0(\cos \theta) = 1, \qquad (13.140)$$
$$l = 1 \quad : \quad P_1^0(\cos \theta) = \cos \theta, \qquad (13.141)$$
$$ \quad : \quad P_1^1(\cos \theta) = \sin \theta, \qquad (13.142)$$
$$l = 2 \quad : \quad P_2^0(\cos \theta) = \tfrac{1}{2}(3 \cos^2 \theta - 1), \qquad (13.143)$$
$$ \quad : \quad P_2^1(\cos \theta) = 3 \sin \theta \cos \theta, \qquad (13.144)$$
$$ \quad : \quad P_2^2(\cos \theta) = 3 \sin^2 \theta, \qquad (13.145)$$
$$l = 3 \quad : \quad P_3^0(\cos \theta) = \tfrac{5}{2} \cos^3 \theta - \tfrac{3}{2} \cos \theta, \qquad (13.146)$$
$$ \quad : \quad P_3^1(\cos \theta) = \tfrac{3}{2} \sin \theta (5 \cos^2 \theta - 1), \qquad (13.147)$$
$$ \quad : \quad P_3^2(\cos \theta) = 15 \sin^2 \theta \cos \theta, \qquad (13.148)$$
$$ \quad : \quad P_3^3(\cos \theta) = 15 \sin^3 \theta. \qquad (13.149)$$

The result is

$$4\pi\varepsilon_0\phi(r) = \sum_{l=0}^{\infty} \sum_{m=-l}^{l} \frac{1}{r^{l+1}} \frac{(l - |m|)!}{(l + |m|)!} \mathcal{M}_l^m Y_l^m(\theta, \phi), \qquad (13.150)$$

where $\mathcal{M}_l^m$ are the $2l + 1$ components of the l-th spherical multipole:

$$\mathcal{M}_l^m = \sum_j q_j r_j^l Y_l^{-m}(\theta_j, \phi_j). \qquad (13.151)$$

These spherical harmonic definitions are related to the cartesian tensor definitions of (13.132) to (13.135).

---

[11] We use simple *non*-normalized spherical harmonics. Our definition of the spherical multipole moments corresponds to Hirschfelder *et al.* (1965). Definitions in the literature may differ as to the normalization factors and sign of the functions for odd m. See, e.g., Weisstein (2005), Abramowitz and Stegun (1965) and Press *et al.* (1992).

[12] See, e.g., Jahnke and Emde (1945), who list Legendre functions up to $l = 6$ and associated functions up to $l = 4$.

13.9 Potentials and fields in non-periodic systems

Given a set of charges, the calculation of potentials, fields, energies and forces by summation of all pairwise interactions is a problem of N^2 complexity that easily runs out of hand for large systems. The use of a cut-off radius reduces the problem to order-N, but produces large and often intolerable artefacts for the fairly long-ranged Coulomb forces. For gravitational forces, lacking the compensation of sources with opposite sign, cut-offs are not allowed at all. Efficient methods that include long-range interactions are of two types:

(i) hierarchical and multipole methods, employing a clustering of sources for interactions at longer distances; and

(ii) grid methods, essentially splitting the interaction into short- and long-range parts, solving the latter by Poisson's equation, generally on a grid.

The second class of methods are the methods of choice for periodic system, which are treated in detail in the next section. They can in principle also be used for non-periodic systems – and still with reasonable efficiency – by extending the system with periodic images. But also without periodicity the same method of solution can be used when the Poisson equation is solved on a grid with given boundary conditions, possibly employing multigrid methods with spatial resolution adjusted to local densities.

As we emphasize molecular simulations where the long-range problem concerns Coulomb rather than gravitational forces, we shall not further consider the hierarchical and "fast multipole" methods, which are essential for astrophysical simulations and are also used in molecular simulation,[13] but have not really survived the competition with methods described in the next section. Whether the fast multipole methods may play a further role in molecular simulation, is a matter of debate (Board and Schulten, 2000).

13.10 Potentials and fields in periodic systems of charges

In periodic systems (see Section 6.2.1) the Coulomb energy of the charges is given by (see (13.42)):

$$U_C = \tfrac{1}{2} \sum_i q_i (\phi(r_i) - \phi_{\text{self}}), \quad i \in \text{unit cell}, \tag{13.152}$$

---

[13] The basic articles on hierarchical and fast multipole methods are Appel (1985), Barnes and Hut (1986) and Greengard and Rokhlin (1987). Niedermeier and Tavan (1994) and Figueirido *et al.* (1997) describe the use of fast multipole methods in molecular simulations. It is indicated that these methods, scaling proportional to N, are computationally more efficient than lattice summation techniques for systems with more than about 20 000 particles.

with

$$\phi(\boldsymbol{r}) = \frac{1}{4\pi\varepsilon_0} \sum_j \sum_{n_1,n_2,n_3 \in \mathbb{Z}} \frac{q_j}{|\boldsymbol{r} - \boldsymbol{r}_j - n_1\boldsymbol{a} - n_2\boldsymbol{b} - n_3\boldsymbol{c}|}$$

$$= \frac{1}{4\pi\varepsilon_0} \sum_j \sum_{\boldsymbol{n} \in \mathbb{Z}^3} \frac{q_j}{|\boldsymbol{r} - \boldsymbol{r}_j - \mathbf{T}\boldsymbol{n}|}, \qquad (13.153)$$

where $\mathbf{T}$ is the transformation matrix from relative coordinates in the unit cell to cartesian coordinates (see (6.3) on page 143), i.e., a matrix of which the columns are the cartesian base vectors of the unit cell $\boldsymbol{a}, \boldsymbol{b}, \boldsymbol{c}$. The last line of (13.153) is in matrix notation; the meaning of $|\mathbf{x}|$ is $(\mathbf{x}^\mathsf{T}\mathbf{x})^{1/2}$. Note that the displacements can be either subtracted (as shown) or added in (13.153). The self-energy contains the diverging interaction of q_i with itself, but not with the images of itself; the images are to be considered as different particles as in a crystal. The interaction of a charge with its images produces zero force, as for every image there is another image at equal distance in the opposite direction; thus the interaction energy of each charge with its own images is a constant, which diverges with the number of periodic images considered. In order to avoid the divergence we may *assume* that every charge has a homogeneous charge distribution of equal magnitude but opposite sign associated with it. If the total charge in a unit cell vanishes, i.e., *for electroneutral systems*, the homogeneous background cancels and need not be invoked.

The direct sum of Coulomb terms is only conditionally convergent (i.e., the convergence depends on the sequence of terms in the summation) and converges very slowly. For an efficient evaluation of the lattice energies and forces it is necessary to split the Coulomb potential into a short-range part that can be directly evaluated by summation in real space, and a long-range part that can be efficiently computed by solving Poisson's equation. The easiest way to accomplish this is to consider each (point) charge as a sum of two charge distributions (see Fig. 13.5):

$$q_i\delta(\boldsymbol{r} - \boldsymbol{r}_i) = q_i[\delta(\boldsymbol{r} - \boldsymbol{r}_i) - w(\boldsymbol{r} - \boldsymbol{r}_i)] + q_i w(\boldsymbol{r} - \boldsymbol{r}_i), \qquad (13.154)$$

where $w(\boldsymbol{r}) = w(r)$ is an isotropic *spread function* which decreases smoothly and rapidly with distance and integrates to 1 over space:

$$\int_0^\infty w(r)4\pi r^2\, dr = 1. \qquad (13.155)$$

For the time being we do not specify the spread function and derive the equations in a general form. Subsequently two specific spread functions will be considered.

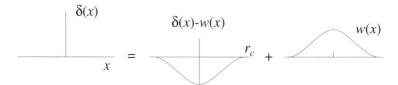

Figure 13.5 A point charge with δ-function distribution (left) is split up into a distribution with short-range potential (middle) and a distribution with long-range potential (right) by a smooth *charge-spread function* $w(\mathbf{r})$.

The total Coulomb energy is split into two contributions:

$$U_C = U_C^s + U_C^l = \frac{1}{2}\sum_i q_i\phi_i^s + \frac{1}{2}\sum_i q_i\phi_i^l, \quad i \in \text{unit cell.} \tag{13.156}$$

Note that we did not split the energy into the sum of energies of the two charge distributions and their interactions, which would require four terms. Each of the contributions should include the self-energy correction. In addition there is a contribution to the energy as a result of the net dipole moment of the unit cell, treated in Section 13.10.5.

13.10.1 Short-range contribution

For the short-range contributions U_C^s to the energy we can write:

$$U_C^s = \tfrac{1}{2}\sum_i q_i\phi_i^s, \quad i \in \text{unit cell.} \tag{13.157}$$

$$\phi_i^s = \frac{1}{4\pi\varepsilon_0}\sideset{}{'}\sum_j\sum_{\mathbf{n}} q_j\varphi^s(r_{ij\mathbf{n}}), \tag{13.158}$$

where the prime in the sum means exclusion of $j = i$ for $\mathbf{n} = \mathbf{0}$, $r_{ij\mathbf{n}} \overset{\text{def}}{=} \mathbf{r}_i - \mathbf{r}_j - \mathbf{Tn}$ and φ^s is a potential function related to the spread function:

$$\varphi^s(r) \overset{\text{def}}{=} \int_r^\infty dr' \frac{1}{r'^2} \int_{r'}^\infty dr'' \, 4\pi r''^2 w(r''). \tag{13.159}$$

The force $\mathbf{F}_i^s$ on particle i due to the short-range potential equals the charge q_i times the electric field $\mathbf{E}^s(\mathbf{r}_i) = -(\nabla\phi^s(\mathbf{r}))_{\mathbf{r}_i}$:

$$\mathbf{F}_i^s = -q_i(\nabla\phi^s(\mathbf{r}))_{\mathbf{r}_i} = \frac{q_i}{4\pi\varepsilon_0}\sideset{}{'}\sum_j\sum_{\mathbf{n}\in\mathbb{Z}^3} q_j f^s(r_{ij\mathbf{n}})\frac{\mathbf{r}_{ij\mathbf{n}}}{r_{ij\mathbf{n}}}, \tag{13.160}$$

where f^s is a force function related to the spread function:

$$f^s(r) \overset{\text{def}}{=} -\frac{d\varphi^s(r)}{dr} = \frac{1}{r^2}\int_r^\infty w(r')\,4\pi r'^2\,dr'. \tag{13.161}$$

One may also evaluate the force on particle i from taking minus the gradient of the total short-range *energy* (13.158). Although the expression for the energy contains a factor $\frac{1}{2}$, particle number i occurs twice in the summation, and one obtains the same equation (13.160) as above. Note that, by omitting $j = i$ for $\mathbf{n} = \mathbf{0}$ from the sum, the short-range terms are corrected for the short-range part of the self-energy. Similarly, Coulomb interactions between specified pairs can be omitted from the short-range evaluation, if so prescribed by the force field. Usually, Coulomb interactions are omitted between atoms that are first or second (and often modified for third) neighbors in a covalent structure because other bond and bond-angle terms take care of the interaction.

When the spread function is such that the potentials and forces are negligible beyond a *cut-off distance* r_c, which does not exceed half the smallest box size, the sums contain only one *nearest image* of each particle pair, which can best be evaluated using a *pair list* that also contains a code for the proper displacement to obtain the nearest image for each pair.

13.10.2 Long-range contribution

The long-range contribution expressed as an explicit particle sum

$$\phi^l(\mathbf{r}_i) = \frac{1}{4\pi\varepsilon_0} \sum_j {}' \sum_\mathbf{n} q_j \left[\frac{1}{r_{ijn}} - \varphi^s(r_{ij\mathbf{n}}) \right] \tag{13.162}$$

converges very slowly because of the $1/r$ nature of the function. The long-range potential can be efficiently evaluated by solving Poisson's equation (see (13.47)):

$$-\varepsilon_0 \nabla^2 \phi^l(\mathbf{r}) = \rho^l(\mathbf{r}) = \sum_i \sum_{\mathbf{n} \in \mathbb{Z}^3} q_i w(\mathbf{r} - \mathbf{r}_i - \mathbf{Tn}). \tag{13.163}$$

The solution is equivalent to (13.162), except that no restrictions, such as $j \neq i$ for $\mathbf{n} = \mathbf{0}$ or any other specified pairs, can be included. The Poisson solution therefore contains a self-energy part (which is a constant, given the unit cell base vectors and spread function), that must be subtracted separately. If Coulomb interactions between specified pairs must be omitted, their contribution included in the long-range interaction must be subtracted. The charge distribution is periodic, and so must be the solution of this equation. The solution is determined up to any additive periodic function satisfying the Laplace equation $\nabla^2 \phi = 0$, which can only be a constant if continuity at the cell boundaries is required. The constant is irrelevant.

There are several ways to solve the Poisson equation for periodic systems, including iterative relaxation on a lattice (see Press *et al.*, 1992), but the obvious solution can be obtained in reciprocal space, because the Laplace operator then transforms to a simple multiplication. We now proceed to formulate this Fourier solution.

First define the discrete set of wave vectors

$$k = 2\pi(m_1 a^* + m_2 b^* + m_3 c^*), \tag{13.164}$$

with $m_1, m_2, m_3 \in \mathbb{Z}$, which enumerate the Fourier terms, and a^*, b^*, c^* the reciprocal lattice vectors. See Section 12.9 on page 331 for a description of the reciprocal lattice and corresponding Fourier transforms. For the evaluation of the scalar product $k \cdot r$ it is generally easier to use the relative coordinates (ξ, η, ζ) of r:

$$k \cdot r = 2\pi(m_1 \xi + m_2 \eta + m_3 \zeta), \tag{13.165}$$

with $\xi = r \cdot a^*$, etc. Now construct the Fourier transform (often called *structure factors*) of the ensemble of point charges:

$$Q_k \overset{\text{def}}{=} \sum_j q_j e^{-ik \cdot r_j}, \quad j \in \text{unit cell}. \tag{13.166}$$

Since we have broadened the point charges with the spread function, the charge density $\rho^l(r)$ (13.163) is the convolution of the point charge distribution and the spread function. The Fourier transform P_k^l of $\rho^l(r)$ therefore is the product of the charge structure factors and the Fourier transform W_k of the spread function:

$$P_k^l \overset{\text{def}}{=} \frac{1}{V} \int_V \rho^l(r) e^{-ik \cdot r} \, dr \tag{13.167}$$

$$= Q_k W_k, \tag{13.168}$$

where

$$W_k \overset{\text{def}}{=} \frac{1}{V} \sum_n \int_V w(r + Tn) e^{-ik \cdot (r+Tn)} \, dr \tag{13.169}$$

$$= \frac{1}{V} \int_{\text{all space}} w(r) e^{-ik \cdot r} \, dr$$

$$= \frac{1}{V} \int_0^\infty dr \int_0^\pi d\theta \, 2\pi r^2 w(r) \sin\theta \, e^{-ikr\cos\theta}$$

$$= \frac{1}{V} \int_0^\infty 4\pi r w(r) \frac{\sin kr}{k} \, dr \tag{13.170}$$

Here $\int_V$ means integration over one unit cell, and we have used the fact that the spread function is isotropic. We see that W_k depends only on the absolute value k of $\mathbf{k}$. The validity of (13.168) can easily be checked by evaluating (13.167) using (13.163).

The Poisson equation (13.163) in reciprocal space reads

$$-k^2 \varepsilon_0 \Phi^{\rm l}_{\mathbf{k}} = -P^{\rm l}_{\mathbf{k}}, \tag{13.171}$$

and thus

$$\Phi^{\rm l}_{\mathbf{k}} = \frac{Q_{\mathbf{k}} W_{\mathbf{k}}}{\varepsilon_0 k^2}; \ \mathbf{k} \neq \mathbf{0}. \tag{13.172}$$

Note that $\mathbf{k} = \mathbf{0}$ must not be allowed in this equation and Φ_0 is therefore not defined; indeed for electroneutral systems $Q_0 = 0$. Electroneutrality therefore is required; if the system is charged the potential does not converge and electroneutrality must be enforced by adding a homogeneous background charge of opposite sign. This enforces $Q_0 = 0$. The real-space potential $\phi^{\rm l}(\mathbf{r})$ follows up to a constant by Fourier transformation:

$$\phi^{\rm l}(\mathbf{r}) = \sum_{\mathbf{k}\neq\mathbf{0}} \Phi^{\rm l}_{\mathbf{k}} e^{i\mathbf{k}\cdot\mathbf{r}} = \sum_{\mathbf{k}\neq\mathbf{0}} \frac{Q_{\mathbf{k}} W_{\mathbf{k}}}{\varepsilon_0 k^2} e^{i\mathbf{k}\cdot\mathbf{r}}. \tag{13.173}$$

The total energy can be expressed in terms of a sum over wave vectors

$$U^{\rm l}_{\rm C} = \frac{1}{2} \sum_i q_i \phi(\mathbf{r}_i) = \frac{1}{2\varepsilon_0} \sum_{\mathbf{k}\neq\mathbf{0}} k^{-2} Q_{\mathbf{k}} Q_{-\mathbf{k}} W_{\mathbf{k}} - U^{\rm l}_{\rm self}. \tag{13.174}$$

The self-energy contained in the long-range energy is a constant given by the $j = i, \mathbf{n} = \mathbf{0}$ part of (13.162):

$$U^{\rm l}_{\rm self} = \frac{1}{4\pi\varepsilon_0} \sum_i q_i^2 \lim_{r\to 0}[r^{-1} - \varphi^{\rm s}(r)] \tag{13.175}$$

(for $\varphi^{\rm s}$ see (13.159)). Similarly, the interaction energy between excluded pairs ij for which the long-range energy must be corrected is

$$U^{\rm l}_{\rm excl} = \frac{q_i q_j}{4\pi\varepsilon_0}[r_{ij}^{-1} - \varphi^{\rm s}(r_{ij})]. \tag{13.176}$$

The long-range *force* on particle i can be evaluated from the gradient of the potential:

$$\mathbf{F}^{\rm l}_i = q_i \mathbf{E}^{\rm l}(\mathbf{r}_i) = -q_i(\nabla\phi^{\rm l}(\mathbf{r}))_{\mathbf{r}_i} = -q_i \sum_{\mathbf{k}\neq\mathbf{0}} \frac{Q_{\mathbf{k}} W_{\mathbf{k}}}{\varepsilon_0 k^2} i\mathbf{k} e^{i\mathbf{k}\cdot\mathbf{r}_i}. \tag{13.177}$$

The sum is real since for every $\mathbf{k}$-term there is a complex conjugate $-\mathbf{k}$-term. The self-energy term does not produce a force; the ij exclusion term produces a long-range force on particle i (and the opposite force on j):

$$\mathbf{F}^{\rm l}_{i,\rm excl} = \frac{q_i q_j}{4\pi\varepsilon_0}[r_{ij}^{-2} - f^{\rm s}(r_{ij})]\frac{\mathbf{r}_{ij}}{r_{ij}} \tag{13.178}$$

(for f^s see (13.161)) which should be subtracted from the long-range force evaluation.

13.10.3 Gaussian spread function

In the special case a Gaussian distribution is chosen for the spread function, the solution is expressed in sums of analytical functions and the classical *Ewald summation* is obtained (Ewald, 1921). The advantage of a Gaussian function is that its Fourier transform is also a Gaussian function, and both the real-space and reciprocal-space functions taper off quickly and can be restricted to a limited range. The Gaussian function contains an inverse width parameter β (the variance of the Gaussian distribution is $1/2\beta^2$); if β is small, the spread function is wide and the real-space summation has many terms. The reciprocal functions, on the other hand, then decrease rapidly with increasing k. For large β the inverse is true. Therefore β can be tuned for the best compromise, minimizing the computational effort for a given error margin.

Noting that

$$\operatorname{erf}(x) \stackrel{\text{def}}{=} \frac{2}{\sqrt{\pi}} \int_0^x e^{-u^2}\,du, \tag{13.179}$$

$$\operatorname{erfc}(x) \stackrel{\text{def}}{=} 1 - \operatorname{erf}(x) = \frac{2}{\sqrt{\pi}} \int_x^\infty e^{-u^2}\,du, \tag{13.180}$$

we summarize the relevant functions:

$$w(\mathbf{r}) = w(r) = \frac{\beta^3}{\pi^{3/2}} e^{-(\beta r)^2}, \tag{13.181}$$

$$(13.159) \quad \varphi^s(r) = \frac{1}{r}\operatorname{erfc}(\beta r), \tag{13.182}$$

$$(13.161) \quad f^s(r) = \frac{1}{r^2}\operatorname{erfc}(\beta r) + \frac{2}{\sqrt{\pi}}\frac{\beta}{r}e^{-\beta^2 r^2}, \tag{13.183}$$

$$(13.170) \quad W_{\mathbf{k}} = \frac{1}{V}\exp\left(-\frac{k^2}{4\beta^2}\right). \tag{13.184}$$

The explicit expressions for the energies and forces are given below. A prime above a sum means that the self-term $i = j$ for $\mathbf{n} = \mathbf{0}$ and the excluded pairs $(i, j, \mathbf{n}) \in$ *exclusion list* are excluded from the sum.

$$(13.158) \quad U_C^s = \frac{1}{4\pi\varepsilon_0}\frac{1}{2}\sideset{}{'}\sum_{i,j}\sum_{\mathbf{n}} q_i q_j \frac{\operatorname{erfc}(\beta r_{ij\mathbf{n}})}{r_{ij\mathbf{n}}}, \tag{13.185}$$

$$(13.160) \qquad U_C^l = \frac{1}{2\varepsilon_0 V} \frac{1}{2} \sum_{i,j} q_i q_j \sum_{\mathbf{k} \neq 0} \frac{1}{k^2} e^{i\mathbf{k}\cdot(\mathbf{r}_i - \mathbf{r}_j)} \exp\left(-\frac{k^2}{4\beta^2}\right)$$

$$-U_{\text{self}}^l - U_{\text{excl}}^l, \qquad (13.186)$$

$$(13.175) \qquad U_{\text{self}}^l = \frac{1}{4\pi\varepsilon_0} \left(\sum_i q_i^2\right) \frac{2\beta}{\sqrt{\pi}}, \qquad (13.187)$$

$$(13.176) \qquad U_{\text{excl}}^l = \frac{1}{4\pi\varepsilon_0} \sum_{i,j,n \in \text{exclusionlist}} q_i q_j \frac{\text{erf}(\beta r_{ijn})}{r_{ijn}}, \qquad (13.188)$$

$$(13.160) \qquad \mathbf{F}_i^s = \frac{q_i}{4\pi\varepsilon_0} \sum_j \sum_{\mathbf{n}}' q_j \left(\frac{\text{erfc}(\beta r)}{r^2} + \frac{2\beta}{r\sqrt{\pi}} e^{-\beta^2 r^2}\right),$$

$$r = r_{ijn} = |\mathbf{r}_i - \mathbf{r}_j - \mathbf{Tn}|, \qquad (13.189)$$

$$((13.177) \qquad \mathbf{F}_i^l = -\frac{q_i}{\varepsilon_0 V} \sum_{\mathbf{k}} \frac{i\mathbf{k}}{k^2} \exp\left(-\frac{k^2}{4\beta^2}\right) \sum_j e^{i\mathbf{k}\cdot(\mathbf{r}_i - \mathbf{r}_j)}, \quad (13.190)$$

$$(13.178) \qquad \mathbf{F}_{i,\text{excl}}^l = -\mathbf{F}_{j,\text{excl}}^l = \frac{q_i q_j}{4\pi\varepsilon_0} \left(\frac{\text{erf}(\beta r)}{r^2} - \frac{2}{\sqrt{\pi}} \frac{\beta}{r} e^{-\beta^2 r^2}\right) \frac{\mathbf{r}}{r},$$

$$\mathbf{r} = \mathbf{r}_i - \mathbf{r}_j - \mathbf{Tn}, \quad (i, j, \mathbf{n}) \in \text{exclusion list}. \qquad (13.191)$$

The exclusion forces $\mathbf{F}_{i,\text{excl}}^l$ must be subtracted from the long-range force $\mathbf{F}_i^l$ calculated from (13.190). There is no force due the self-energy contained in the long-range energy.

Figure 13.6 shows the Gaussian spread function and the corresponding short- and long-range potential functions, the latter adding up to the total potential $1/r$.

13.10.4 Cubic spread function

The Gaussian spread function is by no means the only possible choice.[14] In fact, a spread function that leads to forces which go smoothly to zero at a given cut-off radius r_c and stay exactly zero beyond that radius, have the advantage above Gaussian functions that no cut-off artifacts are introduced in the integration of the equations of motion. Any charge spread function that is exactly zero beyond r_c will produce a short-range force with zero value and zero derivative at r_c. In addition we require that the Fourier transform rapidly decays for large k in order to allow efficient determination of the long-range forces; this implies a smooth spread function. A discontinuous function, and even a discontinuous derivative, will produce wiggles in the

---

[14] Berendsen (1993) lists a number of choices, but does not include the cubic spread function.

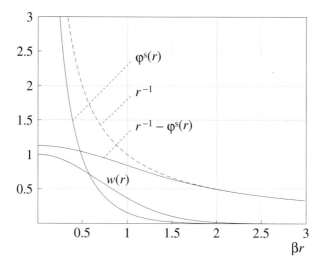

Figure 13.6 Functions for the Ewald sum: $w(r)$ is proportional to the Gaussian spread function; $\varphi^s(r)$ and $r^{-1} - \varphi^s(r)$ are the short-range and long-range potential functions, adding up to the total Coulomb interaction r^{-1}.

Fourier transform. The following cubic polynomial fulfills all requirements; the force function even has a vanishing *second* derivative, allowing the use of higher-order integration algorithms. The functions are all analytical, although tabulation is recommended for efficient implementation. Figure 13.7 shows spread and potential functions for the cubic spread function comparable to Fig. 13.6. Figure 13.8 shows the Fourier transform W_k of both the Gaussian and the cubic spread functions.

The cubic charge spread function is

$$w(r) = \frac{15}{4\pi r_c^3}(1 - \frac{3r^2}{r_c^2} + \frac{2r^3}{r_c^3}) \text{ for } r < r_c,$$
$$= 0 \text{ for } r \geq r_c, \tag{13.192}$$

and its Fourier transform (13.170) is given by

$$W_\kappa = \frac{90}{\kappa^4 V}\left[(1 - \frac{8}{\kappa^2})\cos\kappa - \frac{5}{\kappa}\sin\kappa + \frac{8}{\kappa^2}\right], \tag{13.193}$$

where $\kappa = kr_c$. The short-range force function (13.161) is

$$f^s(r) = \frac{1}{r^2} - \frac{5r}{r_c^3} + \frac{9r^3}{r_c^5} - \frac{5r^4}{r_c^6} \text{ for } r < r_c,$$

$$= 0 \text{ for } r \geq r_c, \tag{13.194}$$

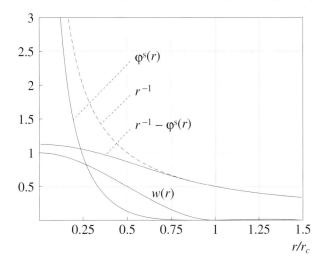

Figure 13.7 Functions for the cubic spread function: $w(r)$ is proportional to the spread function; the potential functions are as in Fig. 13.6, but scaled by 2 to make them comparable.

and the short-range potential function (13.159) is

$$\frac{1}{r} - \frac{9}{4r_c} + \frac{5r^2}{2r_c^3} - \frac{9r^4}{4r_c^5} + \frac{r^5}{r_c^6} \quad \text{for } r < r_c,$$

$$= 0 \quad \text{for } r \geq r_c. \quad (13.195)$$

13.10.5 Net dipolar energy

Special attention needs to be given to the energetic effects of a net non-zero dipole moment, as has been carefully done by de Leeuw *et al.* (1980).[15] The problem is that Coulomb lattice sums over unit cells with non-vanishing total dipole moment converge only *conditionally*, i.e., the sum depends on the sequence of terms in the summation. Consider summation over a chunk of matter containing a (very large, but not infinite) number of unit cells. The total dipole moment of the chunk of matter is proportional to the volume of the chunk. The Coulomb energy, given by the summed dipolar interactions, now depends on the *shape* of the chunk and on its *dielectric environment*. For example, in a flat disc perpendicular to the dipole moment, the interaction is unfavorable (positive), but in a long cylinder parallel to the dipole moment the interaction is favorable (negative). In a sphere of radius R with cubic unit cells the interactions sum to zero, but there will be a *reaction field* E_{RF}

---

[15] See also Caillol (1994), Essmann *et al.* (1995) and Deserno and Holm (1998a).

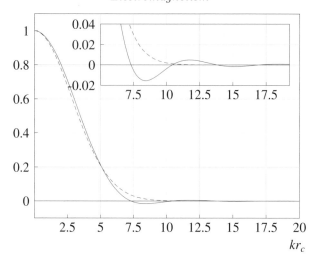

Figure 13.8 Fourier transforms of the cubic (solid line) and Gaussian (dashed line) spread functions. For the Gaussian transform β was set to $2/r_c$. The inset magnifies the tails.

due to the polarizability of the medium in which the sphere is embedded (see (13.83) on page 347):

$$E_{\mathrm{RF}} = \frac{\mu_{\mathrm{tot}}}{4\pi\varepsilon_0 R^3}\frac{2(\varepsilon_r - 1)}{2\varepsilon_r + 1}, \tag{13.196}$$

where ε_r is the relative dielectric constant of the medium. The energy per unit cell $-\mu_{\mathrm{tot}}E_{\mathrm{RF}}/(2N)$ (where N is the number of unit cells in the sphere) in the reaction field can now be written as

$$U_{\mathrm{RF}} = -\frac{\mu^2}{6\varepsilon_0 V}\frac{2(\varepsilon_r - 1)}{2\varepsilon_r + 1}, \tag{13.197}$$

where μ is the unit cell dipole moment and V the unit cell volume. This term does not depend on the size of the system since the R^3 proportionality in the volume just cancels the R^{-3} proportionality of the reaction field. For lower multipoles (i.e., for the total charge) the energy diverges, and the system is therefore required to be electroneutral; for higher multipoles the lattice sum converges unconditionally so that the problem does not arise.

It is clear that the boundary conditions must be specified for periodic systems with non-vanishing total dipole moment. The system behavior, especially the fluctuation of the dipole moment, will depend on the chosen boundary conditions. A special case is the *tin-foil* or *metallic* boundary condition, given by $\varepsilon_r = \infty$, which is equivalent to a conducting boundary.

Applied to a sphere, the RF energy per unit cell then becomes

$$U_{\rm RF} = -\frac{\mu^2}{6\varepsilon_0 V} \quad \text{(spherical tin-foil b.c.).} \tag{13.198}$$

Since the question of the boundary condition did not come up when solving for the long-range Coulomb interaction, leading to (13.174), one wonders whether this equation silently implies a specific boundary condition, and if so, which one. By expanding $\exp(\pm i\mathbf{k}\cdot\mathbf{r})$ in powers of k, we see that $Q_{\mathbf{k}}Q_{-\mathbf{k}} = (\mathbf{k}\cdot\boldsymbol{\mu})^2 + \mathcal{O}(k^4)$, while $W_{\mathbf{k}} = (1/V) + \mathcal{O}(k^2)$. The term $(\mathbf{k}\cdot\boldsymbol{\mu})^2$ equals $\frac{1}{3}\mu^2 k^2$ when averaged over all orientations of the dipole moment. Thus the energy term $k^{-2}Q_{\mathbf{k}}Q_{-\mathbf{k}}W_{\mathbf{k}}/(2\varepsilon_0)$ equals $-\mu^2/(6\varepsilon_0 V) + \mathcal{O}(k^2)$, which is exactly the dipolar energy for the tin-foil boundary conditions. The conclusion is that application of the equations for the Coulomb energy, as derived here based on a splitting between short- and long-range components, and consequently also for the Ewald summation, automatically imply tin-foil boundary conditions.

If one wishes to exert spherical boundary conditions corresponding to a dielectric environment with relative dielectric constant ε_r rather than conducting boundary conditions, an extra term making up the difference between (13.197) and (13.198) must be added to the computed energy. This extra term is

$$U_{\rm dipole} = \frac{\mu^2}{2\varepsilon_0 V}\frac{1}{(2\varepsilon_r + 1)}. \tag{13.199}$$

This term is always positive, as the tin-foil condition (for which the correction is zero) provides the most favorable interaction. In a vacuum environment $(\varepsilon_r = 1)$ it is more unfavorable to develop a net dipole moment, and in a dipolar fluid with fluctuating net dipole moment, the net dipole moment is suppressed compared to tin-foil boundary conditions. The most natural boundary condition for a dipolar fluid would be a dielectric environment with a dielectric constant equal to the actual dielectric constant of the medium.

13.10.6 Particle–mesh methods

The computational effort of the Ewald summation scales as N^2 with the number of charges N and becomes prohibitive for large systems.[16] Fast Fourier transforms (FFT)[17] are computationally attractive although they

---

[16] With optimized truncation of the real and reciprocal sums (Perram *et al.*, 1988) a $N^{3/2}$-scaling can be accomplished. The computation can also be made considerably faster by using tabulated functions (see Chapter 19).

[17] See Section 12.7 of Chapter 12 on page 324 for details on fast Fourier transforms.

restrict the spatial solutions to lattice points. Interpolation is then needed to
obtain the energies and forces acting on charges. They scale as $N \log N$ and
form the long-range methods of choice, e.g., as implemented in the *particle–
mesh–Ewald* (PME) method of Darden *et al.* (1993) who use a Gaussian
spread function and a Lagrange interpolation, or – preferably – the more
accurate *smooth particle–mesh–Ewald* (SPME) method of Essmann *et al.*
(1995), who use a B-spline interpolation.[18] The advantage of using B-spline
interpolation is that the resulting potential function is twice continuously
differentiable if the order of the spline is at least four; smooth forces can
therefore be immediately obtained from the differentiated potential. With
Lagrange interpolation the interpolated potential is only piecewise differentiable and cannot be used to derive the forces. The SPME works as follows,
given a system of N charges q_i at positions r_i within a unit cell of base
vectors $a.b, c$:

- Choose three integers K_1, K_2, K_3 that subdivide the unit cell into small,
  reasonably isotropic, grid cells. Choose an appropriate Ewald parameter β (see (13.181)), a cutoff-radius r_c for the short-range interaction,
  which should not exceed half the length of the smallest base vector, and
  a cut-off radius in reciprocal space. Choose the order p of the B-spline
  interpolation, which should be at least 4 (cubic spline). A cutoff of 4
  times the standard deviation of the Gaussian spread function implies that
  $\beta r_c = 2\sqrt{2}$. A grid size $a/K_1, b/K_2, c/K_3$ of about $0.3/\beta$ and a reciprocal cut-off of 1.5β would be reasonable for a start. The optimal values
  depend on system size and density; they should be adjusted for optimal
  computational efficiency, given a desired overall accuracy.
- Compute the structure factors using the exponential spline interpolation
  as explained in Chapter 19, using (19.87) and (19.88) on page 554. Note
  that for odd order p the value $m = \pm K/2$ must be excluded

For actual programs the reader is referred to the original authors. A full
description of the SPME algorithm can be found in Griebel *et al.* (2003).

  These methods use a splitting between short-range and long-range potentials and solve the long-range potential on a grid; they are really variants
of the PPPM *particle–particle particle–mesh*) method developed earlier by
Hockney and Eastwood (1988). In the PPPM method the charges are distributed over grid points; the Poisson equation is solved and the potentials
and fields are interpolated, using optimized local functions that minimize the
total error. Deserno and Holm (1998a) have compared the accuracies and

---

[18] For details on B-splines see Chapter 19, Section 19.7, on page 548.

efficiencies of various methods and evaluated the error in a PPPM scheme in a second paper (1998b).

It seems that SPME methods have not yet been applied to other charge spread functions than Gaussian ones, although short-range force functions that go exactly and smoothly to zero at the cut-off radius would have the advantage of avoiding cut-off noise in the short-range forces. A suitable candidate would be the cubic function, discussed on page 369.

13.10.7 Potentials and fields in periodic systems of charges and dipoles

Certain force fields describe charge distributions not only with a set of charges, but also with a set of dipoles, or even higher multipoles. The dipoles may be permanent ones, designed to describe the charge distribution with a smaller number of sites. The may also be induced dipoles proportional to the local field, as in certain types of polarizable force fields. In the latter case the induced dipoles are determined in an iterative procedure until consistency, or they may be considered as variables that minimize a free energy functional. In all cases it is necessary to determine potentials and fields, and from those energies and forces, from a given distribution of charges and dipoles.

The methods described above, splitting interactions into short- and long-range parts, can be extended to include dipolar sources as well. One must be aware that such extensions considerably complicate the computation of energies and forces, as the dipolar terms involve higher derivatives than are required for charges. It could well be advantageous to avoid dipoles – and certainly higher multipoles – if the problem at hand allows formulation in terms of charges alone. Here we shall review the methods in order to give the reader a flavor of the additional complexity, referring the reader to the literature for details.

Ewald summations including multipolar sources have been worked out by Smith (1982); based on this work, Toukmaji *et al.* (2000) extended PME methods to dipolar sources. The effect of adding dipole moments $\boldsymbol{\mu}_i$ to the sources q_i is that q_i is replaced by $q_i + \boldsymbol{\mu}_i \cdot \boldsymbol{\nabla}_i$, which has consequences for the structure factors as well as for the short- and long-range force and energy terms. Consider the energy U_{12} between two charges q_1 at $\boldsymbol{r}_1$ and q_2 at $\boldsymbol{r}_2$:

$$U_{12} = q_1 \Phi(\boldsymbol{r}_1) = \frac{1}{4\pi\varepsilon_0} q_1 q_2 \frac{1}{r_{12}}, \tag{13.200}$$

where $r_{12} = |\boldsymbol{r}_1 - \boldsymbol{r}_2|$.[19] When dipoles are present this modifies to (cf (13.123))

$$U_{12} = q_1\Phi(\boldsymbol{r}_1) - \boldsymbol{\mu}_1 \cdot \boldsymbol{E}(\boldsymbol{r}_1) = (q_1 + \boldsymbol{\mu}_1 \cdot \boldsymbol{\nabla}_1)\Phi(\boldsymbol{r}_1), \tag{13.201}$$

with the potential given by (see (13.136))

$$4\pi\varepsilon_0\Phi(\boldsymbol{r}_1) = \frac{q_2}{r_{12}} - \boldsymbol{\mu}_2 \cdot \boldsymbol{\nabla}_1 \frac{1}{r_{12}} = (q_2 + \boldsymbol{\mu}_2 \cdot \boldsymbol{\nabla}_2)\frac{1}{r_{12}}. \tag{13.202}$$

The interaction thus changes from $q_1 q_2/r_{12}$ to

$$U_{12} = (q_1 + \boldsymbol{\mu}_1 \cdot \boldsymbol{\nabla}_1)(q_2 + \boldsymbol{\mu}_2 \cdot \boldsymbol{\nabla}_2)\frac{1}{r_{12}}. \tag{13.203}$$

We note that this represents the total electrostatic interaction; for the energy of polarizable systems one should add the energy it costs to create the induced dipole moments, which is a quadratic form in $\boldsymbol{\mu}$ (like $\sum_i \mu_i^2/2\alpha_i$) for the case of linear polarizability.

This replacement works through all equations; for example, the structure factor $Q_{\mathbf{k}}$ (13.166) now becomes

$$
\begin{aligned}
Q_{\mathbf{k}} &= \sum_j (q_j + \boldsymbol{\mu}_j \cdot \boldsymbol{\nabla}_j)e^{-i\boldsymbol{k}\cdot\boldsymbol{r}_j} \\
&= \sum_j (q_j - 2\pi i\boldsymbol{\mu}_j \cdot \boldsymbol{k})e^{-i\boldsymbol{k}\cdot\boldsymbol{r}_j}, \quad j \in \text{unit cell}, \tag{13.204}
\end{aligned}
$$

with consequences for the spline interpolation procedure. The reader is referred to Toukmaji *et al.* (2000) for details.

Exercises

13.1 Consider an electron as a (classical) sphere with homogeneous charge distribution. What would its radius be when the total field energy equals the relativistic rest energy mc^2?

13.2 What is the amplitude in vacuum of E and B in a circular laser beam of 50 mW monochromatic radiation ($\lambda = 632.8$ nm), when the beam has a diameter of 2 mm?

13.3 Consider an infinite planar surface with vacuum on one side and a dielectric medium with relative dielectric constant ε_r on the other side, with a charge q situated on the vacuum side at a distance d from the plane. Show that the following potential satisfies the boundary conditions (13.56): on the vacuum side the direct Coulomb

---

[19] See the beginning of Section 13.8 on page 353 for a discussion of charge interactions and where a factor 2 should or should not appear.

potential of the charge plus the vacuum potential of an image charge $q_{im} = -q(\varepsilon_r - 1)/(\varepsilon_r + 1)$ at the mirror position; on the medium side the direct Coulomb potential of the charge, divided by a factor ε_{eff}. Express ε_{eff} in ε_r.

13.4   Compare the exact solvation free energy of two charges q_1 and q_2, both at a distance d from the planar surface that separates vacuum and medium (ε_r) as in Exercise 13.3, and separated laterally by a distance r_{12}, with the generalized Born expression (13.103) using Still's expression (13.105).

13.5   Verhoeven and Dymanus (1970) have measured the quadrupole moment of D_2O. They report the values:

$$\theta_a = 2.72(2), \quad \theta_b = -0.32(3), \quad \theta_c = -2.40,$$

in 10^{-26} esu.cm^2, for the diagonal components in a coordinate system with its origin in the center-of-mass, where a and b are in the plane of the molecule and b is in the direction of the molecular symmetry axis. They use the following traceless definition of the quadrupole moment:

$$\theta_{\alpha\beta} = \frac{1}{2} \int \rho(\mathbf{r})[3x_\alpha x_\beta - r^2 \delta_{\alpha\beta}] \, d\mathbf{r}.$$

From these data, derive the quadrupole moment $Q_{\alpha\beta}$ as defined in (13.119) on page 356, expressed in "molecular units" e nm^2 (see Table 8 on page xxv), and in a coordinate system with its origin in the position of the oxygen atom. Use the following data for the transformation: OD-distance: 0.09584 nm, DOD-angle: $104°\,27'$, dipole moment: 1.85 Debye, oxygen mass: 15.999 u, deuterium mass: 2.014 u. An esu (electrostatic unit) of charge equals $3.335\,64 \times 10^{-10}$ C; the elementary charge e equals 4.8032×10^{-10} esu. Give the accuracies as well.

14

Vectors, operators and vector spaces

14.1 Introduction

A vector we know as an arrow in 3D-space with a direction and a length, and we can add any two vectors to produce a new vector in the same space. If we define three coordinate axes, not all in one plane, and define three basis vectors e_1, e_2, e_3 along these axes, then any vector v in 3D-space can be written as a linear combination of the basis vectors:

$$v = v_1 e_1 + v_2 e_2 + v_3 e_3. \qquad (14.1)$$

v_1, v_2, v_3 are the *components* of v on the given basis set. These components form a specific *representation* of the vector, depending on the choice of basis vectors. The components are usually represented as a matrix of one column:

$$\mathbf{v} = \begin{pmatrix} v_1 \\ v_2 \\ v_3 \end{pmatrix}. \qquad (14.2)$$

Note that the *matrix* $\mathbf{v}$ and the *vector* v are different things: v is an entity in space independent of any coordinate system we choose; $\mathbf{v}$ represents v on a specific set of basis vectors. To stress this difference we use a different notation: italic bold for vectors and roman bold for their matrix representations.

Vectors and basis vectors need not be arrows in 3D-space. They can also represent other constructs for which it is meaningful to form linear combinations. For example, they could represent *functions* of one or more variables. Consider all possible real polynomials $f(x)$ of the second degree, which can be written as

$$f(x) = a + bx + cx^2, \qquad (14.3)$$

where a, b, c can be any real number. We could now define the functions $1, x,$

and x^2 as basis vectors (or basis functions) and consider $f(x)$ as a vector with components (a, b, c) on this basis set. These vectors also live in a real 3D-space $\mathbb{R}^3$.

14.2 Definitions

Now we wish to give more general and a bit more precise definitions, without claiming to be mathematically exact.

- A set of elements, called *vectors*, form a **vector space** $\mathcal{V}$ over a *scalar field* $\mathcal{F}$ when:

    (i) $\mathcal{V}$ is an *Abelian group* under the sum operation $+$;
    (ii) for every $\boldsymbol{v} \in \mathcal{V}$ and every $a \in \mathcal{F} : a\boldsymbol{v} \in \mathcal{V}$;
    (iii) for every $\boldsymbol{v}, \boldsymbol{w} \in \mathcal{V}$ and $a, b \in \mathcal{F}$:

$$a(\boldsymbol{v} + \boldsymbol{w}) = a\boldsymbol{v} + b\boldsymbol{w}$$
$$(a + b)\boldsymbol{v} = a\boldsymbol{v} + b\boldsymbol{v}$$
$$(ab)\boldsymbol{v} = a(b\boldsymbol{v})$$
$$1\boldsymbol{v} = \boldsymbol{v}$$
$$0\boldsymbol{v} = \boldsymbol{0}$$

A scalar field is precisely defined in set theory, but for our purpose it suffices to identify $\mathcal{F}$ with the set of real numbers $\mathbb{R}$ or the set of complex numbers $\mathbb{C}$. An Abelian group is a set of elements for which a binary operation $+$ (in this case a summation) is defined such that $\boldsymbol{v} + \boldsymbol{w} = \boldsymbol{w} + \boldsymbol{v}$ is also an element of the set, in which an element $\boldsymbol{0}$ exists for which $\boldsymbol{v} + \boldsymbol{0} = \boldsymbol{0} + \boldsymbol{v}$, and in which for every $\boldsymbol{v}$ an element $-\boldsymbol{v}$ exists with $\boldsymbol{v} + (-\boldsymbol{v}) = \boldsymbol{0}$.

- A vector space is **n-dimensional** if n vectors $\boldsymbol{e}_1, \ldots, \boldsymbol{e}_n$ exist, such that every element $\boldsymbol{v} \in \mathcal{V}$ can be written as $\boldsymbol{v} = \sum_{i=1}^{n} v_i \boldsymbol{e}_i$. The n vectors must be linearly independent, i.e., no non-zero set of numbers $c_1, c_2, \ldots, c_n$ exists for which $\sum_{i=1}^{n} c_i \boldsymbol{e}_i = 0$. The vectors $\boldsymbol{e}_1, \ldots, \boldsymbol{e}_n$ form a *basis* of $\mathcal{V}$.

- A vector space is **real** if $\mathcal{F} = \mathbb{R}$ and **complex** if $\mathcal{F} = \mathbb{C}$.

- A vector space is *normed* if to every $\boldsymbol{v}$ a non-negative real number $||\boldsymbol{v}||$ is associated (called the *norm*), such that for every $\boldsymbol{v}, \boldsymbol{w} \in \mathcal{V}$ and every complex number c:

    (v) $||c\boldsymbol{v}|| = |c| ||\boldsymbol{v}||$;
    (vi) $||\boldsymbol{v} + \boldsymbol{w}|| \leq ||\boldsymbol{v}|| + ||\boldsymbol{w}||$;
    (vii) $||\boldsymbol{v}|| > 0$ for $\boldsymbol{v} \neq \boldsymbol{0}$.

- A vector space is **complete** if:

(viii) for every series v_n with $\lim_{m,n\to\infty}||v_m - v_n|| = 0$ there exists a v such that $\lim_{m,n\to\infty}||v - v_n|| = 0$. Don't worry: all vector spaces we encounter are complete.

- A **Banach space** is a complete, normed vector space.
- A **Hilbert space** $\mathcal{H}$ is a Banach space in which a *scalar product* or *inner product* is defined as follows: to every pair v, w a complex number is associated (often denoted by (v, w) or $\langle v|w\rangle$), such that for every $u, v, w \in \mathcal{H}$ and every complex number c:

  (ix) $\langle cv|w\rangle = c^*\langle v|w\rangle$;
  (x) $\langle u + v|w\rangle = \langle u|w\rangle + \langle v|w\rangle$;
  (xi) $\langle v|w\rangle = \langle w|v\rangle^*$;
  (xii) $\langle v|v\rangle > 0$ if $v \neq 0$;
  (xiii) $||v|| = \langle v|v\rangle^{1/2}$.

- Two vectors are **orthogonal** if $\langle v|w\rangle = 0$. A vector is **normalized** if $||v|| = 1$. A set of vectors is orthogonal if all pairs are orthogonal and the set is **orthonormal** if all vectors are in addition normalized.

14.3 Hilbert spaces of wave functions

We consider functions ψ (it is irrelevant what variables these are functions of) that can be expanded in a set of basis functions ϕ_n:

$$\psi = \sum_n c_n \phi_n, \tag{14.4}$$

where c_n are complex numbers. The functions may also be complex-valued. We define the scalar product of two functions as

$$(\phi, \psi) = \langle \phi|\psi\rangle = \int \phi^* \psi \, d\tau, \tag{14.5}$$

where the integral is over a defined volume of the variables τ. The norm is now defined as

$$||\psi|| = \int \psi^* \psi \, d\tau. \tag{14.6}$$

These definitions comply with requirements (viii) - (xii) of the previous section, as the reader can easily check. Thus the functions are vectors in a Hilbert space; the components $c_1, \ldots, c_n$ form a representation of the vector ψ which we shall denote in matrix notation by the one-column matrix **c**.

The basis set $\{\phi_n\}$ is *orthonormal* if $\langle\phi_n|\phi_m\rangle = \delta_{nm}$. It is not mandatory, but very convenient, to work with orthonormal basis sets. For non-orthonormal basis sets it is useful to define the *overlap matrix* $\mathbf{S}$:

$$S_{nm} = \langle\phi_n|\phi_m\rangle. \tag{14.7}$$

The representation $\mathbf{c}$ of a normalized function satisfies[1]

$$\mathbf{c}^\dagger\mathbf{c} = \sum_n c_n^* c_n = 1 \tag{14.8}$$

on an orthogonal basis set; on an arbitrary basis set $\mathbf{c}^\dagger\mathbf{S}\mathbf{c} = 1$.

14.4 Operators in Hilbert space

An operator acts on a function (or vector) to transform it into another function (or vector) in the same space. We restrict ourselves to *linear operators* which transform a function into a linear combination of other functions and denote operators by a hat, as $\hat{A}$:

$$\psi' = \hat{A}\psi. \tag{14.9}$$

An operator can be represented by a *matrix* on a given orthonormal basis set $\{\phi_n\}$, transforming the representation $\mathbf{c}$ of ψ into $\mathbf{c}'$ of ψ' by an ordinary matrix multiplication

$$\mathbf{c}' = \mathbf{A}\mathbf{c}, \tag{14.10}$$

where

$$A_{nm} = \langle\phi_n|\hat{A}|\phi_m\rangle = \int \phi_n^* \hat{A}\phi_m \, d\tau. \tag{14.11}$$

Proof Expanding ψ' on an orthonormal basis set ϕ_m and applying (14.9) we have:

$$\psi' = \sum_m c_m'\phi_m = \hat{A}\psi = \sum_m c_m \hat{A}\phi_m.$$

Now left-multiply by ϕ_n^* and integrate over coordinates to form the scalar products

$$\sum_m c_m'\langle\phi_n|\phi_m\rangle = \sum_m c_m\langle\phi_n|\hat{A}|\phi_m\rangle = \sum_m A_{nm}c_m,$$

or

$$c_n' = (\mathbf{A}\mathbf{c})_n.$$

---

[1] With the superscript $\dagger$ we denote the *hermitian conjugate*, which is the transpose of the complex conjugate: $(\mathbf{A}^\dagger)_{nm} = A_{mn}^*$. This is the usual notation in physics and chemistry, but in mathematical texts the hermitian conjugate is often denoted by $*$.

□

The **eigenvalue equation** for an operator $\hat{A}$:

$$\hat{A}\psi = \lambda\psi, \tag{14.12}$$

now becomes on an orthonormal basis set an eigenvalue equation for the matrix $\mathbf{A}$:

$$\mathbf{Ac} = \lambda\mathbf{c}. \tag{14.13}$$

Solutions are eigenvectors $\mathbf{c}$ and eigenvalues λ. If the basis set is not orthonormal, the equation becomes

$$\mathbf{Ac} = \lambda\mathbf{Sc}. \tag{14.14}$$

Hermitian operators form an important subclass of operators. An operator $\hat{A}$ is hermitian if

$$\langle f|\hat{A}g\rangle = \langle g|\hat{A}f\rangle^*, \tag{14.15}$$

or

$$\int f^*\hat{A}g\,d\tau = \int(\hat{A}^*f^*)g\,d\tau. \tag{14.16}$$

Hermitian operators have real expectation values ($f = g = \psi$) and real eigenvalues ($f = g; \hat{A}f = \lambda f \rightarrow \lambda = \lambda^*$). The operators of physically meaningful observables are hermitian. The matrix representation of a hermitian operator is a hermitian matrix $\mathbf{A} = \mathbf{A}^\dagger(f = \phi_n, g = \phi_m)$.

Not only do hermitian operators have real eigenvalues, they also have orthogonal eigenfunctions for non-degenerate (different) eigenvalues. The eigenfunctions within the subspace corresponding to a set of degenerate eigenvalues can be chosen to be orthogonal as well,[2] and all eigenfunctions may be normalized: *The eigenfunctions of a hermitian operator (can be chosen to) form an orthonormal set.*

Proof Let λ_n, ψ_n be eigenvalues and eigenfunctions of $\hat{A}$:

$$\hat{A}\psi_n = \lambda_n\psi_n.$$

Then

$$\int\psi^*\hat{A}\psi_m\,d\tau = \lambda_m\int\psi_n^*\psi_m\,d\tau,$$

---

[2] If ϕ_1 and ϕ_2 are two eigenfunctions of the same (degenerate) eigenvalue λ, then any linear combination of ϕ_1 and ϕ_2 is also an eigenfunction.

and

$$\left(\int \psi_m^* \hat{A} \psi_n \, d\tau \right)^* = \lambda_n^* \int \psi_m^* \psi_n \, d\tau.$$

When $\mathbf{A} = \mathbf{A}^\dagger$ then for $n = m : \lambda_n = \lambda_n^* \rightarrow \lambda$ is real; for $m \neq n$ and $\lambda_m \neq \lambda_n : \int \psi_n^* \psi_m \, d\tau = 0.$ $\qquad\square$

The **commutator** $[\hat{A}, \hat{B}]$ of two operators is defined as

$$[\hat{A}, \hat{B}] = \hat{A}\hat{B} - \hat{B}, \hat{A}, \tag{14.17}$$

and we say that $\hat{A}$ and $\hat{B}$ commute if their commutator is zero. If two operators commute, they have the same set of eigenvectors.

14.5 Transformations of the basis set

It is important to clearly distinguish operators that act on functions (vectors) in Hilbert space, changing the vector itself, from *coordinate transformations* which are operators acting on the basis functions, thus changing the representation of a vector, without touching the vector itself.

Consider a linear coordinate transformation Q changing a basis set $\{\phi_n\}$ into a new basis set $\{\phi_n'\}$:

$$\phi_n' = \sum_i Q_{in}\phi_i. \tag{14.18}$$

Let $\mathbf{A}$ be the representation of an operator $\hat{A}$ on $\{\phi_n\}$ and $\mathbf{A}'$ its representation on $\{\phi_n'\}$. Then $\mathbf{A}'$ and $\mathbf{A}$ relate as

$$\mathbf{A}' = \mathbf{Q}^\dagger \mathbf{A} \mathbf{Q}. \tag{14.19}$$

Proof Consider one element of $\mathbf{A}'$ and insert (14.18):

$$(\mathbf{A}')_{nm} = \langle \phi_n' | \hat{A} | \phi_m' \rangle = \sum_{ij} Q_{in}^* Q_{jm} A_{ij} = (\mathbf{Q}^\dagger \mathbf{A} \mathbf{Q})_{nm}.$$

$\qquad\square$

If both basis sets are orthonormal, then the transformation $\mathbf{Q}$ is *unitary*.[3]

Proof Orthonormality implies that

$$\langle \phi_n' | \phi_m' \rangle = \delta_{nm}.$$

---

[3] A transformation (matrix) $\mathbf{U}$ is unitary if $\mathbf{U}^\dagger = \mathbf{U}^{-1}$.

Since

$$\langle \phi'_n | \phi'_m \rangle = \sum_{ij} Q^*_{in} Q_{jm} \langle \phi_i | \phi_j \rangle = \sum_i (\mathbf{Q}^\dagger)_{ni} Q_{im},$$

it follows that

$$\sum_i (\mathbf{Q}^\dagger)_{ni} Q_{im} = \delta_{nm},$$

or

$$\mathbf{Q}^\dagger \mathbf{Q} = 1,$$

which implies that $\mathbf{Q}$ is unitary. □

The representations $\mathbf{c}$ and $\mathbf{c}'$ of a vector are related as

$$\mathbf{c} = \mathbf{Q}\mathbf{c}', \tag{14.20}$$

$$\mathbf{c}' = \mathbf{Q}^{-1}\mathbf{c}, \tag{14.21}$$

$$\mathbf{c}' = \mathbf{Q}^\dagger \mathbf{c}, \tag{14.22}$$

where (14.22) is only valid for unitary transformations.

Let $\hat{A}$ be a hermitian operator with eigenvalues $\lambda_1, \lambda_2, \ldots$ and with orthonormal eigenvectors $\mathbf{c}_1, \mathbf{c}_2, \ldots$. Then, if we construct a matrix $\mathbf{U}$ with columns formed by the eigenvectors, it follows that

$$\mathbf{A}\mathbf{U} = \mathbf{\Lambda}\mathbf{U}, \tag{14.23}$$

where $\mathbf{\Lambda}$ is the diagonal matrix of eigenvalues. Since all columns of $\mathbf{U}$ are orthonormal, $\mathbf{U}$ is a unitary matrix, and thus

$$\mathbf{U}^\dagger \mathbf{A}\mathbf{U} = \mathbf{\Lambda}. \tag{14.24}$$

In other words: $\mathbf{U}$ *is exactly the coordinate transformation that diagonalizes* $\mathbf{A}$.

14.6 Exponential operators and matrices

We shall often encounter operators of the form $\exp(\hat{A}) = e^{\hat{A}}$, e.g., as formal solutions of first-order differential equations. The definition is

$$e^{\hat{A}} = \sum_{k=0}^{\infty} \frac{1}{k!} \hat{A}^k. \tag{14.25}$$

Exponential matrices are similarly defined.

From the definition it follows that

$$\hat{A}e^{\hat{A}} = e^{\hat{A}}\hat{A}, \tag{14.26}$$

and

$$e^{\hat{A}}(f + g) = e^{\hat{A}}f + e^{\hat{A}}g. \tag{14.27}$$

The matrix representation of the operator $\exp(\hat{A})$ is $\exp(\mathbf{A})$:

$$\left\langle n|e^{\hat{A}}|m\right\rangle = \sum_{k=0}^{\infty} \frac{1}{k!}\langle n|\hat{A}^k|m\rangle = \sum_{k=0}^{\infty} \frac{1}{k!}(\mathbf{A}^k)_{nm} = (e^{\mathbf{A}})_{nm}. \tag{14.28}$$

The matrix element $(\exp \mathbf{A})_{nm}$ is in general *not* equal to $\exp(A_{nm})$, unless $\mathbf{A}$ is a diagonal matrix $\mathbf{\Lambda} = \mathrm{diag}(\lambda_1, \ldots)$:

$$(e^{\mathbf{A}})_{nm} = e^{\lambda_n}\delta_{nm}. \tag{14.29}$$

From the definition follows that $\exp(\hat{A})$ or $\exp(\mathbf{A})$ transforms just like any other operator under a unitary transformation:

$$\mathbf{U}^{\dagger}e^{\hat{A}}\mathbf{U} = \sum_{k=0}^{\infty} \frac{1}{k!}\mathbf{U}^{\dagger}\hat{A}^k\mathbf{U} = \sum_{k=0}^{\infty} \frac{1}{k!}(\mathbf{U}^{\dagger}\hat{A}\mathbf{U})^k = e^{\mathbf{U}^{\dagger}\hat{A}\mathbf{U}}. \tag{14.30}$$

This transformation property is true not only for unitary transformations, but for any *similarity transformation* $\mathbf{Q}^{-1}\mathbf{A}\mathbf{Q}$.

Noting that the trace of a matrix is invariant for a similarity transformation, it follows that

$$\det(e^{\mathbf{A}}) = \Pi_n e^{\lambda_n} = \exp\left(\sum_n \lambda_n\right) = \exp(\mathrm{tr}\,\mathbf{A}). \tag{14.31}$$

Some other useful properties of exponential matrices or operators are

$$\left(e^{\mathbf{A}}\right)^{-1} = e^{-\mathbf{A}}, \tag{14.32}$$

and

$$\frac{d}{dt}e^{\mathbf{A}t} = \mathbf{A}e^{\mathbf{A}t} = e^{\mathbf{A}t}\mathbf{A} \quad (t \text{ is a scalar variable}). \tag{14.33}$$

Generally, $\exp(\mathbf{A} + \mathbf{B}) \neq \exp(\mathbf{A})\exp(\mathbf{B})$, unless $\mathbf{A}$ and $\mathbf{B}$ commute. If $\mathbf{A}$ and $\mathbf{B}$ are small (proportional to a smallness parameter ε), the first error term is of order ε^2 and proportional to the commutator $[\mathbf{A}, \mathbf{B}]$:

$$e^{\varepsilon(\mathbf{A}+\mathbf{B})} = e^{\varepsilon\mathbf{A}}e^{\varepsilon\mathbf{B}} - \tfrac{1}{2}\varepsilon^2[\mathbf{A}, \mathbf{B}] + O(\varepsilon^3), \tag{14.34}$$

$$= e^{\varepsilon\mathbf{B}}e^{\varepsilon\mathbf{A}} + \tfrac{1}{2}\varepsilon^2[\mathbf{A}, \mathbf{B}] + O(\varepsilon^3). \tag{14.35}$$

We can approximate $\exp(\mathbf{A} + \mathbf{B})$ in a series of approximations, called the *Lie–Trotter–Suzuki expansion*. These approximations are quite useful for the design of stable algorithms to solve the evolution in time of quantum or classical systems (see de Raedt, 1987, 1996). The basic equation, named the

Trotter formula after Trotter (1959), but based on earlier ideas of Lie (see Lie and Engel, 1888), is

$$e^{(\mathbf{A}+\mathbf{B})} = \lim_{m\to\infty} \left(e^{\mathbf{A}/m} e^{\mathbf{B}/m} \right)^m. \tag{14.36}$$

Let us try to solve the time propagator

$$U(\tau) = e^{-i(\mathbf{A}+\mathbf{B})\tau}, \tag{14.37}$$

where $\mathbf{A}$ and $\mathbf{B}$ are real matrices or operators. The *first-order* solution is obviously

$$U_1(\tau) = e^{-i\mathbf{A}\tau} e^{-i\mathbf{B}\tau} + O(\tau^2). \tag{14.38}$$

Since

$$U_1^\dagger(\tau) = e^{i\mathbf{B}\tau} e^{i\mathbf{A}\tau} = U_1^{-1}(\tau), \tag{14.39}$$

the propagator $U_1(\tau)$ is unitary. Suzuki (1991) gives a recursive recipe to derive higher-order products for the exponential operator. *Symmetric* products are special cases, leading to algorithms with even-order precision. For *second order* precision Suzuki obtains

$$U_2(\tau) = e^{-i\mathbf{B}\tau/2} e^{-i\mathbf{A}\tau} e^{-i\mathbf{B}\tau/2} + O(\tau^3). \tag{14.40}$$

Higher-order precision is obtained by the recursion equation (for symmetric products; $m \geq 2$)

$$U_{2m}(\tau) = [U_{2m-2}(p_m\tau)]^2 U_{2m-2}((1 - 4p_m)\tau)[U_{2m-2}(p_m\tau)]^2 + O(\tau^{2m+1}), \tag{14.41}$$

with

$$p_m = \frac{1}{4 - 4^{1/(2m-1)}}. \tag{14.42}$$

For *fourth-order* precision this works out to

$$U_4(\tau) = U_2(p\tau)U_2(p\tau)U_2((1 - 4p)\tau)U_2(p\tau)U_2(p\tau) + O(\tau^5), \tag{14.43}$$

with

$$p = \frac{1}{4 - 4^{1/3}} = 0.4145. \tag{14.44}$$

All of the product operators are unitary, which means that algorithms based on these product operators – provided they realize the unitary character of each term – are unconditionally stable.

The following relation is very useful as starting point to derive the behavior of reduced systems, which can be viewed as projections of the complete system onto a reduced space (Chapter 8). For any pair of time-independent, non-commuting operators $\hat{A}$ and $\hat{B}$ we can write

$$e^{(\hat{A}+\hat{B})t} = e^{\hat{A}t} + \int_0^t e^{\hat{A}(t-\tau)} \hat{B} e^{(\hat{A}+\hat{B})\tau} \, d\tau. \qquad (14.45)$$

Proof First write

$$e^{(\hat{A}+\hat{B})t} = e^{\hat{A}t} \hat{Q}(t), \qquad (14.46)$$

so that

$$\hat{Q}(t) = e^{-\hat{A}t} e^{(\hat{A}+\hat{B})t}.$$

By differentiating (14.46), using the differentiation rule (14.33), we find

$$(\hat{A} + \hat{B}) e^{(\hat{A}+\hat{B})t} = \hat{A} e^{\hat{A}t} \hat{Q}(t) + e^{\hat{A}t} \frac{d\hat{Q}}{dt},$$

and using the equality

$$(\hat{A} + \hat{B}) e^{(\hat{A}+\hat{B})t} = \hat{A} e^{\hat{A}t} \hat{Q}(t) + \hat{B} e^{(\hat{A}+\hat{B})t},$$

we see that

$$\frac{d\hat{Q}}{dt} = e^{-\hat{A}t} \hat{B} e^{(\hat{A}+\hat{B})t}.$$

Hence, by integration, and noting that $\hat{Q}(0) = 1$:

$$\hat{Q}(t) = 1 + \int_0^t e^{-\hat{A}\tau} \hat{B} e^{(\hat{A}+\hat{B})\tau} \, d\tau.$$

Inserting this $\hat{Q}$ in (14.46) yields the desired expression. ☐

There are two routes to practical computation of the matrix $\exp(\mathbf{A})$. The first is to diagonalize $\mathbf{A}$: $\mathbf{Q}^{-1}\mathbf{A}\mathbf{Q} = \mathbf{\Lambda}$ and construct

$$e^{\mathbf{A}} = \mathbf{Q} \, \text{diag} \, (e^{\lambda_1}, e^{\lambda_2}, \ldots) \, \mathbf{Q}^{-1}. \qquad (14.47)$$

For large matrices diagonalization may not be feasible. Then in favorable cases the matrix may be split up into a sum of block-diagonal matrices, each of which is easy to diagonalize, and the Trotter expansion applied to the exponential of the sum of matrices. It may also prove possible to split the operator into a diagonal part and a part that is diagonal in reciprocal space, and therefore solvable by Fourier transformation, again applying the Trotter expansion.

The second method[4] is an application of the *Caley–Hamilton relation*, which states that every $n \times n$ matrix satisfies its *characteristic equation*

$$\mathbf{A}^n + a_1 \mathbf{A}^{n-1} + a_2 \mathbf{A}^{n-2} + \ldots + a_{n-1} \mathbf{A} + a_n \mathbf{1} = 0. \tag{14.48}$$

Here $a_1, \ldots a_n$ are the coefficients of the characteristic or eigenvalue equation $\det(\mathbf{A} - \lambda \mathbf{1}) = 0$, which is a n-th degree polynomial in λ:

$$\lambda^n + a_1 \lambda^{n-1} + a_2 \lambda^{n-2} + \ldots + a_{n-1} \lambda + a_n = 0. \tag{14.49}$$

Equation (14.49) is valid for each eigenvalue, and therefore for the diagonal matrix $\mathbf{\Lambda}$; (14.48) then follows by applying the similarity transformation $\mathbf{Q} \mathbf{\Lambda} \mathbf{Q}^{-1}$.

According to the Caley–Hamilton relation, $\mathbf{A}^n$ can be expressed as a linear combination of $\mathbf{A}^k$, $k = 0, \ldots, n - 1$, and so can any $\mathbf{A}^m$, $m \geq n$. Therefore, the infinite sum in (14.25) can be replaced by a sum over powers of $\mathbf{A}$ up to $n - 1$:

$$e^{\mathbf{A}} = \mu_0 \mathbf{1} + \mu_1 \mathbf{A} + \cdots + \mu_{n-1} \mathbf{A}^{n-1}. \tag{14.50}$$

The coefficients μ_i can be found by solving the system of equations

$$\mu_0 + \mu_1 \lambda_k + \mu_2 \lambda_k^2 + \cdots + \mu_{n-1} \lambda_k^{n-1} = \exp(\lambda_k), \quad k = 1, \ldots, n \tag{14.51}$$

(which follows immediately from (14.50) by transforming $\exp(\mathbf{A})$ to diagonal form). In the case of degenerate eigenvalues, (14.51) are dependent and the superfluous equations must be replaced by derivatives:

$$\mu_1 + 2\mu_2 \lambda_k + \cdots + (n - 1)\mu_{n-1} \lambda_k^{n-2} = \exp(\lambda_k) \tag{14.52}$$

for a doubly degenerate eigenvalue, and higher derivatives for more than doubly degenerate eigenvalues.

14.6.1 Example of a degenerate case

Find the exponential matrix for

$$\mathbf{A} = \begin{pmatrix} 0 & 1 & 0 \\ 1 & 0 & 0 \\ 0 & 0 & 1 \end{pmatrix}.$$

According to the Caley–Hamilton relation, the exponential matrix can be expressed as

$$\exp(\mathbf{A}) = \mu_0 \mathbf{1} + \mu_1 \mathbf{A} + \mu_2 \mathbf{A}^2.$$

[4] See, e.g., Hiller (1983) for the application of this method in system theory.

Note that the eigenvalues are $+1, +1, -1$ and that $\mathbf{A}^2 = \mathbf{I}$. The equations for μ are (because of the twofold degeneracy of λ_1 the second line is the derivative of the first)

$$\mu_0 + \mu_1 \lambda_1 + \mu_2 \lambda_1^2 = \exp(\lambda_1),$$
$$\mu_1 + 2\mu_2 \lambda_1 = \exp(\lambda_1),$$
$$\mu_0 + \mu_1 \lambda_3 + \mu_2 \lambda_3^2 = \exp(\lambda_3).$$

Solving for μ we find

$$\mu_0 = \mu_2 = \frac{1}{4}\left(e + \frac{1}{e}\right),$$
$$\mu_1 = \frac{1}{2}\left(e - \frac{1}{e}\right),$$

which yields the exponential matrix

$$e^{\mathbf{A}} = \frac{1}{2}\begin{pmatrix} e + 1/e & e - 1/e & 0 \\ e - 1/e & e + 1/e & 0 \\ 0 & 0 & 2e \end{pmatrix}.$$

The reader is invited to check this solution with the first method.

14.7 Equations of motion

In this section we consider solutions of the time-dependent Schrödinger equation, both in terms of the wave function and its vector representations, and in terms of the expectation values of observables.

14.7.1 Equations of motion for the wave function and its representation

The time-dependent Schrödinger equation

$$\frac{\partial}{\partial t}\Psi(\boldsymbol{r}, t) = -\frac{i}{\hbar}\hat{H}\Psi(\boldsymbol{r}, t) \tag{14.53}$$

reads as vector equation in Hilbert space on a stationary orthonormal basis set:

$$\dot{\mathbf{c}} = -\frac{i}{\hbar}\mathbf{H}\mathbf{c}. \tag{14.54}$$

In these equations the Hamiltonian operator or matrix may itself be a function of time, e.g., it could contain time-dependent external potentials.

These equations can be formally solved as

$$\Psi(\mathbf{r}, t) = \exp\left(-\frac{i}{\hbar} \int_0^t \hat{H}(t')\, dt'\right) \Psi(\mathbf{r}, 0), \tag{14.55}$$

$$\mathbf{c}(t) = \exp\left(-\frac{i}{\hbar} \int_0^t \mathbf{H}(t')\, dt'\right) \mathbf{c}(0), \tag{14.56}$$

which reduce in the case that the Hamiltonian does not depend explicitly on time to

$$\Psi(\mathbf{r}, t) = \exp\left(-\frac{i}{\hbar}\hat{H}t\right) \Psi(\mathbf{r}, 0), \tag{14.57}$$

$$\mathbf{c}(t) = \exp\left(-\frac{i}{\hbar}\mathbf{H}t\right) \mathbf{c}(0). \tag{14.58}$$

These exponential operators are *propagators* of the wave function in time, to be written as

$$\Psi(\mathbf{r}, t) = \hat{U}(t)\Psi(\mathbf{r}, 0), \tag{14.59}$$
$$\mathbf{c}(t) = \mathbf{U}(t)\mathbf{c}(0). \tag{14.60}$$

The propagators are *unitary* because they must keep the wave function normalized at all times: $\mathbf{c}^\dagger\mathbf{c}(t) = \mathbf{c}(0)^\dagger \mathbf{U}^\dagger\mathbf{U}\mathbf{c}(0) = 1$ for all times only if $\mathbf{U}^\dagger\mathbf{U} = 1$. We must agree on the interpretation of the role of the time in the exponent: the exponential operator is *time-ordered* in the sense that changes at later times act subsequent to changes at earlier times. This means that, for $t = t_1 + t_2$, where t_1 is first, followed by t_2, the operator factorizes as

$$\exp\left(-\frac{i}{\hbar}\hat{H}t\right) = \exp\left(-\frac{i}{\hbar}\hat{H}t_2\right)\exp\left(-\frac{i}{\hbar}\hat{H}t_1\right). \tag{14.61}$$

Time derivatives must be interpreted as

$$\hat{U}(t + dt) = -\frac{i\, dt}{\hbar}\hat{H}(t)\,\hat{U}(t), \tag{14.62}$$

even when $\hat{U}$ and $\hat{H}$ do not commute.

14.7.2 Equation of motion for observables

The equation of motion for the *expectation* $\langle A \rangle$ of an observable with operator $\hat{A}$,

$$\langle A \rangle = \langle \Psi | A | \Psi \rangle, \tag{14.63}$$

is given by

$$\frac{d}{dt}\langle A \rangle = \frac{i}{\hbar}\langle [\hat{H}, \hat{A}] \rangle + \left\langle \frac{\partial A}{\partial t} \right\rangle. \tag{14.64}$$

Proof

$$\frac{d}{dt}\int \Psi^* \hat{A}\Psi \, dt = \frac{i}{\hbar}\int (\hat{H}^*\Psi^*)\hat{A}\Psi \, d\tau + \langle \frac{\partial A}{\partial t} \rangle - \frac{i}{\hbar}\int \Psi^* \hat{A}\hat{H}\Psi \, d\tau.$$

Because $\hat{H}$ is hermitian:

$$\int (\hat{H}^*\Psi^*)\hat{A}\Psi \, d\tau = \int \Psi^* \hat{H}\hat{A}\Psi \, d\tau,$$

and (14.64) follows. □

Instead of solving the time-dependence for several observables separately by (14.64), it is more convenient to solve for $\mathbf{c}(t)$ and derive the observables from $\mathbf{c}$. When ensemble averages are required, the method of choice is to use the density matrix, which we shall now introduce.

14.8 The density matrix

Let $\mathbf{c}$ be the coefficient vector of the wave function $\Psi(\mathbf{r}, t)$. on a given orthonormal basis set. We define the *density matrix* ρ by

$$\rho_{nm} = c_n c_m^*, \tag{14.65}$$

or, equivalently,

$$\rho = \mathbf{c}\mathbf{c}^\dagger. \tag{14.66}$$

The expectation value of an observable A is given by

$$\langle A \rangle = \int \sum_m c_m^* \phi_m^* \hat{A} \sum_n c_n \phi_n \, d\tau = \sum_{n,m} \rho_{nm} A_{mn} = \sum_n (\rho \mathbf{A})_{nn} \tag{14.67}$$

so that we obtain the simple equation[5]

$$\langle A \rangle = \operatorname{tr} \rho \mathbf{A}. \tag{14.68}$$

So, if we have solved $\mathbf{c}(t)$ then we know $\rho(t)$ and hence $\langle A \rangle(t)$.

The evolution of the density matrix in time can also be solved directly from its equation of motion, called the *Liouville–von Neumann equation*:

$$\dot{\rho} = \frac{i}{\hbar}[\rho, \mathbf{H}]. \tag{14.69}$$

---

[5] The trace of a matrix is the sum of its diagonal elements.

Proof By taking the time derivative of (14.66) and applying (14.54), we see that

$$\dot{\rho} = \dot{\mathbf{c}}\mathbf{c}^\dagger + \mathbf{c}\dot{\mathbf{c}}^\dagger = -\frac{i}{\hbar}\mathbf{H}\mathbf{c}\mathbf{c}^\dagger + \frac{i}{\hbar}\mathbf{c}(\mathbf{H}\mathbf{c})^\dagger.$$

Now $(\mathbf{H}\mathbf{c})^\dagger = \mathbf{c}^\dagger\mathbf{H}^\dagger = \mathbf{c}^\dagger\mathbf{H}$ because $\mathbf{H}$ is hermitian, so that

$$\dot{\rho} = -\frac{i}{\hbar}\mathbf{H}\rho + \frac{i}{\hbar}\rho\mathbf{H} = \frac{i}{\hbar}[\rho, \mathbf{H}].$$

□

This equation also has a formal solution:

$$\rho(t) = \exp\left(-\frac{i}{\hbar}\mathbf{H}t\right)\rho(0)\exp\left(+\frac{i}{\hbar}\mathbf{H}t\right),\qquad(14.70)$$

where, if $\mathbf{H}$ is time-dependent, $\mathbf{H}(t)$ in the exponent is to be replaced by $\int_0^t \mathbf{H}(t')\,dt'$.

Proof We prove that the time derivative of (14.70) is the equation of motion (14.69):

$$\dot{\rho} = -\frac{i}{\hbar}\mathbf{H}\exp\left(-\frac{i}{\hbar}\mathbf{H}t\right)\rho(0)\exp\left(+\frac{i}{\hbar}\mathbf{H}t\right)\qquad(14.71)$$

$$= +\frac{i}{\hbar}\exp\left(-\frac{i}{\hbar}\mathbf{H}t\right)\rho(0)\mathbf{H}\exp\left(+\frac{i}{\hbar}\mathbf{H}t\right)\qquad(14.72)$$

$$= \frac{i}{\hbar}[\rho, \mathbf{H}].\qquad(14.73)$$

Here we have used the fact that $\mathbf{H}$ and $\exp\left(-\frac{i}{\hbar}\mathbf{H}t\right)$ commute.   □

The density matrix transforms as any other matrix under a unitary coordinate transformation $\mathbf{U}$:

$$\rho' = \mathbf{U}^\dagger\rho\mathbf{U}.\qquad(14.74)$$

On a basis on which $\mathbf{H}$ is diagonal (i.e., on a basis of eigenfunctions of $\hat{H}: H_{nm} = E_n\delta_{nm}$) the solution of $\rho(t)$ is

$$\rho_{nm}(t) = \rho_{nm}(0)\exp\left(\frac{i}{\hbar}(E_m - E_n)t\right),\qquad(14.75)$$

implying that ρ_{nn} is constant.

14.8.1 *The ensemble-averaged density matrix*

The density matrix can be averaged over a statistical ensemble of systems without loss of information about ensemble-averaged observables. This is in contrast to the use of $\mathbf{c}(t)$ which contains a phase factor and generally averages out to zero over an ensemble.

In thermodynamic equilibrium (in the canonical ensemble) the probability of a system to be in the n-th eigenstate with energy E_n is proportional to its Boltzmann factor:

$$P_n = \frac{1}{Q} e^{-\beta E_n}, \tag{14.76}$$

where $\beta = 1/k_B T$ and

$$Q = \sum_n e^{-\beta E_n} \tag{14.77}$$

is the partition function (summed over all quantum states). On a basis set of eigenfunctions of $\hat{H}$, in which $\mathbf{H}$ is diagonal,

$$\rho_{eq} = \frac{1}{Q} e^{-\beta \mathbf{H}}, \tag{14.78}$$

$$Q = \operatorname{tr} e^{-\beta \mathbf{H}}, \tag{14.79}$$

implying that off-diagonal elements vanish, which is equivalent to the assumption that the phases of ρ_{nm} are randomly distributed over the ensemble (*random phase approximation*).

But (14.78) and (14.79) are also valid after any unitary coordinate transformation, and thus these equations are generally valid on any orthonormal basis set.

14.8.2 *The density matrix in coordinate representation*

The ensemble-averaged density matrix gives information on the probability of quantum states ϕ_n, but it does not give direct information on the probability of a configuration of the particles in space. In the *coordinate representation* we define the equilibrium density matrix as a function of (multiparticle) spatial coordinates $\mathbf{r}$:

$$\rho(\mathbf{r}, \mathbf{r}'; \beta) = \sum_n \phi_n^*(\mathbf{r}) e^{-\beta E_n} \phi_n(\mathbf{r}'). \tag{14.80}$$

This is a square continuous "matrix" of $\infty \times \infty$ dimensions. The trace of ρ is

$$\operatorname{tr} \rho = \int \rho(\mathbf{r}, \mathbf{r}; \beta) \, d\mathbf{r}, \tag{14.81}$$

which is equal to the partition function Q.

A product of such matrices is in fact an integral, which is itself equal to a density matrix:

$$\int \rho(\boldsymbol{r}, \boldsymbol{r}_1; \beta_1)\rho(\boldsymbol{r}_1, \boldsymbol{r}'; \beta_2)\, d\boldsymbol{r}_1 = \rho(\boldsymbol{r}, \boldsymbol{r}'; \beta_1 + \beta_2), \qquad (14.82)$$

as we can check by working out the l.h.s.:

$$\int \left(\sum_n \phi_n^*(\boldsymbol{r})e^{-\beta_1 E_n} \phi_n(\boldsymbol{r}_1) \right) \left(\sum_m \phi_m^*(\boldsymbol{r}_1)e^{-\beta_2 E_m} \phi_m(\boldsymbol{r}') \right) d\boldsymbol{r}_1$$

$$= \sum_{n,m} \phi_n^*(\boldsymbol{r})e^{-\beta_1 E_n - \beta_2 E_m} \phi_m(\boldsymbol{r}') \int \phi_n(\boldsymbol{r}_1)\phi_m^*(\boldsymbol{r}_1)\, d\boldsymbol{r}_1$$

$$= \sum_n \phi_n^*(\boldsymbol{r})e^{-(\beta_1 + \beta_2) E_n} \phi_n(\boldsymbol{r}') = \rho(\boldsymbol{r}, \boldsymbol{r}'; \beta_1 + \beta_2).$$

A special form of this equality is

$$\rho(\boldsymbol{r}, \boldsymbol{r}'; \beta) = \int \rho(\boldsymbol{r}, \boldsymbol{r}_1; \beta/2)\rho(\boldsymbol{r}_1, \boldsymbol{r}'; \beta/2)\, d\boldsymbol{r}_1, \qquad (14.83)$$

which can be written more generally as

$$\rho(\boldsymbol{r}, \boldsymbol{r}'; \beta) \qquad\qquad\qquad\qquad\qquad\qquad\qquad (14.84)$$

$$= \int \rho(\boldsymbol{r}, \boldsymbol{r}_1; \beta/n)\rho(\boldsymbol{r}_1, \boldsymbol{r}_2; \beta/n) \ldots \rho(\boldsymbol{r}_{n-1}, \boldsymbol{r}'; \beta/n)\, d\boldsymbol{r}_1, \ldots, d\boldsymbol{r}_{n-1}.$$

Applying this to the case $\boldsymbol{r} = \boldsymbol{r}'$, we see that

$$Q = \operatorname{tr}\rho \qquad\qquad\qquad\qquad\qquad\qquad\qquad\qquad (14.85)$$

$$= \int \rho(\boldsymbol{r}, \boldsymbol{r}_1; \beta/n)\rho(\boldsymbol{r}_1, \boldsymbol{r}_2; \beta/n) \ldots \rho(\boldsymbol{r}_{n-1}, \boldsymbol{r}; \beta/n)\, d\boldsymbol{r}\, d\boldsymbol{r}_1, \ldots, d\boldsymbol{r}_{n-1}.$$

Thus the partition function can be obtained by an integral over density matrices with the "high temperature" β/n; such density matrices can be approximated because of the small value in the exponent. This equality is used in *path integral Monte Carlo* methods to incorporate quantum distributions of "heavy" particles into simulations.

15

Lagrangian and Hamiltonian mechanics

15.1 Introduction

Classical mechanics is not only an approximation of quantum mechanics, valid for heavy particles, but historically it also forms the basis on which quantum-mechanical notions are formed. We also need to be able to describe mechanics in generalized coordinates if we wish to treat constraints or introduce other ways to reduce the number of degrees of freedom. The basis for this is Lagrangian mechanics, from which the Hamiltonian description is derived. The latter is not only used in the Schrödinger equation, but forms also the framework in which (classical) statistical mechanics is developed. A background in Lagrangian and Hamiltonian mechanics is therefore required for many subjects treated in this book.

After the derivation of Lagrangian and Hamiltonian dynamics, we shall consider how constraints can be built in. The common type of constraint is a *holonomic constraint* that depends only on coordinates, such as a bond length constraint, or constraints between particles that make them behave as one rigid body. An example of a non-holonomic constraint is the total kinetic energy (to be kept at a constant value or at a prescribed time-dependent value).

We shall only give a concise review; for details the reader is referred to text books on classical mechanics, in particular to Landau and Lifschitz (1982) and to Goldstein *et al.* (2002).

There are several ways to introduce the principles of mechanics, leading to Newton's laws that express the equations of motion of a mechanical system. A powerful and elegant way is to start with *Hamilton's principle of least action* as a postulate. This is the way chosen by Landau and Lifshitz (1982).

15.2 Lagrangian mechanics

Consider a system described by n *degrees of freedom* or coordinates $q = q_1, \ldots, q_n$ (not necessarily the $3N$ cartesian coordinates of N particles) that evolve in time t. A function $\mathcal{L}(q, \dot{q}, t)$ exists with the property that the *action*

$$S \overset{\text{def}}{=} \int_{t_1}^{t_2} \mathcal{L}(q, \dot{q}, t)\, dt \tag{15.1}$$

is minimal for the actual path followed by $q(t)$, given it is at coordinates $q(t_1)$ and $q(t_2)$ at the times t_1 and t_2. $\mathcal{L}$ is called the *Lagrangian* of the system.

This principle (see the proof below) leads to the *Lagrange equations*

$$\frac{d}{dt}\left(\frac{\partial \mathcal{L}}{\partial \dot{q}_i}\right) - \frac{\partial \mathcal{L}}{\partial q_i} = 0 \qquad (i = 1, \cdots, n). \tag{15.2}$$

The Lagrangian is not uniquely determined by this requirement because any (total) time derivative of some function of q and t will have an action independent of the path and can therefore be added to $\mathcal{L}$.

Proof We prove (15.2). The variation of the action (15.1), when the path between $q(t_1)$ and $q(t_2)$ is varied (but the end points are kept constant), must vanish if S is a minimum:

$$\delta S = \int_{t_1}^{t_2} \left(\frac{\partial \mathcal{L}}{\partial q}\delta q + \frac{\partial \mathcal{L}}{\partial \dot{q}}\delta \dot{q}\right) dt = 0.$$

Partial integration of the second term, with $\delta \dot{q} = d\delta q/dt$ and realizing that $\delta q = 0$ at both integration limits t_1 and t_2 because there q is kept constant, converts this second term to

$$-\int_{t_1}^{t_2} \delta q \frac{d}{dt}\left(\frac{\partial \mathcal{L}}{\partial \dot{q}}\right) dt.$$

Now

$$\delta S = \int_{t_1}^{t_2} \left[\frac{\partial \mathcal{L}}{\partial q} - \frac{d}{dt}\left(\frac{\partial \mathcal{L}}{\partial \dot{q}}\right)\right] \delta q\, dt = 0.$$

Since the variation must be zero for any choice of δq, (15.2) follows. $\qquad \square$

For a free particle with position $\boldsymbol{r}$ and velocity $\boldsymbol{v}$ the Lagrangian $\mathcal{L}$ can only be a function of v^2 if we assume *isotropy of space-time*, i.e., that mechanical laws do not depend on the position of the space and time origins and on the orientation in space. In fact, from the requirement that the particle behavior is the same in a coordinate system moving with constant velocity, it follows[1] that $\mathcal{L}$ must be proportional to v^2.

For a system of particles interacting through a position-dependent potential $V(\boldsymbol{r}_1, \ldots, \boldsymbol{r}_N)$, the following Lagrangian:

$$\mathcal{L}(\boldsymbol{r}, \boldsymbol{v}) = \sum_{i=1}^{N} \tfrac{1}{2} m_i v_i^2 - V, \tag{15.3}$$

yields Newtons equations of motion

$$m\dot{\boldsymbol{v}}_i = -\frac{\partial V}{\partial \boldsymbol{r}_i}, \tag{15.4}$$

as the reader can easily verify by applying the Lagrange equations of motion (15.2).

15.3 Hamiltonian mechanics

In many cases a more appropriate description of the equations of motions in generalized coordinates is obtained with the Hamilton formalism. We first define a *generalized momentum p_k, conjugate to the coordinate q_k* from the Lagrangian as

$$p_k \stackrel{\text{def}}{=} \frac{\partial \mathcal{L}}{\partial \dot{q}_k}. \tag{15.5}$$

Then we define a *Hamiltonian* $\mathcal{H}$ in such a way that $d\mathcal{H}$ is a total differential in dp and dq:

$$\mathcal{H} \stackrel{\text{def}}{=} \sum_{k=1}^{n} p_k \dot{q}_k - \mathcal{L}. \tag{15.6}$$

¿From this definition it follows that

$$d\mathcal{H} = \sum_{k=1}^{n} \left(p_k \, d\dot{q}_k + \dot{q}_k \, dp_k - \frac{\partial \mathcal{L}}{\partial q_k} \, dq_k - \frac{\partial \mathcal{L}}{\partial \dot{q}_k} \, d\dot{q}_k \right). \tag{15.7}$$

The first and the last terms cancel, so that a total differential in dp and dq is obtained, with the following derivatives:

$$\frac{\partial \mathcal{H}}{\partial p_k} = \dot{q}_k, \tag{15.8}$$

$$\frac{\partial \mathcal{H}}{\partial q_k} = -\dot{p}_k. \tag{15.9}$$

These are *Hamilton's equations of motion.*

[1] See Landau and Lifshitz, (1982), Chapter 1.

The reader may check that these also lead to Newton's equations of motion for a system of particles interacting through a coordinate-dependent potential V, where

$$\mathcal{H}(p, q) = \frac{1}{2} \sum_{k=1}^{n} \frac{p_k^2}{2m_k} + V(q). \tag{15.10}$$

In this case $\mathcal{H}$ is the *total energy* of the system of particles, composed of the kinetic energy[2] K and potential energy V.

If $\mathcal{H}$ does not depend explicitly on time, it is a *constant of the motion*, since

$$\frac{d\mathcal{H}}{dt} = \sum_{k} \left(\frac{\partial \mathcal{H}}{\partial p_k} \dot{p}_k + \frac{\partial \mathcal{H}}{\partial q_k} \dot{q}_k \right)$$

$$= \sum_{k} (\dot{q}_k \dot{p}_k - \dot{p}_k \dot{q}_k) = 0. \tag{15.11}$$

So Hamiltonian mechanics conserves the value of $\mathcal{H}$, or – in the case of an interaction potential that depends on position only – the total energy. Therefore, in the latter case such a system is also called *conservative*.

15.4 Cyclic coordinates

A coordinate q_k is called *cyclic* if the Lagrangian does not depend on q_k:

$$\frac{\partial \mathcal{L}(q, \dot{q}, t)}{\partial q_k} = 0 \text{ for cyclic } q_k. \tag{15.12}$$

For the momentum $p_k = \partial \mathcal{L}/\partial \dot{q}_k$, conjugate to a cyclic coordinate, the time derivative is zero: $\dot{p}_k = 0$. Therefore: *The momentum conjugate to a cyclic coordinate is conserved, i.e., it is a constant of the motion.*

An example, worked out in detail in Section 15.6, is the center-of-mass motion of a system of mutually interacting particles isolated in space: since the Lagrangian cannot depend on the position of the center of mass, its coordinates are cyclic, and the conjugate momentum, which is the total linear momentum of the system, is conserved. Hence the center of mass can only move with constant velocity.

It is not always true that the cyclic coordinate itself is moving with constant velocity. An example is the motion of a diatomic molecule in free space, where the rotation angle (in a plane) is a cyclic coordinate. The conjugate momentum is the angular velocity multiplied by the moment of

---

[2] We use the notation K for the kinetic energy rather than the usual T in order to avoid confusion with the temperature T.

inertia of the molecule. So, if the bond distance changes, e.g., by vibration, the moment of inertia changes, and the angular velocity changes as well.

A coordinate that is *constrained* to a constant value, by some property of the system itself or by an external action, also acts as a cyclic coordinate, because it is not really a variable any more. However, its time derivative is also zero, and such a coordinate vanishes from the Lagrangian altogether. In Section 15.8 the equations of motion for a system with constraints will be considered in detail.

15.5 Coordinate transformations

Consider a transformation from cartesian coordinates r to general coordinates q:

$$r_i = r_i(q_1, \ldots, q_n), \ i = 1, \ldots, N, \ n = 3N. \tag{15.13}$$

The kinetic energy can be written in terms of q:

$$K = \sum_{i=1}^{N} \frac{1}{2} m_i \dot{r}^2 = \frac{1}{2} \sum_{k,l=1}^{n} \sum_{i=1}^{N} m_i \frac{\partial r_i}{\partial q_k} \cdot \frac{\partial r_i}{\partial q_l} \dot{q}_k \dot{q}_l. \tag{15.14}$$

This is a quadratic form that can be expressed in matrix notation:[3]

$$K(q, \dot{q}) = \tfrac{1}{2} \dot{\mathbf{q}}^\mathsf{T} \mathbf{M}(q) \dot{\mathbf{q}}, \tag{15.15}$$

where

$$M_{kl} = \sum_{i=1}^{N} m_i \frac{\partial r_i}{\partial q_k} \cdot \frac{\partial r_i}{\partial q_l}. \tag{15.16}$$

The tensor $\mathbf{M}$, defined in (15.16), is called the *mass tensor* or sometimes the *mass-metric tensor*.[4] The matrix $\mathbf{M}(q)$ is in general a function of the coordinates; it is symmetric and invertible ($\det \mathbf{M} \neq 0$). Its eigenvalues are the masses m_i, each three-fold degenerate.

Now we consider a conservative system

$$\mathcal{L}(q, \dot{q}) = K(q, \dot{q}) - V(q). \tag{15.17}$$

---

[3] We use roman bold type for matrices, a vector being represented by a column matrix, in contrast to italic bold type for vectors. For example: $v \cdot w = \mathbf{v}^\mathsf{T} \mathbf{w}$. The superscript T denotes the transpose of the matrix.

[4] The latter name refers to the analogy with the metric tensor $g_{kl} = \sum_i [(\partial r_i/\partial q_k) \cdot (\partial r_i/\partial q_l)]$ which defines the *metric* of the generalized coordinate system: the distance ds between q and $q + dq$ is given by $(ds)^2 = \sum_{kl} g_{kl} \, dq_k \, dq_l$.

The conjugate momenta are defined by

$$p_k = \frac{\partial K(q, \dot{q})}{\partial q_k} = \sum_l M_{kl} \dot{q}_l \tag{15.18}$$

or

$$\mathbf{p} = \mathbf{M}\dot{\mathbf{q}}, \tag{15.19}$$

and the Lagrangian equations of motion are

$$\dot{p}_k = \frac{\partial \mathcal{L}}{\partial q_k} = \frac{1}{2}\dot{\mathbf{q}}^\mathsf{T}\frac{\partial \mathbf{M}}{\partial q_k}\dot{\mathbf{q}} - \frac{\partial V}{\partial q_k}. \tag{15.20}$$

By inserting (15.20) into (15.19) a matrix equation is obtained for $\ddot{q}$:

$$\sum_l M_{kl}\ddot{q}_l = -\frac{\partial V}{\partial q_k} + \sum_{\alpha,\beta}\left(\frac{1}{2}\frac{\partial M_{\alpha\beta}}{\partial q_k} - \frac{\partial M_{k\alpha}}{\partial q_\beta}\right)\dot{q}_\alpha\dot{q}_\beta, \tag{15.21}$$

which has the general form

$$\mathbf{M}\ddot{\mathbf{q}} = \mathbf{T}(\mathbf{q}) + \mathbf{C}(\mathbf{q}, \dot{\mathbf{q}}), \tag{15.22}$$

where $\mathbf{T}$ is a generalized force or torque, and $\mathbf{C}$ is a velocity-dependent force that comprises the Coriolis and centrifugal forces. Apart from the fact that these forces are hard to evaluate, we are confronted with a set of equations that require a complexity of order n^3 to solve. Recently more efficient order-n algorithms have been devised as a result of developments in robotics.

By inverting (15.19) to $\dot{\mathbf{q}} = \mathbf{M}^{-1}\mathbf{p}$, the kinetic energy can be written in terms of $\mathbf{p}$ (using the symmetry of $\mathbf{M}$):

$$K = \tfrac{1}{2}(\mathbf{M}^{-1}\mathbf{p})^\mathsf{T}\mathbf{M}(\mathbf{M}^{-1}\mathbf{p}) = \tfrac{1}{2}\mathbf{p}^\mathsf{T}\mathbf{M}^{-1}\mathbf{p}, \tag{15.23}$$

and the Hamiltonian becomes

$$\mathcal{H} = \mathbf{p}^\mathsf{T}\dot{\mathbf{q}} - \mathcal{L} = \mathbf{p}^\mathsf{T}\mathbf{M}^{-1}\mathbf{p} - K + V = \tfrac{1}{2}\mathbf{p}^\mathsf{T}\mathbf{M}^{-1}(q)\mathbf{p} + V(q), \tag{15.24}$$

with the Hamiltonian equations of motion

$$\dot{q}_k = \frac{\partial \mathcal{H}}{\partial p_k} = (\mathbf{M}^{-1}\mathbf{p})_k, \tag{15.25}$$

$$\dot{p}_k = -\frac{\partial \mathcal{H}}{\partial q_k} = -\frac{1}{2}\mathbf{p}^\mathsf{T}\frac{\partial \mathbf{M}^{-1}}{\partial q_k}\mathbf{p} - \frac{\partial V}{\partial q_k}. \tag{15.26}$$

(Parenthetically we note that the equivalence of the kinetic energy in (15.20) and (15.26) implies that

$$\frac{\partial \mathbf{M}^{-1}}{\partial q_k} = -\mathbf{M}^{-1}\frac{\partial \mathbf{M}}{\partial q_k}\mathbf{M}^{-1}, \tag{15.27}$$

which also follows immediately from $\partial \mathbf{M}\mathbf{M}^{-1}/\partial q_k = 0$. We will use this relation in ano ther context.)

The term $-\partial V/\partial q_k$ is a direct transformation of the cartesian forces $\mathbf{F}_i = -\partial V/\partial \mathbf{r}_i$:

$$-\frac{\partial V}{\partial q_k} = \sum_i \mathbf{F}_i \cdot \frac{\partial \mathbf{r}_i}{\partial q_k}, \tag{15.28}$$

and is therefore a kind of generalized force on q_k. Note, however, that in general the time derivative of p_k is not equal to this generalized force! Equality is only valid in the case that the mass tensor is independent of q_k:

$$\text{if } \frac{\partial \mathbf{M}}{\partial q_k} = 0 \text{ then } \dot{p}_k = -\frac{\partial V}{\partial q_k} = \sum_i \mathbf{F}_i \cdot \frac{\partial \mathbf{r}_i}{\partial q_k}. \tag{15.29}$$

15.6 Translation and rotation

Consider a system of N particles that interact mutually under a potential $V^{\text{int}}(\mathbf{r}_1, \ldots, \mathbf{r}_N)$, and are in addition subjected to an external potential $V^{\text{ext}}(\mathbf{r}_1, \ldots, \mathbf{r}_N)$. Homogeneity and isotropy of space dictate that neither the kinetic energy K nor the internal potential V^{int} can depend on the overall position and orientation of the system of particles. As we shall see, these properties give a special meaning to the six generalized coordinates of the overall position and orientation. Their motion is determined exclusively by the external potential. In the absence of an external potential these coordinates are cyclic, and their conjugate moments – which are the total linear and angular momentum – will be conserved. It also follows that a general three-dimensional N-body system has no more than $3N - 6$ internal coordinates.[5]

15.6.1 Translation

Consider[6] a transformation from $\mathbf{r}$ to a coordinate system in which q_1 is a displacement of all coordinates in the direction specified by a unit vector $\mathbf{n}$: $d\mathbf{r}_i = \mathbf{n}\, dq_1$. Hence, for any i,

$$\frac{\partial \mathbf{r}_i}{\partial q_1} = \mathbf{n}, \qquad \frac{\partial \dot{\mathbf{r}}_i}{\partial \dot{q}_1} = \mathbf{n}. \tag{15.30}$$

---

[5] For a system consisting of particles on a straight line, as a diatomic molecule, one of the rotational coordinates does not exist and so there will be at most $3N - 5$ internal coordinates.
[6] We follow the line of reasoning by Goldstein (1980).

Homogeneity of space implies that

$$\frac{\partial K(q, \dot{q})}{\partial q_1} = 0, \qquad \frac{\partial V^{\text{int}}}{\partial q_1} = 0. \tag{15.31}$$

For the momentum p_1 conjugate to q_1 follows

$$p_1 = \frac{\partial K}{\partial \dot{q}_1} = \frac{1}{2} \sum_i m_i \frac{\partial \dot{r}_i^2}{\partial \dot{q}_1} = \sum_i m_i \dot{r}_i \cdot \frac{\partial \dot{r}_i}{\partial \dot{q}_1} = \boldsymbol{n} \cdot \sum_i m_i \boldsymbol{v}_i, \tag{15.32}$$

which is the component of the total linear momentum in the direction $\boldsymbol{n}$. Its equation of motion is

$$\dot{p}_1 = -\frac{\partial V^{\text{ext}}}{\partial q_1} = -\sum_i \frac{\partial V^{\text{ext}}}{\partial \boldsymbol{r}_i} \cdot \frac{\partial \boldsymbol{r}_i}{\partial q_1} = \left(\sum_i \boldsymbol{F}_i^{\text{ext}} \right) \cdot \boldsymbol{n}. \tag{15.33}$$

So the motion is governed by the total external force. In the absence of an external force, the total linear momentum will be conserved.

The direction of $\boldsymbol{n}$ is immaterial: therefore there are three independent general coordinates of this type. These are the components of the *center of mass* (c.o.m.)

$$\boldsymbol{r}_{\text{cm}} \stackrel{\text{def}}{=} \frac{1}{M} \sum_i m_i \boldsymbol{r}_i, \text{ with } M = \sum_i m_i, \tag{15.34}$$

with equation of motion

$$\dot{\boldsymbol{r}}_{\text{cm}} = \frac{1}{M} \sum_i m_i \boldsymbol{v}_i = \frac{1}{M} \sum_i \boldsymbol{F}^{\text{ext}}. \tag{15.35}$$

In other words: the c.o.m. behaves like a single particle with mass M.

15.6.2 Rotation

Consider a transformation from $\boldsymbol{r}$ to a coordinate system in which dq_1 is a rotation of the whole body over an infinitesimal angle around an axis in the direction specified by a unit vector $\boldsymbol{n}$: $d\boldsymbol{r}_i = \boldsymbol{n} \times \boldsymbol{r}_1 \, dq_1$. Hence, for any i

$$\frac{\partial \boldsymbol{r}_i}{\partial q_1} = \frac{\partial \dot{\boldsymbol{r}}_i}{\partial \dot{q}_1} = \boldsymbol{n} \times \boldsymbol{r}_i. \tag{15.36}$$

Isotropy of space implies that

$$\frac{\partial K(q, \dot{q})}{\partial q_1} = 0, \qquad \frac{\partial V^{\text{int}}}{\partial q_1} = 0. \tag{15.37}$$

For the momentum p_1 conjugate to q_1 it follows that[7]

$$p_1 = \frac{\partial K}{\partial \dot{q}_1} = \frac{1}{2} \sum_i m_i \frac{\partial \dot{r}_i^2}{\partial \dot{q}_1} = \sum_i m_i v_i \cdot (n \times r_i)$$

$$= \sum_i m_i (r_i \times v_i) \cdot n = L \cdot n, \tag{15.38}$$

where the *angular momentum* of the system. is given by

$$L \stackrel{\text{def}}{=} \sum_i m_i r_i \times v_i. \tag{15.39}$$

In general, L depends on the choice of the origin of the coordinate system, except when the total linear momentum is zero.

The equation of motion of p_1 is

$$\dot{p}_1 = -\frac{\partial V^{\text{ext}}}{\partial q_1} = \sum_i F_i^{\text{ext}} \cdot (n \times r_i) = \left(\sum_i r_i \times F_i^{\text{ext}} \right) \cdot n = T \cdot n, \tag{15.40}$$

where the *torque* exerted on the system is given by

$$T \stackrel{\text{def}}{=} \sum_i r_i \times F_i^{\text{ext}}. \tag{15.41}$$

Again, n can be chosen in any direction, so that the equation of motion for the angular momentum is

$$\dot{L} = T. \tag{15.42}$$

In general, T depends on the choice of the origin of the coordinate system, except when the total external force is zero.

If there is no external torque, the angular momentum is a constant of the motion.

15.7 Rigid body motion

We now consider the special case of a system of N particles in which the mutual interactions keeping the particles together confine the relative positions of the particles so strongly that the whole system may be considered as one rigid body. We now transform r to generalized coordinates that consist of three c.o.m. coordinates r_{cm}, three variables that define a real orthogonal 3×3 matrix R with determinant $+1$ (a proper rotation matrix) and $3N - 6$ internal coordinates. The latter are all constrained to a constant value and therefore do not figure in the Lagrangian.

---

[7] Use the vector multiplication rule $a \cdot (b \times c) = (a \times b) \cdot c$.

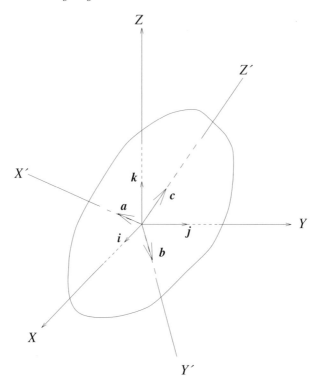

Figure 15.1 Body-fixed coordinate system $X'Y'Z'$ with base vectors $\boldsymbol{a}, \boldsymbol{b}, \boldsymbol{c}$, rotated with respect to a space-fixed coordinate system XYZ with base vectors $\boldsymbol{i}, \boldsymbol{j}, \boldsymbol{k}$.

Using a body-fixed coordinate system X', Y', Z' (see Fig. 15.1, the positions of the particles are specified by $3N$ coordinates $\mathbf{r}'_i$, which are all constants. There are, of course, only $3N - 6$ independent coordinates because six functions of the coordinates will determine the position and orientation of the body-fixed coordinate system. The rotation matrix $\mathbf{R}$ transforms the coordinates of the i-th particle, relative to the center of mass, in the space-fixed system to those in the body-fixed system:[8]

$$\mathbf{r}'_i = \mathbf{R}(\mathbf{r}_i - \mathbf{r}_{\mathrm{cm}}). \tag{15.43}$$

The positions in the space-fixed coordinate system X, Y, Z are given by

$$\mathbf{r}_i = \mathbf{r}_{cm} + \mathbf{R}^\mathsf{T}\mathbf{r}'_i. \tag{15.44}$$

---

[8] When consulting other literature, the reader should be aware of variations in notation that are commonly used. In particular the transformation matrix is often defined as the transpose of our $\mathbf{R}$, i.e., it transforms from the body-fixed to space-fixed coordinates, or the transformation is not defined as a rotation of the system of coordinate axes but rather as the rotation of a vector itself.

In Section 15.6 we have seen that the c.o.m. (defined in (15.34)) behaves independent of all other motions according to (15.35). We disregard the c.o.m. motion from now on, i.e., we choose the c.o.m. as the origin of both coordinate systems, and all vectors are relative to $\mathbf{r}_{cm}$.

The rotation matrix transforms the components $\mathbf{v}$ in the space-fixed system of *any* vector $\mathbf{v}$ to components $\mathbf{v}'$ in the body-fixed system as

$$\mathbf{v}' = \mathbf{R}\mathbf{v}, \qquad \mathbf{v} = \mathbf{R}^{\mathsf{T}}\mathbf{v}'. \tag{15.45}$$

The columns of $\mathbf{R}^{\mathsf{T}}$, or the rows of $\mathbf{R}$, are the components of the three body-fixed unit vectors $(1,0,0)$, $(0,1,0)$ and $(0,0,1)$ in the space-fixed system. They are the *direction cosines* between the axes of the two systems. Denoting the orthogonal unit base vectors of the space-fixed coordinate system by i, j, k and those of the body-fixed system by a, b, c, the rotation matrix is given by

$$\mathbf{R} = \begin{pmatrix} a_x & a_y & a_z \\ b_x & b_y & b_z \\ c_x & c_y & c_z \end{pmatrix} = \begin{pmatrix} a \cdot i & a \cdot j & a \cdot k \\ b \cdot i & b \cdot j & b \cdot k \\ c \cdot i & c \cdot j & c \cdot k \end{pmatrix}. \tag{15.46}$$

The rotational motion is described by the time-dependent behavior of the rotation matrix:

$$\mathbf{r}'_i(t) = \mathbf{R}(t)\mathbf{r}_i. \tag{15.47}$$

Therefore we need differential equations for the matrix elements of $\mathbf{R}(t)$ or for any set of variables that determine $\mathbf{R}$. There are at least three of those (such as the Euler angles that describe the orientation of the body-fixed coordinate axes in the space-fixed system), but one may also choose a redundant set of four to nine variables, which are then subject to internal constraint relations. If one would use the nine components of $\mathbf{R}$ itself, the orthogonality condition would impose six constraints. Expression in terms of Euler angles lead to the awkward *Euler equations*, which involve, for example, division by the sine of an angle leading to numerical problems for small angles in simulations. Another possibility is to use the homomorphism between 3D real orthogonal matrices and 2D complex unitary matrices,[9] leading to the Caley–Klein parameters[10] or to Wigner rotation matrices that are often used to express the effect of rotations on spherical harmonic functions. But all these are too complex for practical simulations. There are two recommended techniques, with the first being the most suitable one for the majority of cases:

---

[9] See, e.g., Jones (1990).
[10] See Goldstein (1980).

- The use of cartesian coordinates of (at least) two (linear system), three (planar system) or four (general 3D system) "particles" in combination with length constraints between them (see Section 15.8). The "particles" are *dummy* particles that have mass and fully represent the motion, while the points that are used to derive the forces (which are likely to be at the position of real atoms) now become *virtual* particles.

- The integration of angular velocity in a principal axes system, preferably combined with the use of *quaternions* to characterize the rotation. This method of solution is described below.

15.7.1 Description in terms of angular velocities

We know that the time derivative of the angular momentum (not the angular velocity!) is equal to the torque exerted on the body (see (15.42)), but that relation does not give us a

straightforward equation for the rate of change of the rotation matrix, which is related to the angular velocity.

First consider an infinitesimal rotation $d\phi = \omega\, dt$ of the body around an axis in the direction of the vector $\boldsymbol{\omega}$. For any point in the body $d\boldsymbol{r} = \boldsymbol{\omega} \times \boldsymbol{r}\, dt$ and hence $\boldsymbol{v} = \boldsymbol{\omega} \times \boldsymbol{r}$. Inserting this into the expression for the angular momentum $\boldsymbol{L}$ and using the vector relation

$$\boldsymbol{a} \times (\boldsymbol{b} \times \boldsymbol{c}) = (\boldsymbol{a} \cdot \boldsymbol{c})\boldsymbol{b} - (\boldsymbol{a} \cdot \boldsymbol{b})\boldsymbol{c}, \tag{15.48}$$

we obtain

$$
\begin{aligned}
\boldsymbol{L} &= \sum_i m_i \boldsymbol{r}_i \times \boldsymbol{v}_i \\
&= \sum_i m_i \boldsymbol{r}_i \times (\boldsymbol{\omega} \times \boldsymbol{r}) \\
&= \sum_i m_i [r_i^2 \boldsymbol{\omega} - \boldsymbol{r}_i(\boldsymbol{r}_i \cdot \boldsymbol{\omega})] \tag{15.49} \\
&= \mathbf{I}\boldsymbol{\omega}, \tag{15.50}
\end{aligned}
$$

where $\mathbf{I}$ is the *moment of inertia* or *inertia tensor*, which is represented by a 3×3 symmetric matrix

$$\mathbf{I} = \sum_i m_i (r_i^2 \mathbf{1} - \mathbf{r}_i \mathbf{r}_i^\mathsf{T}), \tag{15.51}$$

written out as

$$I = \sum_i m_i \begin{pmatrix} y_i^2 + z_i^2 & -x_i y_i & -x_i z_i \\ -y_i x_i & x_i^2 + z_i^2 & -y_i z_i \\ -z_i x_i & -z_i y_i & x_i^2 + y_i^2 \end{pmatrix}. \tag{15.52}$$

Since $\mathbf{I}$ is a tensor and $\boldsymbol{L}$ and $\boldsymbol{\omega}$ are vectors,[11] the relation $\mathbf{L} = \mathbf{I}\boldsymbol{\omega}$ is valid in any (rotated) coordinate system. For example, this relation is also valid for the primed quantities in the body-fixed coordinates:

$$\mathbf{L} = \mathbf{I}\boldsymbol{\omega}, \qquad \mathbf{L}' = \mathbf{I}'\boldsymbol{\omega}', \tag{15.53}$$

$$\mathbf{L}' = \mathbf{R}\mathbf{L}, \qquad \boldsymbol{\omega}' = \mathbf{R}\boldsymbol{\omega}, \qquad \mathbf{I}' = \mathbf{R}\mathbf{I}\mathbf{R}^\mathsf{T}. \tag{15.54}$$

However, we must be careful when we transform a vector $\boldsymbol{v}'$ from a rotating coordinate system: the rotation itself produces an extra term $\boldsymbol{\omega} \times \boldsymbol{v}$ in the time derivative of $\boldsymbol{v}$:

$$\dot{\boldsymbol{v}} = \mathbf{R}^\mathsf{T}\dot{\boldsymbol{v}}' + \boldsymbol{\omega} \times \boldsymbol{v}. \tag{15.55}$$

It is possible to relate the time derivative of the rotation matrix to the angular momentum in the body-fixed system. Since, for an arbitrary vector $\mathbf{v} = \mathbf{R}^\mathsf{T}\mathbf{v}'$, the derivative is

$$\dot{\mathbf{v}} = \frac{d}{dt}\mathbf{R}^\mathsf{T}\mathbf{v}' = \mathbf{R}^\mathsf{T}\dot{\boldsymbol{v}}' + \dot{\mathbf{R}}^\mathsf{T}\boldsymbol{v}', \tag{15.56}$$

which, comparing with (15.55), means that

$$\dot{\mathbf{R}}^\mathsf{T}\boldsymbol{v}' = \boldsymbol{\omega} \times \boldsymbol{v}. \tag{15.57}$$

Hence

$$\dot{\mathbf{R}}^\mathsf{T}\boldsymbol{v}' = \mathbf{R}^\mathsf{T}(\boldsymbol{\omega}' \times \boldsymbol{v}') = \mathbf{R}^\mathsf{T}\boldsymbol{\Omega}'\boldsymbol{v}', \tag{15.58}$$

where $\boldsymbol{\Omega}'$ is a second-rank antisymmetric tensor:

$$\boldsymbol{\Omega}' \overset{\text{def}}{=} \begin{pmatrix} 0 & -\omega_z' & \omega_y' \\ \omega_z' & 0 & -\omega_x' \\ -\omega_y' & \omega_x' & 0 \end{pmatrix}, \tag{15.59}$$

and thus

$$\dot{\mathbf{R}}^\mathsf{T} = \mathbf{R}^\mathsf{T}\boldsymbol{\Omega}'. \tag{15.60}$$

---

[11] Tensors and vectors are defined by their transformation properties under orthogonal transformations. In fact, $\boldsymbol{L}$ and $\boldsymbol{\omega}$ are both *pseudovectors* or *axial vectors* because they change sign under an orthogonal transformation with determinant $-1.$, but this distinction with proper vectors is not relevant in our context.

Other ways of expressing $\dot{\mathbf{R}}$ are

$$\dot{\mathbf{R}} = \mathbf{\Omega}'^{\mathsf{T}}\mathbf{R}, \tag{15.61}$$

$$\dot{\mathbf{R}}^{\mathsf{T}} = \mathbf{\Omega}\mathbf{R}^{\mathsf{T}}, \tag{15.62}$$

$$\dot{\mathbf{R}} = \mathbf{R}\mathbf{\Omega}^{\mathsf{T}}. \tag{15.63}$$

Recall the equation of motion (15.42) for the angular momentum; in matrix notation in the *space-fixed* coordinate system this equation is:

$$\dot{\mathbf{L}} = \frac{d}{dt}(\mathbf{I}\boldsymbol{\omega}) = \mathbf{T}. \tag{15.64}$$

Only in the body-fixed coordinates is $\mathbf{I}$ stationary and constant, so we must refer to those coordinates to avoid time dependent moments of inertia:

$$\frac{d}{dt}\mathbf{R}^{\mathsf{T}}\mathbf{I}'\boldsymbol{\omega}' = \mathbf{T}, \tag{15.65}$$

or

$$\dot{\mathbf{R}}^{\mathsf{T}}\mathbf{I}'\boldsymbol{\omega}' + \mathbf{R}^{\mathsf{T}}\mathbf{I}'\dot{\boldsymbol{\omega}}' = \mathbf{T},$$
$$\mathbf{I}'\dot{\boldsymbol{\omega}}' = \mathbf{R}\mathbf{T} - \mathbf{R}\dot{\mathbf{R}}^{\mathsf{T}}\mathbf{I}'\boldsymbol{\omega}',$$
$$\mathbf{I}'\dot{\boldsymbol{\omega}}' = \mathbf{T}' - \mathbf{\Omega}'\mathbf{I}'\boldsymbol{\omega}'. \tag{15.66}$$

The latter equation enables us to compute the angular accelerations $\dot{\boldsymbol{\omega}}$. This equation becomes particularly simple if the body-fixed, primed, coordinate system is chosen such that the moment of inertia matrix $\mathbf{I}'$ is diagonal. The equation of motion for the rotation around the principal X'-axis then is

$$\dot{\omega}'_x = \frac{T'_x}{I'_{xx}} + \frac{I'_{yy} - I'_{zz}}{I'_{xx}}\omega'_y\omega'_z, \tag{15.67}$$

and the equations for the y- and z-component follow from cyclic permutation of x, y, z in this equation. Thus the angular acceleration in the body-fixed coordinates can be computed, provided both the angular velocity and the torque are known in the body-fixed frame. For the latter we must know the rotation matrix, which means that the time-dependence of the rotation matrix must be solved simultaneously, e.g., from (15.61).

The *kinetic energy* can best be calculated from the angular velocity in the body-fixed coordinate system. It is given by

$$K = \tfrac{1}{2}\sum_i m_i \dot{r}_i^2 = \tfrac{1}{2}\boldsymbol{\omega} \cdot \mathbf{L} = \tfrac{1}{2}\boldsymbol{\omega} \cdot (\mathbf{I} \cdot \boldsymbol{\omega})$$
$$= \tfrac{1}{2}\boldsymbol{\omega}^{\mathsf{T}}\mathbf{I}\boldsymbol{\omega} = \tfrac{1}{2}\boldsymbol{\omega}'^{\mathsf{T}}\mathbf{I}'\boldsymbol{\omega}' = \tfrac{1}{2}\sum_{\alpha=1}^{3} I'_\alpha \omega'^2_\alpha \text{ (principal axes).} \tag{15.68}$$

Proof Since $\dot{\boldsymbol{r}}_i = \boldsymbol{\omega} \times \boldsymbol{r}_i$ the kinetic energy is

$$K = \frac{1}{2} \sum_i m_i \dot{r}_i^2 = \frac{1}{2} \sum_i m_i (\boldsymbol{\omega} \times \boldsymbol{r}_i)^2.$$

With the general vector rule

$$(\boldsymbol{a} \times \boldsymbol{b}) \cdot (\boldsymbol{c} \times \boldsymbol{d}) = (\boldsymbol{a} \cdot \boldsymbol{b})(\boldsymbol{a} \cdot \boldsymbol{b}) - (\boldsymbol{a} \cdot \boldsymbol{d})(\boldsymbol{b} \cdot \boldsymbol{c}), \qquad (15.69)$$

this can be written as

$$K = \frac{1}{2} \sum_i m_i [\omega^2 r_i^2 - (\boldsymbol{\omega} \cdot \boldsymbol{r}_i)^2] = \frac{1}{2} \boldsymbol{\omega} \cdot \boldsymbol{L},$$

using the expression (15.49) for $\boldsymbol{L}$. □

Let us summarize the steps to obtain the equations for simulation of the angular motion of a rigid body:

(i) Determine the center of mass $\boldsymbol{r}_{\mathrm{cm}}$ and the total mass M of the body.
(ii) Determine the moment of inertia of the body and its principal axes. If not obvious from the symmetry of the body, then first obtain $\mathbf{I}$ in an arbitrary body-fixed coordinate system (see (15.52)) and diagonalize it subsequently.
(iii) Assume the initial (cartesian) coordinates and velocities of the constituent particles are known in a given space-fixed coordinate system. From these determine the initial rotation matrix $\mathbf{R}$ and angular velocity $\boldsymbol{\omega}$.
(iv) Determine the forces $\boldsymbol{F}_i$ on the constituent particles, and the total force $\boldsymbol{F}_{\mathrm{tot}}$ on the body.
(v) Separate the c.o.m. motion: $\boldsymbol{F}_{\mathrm{tot}}/M$ is the acceleration of the c.o.m., and thus of all particles. Subtract $m_i \boldsymbol{F}_{\mathrm{tot}}/M$ from every force $\boldsymbol{F}_i$.
(vi) Compute the total torque $\boldsymbol{T}'$ in the body-fixed principal axes coordinate system, best by first transforming the particle forces (after c.o.m. correction) with $\mathbf{R}$ to the body-fixed frame, to obtain $\boldsymbol{F}'_i$. Then apply

$$\boldsymbol{T}' = \sum_i \boldsymbol{F}'_i \times \boldsymbol{r}'_i. \qquad (15.70)$$

(vii) Determine the time derivative of $\boldsymbol{\omega}$ from (15.67) and the time derivative of $\mathbf{R}$ from (15.61). Integrate both equations simultaneously with an appropriate algorithm.

The differential equation for the rotation matrix (last step) can be cast in several forms, depending on the definition of the variables used to describe the matrix. In the next section we consider three different possibilities.

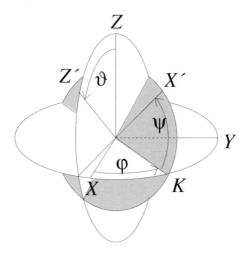

Figure 15.2 Euler angles defining the rotation of coordinate axes XYZ to $X'Y'Z'$.
For explanation see text.

15.7.2 Unit vectors

A straightforward solution of the equation of motion for $\mathbf{R}$ is obtained by
applying (15.61) directly to its nine components, i.e., to the body-fixed basis
vectors a, b, c. There are two possibilities: either one rotates $\boldsymbol{\omega}'$ to the space-
fixed axes and then applies (15.61):

$$\boldsymbol{\omega} = \mathbf{R}^{\mathsf{T}}\boldsymbol{\omega}', \tag{15.71}$$

$$\dot{a} = \boldsymbol{\omega} \times a \tag{15.72}$$

(and likewise for b and c), or one applies (15.63) term for term. Both
methods preserve the orthonormality of the three basis vectors, but in a
numerical simulation the matrix may drift slowly away from orthonormality
by integration errors. Provisions must be made to correct such drift.

15.7.3 Euler angles

The traditional description of rotational motion is in *Euler angles*, of which
there are three and no problem with drifting constraints occur. The Euler
angles do have other severe problems related to the singularity of the Euler
equations when the second Euler angle ϑ has the value zero (see below). For
that reason Euler angles are not popular for simulations and we shall only
give the definitions and equations of motion for the sake of completeness.

The Euler angles φ, ϑ, ψ, defined as follows (see Fig. 15.2). We rotate
XYZ in three consecutive rotations to $X'Y'Z'$. First locate the *line of*

nodes K which is the line of intersection of the XY- and the $X'Y'$-planes (there are two directions for K; the choice is arbitrary). Then:

(i) rotate XYZ around Z over an angle φ until X coincides with K;
(ii) rotate around K over an angle ϑ until Z coincides with Z';
(iii) rotate around Z' over an angle ψ until K coincides with X'.

Rotations are positive in the sense of a right-hand screw (make a fist and point the thumb of your right hand in the direction of the rotation axis; your fingers bend in the direction of positive rotation). The rotation matrix is

$$
\mathbf{R} = \begin{pmatrix}
\cos\varphi\cos\psi+ & \sin\varphi\cos\psi+ & +\sin\vartheta\sin\psi \\
-\sin\varphi\cos\vartheta\sin\psi & \cos\varphi\cos\vartheta\sin\psi & \\
-\cos\varphi\sin\psi+ & -\sin\varphi\sin\psi+ & +\sin\vartheta\cos\psi \\
-\sin\varphi\cos\vartheta\cos\psi & \cos\varphi\cos\vartheta\cos\psi & \\
\sin\varphi\sin\vartheta & -\cos\varphi\sin\vartheta & \cos\vartheta
\end{pmatrix}. \tag{15.73}
$$

The equations of motion, relating the angular derivatives to the body-fixed angular velocities, can be derived from (15.61). They are

$$
\begin{pmatrix} \dot\varphi \\ \dot\vartheta \\ \dot\psi \end{pmatrix} = \begin{pmatrix}
\frac{\sin\varphi\cos\vartheta}{\sin\vartheta} & \frac{\cos\varphi\cos\vartheta}{\sin\vartheta} & 1 \\
\cos\varphi & \sin\varphi & \\
\frac{\sin\varphi}{\sin\vartheta} & -\frac{\cos\varphi}{\sin\vartheta} &
\end{pmatrix} \begin{pmatrix} \omega'_x \\ \omega'_y \\ \omega'_z \end{pmatrix}. \tag{15.74}
$$

15.7.4 Quaternions

Quaternions[12] $[q_0, q_1, q_2, q_3] = q_0 + q_1 i + q_2 j + q_3 k$ are hypercomplex numbers with four real components that can be viewed as an extension of the complex numbers $a + bi$. They were invented by Hamilton (1844). The *normalized quaternions*, with $\sum_i q_i^2 = 1$, can be conveniently used to describe 3D rotations; these are known as the *Euler–Rodrigues parameters*, described a few years before Hamilton's quaternions by Rodrigues (1840). Subsequently the quaternions have received almost no attention in the mechanics of molecules,[13] until they were revived by Evans (1977). Because equations of motion using quaternions do not suffer from singularities as

---

[12] See Altmann (1986) for an extensive review of quaternions and their use in expressing 3D rotations, including a survey of the historical development. Another general reference is Kyrala (1967).

[13] In the dynamics of macroscopic multi-body systems the Euler–Rodrigues parameters are well-known, but not by the name "quaternion". See for example Shabana (1989). They are commonly named *Euler parameters*; they differ from the three *Rodrigues* parameters, which are defined as the component of the unit vector $\mathbf{N}$ multiplied by $\tan(\frac{1}{2}\phi)$ (see (15.79)).

those with Euler angles do, and do not involve the computation of gonio-metric functions, they have become popular in simulations involving rigid bodies (Allen and Tildesley, 1987; Fincham, 1992).

The unit quaternions[14] $1 = [1,0,0,0]$, $i = [0,1,0,0]$, $j = [0,0,1,0]$ and $k = [0,0,0,1]$ obey the following multiplication rules ($[q]$ is any quaternion):

$$1[q] = [q]1 = [q], \tag{15.75}$$

$$i^2 = j^2 = k^2 = -1, \tag{15.76}$$

$$ij = -ji = k \quad \text{and cyclic permutations.} \tag{15.77}$$

A general quaternion can be considered as the combination $[q_0, \boldsymbol{Q}]$ of a scalar q_0 and a vector $\boldsymbol{Q}$ with components (q_1, q_2, q_3). The multiplication rules then imply (as the reader is invited to check) that

$$[a, \boldsymbol{A}][b, \boldsymbol{B}] = [ab - \boldsymbol{A} \cdot \boldsymbol{B}, a\boldsymbol{B} + b\boldsymbol{A} + \boldsymbol{A} \times \boldsymbol{B}]. \tag{15.78}$$

According to Euler's famous theorem,[15] any 3D rotation with fixed origin can be characterized as single rotation about an axis $\boldsymbol{n}$ over an angle ϕ: $\mathbf{R}(\phi, \boldsymbol{n})$. Thus four parameters ϕ, n_x, n_y, n_z describe a rotation, with the constraint that $|\boldsymbol{n}| = 1$. Note that $\mathbf{R}(-\phi, -\boldsymbol{n})$ and $\mathbf{R}(\phi, \boldsymbol{n})$ represent the same rotation. The Euler–Rodrigues parameters are expressions of these four parameters:

$$[q] = [\cos \tfrac{1}{2}\phi, \boldsymbol{n} \sin \tfrac{1}{2}\phi]. \tag{15.79}$$

They are indeed quaternions that obey the multiplication rule (15.78). They should be viewed as rotation operators, the product [a][b] meaning the sequential operation of first [b], then [a]. Beware that $[q]$ rotates a vector in a given coordinate system and not the coordinate system itself, which implies that $[q]$ is to be identified with $\mathbf{R}^\mathsf{T}$ and not with $\mathbf{R}$. The unit quaternions now have the meaning of the identity operator ($[1, 0, 0, 0]$), and rotations by π about the x-, y- and z-axes, respectively. Such unit rotations are called *binary rotations*. Note that the equivalent rotations $(-\phi, -\boldsymbol{n})$ and $(\phi, \boldsymbol{n})$ are given by the same quaternion. Note also that the full range of all normalized quaternions $(-1 < q_i \leq +1)$ includes all possible rotations twice. The inverse of a normalized quaternion obviously is a rotation about the same axis but in opposite direction:

$$[q_0, \boldsymbol{Q}]^{-1} = [q_0, -\boldsymbol{Q}]. \tag{15.80}$$

---

[14] We shall use the notation $[q]$ or $[q_0, q_1, q_2, q_3]$ or $[\mathrm{q}, \boldsymbol{Q}]$ for quaternions.

[15] Euler's theorem: "Two arbitrarily oriented orthonormal bases with common origin P can be made to coincide with one another by rotating one of them through a certain angle about an axis which is passing through P and which has the direction of the eigenvector $\boldsymbol{n}$ of the rotation matrix" (Wittenburg, 1977).

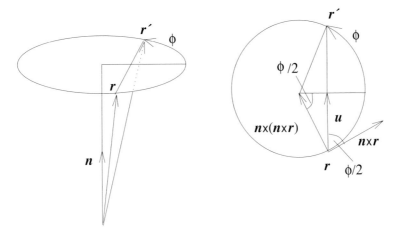

Figure 15.3 Rotation of the vector r to r' by a rotation over an angle ϕ about an axis n.

We now seek the relation between quaternions and the rotation matrix or the Euler angles. The latter is rather simple if we realize that the rotation expressed in Euler angles $(\varphi, \vartheta, \psi)$ is given by

$$\mathbf{R} = \mathbf{R}(\psi, k)\mathbf{R}(\vartheta, i)\mathbf{R}(\varphi, k) \tag{15.81}$$
$$= [\cos\tfrac{1}{2}\psi, 0, 0, \sin\tfrac{1}{2}\psi]\,[\cos\tfrac{1}{2}\vartheta, \sin\tfrac{1}{2}\vartheta, 0, 0]\,[\cos\tfrac{1}{2}\varphi, 0, 0, \sin\tfrac{1}{2}\varphi]$$
$$= [q_0, q_1, q_2, q_3] \tag{15.82}$$

with

$$q_0 = \cos\tfrac{1}{2}\vartheta \cos\tfrac{1}{2}(\varphi + \psi), \tag{15.83}$$
$$q_1 = \sin\tfrac{1}{2}\vartheta \cos\tfrac{1}{2}(\varphi - \psi), \tag{15.84}$$
$$q_2 = \sin\tfrac{1}{2}\vartheta \cos\tfrac{1}{2}(\varphi - \psi), \tag{15.85}$$
$$q_3 = \cos\tfrac{1}{2}\vartheta \cos\tfrac{1}{2}(\varphi + \psi). \tag{15.86}$$

The relation with the rotation matrix itself can be read from the transformation of an arbitrary vector r to r' by a general rotation over an angle ϕ about an axis n, given by the quaternion (15.79). We must be careful with the definition of rotation: here we consider a vector to be rotated in a fixed coordinate system, while $\mathbf{R}$ defines a transformation of the components of a fixed vector to a rotated system of coordinate axes. Each rotation is the transpose of the other, and the present vector rotation is given by $r' = \mathbf{R}^T r$. Refer to Fig. 15.3. Define a vector u, which can be constructed from the two perpendicular vectors $n \times (n \times r)$ and $n \times r$:

$$u = \sin\tfrac{1}{2}\phi \left[(n \times (n \times r))\sin\tfrac{1}{2}\phi + (n \times r)\cos\tfrac{1}{2}\phi\right].$$

With

$$r' = r + 2u = \mathbf{R}^\mathsf{T} r$$

the rotation matrix can be written out and we obtain

$$\mathbf{R} = \begin{pmatrix} q_0^2 + q_1^2 - q_2^2 - q_3^2 & 2(q_1 q_2 + q_0 q_3) & 2(q_1 q_3 - q_0 q_2) \\ 2(q_1 q_2 - q_0 q_3) & q_0^2 - q_1^2 + q_2^2 - q_3^2 & 2(q_2 q_3 + q_0 q_1) \\ 2(q_3 q_1 + q_0 q_2) & 2(q_3 q_2 - q_0 q_1) & q_0^2 - q_1^2 - q_2^2 + q_3^2 \end{pmatrix}.$$
$$(15.87)$$

Note Two remarks will be made. The first concerns the *reverse* relation, from $\mathbf{R}$ to $[q]$, and the second concerns the symmetry of $[q]$ and its use in generating random rotations.

The reverse relation is determined by the fact that $\mathbf{R}$ has the eigenvalues 1, $\exp(i\phi)$ and $\exp(-i\phi)$, and that the eigenvector belonging to the eigenvalue 1 is n. So, once these have been determined, $[q]$ follows from (15.79).

As can be seen from (15.87), the four quantities in $[q]$ play an almost, but not quite symmetric role: q_0 differs from q_1, q_2, q_3. This is important if we would fancy to generate a *random* rotation. The correct procedure, using quaternions, would be to generate a vector n randomly distributed over the unit sphere, then generate a random angle ϕ in the range $(0, 2\pi)$, and construct the quaternion from (15.79). The vector n could be obtained[16] by choosing three random numbers x, y, z from a homogeneous distribution between -1 and $+1$, computing $r^2 = x^2 + y^2 + z^2$, discarding all triples for which $r^2 > 1$, and scaling all accepted triples to unit length by division by r. Such a procedure cannot be used for the four dimensions of $[q]$, simply because q_0 is different, e.g., for a random 3D rotation $\langle q_0^2 \rangle = 1/2$ while $\langle q_1^2 \rangle = \langle q_2^2 \rangle = \langle q_3^2 \rangle = 1/6$. To generate a small random rotation in a given restricted range, as might be required in a Monte Carlo procedure, first generate a random n, then choose a value of ϕ in the range $\Delta\phi$ as required (both $(0, \Delta\phi)$ and $(-\Delta\phi/2, \Delta\phi/2)$ will do), and construct $[q]$ from (15.79).

What we finally need is the rate of change of $[q]$ due to the angular velocity ω. Noting that $[q]$ rotates vectors rather than coordinate axes, so that $[\dot{q}]$ is equal to $\dot{\mathbf{R}}^\mathsf{T}$, which is given by (15.60):

$$\dot{\mathbf{R}}^\mathsf{T} = \mathbf{R}^\mathsf{T} \Omega', \tag{15.88}$$

we can cast this into a quaternion multiplication

$$[\dot{q}] = [q]\,[0, \tfrac{1}{2}\omega']. \tag{15.89}$$

Here we have used the fact that Ω' is the time derivative of an angular rotation around an axis in the direction of ω; the factor $1/2$ comes from the derivative of $\sin \frac{1}{2}\phi$. After working this out (reader, please check!) we find

---

[16] See Allen and Tildesley (1987), Appendix G4.

the time derivative of $[q]$:

$$
\begin{pmatrix} \dot{q}_0 \\ \dot{q}_1 \\ \dot{q}_2 \\ \dot{q}_3 \end{pmatrix} = \frac{1}{2} \begin{pmatrix} -q_1 & -q_2 & -q_3 \\ q_0 & -q_3 & q_2 \\ q_3 & q_0 & -q_1 \\ -q_2 & q_1 & q_0 \end{pmatrix} \begin{pmatrix} \omega'_x \\ \omega'_y \\ \omega'_z \end{pmatrix}.
\tag{15.90}
$$

In simulations numerical and integration errors may produce a slow drift in the constraint $q_0^2 + q_1^2 + q_2^2 + q_3^2 = 1$, which is usually compensated by regular scaling of the q's.

15.8 Holonomic constraints

Holonomic constraints depend only on coordinates and can be described by a *constraint equation* $\sigma(\mathbf{r}) = 0$ that should be satisfied at all times. For every constraint there is such an equation. Examples are (we use the notation $\mathbf{r}_{ij} = \mathbf{r}_i - \mathbf{r}_j$):

- *distance constraint between two particles:* $|\mathbf{r}_{12}| - d_{12} = 0$, or, alternatively, $(\mathbf{r}_{12})^2 - d_{12}^2 = 0$;
- *angle 1-2-3 constraint between two constrained bonds:* $\mathbf{r}_{12} \cdot \mathbf{r}_{32} - c = 0$, where $c = d_{12}d_{32} \cos \phi$, or, alternatively, $r_{13}^2 - d_{13}^2 = 0$.

The way to introduce holonomic constraints into the equations of motion is by minimizing the action while preserving the constraints, using Lagrange multipliers. Thus we add each of the m constraints $\sigma_s(q)$, $s = 1, \ldots, m$ with their undetermined Lagrange multipliers λ_s to the Lagrangian and minimize the action. As can be easily verified, this results in the modified Lagrange equations (compare (15.2))

$$
\frac{d}{dt} \left(\frac{\partial \mathcal{L}'}{\partial \dot{q}_k} \right) - \frac{\partial \mathcal{L}'}{\partial q_k} = 0 \qquad i = 1, \ldots, n,
\tag{15.91}
$$

where

$$
\mathcal{L}' \stackrel{\text{def}}{=} \mathcal{L} + \sum_{s=1}^{m} \lambda_s \sigma_s(q),
\tag{15.92}
$$

while for all q along the path

$$
\sigma_s(q) = 0, \ s = 1, \ldots, m
\tag{15.93}
$$

Equations (15.91) and (15.93) fully determine the path, i.e., both $q(t)$ and $\lambda(t)$. In fact the path is restricted to a *hypersurface* determined by the constraint equations. There are $n+m$ variables (the n q's and the m λ's) and an equal number of equations (n Lagrange equations and m constraint equations). Note that the generalized momenta are not modified by holonomic constraints[17] because the constraints are not functions of $\dot{q}$, but the forces are. The total generalized force is built up from an *unconstrained force* and a *constraint force*:

$$\frac{\partial \mathcal{L}'}{\partial q_k} = \frac{\partial \mathcal{L}}{\partial q_k} + \sum_{s=1}^{m} \lambda_s \frac{\partial \sigma_s}{\partial q_k}. \tag{15.94}$$

If the constraints are not eliminated by the use of generalized coordinates, the λ's must be solved from the constraint equations. We can distinguish two ways to obtain the solution, both of which will be worked out in more detail in the following subsections. The first method, which is historically the oldest and in practice the most popular one, was devised by Ryckaert *et al.* (1977). It resets the coordinates after an unconstrained time step, so as to satisfy the constraints to within a given numerical precision, and therefore prevents the propagation of errors. The method is most suitable in conjunction with integration algorithms that do not contain explicit velocities, although a velocity variant is also available (Andersen, 1983). The second method rewrites the equations of motion to include the constraints by solving the λ's from the fact that the σ_s's are zero at all times: hence all time derivatives of σ_s are also zero. This allows an explicit solution for the Lagrange multipliers, but the solutions contain the velocities, and since only derivatives of the constraints appear, errors may propagate. We shall call this class of solutions *projection methods* because in fact the accelerations are projected onto the hypersurface of constraint. The first algorithm using this method for molecules was published by Edberg *et al.* (1986); the solution was cast in more general terms by de Leeuw *et al.* (1990) and discussed in the context of various types of differential equations by Bekker (1996). A similar method was devised by Yoneya *et al.* (1994), and an efficient algorithm (LINCS) was published by Hess *et al.* (1997). We note that matrix methods to solve for holonomic as well as non-holonomic (velocity-dependent) constraints were already known in the field of macroscopic rigid body dynamics.[18]

---

[17] That is true for our definition of constraints; de Leeuw *et al.* (1990) also consider a definition based on the constancy of the time derivative of σ, in which case the generalized momenta are modified, but the final equations are the same.

[18] See, e.g., Section 5.3 of Wittenburg (1977).

15.8.1 Generalized coordinates

A straightforward, but often not feasible, method to implement constraints
is the use of generalized coordinates in such a way that the constraints are
themselves equivalent to generalized coordinates. Suppose that we transform

$$(\boldsymbol{r}_1 \ldots \boldsymbol{r}_N) \rightarrow (q', q''), \qquad (15.95)$$

where $q' = q_1, \ldots, q_{n=3N-m}$ and $q'' = q_{n+1}, \ldots q_{n+m=3N}$ are the free and
constrained coordinates, respectively. The constrained coordinates fulfill
the m constraint equations:

$$\sigma_s = q_{n+s} - c_s = 0. \qquad (15.96)$$

Because $q'' = c$, $\dot{q}'' = 0$ and the kinetic energy does not contain $\dot{q}''$. The
Lagrangian does also not depend on q'' as variables, but contains the c's
only as fixed parameters. Thus the q'' do not figure at all in the equations
of motion and the Lagrangian or Hamiltonian mechanics simply preserves
the constraints. The dynamics is equivalent to that produced by the use
of Lagrange multipliers. The difficulty with this method is that only in
simple cases the equations of motion can be conveniently written in such
generalized coordinates.

15.8.2 Coordinate resetting

One popular method of solving the constraint equations (Ryckaert *et al.*,
1977) can be used in conjunction with the Verlet algorithm, usually applied
with Cartesian coordinates:

$$\boldsymbol{r}_i(t + \Delta t) = 2\boldsymbol{r}_i(t) - \boldsymbol{r}_i(t - \Delta t) + \frac{(\Delta t)^2}{m_i} [\boldsymbol{F}_i^u(t) + \boldsymbol{F}_i^c(t)], \qquad (15.97)$$

where $\boldsymbol{F}^u$ are the forces disregarding the constraints, and the constraint
force on particle i at time t is given by

$$\boldsymbol{F}_i^c(t) = \sum_s \lambda_s(t) \frac{\partial \sigma_s}{\partial \boldsymbol{r}_i}. \qquad (15.98)$$

The effect of the constraint force is to add a second contribution to the displacement of the particles. The algorithm first computes the new positions
$\boldsymbol{r}_i'$ disregarding the constraints:

$$\boldsymbol{r}_i' = 2\boldsymbol{r}_i(t) - \boldsymbol{r}_i(t - \Delta t) + \frac{(\Delta t)^2}{m_i} \boldsymbol{F}_i^u(t), \qquad (15.99)$$

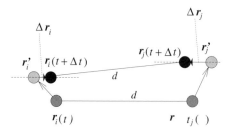

Figure 15.4 Coordinate resetting to realize a bond length constraint.

and then corrects the positions with Δr_i such that

$$\sigma_s(r' + (\Delta r) = 0, \ s = 1, \ldots, m, \tag{15.100}$$

where

$$\Delta r_i = \frac{(\Delta t)^2}{m_i} \sum_s \lambda_s(t) \frac{\partial \sigma_s(r(t))}{\partial r_i}. \tag{15.101}$$

These equations represent a set of m (generally non-linear) coupled equations for the m λ's, which can be solved in several ways, but as a result of the nonlinear character always requiring iteration. They can be either linearized and then solved as a set of linear equations, or the constraints can be solved sequentially and the whole procedure iterated to convergence. The latter method is easy to implement and is used in the routine SHAKE (Ryckaert *et al.*, 1977).

Let us illustrate how one distance constraint between particles i and j:

$$\sigma = r_{ij}^2 - d^2 = 0 \tag{15.102}$$

will be reset in a partial iteration step of SHAKE. See Fig. 15.4. The particle positions are first displaced to r'. Because $\partial \sigma / \partial r_i = -\partial \sigma / \partial r_j = 2r_{ij}$, the displacements must be in the direction of $r_{ij}(t)$ and proportional to the inverse mass of each particle:

$$\Delta r_i = \frac{2(\Delta t)^2}{m_i} \lambda \, r_{ij}(t),$$

$$\Delta r_j = -\frac{2(\Delta t)^2}{m_j} \lambda r_{ij}(t). \tag{15.103}$$

The variable λ is determined such that the distance between $r_i + \Delta r_i$ and $r_j + \Delta r_j$ is equal to d. This procedure is repeated for all constraints until all constraints have converged within a given tolerance.

It is illustrative to consider the motion of a particle that is constrained to

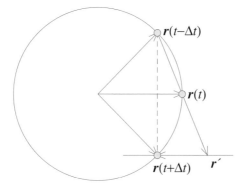

Figure 15.5 The action of SHAKE for a rotating particle constrained to move on a circle.

move on a circle by a distance constraint to a fixed origin (Fig. 15.5). There is no external force acting on the particle, so it should move with constant angular velocity. The positions at times $t - \Delta t$ and t are given. According to (15.99), the position is first linearly extrapolated to r'. Subsequently the position is reset in the direction of the constraint $r(t)$ until the constraint is satisfied. Thus $r(t + \Delta t)$ is obtained. It is easily seen that this algorithm gives *exact* results up to an angular displacement per step of close to 90° (four steps per period), beyond which the algorithm is unstable and fails to find a solution. It is also seen that resetting in the direction of the constraint at time t is correct.

15.8.3 Projection methods

Consider a cartesian system of point masses with equations of motion

$$m_i \ddot{r}_i = F_i^u + \sum_{s=1}^{n_c} \frac{\partial \sigma_s}{\partial r_i}, \quad i = 1, \dots, N, \tag{15.104}$$

which we shall write in matrix notation as

$$\mathbf{M}\ddot{\mathbf{x}} = \mathbf{f} + \mathbf{C}^{\mathsf{T}}\boldsymbol{\lambda}, \tag{15.105}$$

where we use $\mathbf{x}$ for the $3N \times 1$ column matrix $(x_1, y_1, z_1, x_2, \dots, y_N, z_N)^{\mathsf{T}}$, similarly $\mathbf{f}$ for $\boldsymbol{F}^u$, $\mathbf{M}$ is the $3N \times 3N$ diagonal matrix of masses, and the *constraint matrix* $\mathbf{C}$ is defined by

$$C_{si} = \frac{\partial \sigma_s}{\partial x_i}. \tag{15.106}$$

By taking the time derivative of

$$\dot{\sigma}_s = (\mathbf{C}\dot{\mathbf{x}})_s = 0 \tag{15.107}$$

the following relation is found:

$$\mathbf{C}\ddot{\mathbf{x}} = -\dot{\mathbf{C}}\dot{\mathbf{x}}. \tag{15.108}$$

By left-multiplying (15.105) first by $\mathbf{M}^{-1}$ and then by $\mathbf{C}$, and substituting (15.108), $\boldsymbol{\lambda}$ can be solved and we obtain

$$\boldsymbol{\lambda} = -(\mathbf{CM}^{-1}\mathbf{C}^{\mathsf{T}})^{-1}(\mathbf{CM}^{-1}\mathbf{f} + \dot{\mathbf{C}}\dot{\mathbf{x}}). \tag{15.109}$$

The matrix $\mathbf{CM}^{-1}\mathbf{C}^{\mathsf{T}}$ is non-singular and can be inverted if the constraints are independent.[19] Substituting (15.109) into (15.105) we obtain the following equation of motion for $\mathbf{x}$:

$$\ddot{\mathbf{x}} = (\mathbf{1} - \mathbf{TC})\mathbf{M}^{-1}\mathbf{f} - \mathbf{T}\dot{\mathbf{C}}\dot{\mathbf{x}}, \tag{15.110}$$

where

$$\mathbf{T} \stackrel{\text{def}}{=} \mathbf{M}^{-1}\mathbf{C}\mathbf{T}(\mathbf{CM}^{-1}\mathbf{C}^{\mathsf{T}})^{-1}. \tag{15.111}$$

The matrix $\mathbf{1} - \mathbf{TC}$ projects the accelerations due to the unconstrained forces onto the constraint hypersurface. The first term in (15.110) gives the constrained accelerations due to the systematic forces (derivatives of the potential) and the second term gives the constrained accelerations due to centripetal forces.

Equation (15.110) contains the velocities at the same time as when the forces are evaluated, but these velocities are not known at that time. Therefore this equation is in principle implicit and needs an iterative solution. In practice this is not done and not necessary. Hess *et al.* (1997) show that the velocities at the previous half step can be used in conjunction with appropriate corrections. The corrections are made in such a way that a stable algorithm results without drift.

Although the SHAKE algorithm of Ryckaert *et al.* (1977) is easier to implement, the LINCS (LINear Constraint Solver) algorithm of Hess *et al.* (1997) is faster, more robust, more accurate and more suitable for parallel computers. For special cases it can be advantageous to solve the equations analytically, as in the SETTLE algorithm of Miyamoto and Kollman (1992) for water molecules.

---

[19] De Leeuw *et al.* (1990) show that the matrix is not only non-singular but also positive definite.

16

Review of thermodynamics

16.1 Introduction and history

This book is not a textbook on thermodynamics or statistical mechanics. The reason to incorporate these topics nevertheless is to establish a common frame of reference for the readers of this book, including a common nomenclature and notation. For details, commentaries, proofs and discussions, the reader is referred to any of the numerous textbooks on these topics.

Thermodynamics describes the macroscopic behavior of systems in equilibrium, in terms of macroscopic measurable quantities that do not refer at all to atomic details. Statistical mechanics links the thermodynamic quantities to appropriate averages over atomic details, thus establishing the ultimate *coarse-graining* approach. Both theories have something to say about non-equilibrium systems as well. The logical exposition of the link between atomic and macroscopic behavior would be in the sequence:

(i) describe atomic behavior on a quantum-mechanical basis;
(ii) simplify to classical behavior where possible;
(iii) apply statistical mechanics to average over details;
(iv) for systems in equilibrium: derive thermodynamic; quantities and phase behavior; for non-equilibrium systems: derive macroscopic rate processes and transport properties.

The historical development has followed a quite different sequence. Equilibrium thermodynamics was developed around the middle of the nineteenth century, with the definition of *entropy* as a state function by Clausius forming the crucial step to completion of the theory. No detailed knowledge of atomic interactions existed at the time and hence no connection between atomic interactions and macroscopic behavior (the realm of statistical mechanics) could be made. Neither was such knowledge needed to define the state functions and their relations.

Thermodynamics describes equilibrium systems. Entropy is really only defined in equilibrium. Still, thermodynamics is most useful when *processes* are considered, as phase changes and chemical reactions. But equilibrium implies reversibility of processes; processes that involve changes of state or constitution cannot take place in equilibrium, unless they are infinitely slow. Processes that take place at a finite rate always involve some degree of irreversibility, about which traditional thermodynamics has nothing to say. Still, the *second law* of thermodynamics makes a *qualitative* statement about the *direction* of processes: a system will spontaneously evolve in the direction of increasing excess entropy (i.e., entropy in excess of the reversible exchange with the environment). This is sometimes formulated as: *the entropy of the universe (= system plus its environment) can only increase.* Such a statement cannot be made without a definition of entropy in a non-equilibrium system, which is naturally not provided by equilibrium thermodynamics! The second law is therefore not precise within the bounds of thermodynamics proper; the notion of entropy of a non-equilibrium system rests on the assumption that the total system can be considered as the sum of smaller systems that are locally in equilibrium. The smaller systems must still contain a macroscopic number of particles.

This somewhat uneasy situation, given the practical importance of the second law, gave rise to deeper consideration of irreversible processes in the thirties and later. The crucial contributions came from Onsager (1931a, 1931b) who considered the behavior of systems that deviate slightly from equilibrium and in which irreversible fluxes occur proportional to the deviation from equilibrium. In fact, the *thermodynamics of irreversible processes*, treating the linear regime, was born. It was more fully developed in the fifties by Prigogine (1961) and others.[1] In the mean time, and extending into the sixties, also the statistical mechanics of irreversible processes had been worked out, and relations were established between transport coefficients (in the linear regime of irreversible processes) and fluctuations occurring in equilibrium. Seminal contributions came from Kubo and Zwanzig.

Systems that deviate from equilibrium beyond the linear regime have been studied extensively in the second half of the twentieth century, notably by the Brussels school of Prigogine. Such systems present new challenges: different quasi-stationary regimes can emerge with structured behavior (in time and/or in space), or with chaotic behavior. Transitions between regimes often involve bifurcation points with non-deterministic behavior. A whole new line of theoretical development has taken place since and is still active,

---

[1] see, for example, de Groot and Mazur (1962)

including chaos theory, complexity theory, and the study of emergent behavior and self-organization in complex systems. In biology, studies of this
type, linking system behavior to detailed pathways and genetic make-up,
are making headway under the term *systems biology.*

In the rest of this chapter we shall summarize equilibrium thermodynamics
based on Clausius' entropy definition, without referring to the statistical interpretation of entropy. This is the traditional thermodynamics, which is an
established part of both physics and chemistry. We emphasize the thermodynamic quantities related to molecular components in mixtures, traditionally treated more extensively in a chemical context. Then in Section 16.10
we review the non-equilibrium extensions of thermodynamics in the linear
regime. Time-dependent linear response theory is deferred to another chapter (18). Chapter 17 (statistical mechanics) starts with the principles of
quantum statistics, where entropy is given a statistical meaning.

16.2 Definitions

We consider systems in equilibrium. It suffices to define equilibrium as the
situation prevailing after the system has been allowed to relax under constant external conditions for a long time t, in the limit $t \to \infty$. Processes that
occur so slowly that in all intermediate states the system can be assumed to
be in equilibrium are called *reversible. State functions* are properties of the
system that depend only on the state of the system and not on its history.
The state of a system is determined by a description of its composition, usually in terms of the number of moles[2] n_i of each constituent chemical species
i, plus two other *independent* state variables, e.g., volume and temperature.
State functions are *extensive* if they are proportional to the size of the system (such as n_i); *intensive* properties are independent of system size (such
as the concentration of the ith species $c_i = n_i/V$). An important intensive
state function is the *pressure* p, which is homogeneous and isotropic in a
fluid in equilibrium and which can be defined by the force acting per unit
area on the wall of the system. Another important intensive thermodynamic
state function is the *temperature*, which is also homogeneous in the system
and can be measured by the pressure of an ideal gas in a (small, with respect
to the system) rigid container in thermal contact with the system.

Since a state function (say, f) depends on the independent variables (say,
x and y) only and not on processes in the past, the differential of f is a *total*

---

[2] Physicists sometimes express the quantity of each constituent in mass units, but that turns out
to be very inconvenient when chemical reactions are considered.

or *exact* differential

$$df = \frac{\partial f}{\partial x}\,dx + \frac{\partial f}{\partial y}\,dy. \tag{16.1}$$

The line integral over a path from point A to point B does not depend on the path, and the integral over any *closed path* is zero:

$$\int_A^B df = f(B) - f(A), \tag{16.2}$$

$$\oint df = 0. \tag{16.3}$$

If the second derivatives are continuous, then the order of differentiation does not matter:

$$\frac{\partial^2 f}{\partial x \partial y} = \frac{\partial^2 f}{\partial y \partial x}. \tag{16.4}$$

As we shall see, this equality leads to several relations between thermodynamics variables.

A thermodynamic system may exchange heat dq, work dw (mechanical work as well as electrical energy) and/or particles dn_i with its environment. We shall adopt the sign convention that dq, dw and dn_i are *positive* if heat is absorbed by the system, work is exerted *on* the system or particles enter the system. Both dq and dw increase the internal energy of the system. If the work is due to a volume change, it follows that

$$dw = -p\,dV. \tag{16.5}$$

Neither dq nor dw is an exact differential: it is possible to extract net heat or work over a closed reversible path. We see, however, that $-1/p$ is an *integrating factor* of dw, yielding the exact differential dV of a state function V. Similarly it can be shown that a function β exists, such that βdq is an exact differential. Thus the function β is an *integrating factor* of dq. It can be identified with the inverse absolute temperature, so that a state function S exists with

$$dS = \frac{dq}{T}. \tag{16.6}$$

The function S is called the *entropy*; it is an extensive state function. The entropy is only defined up to a constant and the unit for S depends on the unit agreed for T. The zero point choice for the entropy is of no consequence for any process and is usually taken as the value at $T = 0$.

In Table 16.1 the important state functions are summarized, with their S.I. units.

Table 16.1 *Thermodynamic state functions and their definitions and units.*
All variables with upper case symbols are extensive and all variables with
lower case symbols are intensive, with the following exceptions: n and m
are extensive; T and M_i are intensive.

	Definition	Name	S.I. unit
V		volume	m^3
p	*see text*	pressure	Pa
T	*see text*	temperature	K
n		total amount of moles	-
m		total mass of system	kg
ρ	m/V	density[a]	kg/m^3
U	*see text*	internal energy	J
S	*see text*	entropy	J/K
H	$U + pV$	enthalpy	J
A	$U - TS$	Helmholtz free energy	J
G	$A + pV$	Gibbs free energy	J
	$= H - TS$	or Gibbs function	
n_i		moles of ith component	mol
M_i		molar mass of ith component	kg/mol
x_i	n_i/n	mole fraction of ith component	-
c_i	n_i/V	concentration[b] of ith componenet	mol/m^3
m_i	n_i/m_s	molality of ith componenet	mol/kg
		(mol solute per kg solvent)	= molal
C_V	$(\partial U/\partial T)_{V,n_i}$	isochoric heat capacity	J/K
C_p	$(\partial H/\partial T)_{p,n_i}$	isobaric heat capacity	J/K
c_V	C_V/n	molar isochoric heat capacity	$J\ mol^{-1}\ K^{-1}$
c_p	C_p/n	molar isobaric heat capacity	$J\ mol^{-1}\ K^{-1}$
$\bar{c}_V$	C_V/m	isochoric specific heat	$J\ kg^{-1}\ K^{-1}$
$\bar{c}_p$	C_p/m	isobaric specific heat	$J\ kg^{-1}\ K^{-1}$
α	$(1/V)(\partial V/\partial T)_{p,n_i}$	volume expansion coeff.	K^{-1}
κ_T	$-(1/V)(\partial V/\partial p)_{T,n_i}$	isothermal compressibility	Pa^{-1}
κ_S	$-(1/V)(\partial V/\partial p)_{S,n_i}$	adiabatic compressibility	Pa^{-1}
μ_{JT}	$(\partial T/\partial p)_{H,n_i}$	Joule–Thomson coefficient	K/Pa

[a] The symbol ρ is sometimes also used for molar density or concentration, or for number density: particles per m^3.
[b] The unit "molar", symbol M, for $mol/dm^3 = 1000\ mol/m^3$ is usual in chemistry.

16.2.1 Partial molar quantities

In Table 16.1 derivatives with respect to composition have not been included. The partial derivative y_i of an extensive state function $Y(p, T, n_i)$, with respect to the number of moles of each component, is called the *partial molar Y*:

$$y_i = \left(\frac{\partial Y}{\partial n_i}\right)_{p,T,n_{j\neq i}}. \tag{16.7}$$

For example, if $Y = G$, we obtain the *partial molar Gibbs free energy*, which is usually called the *thermodynamic potential* or the *chemical potential*:

$$\mu_i = \left(\frac{\partial G}{\partial n_i}\right)_{p,T,n_{j\neq i}}, \tag{16.8}$$

and with the volume V we obtain the *partial molar volume* v_i. Without further specification partial molar quantities are defined at constant pressure and temperature, but any other variables may be specified. For simplicity of notation we shall from now on implicitly assume the condition $n_{j\neq i} = $ constant in derivatives with respect to n_i.

If we enlarge the whole system (under constant p, T) by dn, keeping the mole fractions of all components the same (i.e., $dn_i = x_i dn$), then the system enlarges by a fraction dn/n and all extensive quantities Y will enlarge by a fraction dn/n as well:

$$dY = \frac{Y}{n}\, dn. \tag{16.9}$$

But also:

$$dY = \sum_i \left(\frac{\partial Y}{\partial n_i}\right)_{p,T,n_{j\neq i}} dn_i = \sum_i y_i x_i dn. \tag{16.10}$$

Hence

$$Y = n \sum_i x_i y_i = \sum_i n_i y_i. \tag{16.11}$$

Note that this equality is only valid if the other independent variables are intensive state functions (as p and T) and not for, e.g., V and T. The most important application is $Y = G$:

$$G = \sum_i n_i \mu_i. \tag{16.12}$$

This has a remarkable consequence: since

$$dG = \sum_i \mu_i\, dn_i + \sum_i n_i\, d\mu_i, \tag{16.13}$$

but also, as a result of (16.8),

$$(dG)_{p,T} = \sum_i \mu_i\, dn_i, \tag{16.14}$$

it follows that

$$\sum_i n_i (d\mu_i)_{p,T} = 0. \tag{16.15}$$

This equation is the *Gibbs–Duhem relation*, which is most conveniently expressed in the form

$$\sum_i x_i (d\mu_i)_{p,T} = 0. \tag{16.16}$$

The Gibbs–Duhem relation implies that not all chemical potentials in a mixture are independent. For example, consider a solution of component s (solute) in solvent w (water). If the mole fraction of the solute is x, then

$$x_w = (1 - x) \quad \text{and} \quad x_s = x, \tag{16.17}$$

and

$$x \, d\mu_s + (1 - x) \, d\mu_w = 0. \tag{16.18}$$

This relation allows derivation of the concentration dependence of μ_s from the concentration dependence of μ_w. The latter may be determined from the osmotic pressure as a function of concentration.

There are numerous other partial molar quantities. The most important ones are

$$v_i = \left(\frac{\partial V}{\partial n_i} \right)_{p,T}, \tag{16.19}$$

$$u_i = \left(\frac{\partial U}{\partial n_i} \right)_{p,T}, \tag{16.20}$$

$$h_i = \left(\frac{\partial H}{\partial n_i} \right)_{p,T}, \tag{16.21}$$

$$s_i = \left(\frac{\partial S}{\partial n_i} \right)_{p,T}, \tag{16.22}$$

which are related by

$$\mu_i = h_i - T s_i \quad \text{and} \quad h_i = u_i + p \, v_i. \tag{16.23}$$

16.3 Thermodynamic equilibrium relations

The *first law* of thermodynamics is the conservation of energy. If the number of particles and the composition does not change, the change in internal energy dU is due to absorbed heat dq and to work exerted on the system dw. In equilibrium, the former equals $T \, dS$ and the latter $-p \, dV$, when other types of work as electrical work, nuclear reactions and radiation are disregarded. Hence

$$dU = T\,dS - p\,dV. \tag{16.24}$$

With the definitions given in Table 16.1 and (16.8), we arrive at the following differential relations:

$$dU = T\,dS - p\,dV + \sum_i \mu_i\,dn_i, \tag{16.25}$$

$$dH = T\,dS + V\,dp + \sum_i \mu_i\,dn_i, \tag{16.26}$$

$$dA = -S\,dT - p\,dV + \sum_i \mu_i\,dn_i, \tag{16.27}$$

$$dG = -S\,dT + V\,dp + \sum_i \mu_i\,dn_i. \tag{16.28}$$

Each of the differentials on the left-hand side are total differentials, defining 12 partial differentials such as (from (16.28)):

$$\left(\frac{\partial G}{\partial T}\right)_{p,n_i} = -S, \tag{16.29}$$

$$\left(\frac{\partial G}{\partial p}\right)_{T,n_i} = V, \tag{16.30}$$

$$\left(\frac{\partial G}{\partial n_i}\right)_{p,T} = \mu_i. \tag{16.31}$$

These are among the most important thermodynamic relations. The reader is invited to write down the other nine equations of this type. Note that the entropy follows from the temperature dependence of G. However, one can also use the temperature dependence of G/T (e.g. from an equilibrium constant), to obtain the enthalpy rather than the entropy:

$$\frac{\partial(G/T)}{\partial(1/T)} = H. \tag{16.32}$$

This is the very useful *Gibbs–Helmholtz relation*.

Being total differentials, the second derivatives of mixed type do not depend on the sequence of differentiation. For example, from (16.28):

$$\frac{\partial^2 G}{\partial p\,\partial T} = \frac{\partial^2 G}{\partial T\,\partial p}, \tag{16.33}$$

implies that

$$-\left(\frac{\partial S}{\partial p}\right)_{T,n_i} = \left(\frac{\partial V}{\partial T}\right)_{p,n_i}. \qquad (16.34)$$

This is one of the *Maxwell relations*. The reader is invited to write down the other 11 Maxwell relations of this type.

16.3.1 Relations between partial differentials

The equations given above, and the differential relations that follow from them, are a selection of the possible thermodynamic relations. They may not include a required derivative. For example, what is the relation between C_V and C_p or between κ_S and κ_T? Instead of listing all possible relations, it is much more effective and concise to list the basic mathematical relations from which such relations follow. The three basic rules are given below.

Relations between partial differentials

f is a differentiable function of two variables. There are three variables x, y, z, which are related to each other and all partial differentials of the type $(\partial x/\partial y)_z$ exist. Then the following rules apply:

Rule 1:

$$\left(\frac{\partial f}{\partial x}\right)_z = \left(\frac{\partial f}{\partial x}\right)_y + \left(\frac{\partial f}{\partial y}\right)_x \left(\frac{\partial y}{\partial x}\right)_z. \qquad (16.35)$$

Rule 2:

$$\left(\frac{\partial x}{\partial y}\right)_z = \left[\left(\frac{\partial y}{\partial x}\right)_z\right]^{-1} \quad \text{(inversion)}. \qquad (16.36)$$

Rule 3:

$$\left(\frac{\partial x}{\partial y}\right)_z \left(\frac{\partial y}{\partial z}\right)_x \left(\frac{\partial z}{\partial x}\right)_y = -1 \quad \text{(cyclic chain rule)}. \qquad (16.37)$$

From these rules several relations can be derived. For example, in order to relate general dependencies on volume with those on pressure, one can apply Rule 1:

$$\left(\frac{\partial f}{\partial T}\right)_V = \left(\frac{\partial f}{\partial T}\right)_p + \frac{\alpha}{\kappa_T}\left(\frac{\partial f}{\partial p}\right)_T, \qquad (16.38)$$

or Rule 3:

$$\left(\frac{\partial f}{\partial V}\right)_T = -\frac{1}{\kappa_T V}\left(\frac{\partial f}{\partial p}\right)_T. \qquad (16.39)$$

Useful relations are

$$C_{\mathrm{p}} = C_{\mathrm{V}} + \frac{\alpha^2 V T}{\kappa_T}, \tag{16.40}$$

$$\kappa_T = \kappa_S + \frac{\alpha^2 V T}{C_{\mathrm{p}}}, \tag{16.41}$$

$$p = T \left(\frac{\partial p}{\partial T} \right)_V - \left(\frac{\partial U}{\partial V} \right)_T. \tag{16.42}$$

Equation (16.42) splits pressure into an *ideal gas kinetic* part and an *internal* part due to internal interactions. The term $(\partial U/\partial V)_T$ indicates deviation from ideal gas behavior. The proofs are left to the exercises at the end of this chapter.

16.4 The second law

Thus far we have used the second law of thermodynamics in the form $dS = dq/T$, valid for systems in equilibrium. The full second law, however, states that

$$dS \geq \frac{dq}{T} \tag{16.43}$$

for any system, including non-equilibrium states. It tells us in what direction spontaneous processes will take place. When the system has reached full equilibrium, the equality holds.

This qualitative law can be formulated for closed systems for three different cases:

- **Closed, adiabatic system**: when neither material nor heat is exchanged with the environment $(dq = 0)$, *the system will spontaneously evolve in the direction of maximum entropy*:

$$dS \geq 0. \tag{16.44}$$

In equilibrium S is a maximum.
- **Closed, isothermal and isochoric system**: when volume and temperature are kept constant $(dV = 0, dT = 0)$, then $dq = dU$ and $T\,dS \geq dU$. This implies that

$$dA \leq 0. \tag{16.45}$$

The system will spontaneously evolve in the direction of lowest Helmholtz free energy. In equilibrium A is a minimum.

- **Closed, isothermal and isobaric system**: When pressure and temperature are kept constant ($dp = 0$, $dT = 0$), then $dq = dH$ and $T\,dS \geq dH$. This implies that

$$dG \leq 0. \tag{16.46}$$

The system will spontaneously involve in the direction of lowest Gibbs free energy. In equilibrium G is a minimum.

For *open systems* under constant p and T that are able to exchange material, we can formulate the second law as follows: *the system will spontaneously evolve such that the thermodynamic potential of each component becomes homogeneous.* Since $G = \sum_i n_i \mu_i$, the total G would decrease if particles would move from a region where their thermodynamic potential is high to a region where it is lower. Therefore particles would spontaneously move until their μ would be the same everywhere. One consequence of this is that the thermodynamic potential of any component is the same in two (or more) coexisting phases.

16.5 Phase behavior

A closed system with a one-component homogeneous phase (containing n moles) has *two* independent variables or *degrees of freedom*, e.g., p and T. All other state functions, including V, are now determined. Hence there is relation between p, V, T:

$$\Phi(p, V, T) = 0, \tag{16.47}$$

which is called the *equation of state* (EOS). Examples are the ideal gas EOS: $pv = RT$ (where v is the molar volume V/n), or the van der Waals gas $(p + a/v^2)(v - b) = RT$. If two phases, as liquid and gas, coexist, there is the additional restriction that the thermodynamic potential must be equal in both phases, and only one degree of freedom (either p or T) is left. Thus there is a relation between p and T along the phase boundary; for the liquid–vapor boundary boiling point and vapor pressure are related. When three phases coexist, as solid, liquid and gas, there is yet another restriction which leaves no degrees of freedom. Thus the *triple point* has a fixed temperature and pressure.

Any additional component in a mixture adds another degree of freedom, viz. the mole fraction of the additional component. The number of degrees of freedom F is related to the number of components C and the number of coexisting phases P by *Gibbs' phase rule*:

$$F = C - P + 2, \tag{16.48}$$

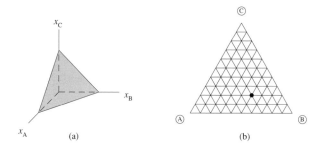

Figure 16.1 (*a*) points representing mole fractions of 3 components A,B,C in a cartesian coordinate system end up in the shaded triangle (*b*) Each vertex represents a pure component and each mole fraction is on a linear 0-1 scale starting at the opposite side. The dot represents the composition $x_A = 0.3$, $x_B = 0.5$, $x_C = 0.2$.

as the reader can easily verify.

A *phase diagram* depicts the phases and phase boundaries. For a single component with two degrees of freedom, a two-dimensional plot suffices, and one may choose as independent variables any pair of p, T, and either V or molar density $\rho = n/V$. Temperature–density phase diagrams contain a *coexistence region* where a single phase does not exist and where two phases (gas and liquid, or solid and liquid) are in equilibrium, one with low density and one with high density. In simulations on a small system a density-temperature combination in the coexistence region may still yield a stable fluid with negative pressure. A large amount of real fluid would separate because the total free energy would then be lower, but the separation is a slow process, and in a small system the free energy cost to create a phase boundary (surface pressure) counteracts the separation.

For mixtures the composition comes in as extra degrees of freedom. Phase diagrams of binary mixtures are often depicted as x, T diagrams. Ternary mixtures have two independent mole fractions; if each of the three mole fractions are plotted along three axes of a 3D cartesian coordinate system, the condition $\sum_i x_i = 1$ implies that all possible mixtures lie on the plane through the points (1,0,0), (0,1,0) and (0,0,1) (Fig. 16.1a). Thus any composition can be depicted in the corresponding triangle (Fig. 16.1b).

Along a phase boundary between two phases 1 and 2 in the T, p plane we know that the thermodynamic potential at every point is equal on both sides of the boundary. Hence, stepping dT, dp along the boundary, $d\mu_1 = d\mu_2$:

$$d\mu_1 = v_1\,dp - s_1\,dT = d\mu_2 = v_2\,dp - s_2\,dT. \tag{16.49}$$

Therefore, along the boundary the following relation holds

$$\frac{dp}{dT} = \frac{\Delta s}{\Delta v} = \frac{1}{T}\frac{\Delta h}{\Delta v}, \tag{16.50}$$

where Δ indicates a difference between the two phases. These relations are exact. If we consider the boiling or sublimation line, and one phase can be approximated by an ideal gas and the other condensed phase has a negligible volume compared to the gas, we may set $\Delta v = RT/p$, and we arrive at the *Clausius–Clapeyron equation* for the temperature dependence of the saturation pressure

$$\frac{d \ln p}{dT} = \frac{\Delta h_{\text{vap}}}{RT^2}. \tag{16.51}$$

This equation implies that the saturation pressure increases with temperature as

$$p(T) \propto \exp\left(\frac{-\Delta h_{\text{vap}}}{RT}\right). \tag{16.52}$$

16.6 Activities and standard states

The thermodynamic potential for a gas at low pressure, which approaches the ideal gas for which the molar volume $v = RT/p$, is given by

$$\mu(p) = \mu^0(p^0) + \int_{p^0}^{p} v\, dp = \mu^0(p^0) + RT \ln \frac{p}{p^0}. \tag{16.53}$$

Here μ^0 is the *standard* thermodynamic potential at some standard pressure p^0. For real gases at non-zero pressure the thermodynamic potential does not follow this dependence exactly. One writes

$$\mu(p) = \mu^0 + RT \ln \frac{\gamma p}{p^0}, \tag{16.54}$$

where γ is the *activity coefficient* and $f = \gamma p$ is called the *fugacity* of the gas. It is the pressure the gas would have had it been ideal. For $p \rightarrow 0$, $\gamma \rightarrow 1$.

For solutions the thermodynamic potential of a dilute solute behaves in a similar way. When the concentration (or the mole fraction or molality) of a component approaches zero, the thermodynamic potential of that component becomes linear in the logarithm of the concentration (mole fraction, molality):

$$\mu(c) = \mu_c^0 + RT \ln(\gamma_c c/c^0), \tag{16.55}$$
$$\mu(m) = \mu_m^0 + RT \ln(\gamma_m m/m^0), \tag{16.56}$$
$$\mu(x) = \mu_x^0 + RT \ln(\gamma_x x). \tag{16.57}$$

The standard concentration c^0 is usually 1 M (molar $=$ mol dm^{-3}), and the standard molality m_0 is 1 mole per kg solvent. For mole fraction x the standard reference is the pure substance, $x = 1$. The γ's are *activity coefficients* and the products $\gamma_c c / c^0, \gamma_m m / m^0, \gamma_x x$ are called *activities*; one should clearly distinguish these three different kinds of activities. They are, of course, related through the densities and molar masses.

Note that $\mu(c^0) \neq \mu_c^0$ and $\mu(m^0) \neq \mu_m^0$, unless the activity coefficients happen to be zero at the standard concentration or molality. The definition of μ^0 is

$$\mu_c^0 \stackrel{\text{def}}{=} \lim_{c \to 0} \left[\mu(c) - RT \ln \frac{c}{c^0} \right], \tag{16.58}$$

and similarly for molalities and for mole fractions of solutes. For mole fractions of *solvents* the standard state $x = 1$ represents the pure solvent, and μ_x^0 is now defined as $\mu(x = 1)$, which is usually indicated by μ^*. For $x = 1$ the activity coefficient equals 1.

Solutions that have $\gamma_x = 1$ for any composition are called *ideal*.

The reader is warned about the inaccurate use, or incomplete definitions, of standard states and activities in the literature. Standard entropies and free energies of transfer from gas phase to solution require proper definition of the standard states in both phases. It is very common to see $\ln c$ in equations, meaning $\ln(c/c^0)$. The logarithm of a concentration is mathematically undefined.

16.6.1 Virial expansion

For dilute gases the deviation from ideal behavior can be expressed in the *virial expansion*, i.e., the expansion of p/RT in the molar density $\rho = n/V$:

$$\frac{p}{k_{\rm B} T} = \rho + B_2(T)\rho^2 + B_3(T)\rho^3 + \cdots . \tag{16.59}$$

This is in fact an equation of state for the dilute gas phase. The *second virial coefficient* B_2 is expressed in m^3/mol. It is temperature dependent with usually negative values at low temperatures and tending towards a limiting positive value at high temperature. The second virial coefficient can be calculated on the basis of pair interactions and is therefore an important experimental quantity against which an interaction function can be calibrated.[3] For dilute *solutions* a similar expansion can be made of the

---

[3] See Hirschfelder *et al.* (1954) for many details on determination and computation of virial coefficients.

osmotic pressure (see (16.85)) versus the concentration:

$$\frac{\Pi}{k_B T} = c + B_2(T)c^2 + B_3(T)c^3 + \cdots. \tag{16.60}$$

The activity coefficient is related to the virial coefficients: using the expression

$$\left(\frac{\partial \mu}{\partial \rho}\right)_T = \frac{1}{\rho}\left(\frac{\partial p}{\partial \rho}\right)_T, \tag{16.61}$$

we find that

$$\mu(\rho, T) = \mu_{\text{ideal}} + 2RTB_2(T)\rho + \frac{3}{2}RTB_3(T)\rho^2 + \cdots. \tag{16.62}$$

This implies that

$$\ln \gamma_c = 2B_2\rho + \frac{3}{2}B_3\rho^2 + \cdots. \tag{16.63}$$

Similar expressions apply to the fugacity and the osmotic coefficient.

16.7 Reaction equilibria

Consider *reaction equilibria* like

$$A + 2B \rightleftharpoons AB_2,$$

which is a special case of the general reaction equilibrium

$$0 \rightleftharpoons \sum_i \nu_i C_i \tag{16.64}$$

Here, C_i are the components, and ν_i the *stoichiometric coefficients*, positive on the right-hand side and negative on the left-hand side of the reaction. For the example above, $\nu_A = -1$, $\nu_B = -2$ and $\nu_{AB_2} = +1$. In equilibrium (under constant temperature and pressure) the total change in Gibbs free energy must be zero, because otherwise the reaction would still proceed in the direction of decreasing G:

$$\sum_i \nu_i \mu_i = 0. \tag{16.65}$$

Now we can write

$$\mu_i = \mu_i^0 + RT \ln a_i, \tag{16.66}$$

where we can fill in any consistent standard state and activity definition we desire. Hence

$$\sum_i \nu_i \mu_i^0 = -RT \sum_i \nu_i \ln a_i. \tag{16.67}$$

The left-hand side is a thermodynamic property of the combined reactants, usually indicated by ΔG^0 of the reaction, and the right-hand side can also be expressed in terms of the *equilibrium constant* K:

$$\Delta G^0 \overset{\text{def}}{=} \sum_i \nu_i \mu_i^0 = -RT \ln K, \tag{16.68}$$

$$K \overset{\text{def}}{=} \Pi_i a_i^{\nu_i}. \tag{16.69}$$

The equilibrium constant depends obviously on the definitions used for the activities and standard states. In dilute solutions concentrations or molalities are often used instead of activities; note that such equilibrium "constants" are not constant if activity coefficients deviate from 1.

Dimerization $2\,A \rightleftharpoons A_2$ in a gas diminishes the number of "active" particles and therefore reduces the pressure. In a solution similarly the osmotic pressure is reduced. This leads to a negative second virial coefficient $B_2(T) = -K_p RT$ for dilute gases ($K_p = p_{A_2}/p_A^2$ being the dimerization constant on the basis of pressures) and $B_2(T) = -K_c$ for dilute solutions ($K_c = c_{A_2}/c_A^2$ being the dimerization constant on the basis of concentrations).

A special case of an equilibrium constant is *Henry's constant*, being the ratio between pressure in the gas phase of a substance A, and mole fraction x_A in dilute solution.[4] It is an inverse solubility measure. The reaction is

$$A(\text{sol}, x) \rightleftharpoons A(\text{gas}, p)$$

with standard state $x = 1$ in solution and $p = p^0$ in the gas phase. Henry's constant $K_H = p/x$ relates to the standard Gibbs free energy change as

$$\Delta G^0 = \mu^0(\text{gas}; p^0) - \mu^0(\text{sol}; x = 1) = -RT \ln K_H. \tag{16.70}$$

Other special cases are equilibrium constants for acid–base reactions involving proton transfer, and for reduction–oxidation reactions involving electron transfer, both of which will be detailed below.

16.7.1 Proton transfer reactions

The general proton transfer reaction is

$$HA \rightleftharpoons H^+ + A^-,$$

where the proton donor or acid HA may also be a charged ion (like NH_4^+ or HCO_3^-) and H^+ stands for any form in which the proton may appear in

---

[4] The inverse of K_H, expressed not as mole fraction but as molality per bar, is often tabulated (e.g., by NIST). It is also called Henry's law constant and denoted by k_H.

solution (in aqueous solutions most likely as H_3O^+). A^- is the corresponding proton acceptor or base (like NH_3 or $(CO_3)^{2-}$). The equilibrium constant in terms of activities based on molar concentrations is the *acid dissociation constant* K_a:

$$K_a = \frac{a_{H^+} a_{A^-}}{a_{HA}} \approx \frac{[H^+][A^-]}{[HA]c^0}, \tag{16.71}$$

where the brackets denote concentrations in molar, and $c^0 = 1$ M.[5] When the acid is the solvent, as in the dissociation reaction of water itself:

$$H_2O \rightleftharpoons H^+ + OH^-,$$

the standard state is mole fraction $x = 1$ and the dissociation constant $K_w = 10^{-14}$ is simply the product of ionic concentrations in molar.

With the two definitions

$$pH \stackrel{\text{def}}{=} -\log_{10} a_{H^+} \approx -\log_{10} [H^+] \tag{16.72}$$

$$pK_a \stackrel{\text{def}}{=} -\log_{10} K_a, \tag{16.73}$$

we find that

$$\Delta G^0 = -RT \ln K_a = 2.3026\, RT\, pK_a. \tag{16.74}$$

It is easily seen that the acid is halfway dissociated (activities of acid and base are equal) when the *pH* equals *pK_a*.

16.7.2 Electron transfer reactions

The general electron transfer reaction involves two molecules (or ions): a *donor* D and an *acceptor* A:

$$D + A \rightleftharpoons D^+ + A^- \qquad .$$

In this process the electron donor is the *reductant* that gets *oxidized* and the electron acceptor is the *oxidant* that gets *reduced*. Such reactions can be formally built up from two separate *half-reactions*, both written as a reduction:

$$D^+ + e^- \rightleftharpoons D \qquad ,$$
$$A + e^- \rightleftharpoons A^- \qquad .$$

The second minus the first reaction yields the overall electron transfer reaction. Since the free electron in solution is not a measurable intermediate,[6]

---

[5] Usually, c^0 is not included in the definition of K, endowing K with a dimension (mol dm^{-3}), and causing a formal inconsistency when the logarithm of K is needed.

[6] Solvated electrons do exist; they can be produced by radiation or electron bombardment. They have a high energy and a high reductive potential. In donor-acceptor reactions electrons are transferred through contact, through material electron transfer paths or through vacuum over very small distances, but not via solvated electrons.

one cannot attach a meaning to the absolute value of the chemical potential of the electron, and consequently to the equilibrium constant or the ΔG^0 of half-reactions. However, in practice all reactions involve the difference between two half-reactions and any measurable thermodynamic quantities involve differences in the chemical potential of the electron. Therefore such quantities as μ_e and ΔG^0 are still meaningful if a proper reference state is defined. The same problem arises if one wishes to split the potential difference of an electrochemical cell (between two metallic electrodes) into two contributions of each electrode. Although one may consider the potential difference between two electrodes as the difference between the potential of each electrode with respect to the solution, there is no way to measure the "potential of the solution." Any measurement would involve an electrode again.

The required standard is internationally agreed as the *potential of the standard hydrogen electrode*, defined as zero with respect to the solution (at any temperature). The standard hydrogen electrode is a platinum electrode in a solution with pH $= 0$ and in contact with gaseous hydrogen at a pressure of 1 bar. The electrode reduction half-reaction is

$$2\,H^+ + 2\,e^- \rightleftharpoons H_2$$

As the electron is in equilibrium with the electrons in a metallic electrode at a given electrical potential Φ, under conditions of zero current, the thermodynamic potential of the electron is given by

$$\mu_e = -F\Phi, \tag{16.75}$$

where F is the Faraday, which is the absolute value of the charge of a mole electrons ($96\,485$ C). 1 Volt corresponds to almost 100 kJ/mol. Here, the electrical potential is defined with respect to the "potential of the solution" according to the standard hydrogen electrode convention.

We can now summarize the thermodynamic electrochemical relations for the general half-reaction

$$ox + \nu\,e^- \rightleftharpoons red$$

as follows:

$$\Delta G = \mu_{red} - \mu_{ox} - \nu\mu_e = 0, \tag{16.76}$$

implying that

$$\mu_{red}^0 + RT \ln a_{red} - \mu_{ox}^0 - RT \ln a_{ox} + \nu F\Phi = 0. \tag{16.77}$$

With the definition of the *standard reduction potential* E^0:

$$-\nu F E^0 = \Delta G^0 = \mu^0_{\text{red}} - \mu^0_{\text{ox}}, \qquad (16.78)$$

we arrive at an expression for the equilibrium, i.e., current-free, potential[7] of a (platinum) electrode with respect to the "potential of the solution" (defined through the standard hydrogen electrode)

$$\Phi = E^0 - \frac{RT}{\nu F} \ln \frac{a_{\text{red}}}{a_{\text{ox}}}. \qquad (16.79)$$

Values of E^0 have been tabulated for a variety of reduction–oxidation couples, including redox couples of biological importance. When a metal or other solid is involved, the activity is meant with respect to mole fraction, which is equal to 1. The convention to tabulate reduction potentials is now universally accepted, meaning that a couple with a more negative standard potential has – depending on concentrations – the tendency to reduce a couple with a more positive E^0. A concentration ratio of 10 corresponds to almost 60 mV for a single electron transfer reaction.

16.8 Colligative properties

Colligative properties of solutions are properties that relate to the combined influence of all solutes on the thermodynamic potential of the solvent. This causes the solvent to have an osmotic pressure against the pure solvent and to show shifts in vapor pressure, melting point and boiling point. For *dilute* solutions the thermodynamic potential of the solvent is given by

$$\mu = \mu^* + RT \ln x_{\text{solv}} = \mu^* + RT \ln(1 - \sum_j{}' x_j) \qquad (16.80)$$

$$\approx \mu^* - RT \sum_j{}' x_j \approx \mu^* - M_{\text{solv}} RT \sum_j{}' m_j, \qquad (16.81)$$

where μ^* is the thermodynamic potential of the pure solvent at the same pressure and temperature, and the prime in the sum means omitting the solvent itself. M_{solv} is the molar mass of the solvent. A solution with a thermodynamic potential of the solvent equal to an ideal dilute solution of m molal has an *osmolality* of m.

The consequences of a reduced thermodynamic potential of the solvent $\mu^* - \Delta\mu$ are the following:

---

[7] The current-free equilibrium potential has long been called the *electromotive force* (EMF), but this historical and confusing nomenclature is now obsolete. The inconsistent notation E (usual for electric field, not potential) for the standard potential has persisted, however.

(i) The vapor pressure p is reduced with respect to the saturation vapor pressure p^* of the pure solvent (assuming ideal gas behavior) according to

$$-\Delta\mu = RT \ln \frac{p}{p^*}, \tag{16.82}$$

or

$$p = p^*(1 - \sum_j' x_j). \tag{16.83}$$

This is a form of *Raoult's law* stating that the vapor pressure of a volatile component in an ideal mixture is proportional to the mole fraction of that component.

(ii) The solution has an osmotic pressure Π (to be realized as a real pressure increase after the pure solvent is equilibrated with the solution, from which it is separated by a *semipermeable membrane* that is exclusively permeable to the solvent), determined by the equality

$$\mu^* = \mu^* - \Delta\mu + \Pi v_{\text{solv}}, \tag{16.84}$$

or

$$\Pi = \rho_{\text{solv}} RT m_{\text{osmol}} \approx \frac{RT \sum_j' x_j}{v_{\text{solv}}}. \tag{16.85}$$

The latter equation is the *van 't Hoff equation* stating that the osmotic pressure equals the pressure that the solute would have if the solute particles would behave as in an ideal gas.

(iii) The solution has a *boiling point elevation* ΔT_{b} equal to

$$\Delta T_{\text{b}} = \frac{\Delta\mu}{s_{\text{g}} - s_{\text{l}}}, \tag{16.86}$$

or

$$\frac{\Delta T_{\text{b}}}{T_{\text{b}}} = \frac{\Delta\mu}{\Delta h_{\text{vap}}}. \tag{16.87}$$

In Fig. 16.2 the boiling point elevation is graphically shown to be related to the reduction in thermodynamic potential of the solvent. Since the slopes of the lines are equal to the molar entropies, equation (16.86) follows directly from this plot, under the assumption that the curvature in the range $(T_{\text{b}}^0, T_{\text{b}})$ is negligible,

(iv) Similarly, the solution has a *freezing point depression* (or *melting point depression*) ΔT_{m} (see Fig. 16.2), given by

$$\Delta T_{\text{m}} = \frac{\Delta\mu}{s_{\text{l}} - s_{\text{s}}}, \tag{16.88}$$

Thermodynamic potential

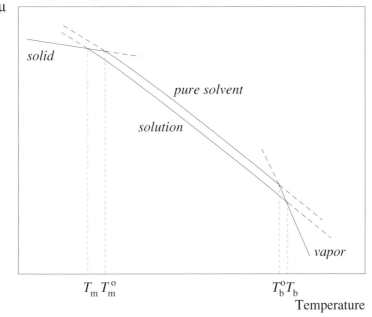

Figure 16.2 Thermodynamic potential of the solvent for solid, liquid and vapor phases, both for the pure liquid (thick lines) and for a solution (thin line), as a function of temperature. The negative slope of the curves is given by the molar entropy of each state. The state with lowest μ is stable. The solution shows a melting point depression and boiling point elevation, proportional to the reduction of the solvent thermodynamic potential.

or

$$\frac{\Delta T_\mathrm{m}}{T_\mathrm{m}} = \frac{\Delta \mu}{\Delta h_\mathrm{fusion}}. \tag{16.89}$$

16.9 Tabulated thermodynamic quantities

Thermodynamic quantities, derived from experiments, need unambiguous specification when tabulated for general use. This applies to standard states and activities and to standard conditions, if applicable. Standard states should include the quantity and the unit in which the quantity is expressed. In the case of pressure, the now preferred standard state is 1 bar (10^5 Pa), but the reader should be aware that some tables still use the standard atmosphere, which equals 101 325 Pa. In all cases the temperature and pressure, and other relevant environmental parameters, should be specified. Good tables include error estimates, give full references and are publicly accessible.

Quantities *of formation* (usually enthalpy, entropy and Gibbs free energy) of a substance refer to the formation of the substance from its elements, all components taken at the reference temperature of 273.15 K and reference pressure of 1 bar, in the phase in which the substance or element is stable under those conditions (if not specified otherwise). The *absolute entropy* refers to the third-law entropy, setting $S = 0$ at $T = 0$.

The international organization CODATA maintains a list of key values for thermodynamics.[8] Other sources for data are the *Handbook of Chemistry and Physics*,[9] and various databases[10] of which many are freely available through the (US) National Institute of Standards and Technology (NIST).[11]

16.10 Thermodynamics of irreversible processes

Consider a system that is slightly off-equilibrium because small gradients exist of temperature, pressure, concentrations, and/or electric potential. The gradients are so small that the system may be considered locally in equilibrium over volumes small enough to have negligible gradients and still large enough to contain a macroscopic number of particles for which thermodynamic relations apply. We now look at two adjacent compartments (see Fig. 16.3), each in local equilibrium, but with slightly different values of parameters such as temperature, pressure, concentration and/or electric potential. There are possibly fluxes from the left to the right compartment or vice-versa, both of heat and of particles.

16.10.1 Irreversible entropy production

The crucial step is to consider the total entropy production in both compartments together. We shall then see that the fluxes cause an entropy production. This *irreversible* entropy production, which is always positive, can be related to the product of properly defined fluxes and *thermodynamic forces* which are the gradients that cause the fluxes.

Start with (16.25). Since the volume of the compartments is constant,

---

[8] CODATA, the Committee on Data for Science and Technology, is based in Paris and acts under the auspices of the International Council for Science ICSU,. See http://www.codata.org. For the key values see http://www.codata.org/codata/databases/key1.html.
[9] Published by CRC Press, Baco Raton, FL, USA. It comes out in a yearly printed edition and in a Web version available at http://www.hbcpnetbase.com
[10] A useful list of links to data sites is provided by
              http://tigger.uic.edu/~mansoori/Thermodynamic.Data.and.Property.html.
[11] http://www.nist gov.

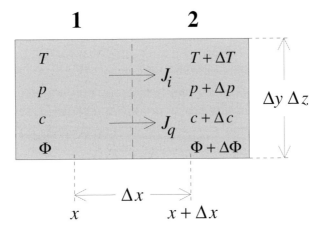

Figure 16.3 Two adjacent compartments in a system in which irreversible fluxes of particles and/or heat are present. Each compartment of size $\Delta x \Delta y \Delta z$ is in local equilibrium, but adjacent compartments have slightly different values of temperature, pressure, concentrations and/or electric potential representing gradients in the x-direction.

(16.25) can be rewritten as

$$dS = \frac{dU}{T} - \sum_i \frac{\mu_i}{T} \, dn_i. \tag{16.90}$$

Numbering the compartments 1 and 2, the total entropy change equals

$$dS_1 + dS_2 = \frac{dU_1}{T_1} + \frac{dU_2}{T_2} - \sum_i \frac{\mu_{1i}}{T_1} \, dn_1 - \sum_i \frac{\mu_{2i}}{T_2} \, dn_2. \tag{16.91}$$

Now call the entropy production per unit volume and per unit time σ. If we call the energy flux per unit time and per unit of surface area J_u (in J m^{-2} s^{-1}) and the particle flux J_i (in mol m^{-2} s^{-1}), then

$$J_u \, \Delta y \, \Delta z = \frac{dU_2}{dt} = -\frac{dU_1}{dt}, \tag{16.92}$$

$$J_i \, \Delta y \, \Delta z = \frac{dn_{2i}}{dt} = -\frac{dn_{1i}}{dt}, \tag{16.93}$$

so that

$$\sigma \, \Delta x \, \Delta y \, \Delta z = \left[J_u \left(\frac{1}{T_2} - \frac{1}{T_1} \right) - \sum_i J_i \left(\frac{\mu_{2i}}{T_2} - \frac{\mu_{1i}}{T_1} \right) \right] \Delta y \, \Delta z. \tag{16.94}$$

Equating the differences between 1 and 2 to the gradient multiplied by Δx, and then extending to three dimensions, we find

$$\sigma = \boldsymbol{J}_{\mathrm{u}} \cdot \boldsymbol{\nabla}\left(\frac{1}{T}\right) - \sum_i \boldsymbol{J}_i \cdot \boldsymbol{\nabla}\left(\frac{\mu_i}{T}\right). \tag{16.95}$$

This equation formulates the irreversible entropy production per unit volume and per unit time as a sum of scalar products of *fluxes* $\boldsymbol{J}_\alpha$ and conjugated *forces* $\boldsymbol{X}_\alpha$:

$$\sigma = \sum_\alpha \boldsymbol{J}_\alpha \cdot \boldsymbol{X}_\alpha = \mathbf{J}^{\mathsf{T}}\mathbf{X}, \tag{16.96}$$

where the gradient of the inverse temperature is conjugate to the flux of internal energy, and minus the gradient of the thermodynamic potential divided by the temperature, is conjugate to the particle flux. The last form in (16.96) is a matrix notation, with $\mathbf{J}$ and $\mathbf{X}$ being column vector representations of all fluxes and forces.

It is possible and meaningful to transform both fluxes and forces such that (16.96) still holds. The formulation of (16.94) is very inconvenient: for example, a temperature gradient does not only work through the first term but also through the second term, including the temperature dependence of the thermodynamic potential. Consider as independent variables: $p, T, \Phi, x_i, i = 1, \ldots, n - 1$[12] and use the following derivatives:

$$\frac{\partial \mu_i}{\partial p} = v_i, \qquad \frac{\partial \mu_i}{\partial T} = -s_i, \qquad \frac{\partial \mu_i}{\partial \Phi} = z_i F, \tag{16.97}$$

where z_i is the charge (including sign) in units of the elementary charge of species i. Equation (16.94) then transforms to

$$\sigma = \boldsymbol{J}_{\mathrm{q}} \cdot \boldsymbol{\nabla}\left(\frac{1}{T}\right) - \frac{1}{T}\boldsymbol{J}_{\mathrm{v}} \cdot \boldsymbol{\nabla}p + \frac{1}{T}\boldsymbol{j} \cdot \boldsymbol{E} - \frac{1}{T}\sum_i \boldsymbol{J}_i \cdot (\boldsymbol{\nabla}\mu_i)_{p,T}. \tag{16.98}$$

Here we have used the relation $\mu_i = h_i - T s_i$ and the following definitions:

- $\boldsymbol{J}_{\mathrm{q}}$ is the *heat flux* (in J m^{-2} s^{-1}), being the energy flux from which the contribution as a result of particle transport has been subtracted:

$$\boldsymbol{J}_{\mathrm{q}} \stackrel{\text{def}}{=} \boldsymbol{J}_{\mathrm{u}} - \sum_i \boldsymbol{J}_i h_i. \tag{16.99}$$

Note that the energy transported by particles is the partial molar enthalpy, not internal energy, because there is also work performed against the pressure when the partial molar volume changes.

---

[12] With n components there are $n - 1$ independent mole fractions. One may also choose $n - 1$ concentrations or molalities.

- $\boldsymbol{J}_{\mathrm{v}}$ is the total *volume flux* (in m/s), which is a complicated way to express the *bulk velocity* of the material:

$$\boldsymbol{J}_{\mathrm{v}} \overset{\text{def}}{=} \sum_i \boldsymbol{J}_i v_i. \tag{16.100}$$

- $\boldsymbol{j}$ is the *electric current density* (in A/m^2):

$$\boldsymbol{j} \overset{\text{def}}{=} \sum_i \boldsymbol{J}_i z_i F. \tag{16.101}$$

Note that the irreversible entropy production due to an electric current is the Joule heat divided by the temperature.

The last term in (16.98) is related to gradients of composition and needs to be worked out. If we have n species in the mixture, there are only $n - 1$ independent composition variables. It is convenient to number the species with $i = 0, 1, \ldots, n - 1$, with $i = 0$ representing the solvent, and $x_i, i = 1, \ldots, n - 1$ the independent variables. Then

$$x_0 = 1 - \sideset{}{'}\sum_i x_i, \tag{16.102}$$

where the prime in the summation means omission of $i = 0$. The Gibbs–Duhem relation (16.16) relates the changes in chemical potential of the different species. It can be written in the form

$$x_0 (\boldsymbol{\nabla} \mu_0)_{p,T} + \sideset{}{'}\sum_i x_i (\boldsymbol{\nabla} \mu_i)_{p,T} = 0. \tag{16.103}$$

Using this relation, the last term in (16.98) can be rewritten as

$$-\frac{1}{T} \sum_i \boldsymbol{J}_i \cdot (\boldsymbol{\nabla} \mu_i)_{p,T} = -\frac{1}{T} \left[\sideset{}{'}\sum_i \left(\boldsymbol{J}_i - \frac{x_i}{x_0} \boldsymbol{J}_0 \right) \cdot (\boldsymbol{\nabla} \mu_i)_{p,T} \right]. \tag{16.104}$$

Here a *difference flux* with respect to flow of the solvent appears in the equation for irreversible entropy production. If all particles move together at the same bulk flux J (total moles per m^2 and per s), then $J_i = x_i J$ for all i, and $J_i^d = 0$. So this definition makes sense: a concentration gradient produces a difference flux by irreversible diffusion processes.

Note that relation (16.98), including (16.104), has the form of a product of fluxes and forces as in (16.96). Hence also for these fluxes and forces linear and symmetric Onsager relations (Section 16.10.3) are expected to be valid.

16.10.2 Chemical reactions

Whenever chemical reactions proceed spontaneously, there is also an irreversible entropy production. According to (16.90), entropy is generated when the composition changes according to $-(1/T)\sum_i \mu_i\, dn_i$. In the previous subsection this term was evaluated when the number of molecules changed due to fluxes. We have not yet considered what happens if the numbers change due to a chemical reaction.

Assume the reaction[13]

$$0 \rightarrow \sum_i \nu_i C_i \tag{16.105}$$

proceeds for a small amount, Δn mole, as written. This means that $\nu_i \Delta n$ moles of species C_i are formed (or removed when ν_i is negative). The irreversible entropy production ΔS is

$$\Delta S = -\frac{1}{T}\left(\sum_i \nu_i\mu_i\right)\Delta n. \tag{16.106}$$

With the following definitions:

- the *affinity* of the reaction $A \overset{\text{def}}{=} -\sum_i \nu_i\mu_i$;
- the *degree of advancement* of the reaction $\xi \overset{\text{def}}{=}$ number of moles per unit volume the reaction (as written) has proceeded,

the irreversible entropy production per unit volume and per unit time due to the advancement of the reaction can be written as

$$\sigma = \frac{1}{T}A\frac{d\xi}{dt}. \tag{16.107}$$

The rate of advancement $d\xi/dt$ can be viewed as the *reaction flux* and A/T then is the *driving force for the reaction*. Note that reaction fluxes and forces are *scalar* quantities, contrary to the vector quantities we encountered thus far. The rate of advancement is equal to the usual net *velocity* of the reaction

$$v_{\text{react}} = \frac{1}{\nu_i}\frac{d[C_i]}{dt}. \tag{16.108}$$

In equilibrium, the affinity is zero. For reactions that deviate only slightly from equilibrium, the velocity is linear in the affinity; far from equilibrium no such relation with affinity exists.

---

[13] We use the same notation as in (16.64).

16.10.3 Phenomenological and Onsager relations

For small deviations from equilibrium, i.e., small values of the driving forces
X, one may *assume* that the fluxes are proportional to the driving forces.
Such linear relations are the first term in a Taylor expansion in the driving
forces,[14] and they are justified on a *phenomenological* basis. The main
driving force for a flux is its conjugated force, but in general the linear
relations may involve any other forces as well:

$$J_k = \sum_l L_{kl} X_l \quad \text{or} \quad \mathbf{J} = \mathbf{L}\mathbf{X}. \tag{16.109}$$

Here L_{kl} are the *phenomenological coefficients*. This implies that for the
entropy production

$$\sigma = \mathbf{X}^\mathsf{T} \mathbf{L}^\mathsf{T} \mathbf{X}. \tag{16.110}$$

From the second law we know that the irreversible entropy production must
always be positive for any combination of driving forces. Mathematically
this means that the matrix $\mathbf{L}$ must be *positive definite* with only positive
eigenvalues.

The diagonal phenomenological coefficients relate to fluxes resulting from
their conjugate forces, such as heat conduction (heat flow due to temperature gradient), viscous flow, e.g., through a membrane or through porous
material (fluid flow due to hydrostatic pressure difference), electrical conduction (current due to electric field), and diffusion (particle flow with respect
to solvent due to concentration gradient). Off-diagonal coefficients relate to
fluxes that result from other than their conjugate forces, such as:

- thermoelectric effect (current due to temperature gradient) and Peltier
  effect (heat flow due to electric field);
- thermal diffusion or Soret effect (particle separation due to temperature
  gradient) and Dufour effect (heat flow due to concentration gradient);
- osmosis (volume flow due to concentration gradient) and reverse osmosis
  (particle separation due to pressure gradient);
- electro-osmosis (volume flow due to electric field) and streaming potential
  (current due to pressure gradient);
- diffusion potential (current due to concentration gradient) and electro-
  phoresis (particle separation due to electric field).

On the basis of microscopic reversibility and by relating the phenomenological coefficients to fluctuations, Onsager (1931b) came to the conclusion

---

[14] Note that this does not imply that the coefficients themselves are constants; they may depend
on temperature, pressure and concentration.

that the matrix $\mathbf{L}$ must be symmetric:

$$L_{kl} = L_{lk}. \tag{16.111}$$

These are Onager's *reciprocal relations* that relate cross effects in pairs. For example, the thermoelectric effect is related to the Peltier effect as follows:

$$J_q = L_{qq} \, \mathbf{grad} \, \frac{1}{T} + L_{qe} \frac{E}{T}, \tag{16.112}$$

$$j = L_{eq} \, \mathbf{grad} \, \frac{1}{T} + L_{ee} \frac{E}{T}. \tag{16.113}$$

The thermoelectric coefficient is defined as the ratio of the potential difference, arising under current-free conditions, to the externally maintained temperature difference (in V/K). This is equal to

$$\left(\frac{\mathbf{grad} \, \Phi}{\mathbf{grad} \, T} \right)_{j=0} = -\frac{L_{eq}}{T L_{ee}}. \tag{16.114}$$

The Peltier effect is defined as the heat flux carried per unit current density under conditions of constant temperature (in J/C). This is equal to

$$\left(\frac{J_q}{j} \right)_{\mathbf{grad} \, T = 0} = -\frac{L_{qe}}{L_{ee}}. \tag{16.115}$$

Onsager's relation $L_{eq} = L_{qe}$ implies that the thermoelectric coefficient equals the Peltier coefficient, divided by the absolute temperature.

16.10.4 Stationary states

When there are no constraints on a system, it will evolve spontaneously into an equilibrium state, in which the irreversible entropy production becomes zero. The direction of this process is dictated by the second law; a flow J_i diminishes the conjugate force, and in course of time the entropy production *decreases*. With external constraints, either on forces or on fluxes, or on combinations thereof, the system will evolve into a *stationary state* (or *steady state*), in which the entropy production becomes *minimal*. This is easily seen as follows.

Assume that forces X_α are constrained by the environment. X_α are a subset of all forces X_i. The system develops in the direction of decreasing entropy production, until the entropy production is minimal. Minimizing $\mathbf{X}^\mathsf{T} \mathbf{L} \mathbf{X}$ under constraints $X_k = constant$, requires that

$$\frac{\partial}{\partial X_k} \left(\sum_{ij} L_{ij} X_i X_j + \sum_{\alpha} \lambda_\alpha X_\alpha \right) = 0, \tag{16.116}$$

for all k, and where λ_α are Lagrange undetermined multipliers. This implies that:

- $J_\alpha = $ constant;
- $J_k = 0$ for $k \neq \alpha$.

Thus, the system evolves into a steady state with constant fluxes; fluxes conjugate to unconstrained forces vanish.

Exercises

16.1   Show that the following concentration dependencies, valid for ideal solutions, satisfy the Gibbs–Duhem relation:

$$\mu_w = \mu_w^0 + RT \ln(1 - x),$$

$$\mu_s = \mu_s^0 + RT \ln x.$$

16.2   Prove the Gibbs–Helmholtz equation, (16.32).

16.3   Show that another Gibbs–Helmholtz equation exists, replacing G by A and H by U.

16.4   At first sight you may wonder why the term $\sum_i \mu_i \, dn_i$ occurs in (16.25) to (16.28). Prove, for example, that $(\partial H / \partial n_i)_{p,T} = \mu_i$.

16.5   Prove (16.40) and (16.41).

16.6   Prove (16.42) by starting from the pressure definition from (16.27), and then using the Maxwell relation derived from (16.27).

16.7   Estimate the boiling point of water from the Clausius–Clapeyron equation at an elevation of 5500 m, where the pressure is 0.5 bar. The heat of vaporization is 40 kJ/mol.

16.8   Estimate the melting point of ice under a pressure of 1 kbar. The densities of ice and water are 917 and 1000 kg/m^3; the heat of fusion is 6 kJ/mol.

16.9   Rationalize that the *cryoscopic constant* (the ratio between freezing point depression and molality of the solution) equals 1.86 K kg mol^{-1} for water.

17

Review of statistical mechanics

17.1 Introduction

Equilibrium statistical mechanics was developed shortly after the introduction of thermodynamic entropy by Clausius, with Boltzmann and Gibbs as the main innovators near the end of the nineteenth century. The concepts of atoms and molecules already existed but there was no notion of quantum theory. The link to thermodynamics was properly made, including the interpretation of entropy in terms of probability distributions over ensembles of particle configurations, but the quantitative counting of the number of possibilities required an unknown elementary volume in phase space that could only later be identified with Planck's constant h. The indistinguishability of particles of the same kind, which had to be introduced in order to avoid the *Gibbs' paradox*,[1] got a firm logical basis only after the invention of quantum theory. The observed distribution of black-body radiation could not be explained by statistical mechanics of the time; discrepancies of this kind have been catalysts for the development of quantum mechanics in the beginning of the twentieth century. Finally, only after the completion of basic quantum mechanics around 1930 could quantum statistical mechanics – in principle – make the proper link between microscopic properties at the atomic level and macroscopic thermodynamics. The classical statistical mechanics of Gibbs is an approximation to quantum statistics.

In this review we shall reverse history and start with quantum statistics, proceeding to classical statistical mechanics as an approximation to quantum

---

[1] In a configuration of N particles there are $N!$ ways to order the particles. If each of these ways are counted as separate realizations of the configuration, a puzzling paradox arises: thermodynamic quantities that involve entropy appear not to be proportional to the size of the system. This paradox does not arise if the number of realizations is divided by $N!$. In quantum statistics one does not count configurations, but quantum eigenstates which incorporate the indistinguishability of particles of the same kind in a natural way.

statistics. This will enable us to see the limitations of classical computational approaches and develop appropriate quantum corrections where necessary.

Consider an equilibrium system of particles, like nuclei and electrons, possibly already pre-organized in a set of molecules (atoms, ions). Suppose that we know all the rules by which the particles interact, i.e., the Hamiltonian describing the interactions. Then we can proceed to solve the time-independent Schrödinger equation to obtain a set of wave functions and corresponding energies, or – by classical approximation – a set of configurations in *phase space*, i.e., the multidimensional space of coordinates and momenta with their corresponding energies. The aim of statistical mechanics is to provide a recipe for the proper averaging over these detailed solutions in order to obtain thermodynamic quantities. Hopefully the averaging is such that the (very difficult and often impossible) computation of the detailed quantities can be avoided. We must be prepared to handle thermodynamic quantities such as temperature and entropy which cannot be obtained as simple averages over microscopic variables.

17.2 Ensembles and the postulates of statistical mechanics

The basic idea, originating from Gibbs,[2] is to consider a hypothetical *ensemble* of a large number of replicas of the system, with the same thermodynamic conditions but different microscopic details. Properties are then obtained by averaging over the ensemble, taken in the limit of an infinite number of replicates. The ensemble is supposed to contain all possible states of the system and be representative for the single system considered over a long time. This latter assumption is the *ergodic* postulate. Whether a realistic system is in practice ergodic (i.e., are all microscopic possibilities indeed realized in the course of time?) is a matter of time scale: often at low temperatures internal processes may become so slow that not all possibilities are realized in the experimental observation time, and the system is not ergodic and in fact not in complete equilibrium. Examples are metastable crystal modifications, glassy states, polymer condensates and computer simulations that provide incomplete sampling or insufficient time scales.

Let us try to set up an appropriate ensemble. Suppose that we can describe discrete *states* of the system, numbered by $i = 1, 2, \ldots$ with energies E_i. In quantum mechanics, these states are the solutions of the Schrödinger

---

[2] Josiah Willard Gibbs (1839–1903) studied mathematics and engineering in Yale, Paris, Berlin and Heidelberg, and was professor at Yale University. His major work on statistical mechanics dates from 1876 and later; his collected works are available (Gibbs, 1957). See also http://www-gap.dcs.st-and.ac.uk/~history/Mathematicians/Gibbs.html.

equation for the whole system.[3] Note that energy levels may be *degenerate*, i.e., many states can have the same energy; i numbers the states and not the distinct energy levels. In classical mechanics a state may be a point in phase space, discretized by subdividing phase space into elementary volumes. Now envisage an ensemble of $\mathcal{N}$ replicas. Let $\mathcal{N}_i = w_i \mathcal{N}$ copies be in state i with energy E_i. Under the ergodicity assumption, the fraction w_i is the probability that the (single) system is to be found in state i. Note that

$$\sum_i w_i = 1. \tag{17.1}$$

The number of ways $\mathcal{W}$ the ensemble can be made up with the restriction of given $\mathcal{N}_i$'s equals

$$\mathcal{W} = \frac{\mathcal{N}!}{\prod_i \mathcal{N}_i!} \tag{17.2}$$

because we can order the systems in $\mathcal{N}!$ ways, but should not count permutations among $\mathcal{N}_i$ as distinct. Using the Stirling approximation for the factorial,[4] we find that

$$\ln \mathcal{W} = -\mathcal{N} \sum_i w_i \ln w_i. \tag{17.3}$$

If the assumption is made (and this is the second *postulate* of statistical mechanics) that all possible ways to realize the ensemble are equally probable, the set of probabilities $\{w_i\}$ that *maximizes* $\mathcal{W}$ is the most probable distribution. It can be shown that in the limit of large $\mathcal{N}$, the number of ways the most probable distribution can be realized approaches the total number of ways that *any* distribution can be realized, i.e., the most probable distribution dominates all others.[5] Therefore our task is to find the set of probabilities $\{w_i\}$ that *maximizes* the function

$$H = -\sum_i w_i \ln w_i. \tag{17.4}$$

This function is equivalent to Shannon's definition of *information* or *uncertainty* over a discrete probability distribution (Shannon, 1948).[6] It is

---

[3] The states as defined here are *microscopic* states unrelated to the *thermodynamic states* defined in Chapter 16.

[4] $N! \approx N^N e^{-N} \sqrt{2\pi N} \{1 + (O)(1/N)\}$. For the present application and with $\mathcal{N} \to \infty$, the approximation $N! \approx N^N e^{-N}$ suffices.

[5] More precisely: the logarithm of the number of realizations of the maximum distribution approaches the logarithm of the number of all realizations in the limit of $\mathcal{N} \to \infty$.

[6] The relation with information theory has led Jaynes (1957a, 1957b) to propose a new foundation for statistical mechanics: from the viewpoint of an observer the most unbiased guess he can make about the distribution $\{w_i\}$ is the one that maximizes the uncertainty H under the constraints of whatever knowledge we have about the system. Any other distribution would

also closely related (but with opposite sign) to the H-function defined by Boltzmann for the classical distribution function for a system of particles in coordinate and velocity space. We shall see that this function is proportional to the entropy of thermodynamics.

17.2.1 Conditional maximization of H

The distribution with maximal H depends on further conditions we may impose on the system and the ensemble. Several cases can be considered, but for now we shall concentrate on the N, V, T or *canonical* ensemble. Here the particle number and volume of the system and the *expectation* of the energy, i.e., the *ensemble-averaged* energy $\sum_i w_i E_i$, are fixed. The systems are allowed to interact weakly and exchange energy. Hence the energy per system is not constant, but the systems in the ensemble belong to the same equilibrium conditions. Thus H is maximized under the conditions

$$\sum_i w_i = 1, \tag{17.5}$$

$$\sum_i w_i E_i = U. \tag{17.6}$$

Using the method of Lagrange multipliers,[7] the function

$$-\sum_j w_j \ln w_j + \alpha \sum_j w_j - \beta \sum_j w_j E_j \tag{17.7}$$

(the minus sign before β is for later convenience) is maximized by equating all partial derivatives to w_i to zero:

$$-1 - \ln w_i + \alpha - \beta E_i = 0, \tag{17.8}$$

or

$$w_i \propto e^{-\beta E_i}. \tag{17.9}$$

be a biased choice that is only justified by additional knowledge. Although this principle leads to exactly the same results as the Gibbs postulate that all realizations of the ensemble are equally probable, it introduces a subjective flavor into physics that is certainly not universally embraced.

[7] Lagrange undetermined multipliers are used to find the optimum of a function $f(\mathbf{x})$ of n variables $\mathbf{x}$ under s constraint conditions of the form $g_k(\mathbf{x}) = 0$, $k = 1, \ldots, s$. One constructs the function $f + \sum_{k=1}^{s} \lambda_k g_k$, where λ_k are as yet undetermined multipliers. The optimum of this function is found by equating all partial derivatives to zero. Then the multipliers are solved from the constraint equations.

The proportionality constant (containing the multiplier α) is determined by normalization condition (17.5), yielding

$$w_i = \frac{1}{Q} e^{-\beta E_i}, \tag{17.10}$$

$$Q = \sum_i e^{-\beta E_i}. \tag{17.11}$$

Q is called the *canonical partition function*. The multiplier β follows from the implicit relation

$$U = \frac{1}{Q} \sum_i E_i e^{-\beta E_i}. \tag{17.12}$$

As we shall see next, β is related to the temperature and identified as $1/k_{\mathrm{B}}T$.

17.3 Identification of thermodynamical variables

Consider a canonical ensemble of systems with given number of particles and fixed volume. The system is completely determined by its microstates labeled i with energy E_i, and its thermodynamic state is determined by the probability distribution $\{w_i\}$, which in equilibrium is given by the canonical distribution (17.10). The distribution depends on one (and only one) parameter β. We have not introduced the temperature yet, but it must be clear that the temperature is somehow related to the distribution $\{w_i\}$, and hence to β. Supplying *heat* to the system has the consequence that the distribution $\{w_i\}$ will change. In the following we shall first identify the relation between β and temperature and between the distribution $\{w_i\}$ and entropy. Then we shall show how the partition function relates to the Helmholtz free energy, and – through its derivatives – to all other thermodynamic functions.

17.3.1 Temperature and entropy

Now consider the ensemble-averaged energy, which is equal to the thermodynamic internal energy U:

$$U = \sum_i w_i E_i. \tag{17.13}$$

The ensemble-averaged energy changes by changing the distribution $\{w_i\}$, corresponding to heat exchange dq:

$$dU = dq = \sum_i E_i \, dw_i. \tag{17.14}$$

Note that, as a result of the normalization of the probabilities,

$$\sum_i dw_i = 0. \tag{17.15}$$

At constant volume, when no work is done on the system, the internal energy can only change by absorption of heat dq, which in equilibrium equals $T\,dS$:

$$dU = dq = T\,dS \quad \text{or} \quad \frac{1}{T}dq = dS. \tag{17.16}$$

So, in thermodynamics the temperature is defined as the inverse of the integrating factor of dq that produces the differential dS of a state function S (see the discussion on page 426). Can we find an integrating factor for dq in terms of the probabilities w_i?

From (17.10) follows that

$$E_i = -\beta^{-1}\ln w_i - \beta^{-1}\ln Q, \tag{17.17}$$

which can be inserted in (17.14) to yield

$$dq = -\beta^{-1}\sum_i \ln w_i\,dw_i. \tag{17.18}$$

Here use has been made of the fact that $\sum_i dw_i = 0$. Using this fact again, it follows that

$$\beta\,dq = d(-\sum_i w_i \ln w_i). \tag{17.19}$$

So we see that β is an integrating factor for dq, yielding a total differential of a thermodynamic state function $-\sum_i w_i \ln w_i$. Therefore this state function can be identified with the entropy S and β with the inverse temperature $1/T$. Both functions can be scaled with an arbitrary constant, which is determined by the convention about units in the definition of temperature. Including the proper constant we conclude that

$$\beta = \frac{1}{k_B T}, \tag{17.20}$$

$$S = -k_B \sum_i w_i \ln w_i. \tag{17.21}$$

These are the fundamental relations that couple statistical mechanics and thermodynamics.[8] Note that the entropy is simply equal to the information function H introduced in (17.4), multiplied by Boltzmann's constant.

---

[8] Several textbooks use these equations as *definitions* for temperature and entropy, thus ignoring the beautiful foundations of classical thermodynamics.

Strictly, the entropy is only defined by (17.21) in the case that $\{w_i\}$ represents a canonical equilibrium distribution. We may, however, extend the definition of entropy by (17.21) for any distribution; in that case finding the equilibrium distribution is equivalent to maximizing the entropy under the constraint that $\sum_i w_i = 1$ and the additional constraints given by the definition of the ensemble (for the canonical ensemble: constant N and V and given expectation for the energy U : $\langle U \rangle = \sum_i w_i E_i$).

17.3.2 Free energy and other thermodynamic variables

The entropy is proportional to the *expectation* of $\ln w_i$, i.e., the average of $\ln w_i$ over the distribution $\{w_i\}$:

$$S = -k_B \langle \ln w_i \rangle. \tag{17.22}$$

From the canonical distribution (17.10), it follows that

$$\ln w_i = -\ln Q - \beta E_i, \tag{17.23}$$

and taking the expectation over both sides, we obtain

$$-\frac{S}{k_B} = -\ln Q - \frac{U}{k_B T}, \tag{17.24}$$

which reshuffles to

$$-k_B T \ln Q = U - TS = A. \tag{17.25}$$

This simple relation between Q and the Helmholtz free energy A is all we need to connect statistical and thermodynamic quantities: if we know Q as a function of V and β, we know A as a function of V and T, from which all other thermodynamic quantities follow.

17.4 Other ensembles

Thus far we have considered the canonical ensemble with constant N and V, and given expectation U of the energy over the ensemble. It appeared that the latter requirement implied the existence of a constant β, identified with the inverse temperature. Thus the canonical ensemble is also called the N, V, T ensemble.

Although the canonical ensemble is often the most useful one, it is by no means the only possible ensemble. For example, we can constrain not only N and V, but also the energy E for each system and obtain the *microcanonical ensemble*; the ensemble then consists of systems in different microstates (wave functions) with the same degenerate energy.[9] Instead of constraining the volume for each system, we can prescribe the ensemble average of the volume, which introduces another constant that appears to be related to the pressure. This is the N, p, T ensemble. Finally, if we do not fix the particle number for each system, but only fix the ensemble average of the particle number, a constant appears that is related to the chemical potential. This produces the *grand-canonical* or μ, V, T ensemble if the volume is fixed, or the μ, p, T ensemble if the ensemble-averaged volume is fixed.

The recipe is always the same: Let w_i be the probability that the system is in state i (numbering each of all possible states, given the freedom we give the various parameters), and maximize $-\sum_i w_i \ln w_i$ under the conditions we impose on the ensemble. Each condition introduces one Lagrange multiplier, which can be identified with a thermodynamic quantity. This can be summarized as follows:

- The N, V, E or *microcanonical* ensemble. The system will have a degeneracy Ω, being the number of states with energy E (or within a very small, fixed energy interval). $-\sum_i w_i \ln w_i$ is maximized under the only condition that $\sum_i w_i = 1$, which implies that all probabilities w_i are equal and equal to $1/\Omega$; it follows that $S = k_B \ln \Omega$. Knowledge of the "partition function" Ω (and hence S) as a function of V and $E = U$ then generates all thermodynamic quantities. For example, $T = \partial S / \partial E$.
- The N, V, T or *canonical* ensemble. See above. The partition function is $Q(N, V, \beta) = \sum_i \exp(-\beta E_i)$ and thermodynamic functions follow from $\beta = (k_B T)^{-1}$ and $A = -k_B T \ln Q$.
- The N, p, T or *isobaric-isothermal* ensemble. Here the particle number of the system and the *ensemble-averaged* energy and volume are fixed. The w_i are now a function of volume (a continuous variable), and we look for the probability $w_i(V) \, dV$ that the system is in state i with volume between V and $V + dV$. Thus[10] $H = -\int dV \sum_i w_i(V) \ln w_i(V)$ is maximized under the conditions

$$\int dV \sum_i w_i(V) = 1, \qquad (17.26)$$

[9] Neither experimentally, nor in simulations is it possible to generate an exact microcanonical ensemble. The spacing between energy levels becomes very small for macroscopic systems and complete thermal isolation including radiation exchange is virtually impossible; algorithms usually do not conserve energy exactly. But the microcanonical ensemble can be defined as having an energy in a small interval $(E, E + \Delta E)$.

[10] This is a somewhat sloppy extension of the H-function of (17.4) with an integral. The H-function becomes infinite when a continuous variable is introduced, because the number of choices in the continuous variable is infinite. The way out is to discretize the variable V in small intervals ΔV. The equation for H then contains $\ln[w_i(V) \Delta V]$. But for maximization the introduction of ΔV is immaterial.

$$\int dV \sum_i w_i(V) E_i(V) = U, \tag{17.27}$$

$$\int dV' V' \sum_i w_i(V') = V. \tag{17.28}$$

The Lagrange optimization yields

$$\ln w_i(V) \propto -\beta E_i(V) - \gamma V, \tag{17.29}$$

or

$$w_i(V) = \frac{1}{\Delta} e^{-\beta E_i(V) - \gamma V}, \tag{17.30}$$

$$\Delta = \int dV \sum_i e^{-\beta E_i(V) - \gamma V} \tag{17.31}$$

Δ is the *isothermal-isobaric partition function*. Identifying $-k_B \langle \ln w_i(V) \rangle$ with the entropy S, we find that

$$\frac{S}{k_B} = \ln \Delta + \beta U + \gamma V. \tag{17.32}$$

Hence, $\beta = (k_B T)^{-1}$, $\gamma = \beta p$, and the thermodynamic functions follow from

$$G = U - TS + pV = -k_B T \ln \Delta. \tag{17.33}$$

- The μ, V, T or *grand-canonical* ensemble.[11] Here the ensemble averages of E and N are fixed, and a Lagrange multiplier δ is introduced, related to the condition $\langle N \rangle = N_A n$, where N_A is Avogadro's number and n is the (average) number of moles in the system.[12] The microstates now involve every possible particle number N and all quantum states for every N. The probabilities and partition function are then given by

$$w_{N,i} = \frac{1}{\Xi} e^{-\beta E_{N,i} + \delta N}, \tag{17.34}$$

$$\Xi = \sum_N e^{\delta N} \sum_i e^{-\beta E_{N,i}}. \tag{17.35}$$

Working out the expression for the entropy $S = -k_B \langle \ln w_{N,i} \rangle$ and comparing with the thermodynamic relation

$$TS = U + pV - n\mu, \tag{17.36}$$

---

[11] Often just called "the grand ensemble."
[12] For a multi-component system, there is a given average number of particles and a Lagrange multiplier for each species. Many textbooks do not introduce Avogadro's number here, with the consequence that the thermodynamic potential is defined per particle and not per mole as is usual in chemistry.

one finds the identifications $\beta = (k_B T)^{-1}$, $\delta = \mu/RT$ and

$$pV = k_B T \ln \Xi(\mu, V, T). \tag{17.37}$$

This equation relates the grand partition function to thermodynamics. Note that

$$\Xi = \sum_{N=0}^{\infty} Q_N \exp\left(\frac{\mu N}{RT}\right). \tag{17.38}$$

If we define the *absolute activity* λ as:[13]

$$\lambda \overset{\text{def}}{=} \exp\left(\left(\frac{\mu}{RT}\right)\right), \tag{17.39}$$

then the grand partition function can be written as

$$\Xi = \sum_{N=0}^{\infty} \lambda^N Q_N. \tag{17.40}$$

Partial derivatives yield further thermodynamic functions:

$$\lambda \frac{\partial \ln \Xi}{\partial \lambda} = \langle N \rangle = N_A n, \tag{17.41}$$

$$\frac{\partial \ln \Xi}{\partial V} = \frac{p}{k_B T}, \tag{17.42}$$

$$\frac{\partial \ln \Xi}{\partial \beta} = -U. \tag{17.43}$$

This ends our review of the most important ensembles. In simulations one strives for realization of one of these ensembles, although integration errors and deviations from pure Hamiltonian behavior may cause distributions that are not exactly equal to those of a pure ensemble. If that is the case, one may in general still trust observed averages, but observed *fluctuations* may deviate significantly from those predicted by theory.

17.4.1 Ensemble and size dependency

One may wonder if and if so, why, the different ensembles yield the same thermodynamic functions. After all, the various ensembles differ in the freedom that we allow for the system and therefore their entropies are different as well. It is quite clear that the entropy of a given system is larger in a canonical than in a microcanonical ensemble, and larger still in a grand ensemble, because there are more microstates allowed. This would seemingly

---

[13] The name "absolute activity" is logical if we compare $\mu = RT \ln \lambda$ with the definition of activity a (e.g., (16.66)): $\mu = \mu^0 + RT \ln a$.

lead to different values for the entropy, as well as for other thermodynamic functions.

The point is that, although the entropies are not strictly the same, they tend to the same value when the system is macroscopic and contains a large number of particles. Each of the thermodynamic variables that is not fixed per system has a probability distribution over the ensemble that tends to a delta function in the limit of an infinite system, with the same value for each kind of ensemble. The ensembles do differ, however, both in the values of averages and in fluctuations, for systems of finite size with a finite number of particles. In numerical simulations in particular one deals with systems of finite size, and one should be aware of (and correct for) the finite-size effects of the various ensembles.

Let us, just for demonstration, consider the finite-size effects in a very simple example: a system of N non-interacting spins, each of which can be either "up" or "down". In a magnetic field the total energy will be proportional to $\sum_i m_i$. Compare the microcanonical ensemble with exact energy $E = 0$, requiring $\frac{1}{2}N$ spins up and $\frac{1}{2}N$ spins down, with the canonical ensemble at such high temperature that all 2^N possible configurations are equally likely (the Boltzmann factor for any configuration equals 1). The entropy in units of k_B is given by

$$\text{microcanonical}: \quad S = \ln \frac{N!}{[(\frac{1}{2}N)!]^2} = \ln N! - 2\ln(\tfrac{1}{2}N)!, \quad (17.44)$$

$$\text{canonical}: \quad S = N \ln 2. \quad (17.45)$$

We see that for large N, in the limit where the Stirling approximation $\ln N! \approx N \ln N - N$ is valid, the two entropies are equal. For smaller N this is not the case, as Fig. 17.1 shows. Plotting the "observed" entropy versus $N^{-1} \ln N$ allows extrapolation to infinite N.

17.5 Fermi–Dirac, Bose–Einstein and Boltzmann statistics

In this section first a more general formulation for the canonical partition function will be given in terms of the trace of an exponential matrix in Hilbert space. The reader may wish to review parts of Chapter 14 as an introduction. Then we shall look at a system of non-interacting particles (i.e., an ideal gas) where the symmetry properties of the total wave function appear to play a role. For fermions this leads to *Fermi–Dirac* (FD) statistics, while bosons obey *Bose–Einstein* (BE) statistics. In the limit of low density or high temperature both kinds of statistics merge into the *Boltzmann approximation*.

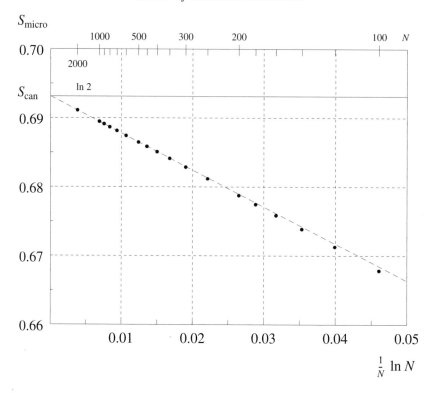

Figure 17.1 Difference between canonical and microcanonical ensemble entropies for finite systems. Case: N non-interacting spins in a magnetic field at high temperature. The canonical entropy per mole in units of R (or per spin in units of k_B) equals $\ln 2$, independent of system size (solid line); the microcanonical (at $E = 0$) molar entropy depends on N and tends to $\ln 2$ for large N. Extrapolation is nearly linear if S is plotted against $N^{-1}\ln N$: the dashed line is a linear connection between the data points at $N = 1000$ and $N = 2000$.

17.5.1 Canonical partition function as trace of matrix

Consider a system of N particles in a box of volume V. For simplicity we take a cubic box and assume the system to be infinite and periodic with the box as repeating unit cell. These restrictions are convenient but not essential: the system may contain various types of different particles, or have another shape. Although not rigorously proved, it is assumed that effects due to particular choices of boundary conditions vanish for large system sizes since these effects scale with the surface area and hence the effect per particle scales as $N^{-1/3}$. The wave function $\Psi(r_1, r_2, \ldots)$ of the N-particle system can be constructed as a linear combination of properly (anti)symmetrized products of single-particle functions. The N-particle wave function must

change sign if two identical fermions are interchanged, and remain the same
if two identical bosons are interchanged (see Chapter 2, page 37, for details).

In equilibrium we look for stationary quantum states involving all particles. There will be stationary solutions with wave functions Ψ_i and energies
E_i and the canonical partition function is given by

$$Q = \sum_i \exp(-\beta E_i).$$
(17.46)

Consider a Hilbert space spanned by all the stationary solutions Ψ_i of the
Hamiltonian (see Chapter 14 for details of vectors and transformations in
Hilbert spaces). Then the matrix H is diagonal and $E_i = H_{ii}$. Thus we can
also write

$$Q = \text{tr} \exp(-\beta H).$$
(17.47)

This equality is quite general and also valid on any complete basis set on
which the Hamiltonian is *not* diagonal. This is easily seen by applying a
unitary transformation U that diagonalizes H, so that $U^\dagger H U$ is diagonal,
and realizing that (see (14.30) on page 386)

$$\exp(-\beta U^\dagger H U) = U^\dagger \exp(-\beta H) U,$$
(17.48)

and that, because the trace of a product is invariant for cyclic exchange of
the elements in the product,

$$\text{tr} \, U^\dagger A U = \text{tr} \, U U^\dagger A = \text{tr} \, A.$$
(17.49)

Solving Q would now require the computation of all diagonal elements
of the Hamiltonian (on a complete basis set). This seems simpler than
solving the Schrödinger equation for the whole system, but is still in practice
impossible for all but the simplest systems.

17.5.2 Ideal gas: FD and BE distributions

In order to get insight into the effect of the symmetry requirements of the
wave function on the partition function we shall now turn to a system which
is solvable: the *ideal gas*. In the ideal gas there are no interactions between
particles. We shall also, for convenience, but without loss of generality, assume that the system contains identical particles, which are either fermions
or bosons. Let the *single-particle wave functions* be given by $\phi_k(\boldsymbol{r})$ with
energy ε_k (Fig. 17.2). The ϕ_k form an orthonormal set of functions, and the

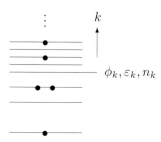

Figure 17.2 Single-particle quantum states k, with wave function ϕ_k, energy ε_k and occupation number n_k. For fermions n_k is restricted to 0 or 1. The shown double occupation of the third level is only possible for bosons.

total wave function is an (anti)symmetrized sum of product states

$$\Psi_i(\boldsymbol{r}_1, \boldsymbol{r}_2, \ldots) = \frac{1}{\sqrt{N!}}(\mp 1)^P \sum_P P[\phi_{k_1}(\boldsymbol{r}_1)\phi_{k_2}(\boldsymbol{r}_2)\ldots]. \tag{17.50}$$

Here the sum is over all possible $N!$ permutations of the N particles, and $(\mp 1)^P$ is a shorthand notation for -1 in case of an odd number of permutations of two fermions (the upper sign) and $+1$ in case of bosons (lower sign). The factor $1/\sqrt{N!}$ is needed to normalize the total wave function again.

It is clear that the total wave function vanishes if two fermions occupy the same single-particle wave function ϕ_k. Therefore the number of particles n_k occupying wave function k is restricted to 0 or 1, while no such restriction exists for bosons:

$$n_k = 0, 1 \quad \text{(fermions).}, \tag{17.51}$$
$$n_k = 0, 1, 2, \ldots \quad \text{(bosons).} \tag{17.52}$$

A N-particle wave function is characterized by the set of occupation numbers $n = \{n_1, n_2, \ldots, n_k, \ldots\}$ with the restriction

$$N = \sum_k n_k. \tag{17.53}$$

The energy E_n is given by

$$E_n = \sum_k n_k \varepsilon_k. \tag{17.54}$$

All possible states with all possible numbers of particles are generated by all possible sets n of numbers subject to the condition (17.51) for fermions.

Thus the grand partition function equals

$$\Xi = \sum_{N=0}^{\infty} \lambda^N \sum_n e^{-\beta E_n}$$

$$= \sum_{n_1} \sum_{n_2} \cdots \lambda^{\sum_k n_k} e^{-\beta \sum_k n_k \varepsilon_k}$$

$$= \sum_{n_1} \sum_{n_2} \cdots \Pi_k \left(\lambda e^{-\beta \varepsilon_k} \right)^{n_k}$$

$$= \sum_{n_1} \left(\lambda e^{-\beta \varepsilon_1} \right)^{n_1} \sum_{n_2} \left(\lambda e^{-\beta \varepsilon_2} \right)^{n_2} \cdots$$

$$= \Pi_k \sum_{n_k} \left(\lambda e^{-\beta \varepsilon_k} \right)^{n_k}. \tag{17.55}$$

where each sum over n_k runs over the allowed values. For fermions only the values 0 or 1 are allowed, yielding the *Fermi–Dirac* statistics:

$$\Xi_{\mathrm{FD}} = \Pi_k \left(1 + \lambda e^{-\beta \varepsilon_k} \right). \tag{17.56}$$

For bosons all values of n_k are allowed, yielding the *Bose–Einstein* statistics:

$$\Xi_{\mathrm{BE}} = \Pi_k \left(1 + \lambda e^{-\beta \varepsilon_k} + \lambda^2 e^{-2\beta \varepsilon_k} + \ldots \right) = \Pi_k \left(1 - \lambda e^{-\beta \varepsilon_k} \right)^{-1}. \tag{17.57}$$

Equations (17.56) and (17.57) can be combined as

$$\Xi_{\mathrm{BE}}^{\mathrm{FD}} = \Pi_k \left(1 \pm \lambda e^{-\beta \varepsilon_k} \right)^{\pm 1} \tag{17.58}$$

with thermodynamic relation

$$\beta p V = \ln \Xi = \pm \sum_k \ln \left(1 \pm \lambda e^{-\beta \varepsilon_k} \right) \tag{17.59}$$

and occupancy numbers given by

$$n_k = \frac{\lambda \exp(-\beta \varepsilon_k)}{1 \pm \lambda \exp(-\beta \varepsilon_k)} = \frac{\exp[-\beta(\varepsilon_k - \mu/N_{\mathrm{A}})]}{1 \pm \exp[-\beta(\varepsilon_k - \mu/N_{\mathrm{A}})]} \tag{17.60}$$

(upper sign: FD; lower sign: BE). Figure 17.3 shows that fermions will fill low-lying energy levels approximately until the thermodynamic potential per particle (which is called the *Fermi level*) is reached; one example is electrons in metals (see exercises). Bosons tend to accumulate on the lowest levels; the thermodynamic potential of bosons is always lower than the lowest level. This *Bose condensation* phenomenon is only apparent at very low temperatures.

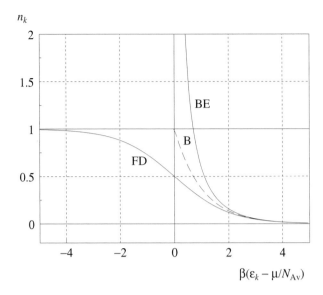

Figure 17.3 Occupation number n_k of kth single-particle quantum state, in an ideal quantum gas, as function of the energy level ε_k above the thermodynamic potential μ, for fermions (FD) and bosons (BE). The dashed line indicates the classical approximation (Boltzmann statistics).

17.5.3 The Boltzmann limit

In gases at high temperature or low density the number of available quantum states considerably exceeds the number of particles. In a periodic box of dimensions $a \times a \times a$; $V = a^3$, the functions

$$\phi_{\boldsymbol{k}} = V^{-1/2} \exp(i\boldsymbol{k}\boldsymbol{r}), \quad \boldsymbol{k} = \frac{2\pi}{a}\boldsymbol{n}, \quad \boldsymbol{n} \in \mathbb{Z}^3, \qquad (17.61)$$

are eigenfunctions of the kinetic energy operator, with eigenvalues

$$\varepsilon_k = \frac{\hbar^2 k^2}{2m}. \qquad (17.62)$$

This implies (see Exercise 17.3) that the number of single-particle translational quantum states between ε and $\varepsilon + d\varepsilon$ is given by

$$g(\varepsilon)\, d\varepsilon = 4\pi\sqrt{2}\frac{m^{3/2}V}{h^3}\varepsilon^{1/2}\, d\varepsilon. \qquad (17.63)$$

Consider 1 cm^3 of neon gas at a pressure of 1 bar and temperature of 300 K, containing 2.4×10^{19} atoms. By integrating (17.63) from 0 up to $k_{\mathrm{B}}T$ we find that there are 6.7×10^{25} quantum states with energies up to $k_{\mathrm{B}}T$. Thus the probability that any particular quantum state is occupied is much smaller than one, and the probability that any state is *doubly* occupied is negligible.

Therefore there will be no distinction between fermions and bosons and the system will behave as in the *classical* or *Boltzmann* limit. In this limit the occupancies $n_k \ll 1$ and hence

$$\lambda e^{-\beta \varepsilon_k} \ll 1. \tag{17.64}$$

In the Boltzmann limit, including the lowest order deviation from the limit, the occupancies and grand partition function are given by (upper sign: FD; lower sign: BE)

$$n_k \approx \lambda e^{-\beta \varepsilon_k} \mp \lambda^2 e^{-2\beta \varepsilon_k} + \cdots, \tag{17.65}$$

$$\beta p V = \ln \Xi \approx \lambda \sum_k e^{-\beta \varepsilon_k} \mp \tfrac{1}{2}\lambda^2 \sum_k e^{-2\beta \varepsilon_k} + \cdots. \tag{17.66}$$

Since $N = \sum_k n_k$, the ideal gas law $pV = Nk_\mathrm{B}T$ is recovered in the Boltzmann limit. The first-order deviation from ideal-gas behavior can be expressed as a second virial coefficient $B_2(T)$:

$$\frac{p}{k_\mathrm{B}T} \approx \frac{N}{V} + N_\mathrm{A}^{-1} B_2(T) \left(\frac{N}{V}\right)^2 + \cdots, \tag{17.67}$$

$$N_\mathrm{A}^{-1} B_2(T)_{\mathrm{BE}}^{\mathrm{FD}} = \pm \frac{h^3}{2(4\pi m k_\mathrm{B}T)^{3/2}} = \pm \frac{\Lambda^3}{2^{5/2}}, \tag{17.68}$$

where Λ is the *de Broglie thermal wavelength*

$$\Lambda \overset{\mathrm{def}}{=} \frac{h}{\sqrt{2\pi m k_\mathrm{B}T}}. \tag{17.69}$$

Avogadro's number comes in because B_2 is defined per mole and not per molecule (see (16.59)). This equation is proved as follows: first show that $N_\mathrm{A}^{-1} B_2 = \pm q_2 V/(2q^2)$, where $q = \sum_k \exp(-\beta \varepsilon_k)$ and $q_2 = \sum_k \exp(-2\beta \varepsilon_k)$, and then solve q and q_2 by approximating the sum by an integral:

$$q = \sum_k e^{-\beta \varepsilon_k}, \qquad \varepsilon_k = \frac{\hbar^2 k^2}{2m} = \frac{h^2 n^2}{2ma^2}. \tag{17.70}$$

Here, $n^2 = n_x^2 + n_y^2 + n_z^2$ with $n_x, n_y, n_z \in \{0, \pm 1, \pm 2, \ldots\}$ (see (17.61)). Use has been made of the periodicity requiring that $\boldsymbol{k} = (2\pi/a)\boldsymbol{n}$. Since the occupation numbers are high (n is large), the sum can be approximated by

$$q \approx \int_{-\infty}^{\infty} \int_{-\infty}^{\infty} \int_{-\infty}^{\infty} \exp\left(-\frac{\beta h^2}{2ma^2} n^2\right) d\boldsymbol{n} = \frac{a^3}{h^3} (2\pi m k_\mathrm{B}T)^{3/2} = \frac{V}{\Lambda^3}. \tag{17.71}$$

Note that this q is the *single particle canonical translational partition function* of a particle in a periodic box.[14] Also note that the quantum deviation from the Boltzmann limit,[15] due to particle symmetry, is of the order h^3.

In the Boltzmann limit, the grand partition function Ξ and the single-particle canonical partition function q are related by

$$\ln \Xi = \lambda q \tag{17.72}$$

and thus

$$\Xi = e^{\lambda q} = \sum_N \lambda^N \frac{q^N}{N!}. \tag{17.73}$$

Since $\Xi = \sum_N \lambda^N Q_N$ (see (17.40)), it follows that the N-particle *canonical partition function* Q_N for non-interacting particles equals

$$Q_N = \frac{q^N}{N!}. \tag{17.74}$$

The $N!$ means that any state obtained by interchanging particles should not be counted as a new microstate, as we expect from the indistinguishability of identical quantum particles. It is a result of quantum symmetry that persists in the classical limit. It's omission would lead to thermodynamic functions that are neither intensive nor extensive (the Gibbs' paradox) as the following will show.

Consider a gas of N non-interacting atoms in a periodic box of volume V, with translational single-atom partition function (17.71)

$$q = \frac{V}{\Lambda^3} \tag{17.75}$$

Using (17.74) and the Stirling approximation (see footnote on page 455) for $N!$, the Helmholtz free energy is given by

$$\begin{aligned} A &= -k_{\mathrm{B}} T \ln Q_N = -k_{\mathrm{B}} T \ln \frac{q^N}{N^N e^{-N}} \\ &= -N k_{\mathrm{B}} T \ln \frac{q}{N} - N k_{\mathrm{B}} T = -N k_{\mathrm{B}} T \ln \frac{q}{N} - pV. \end{aligned} \tag{17.76}$$

From this follows the absolute thermodynamic potential of the gas

$$\mu = \frac{G}{n} = \frac{A + pV}{n} = -RT \ln \frac{q}{N} \tag{17.77}$$

[14] The same result is obtained if the box is not periodic, but closed with infinitely high walls. The wave functions must than vanish at the boundaries and thus be composed of sine waves with wave lengths that are whole fractions of twice the box length. This leads to $8\times$ higher density of points in *n*-space, of which only one octant (positive n) is valid, and thus to the same value of the integral.

[15] Any quantum correction to classical behavior should contain h; the classical limit is often viewed as the limit for $h \to 0$, which is nonsensical for a physical constant, but useful.

$$= RT \ln \frac{\Lambda^3 p^0}{k_B T} + RT \ln \frac{p}{p^0}, \tag{17.78}$$

where p^0 is (any) standard pressure. We recover the linear dependence of μ of the logarithm of the pressure. Without the $N!$ we would have found μ to be proportional to the logarithm of the pressure divided by the number of particles: a nonsensical result.

The single-particle partition function q is still fully quantum-mechanical. It consists of a product of the translational partition function q_{trans} (computed above) and the internal partition function, which – in good approximation – consists of a product of the rotational partition function q_{rot} and the internal vibrational partition function q_{vib}, all for the electronic ground state. If there are low-lying excited electronic states that could be occupied at the temperatures considered, the internal partition function consists of a sum of vibro-rotational partition functions, if applicable multiplied by the degeneracy of the electronic state, for each of the relevant electronic states.

The *rotational partition function* for a linear molecule with moment of inertia I (rotating in 2 dimension) equals

$$q_{rot} = \sum_J (2J+1) \exp\left[-\frac{\hbar^2 J(J+1)}{2 I k_B T} \right] \tag{17.79}$$

$$= \frac{T}{\sigma \Theta} \left[1 + \frac{\Theta}{3T} + \frac{1}{15}\left(\frac{\Theta}{T}\right)^2 + \dots \right], \tag{17.80}$$

where $\Theta \overset{\text{def}}{=} \hbar^2/(2I k_B)$, and σ is the *symmetry factor*. The summation is over the symmetry-allowed values of the quantum number J: for a homonuclear diatomic molecule $\sigma = 2$ because J can be either even or odd, depending on the symmetry of the wave function on interchange of nuclei, and on the symmetry of the spin state of the nuclei. The high-temperature limit for the linear rotator, valid for most molecules at ordinary temperatures, is

$$q_{rot} = \frac{2 I k_B T}{\sigma \hbar^2}. \tag{17.81}$$

For a general non-linear molecule rotating in three dimensions, with moment of inertia tensor[16] $\boldsymbol{I}$, the high-temperature limit for the partition function is given by

$$q_{rot} = \frac{(2k_B T)^{3/2}}{\sigma \hbar^3} \sqrt{\pi \det(\boldsymbol{I})}. \tag{17.82}$$

In contrast to the rotational partition function, the *vibrational partition*

[16] For the definition of the inertia tensor see (15.52) on page 409.

function can in general not be approximated by its classical high-temperature limit. Low-temperature molecular vibrations can be approximated by
a set of independent harmonic oscillators (*normal modes* with frequency ν_i)
and the vibrational partition function is a product of the p.f. of each normal
mode. A harmonic oscillator with frequency ν has equidistant energy levels
(if the minimum of the potential well is taken as zero energy):

$$\varepsilon_n = (n + \tfrac{1}{2})h\nu, \quad n = 0, 1, 2, \ldots, \tag{17.83}$$

and the partition function is

$$q_{\text{ho}} = \frac{\exp(-\tfrac{1}{2}\xi)}{1 - \exp(-\xi)} = \tfrac{1}{2}[\sinh(\xi/2)]^{-1}, \tag{17.84}$$

where $\xi = h\nu/k_{\text{B}}T$. In the low-temperature limit only the first level is occupied and q tends to $\exp(-\tfrac{1}{2}\xi)$ (or one if the lowest level is taken as zero
energy); in the high-temperature (classical) limit q tends to $k_{\text{B}}T/h\nu$. Figure 17.4 compares the quantum-statistical partition function, (free) energy
and entropy with the classical limit: although the difference in Q is not
large, the difference in S is dramatic. The classical entropy tends to $-\infty$
as $T \to 0$, which is a worrying result! For temperatures above $h\nu/k_{\text{B}}$ the
classical limit is a good approximation.

17.6 The classical approximation

A full quantum calculation of the partition function of a multidimensional
system is in general impossible, also if the Hamiltonian can be accurately
specified. But, as quantum dynamics for "heavy" particles can be approximated by classical mechanics, quantum statistics can be approximated by a
classical version of statistical mechanics. In the previous section we considered the classical limit for an ideal gas of (quantum) molecules, and found
$q^N/N!$ for the classical or Boltzmann limit of the partition function of N
indistinguishable molecules (see (17.74) on page 470). We also found the
first-order correction for either Fermi–Dirac or Bose–Einstein particles in
terms of a virial coefficient proportional to h^3 (Eq. (17.68) on page 469).
But these equations are only valid in the ideal gas case when the interaction
between the particles can be ignored. In this section we shall consider a
system of interacting particles and try to expand the partition function in
powers of $\hbar$. We expect the zero-order term to be the classical limit, and
we expect at least a third-order term to distinguish between FD and BE
statistics.

The approach to the classical limit of the quantum partition function

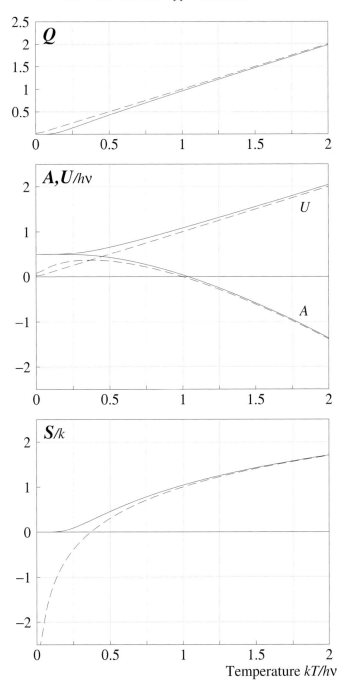

Figure 17.4 Partition function Q, Helmholtz free energy A, energy U and entropy S for the harmonic oscillator as a function of temperature, for the quantum oscillator (drawn lines) and the classical oscillator (dashed lines). Temperature is expressed as $k_{\mathrm{B}}T/h\nu$, energies are in units of $h\nu$ and entropy is in units of k_{B} (ν being the oscillator frequency).

was effectively solved in the early 1930s. The original expansion in powers of $\hbar$ was done by Wigner (1932), but without considering the symmetry properties that distinguish FD and BE statistics. The latter was solved separately by Uhlenbeck and Gropper (1932). Kirkwood (1933) gave a lucid combined derivation that found its way into most textbooks on quantum statistics. We shall not copy the derivation, but only give Kirkwood's line of reasoning and the main result.

We start with the expression of the quantum partition function (17.47) for a system of N identical particles:[17]

$$Q = \text{tr } \exp(-\beta \mathbf{H}), \tag{17.85}$$

where $\mathbf{H}$ is the Hamiltonian matrix in an arbitrary orthonormal complete set of basis functions. The basis functions must have the symmetry imposed by the particle characters, such as specified in (17.50). One convenient choice of basis function is the product of the wave functions for a single particle in a cubic periodic box with edge a and volume V, see (17.61). For this choice of *plane wave functions*

$$Q = \frac{1}{N!} \sum_{\mathbf{k}} \int \psi_{\mathbf{k}}^* e^{-\beta \hat{H}(\mathbf{r})} \psi_{\mathbf{k}} \, d\mathbf{r}, \tag{17.86}$$

with

$$\mathbf{k} = \{\mathbf{k}_1, \mathbf{k}_2, \ldots, \mathbf{k}_N\}, \quad \mathbf{k}_j = \frac{2\pi}{a} \mathbf{n}_j, \quad \mathbf{n}_j \in \mathbb{Z}^3, \tag{17.87}$$

$$\mathbf{r} = \{\mathbf{r}_1, \mathbf{r}_2, \ldots, \mathbf{r}_N\}, \quad \mathbf{r}_i \in box, \tag{17.88}$$

$$\psi_{\mathbf{k}} = \frac{1}{\sqrt{N!}} V^{-N/2} \sum_{\mathcal{P}} (\mp 1)^{\mathcal{P}} e^{i\mathcal{P}[\sum_j \mathbf{k}_j \cdot \mathbf{r}_j]}. \tag{17.89}$$

The permutation $\mathcal{P}$ permutes indexes of identical particles, such as exchanging $\mathbf{r}_i$ and $\mathbf{r}_j$, but it does not touch the indexing of $\mathbf{k}$. The sign $(\mp 1)^{\mathcal{P}}$ is negative when an odd number of fermion pairs are interchanged, and positive otherwise. The factor $1/N!$ in the partition function needs some discussion. The product of plane waves does *not* form an orthogonal set, as each permutation within the set $\{\mathbf{k}\}$ produces an identical (except for the sign) wave function. Therefore sequences obtained by permutation should be omitted from the set $\{\mathbf{k}\}$, which can be done, for example, by only allowing sequences for which the component k's are ordered in non-decreasing order. If we allow all sequences, as we do, we overcount the sum by $N!$ and we should therefore divide the sum by $N!$.[18] Note that this $N!$ has nothing to do with

---

[17] For a mixture of different types of particles, the equations are trivially modified.
[18] This is a rather subtle and usually disregarded consideration. The reader may check the

the $1/\sqrt{N!}$ in the definition of $\psi(\boldsymbol{k})$, which is meant to normalize the wave function consisting of a sum of $N!$ terms.

Since the box is large, the distance $\Delta k = 2\pi/a$ between successive levels of each component of $\boldsymbol{k}$ is small and the sum over $\boldsymbol{k}$ can be replaced by an integral. When we also write $\boldsymbol{p}$ for $\hbar\boldsymbol{k}$, we can replace the sum as:

$$\sum_{\boldsymbol{k}} \rightarrow \frac{V^N}{(2\pi)^{3N}} \int d\boldsymbol{k} = \frac{V^N}{h^{3N}} \int d\boldsymbol{p}, \qquad (17.90)$$

and obtain

$$Q = \frac{1}{N!h^{3N}} \frac{1}{N!} \sum_{\mathcal{P}',\mathcal{P}} (\mp 1)^{\mathcal{P}'+\mathcal{P}} \int d\boldsymbol{p} \int d\boldsymbol{r}\, \phi_0^*(\mathcal{P})\, e^{-\beta\hat{H}(\boldsymbol{r})}\, \phi_0(\mathcal{P}), \qquad (17.91)$$

$$\phi_0(\mathcal{P}) \overset{\text{def}}{=} e^{(i/\hbar)\mathcal{P}[\sum_j \boldsymbol{P}_j \cdot \boldsymbol{r}_j]}. \qquad (17.92)$$

The problem is to evaluate the function

$$u(\boldsymbol{r};\mathcal{P}) \overset{\text{def}}{=} e^{-\beta\hat{H}(\boldsymbol{r})}\, \phi_0(\mathcal{P}), \qquad (17.93)$$

which is in general impossible because the two constituents of the hamiltonian $\hat{K} = -(\hbar^2/2m)\sum_j \nabla_j^2$ and $V(\boldsymbol{r})$ do not commute. If they would commute, we could write (see Section 14.6 on page 385)

$$e^{-\beta\hat{H}} = e^{-\beta V} e^{-\beta\hat{K}}, \qquad (17.94)$$

and evaluate u as

$$u(\boldsymbol{r};\mathcal{P}) = e^{-\beta V(\boldsymbol{r})} e^{-\beta\sum_j(\boldsymbol{P}_j^2/2m)}\, \phi_0(\mathcal{P}). \qquad (17.95)$$

In (17.91) only the $N!$ terms with $\mathcal{P}' = \mathcal{P}$ survive and we would obtain the *classical limit*

$$Q^{\text{cl}} = \frac{1}{N!h^{3N}} \int d\boldsymbol{p} \int d\boldsymbol{r}\, e^{-\beta H(\boldsymbol{p},\boldsymbol{q})}. \qquad (17.96)$$

In general, we need to solve $u(\boldsymbol{r};\mathcal{P})$ (see (17.93)). By differentiating with respect to β, it s found that u satisfies the equation

$$\frac{\partial u}{\partial\beta} = -\hat{H}u, \qquad (17.97)$$

which is the Schrödinger equation in imaginary time: $it/\hbar$ being replaced by β.

correctness for the case of two one-dimensional particles with plane waves k_1, k_2 and observe that not only $\langle k_1 k_2|k_1 k_2\rangle = 1$, but also $\langle k_1 k_2|k_2 k_1\rangle = \mp 1$. In fact, Kirkwood's original article (1933) omitted this $N!$, which led to an incorrect omission of $1/N!$ in the final equations. Therefore his equations were still troubled by the Gibbs paradox. In a correction published in 1934 he corrected this error, but did not indicate what was wrong in the original article. The term is properly included by Huang (1987).

Kirkwood proceeded by writing u as

$$u = w\,\phi_0(\mathcal{P})e^{-\beta H(\mathbf{p},\mathbf{r})}, \tag{17.98}$$

formulating a differential equation for w and then expanding w in a power series in $\hbar$. This yields

$$w = 1 + \hbar w_1 + \hbar^2 w_2 + \mathcal{O}(\hbar^3), \tag{17.99}$$

$$w_1 = -\frac{i\beta^2}{2m}\mathcal{P}\left[\sum_j \mathbf{p}_j \cdot \boldsymbol{\nabla}_j V\right], \tag{17.100}$$

$$w_2 = -\frac{\beta^2}{4m}\sum_j \boldsymbol{\nabla}_j^2 V - \frac{\beta^3}{6m}\left[\sum_j (\boldsymbol{\nabla}_j V)^2 + \frac{1}{m}(\mathcal{P}\sum_j \mathbf{p}_j \cdot \boldsymbol{\nabla}_j)^2 V\right]$$

$$+ \frac{\beta^4}{8m^2}(\mathcal{P}\sum_j \mathbf{p}_j \cdot \boldsymbol{\nabla}_j V)^2. \tag{17.101}$$

Inserting this in (17.91) and integrating all correction terms over $d\mathbf{p}$, first the terms in the sum over permutations for which $\mathcal{P}' = \mathcal{P}$ ($N!$ terms) can be separated. All odd-order terms, which are antisymmetric in $\mathbf{p}$, disappear by integration over $\mathbf{p}$. Then the $\frac{1}{2}N!$ terms for which $\mathcal{P}'$ and $\mathcal{P}$ differ by the exchange of one pair of particles can be integrated. This is the leading exchange term; higher-order exchange terms will be neglected. The end result is

$$Q = \frac{1}{N!}\left(\frac{2\pi m k_B T}{h^2}\right)^{3N/2}\int d\mathbf{r}\,e^{-\beta V(\mathbf{r})}\,(1 + f_{\text{cor}}),$$

$$f_{\text{cor}} = -\frac{\hbar^2 \beta^2}{12}\sum_j \frac{1}{m_j}\left[\boldsymbol{\nabla}_j^2 V - \frac{\beta}{2}(\boldsymbol{\nabla}_j V)^2\right] + \mathcal{O}(\hbar^4)$$

$$\mp \sum_{j\neq k} e^{-m_j r_{jk}^2/\beta\hbar^2}\left[1 + \frac{\beta}{2}r_{jk}\cdot(\boldsymbol{\nabla}_j V - \boldsymbol{\nabla}_k V) + \ldots\right] \tag{17.102}$$

We note that the $\hbar^2$ correction term was earlier derived by Wigner (1932); the exchange term had been derived by Uhlenbeck and Gropper (1932) in the slightly different form an ensemble average of the product over particle pairs (i,j) of $(1\mp\exp(-mk_B T r_{ij}^2/\hbar^2))$, and without the first-order term in β in (17.102).

What does this result mean in practice? First we see that the *classical canonical partition function* is given by

$$Q^{\text{cl}} = \frac{1}{N!}\left(\frac{2\pi m k_B T}{h^2}\right)^{3N/2}\int d\mathbf{r}\,e^{-\beta V(\mathbf{p},\mathbf{r})} \tag{17.103}$$

$$= \frac{1}{N!h^{3N}} \int d\boldsymbol{p} \int d\boldsymbol{r} e^{-\beta H(\boldsymbol{p},\boldsymbol{r})}. \tag{17.104}$$

This is the starting point for the application, in the following section, of statistical mechanics to systems of particles that follow classical equations of motion. We observe that this equation is consistent with an equilibrium *probability density* proportional to $\exp -\beta H(\boldsymbol{p}, \boldsymbol{r})$ in an isotropic phase space $\boldsymbol{p}, \boldsymbol{r}$, divided into elementary units of area h with equal a priori statistical weights. There is one single-particle quantum state per unit of 6D volume h^3. The $N!$ means the following: If two identical particles (1) and (2) exchange places in phase space, the two occupations $\boldsymbol{p}(1)\boldsymbol{p}'(2)\boldsymbol{r}(1)\boldsymbol{r}'(2)$ and $\boldsymbol{p}(2)\boldsymbol{p}'(1)\boldsymbol{r}(2)\boldsymbol{r}'(1)$ should statistically be counted as one.

The quantum corrections to the classical partition function can be expressed in several ways. The effect of quantum corrections on thermodynamical quantities is best evaluated through the quantum corrections to the Helmholtz free energy A. Another view is obtained by expressing quantum corrections as corrections to the classical Hamiltonian. These can then be used to generate modified equations of motion, although one should realize that in this way we do not generate true quantum corrections to classical dynamics, but only generate some kind of modified dynamics that happens to produce proper quantum corrections to *equilibrium* phase-space distributions.

First look at the quantum correction to the free energy $A = -k_{\mathrm{B}}T \ln Q$. Noting that

$$Q = Q^{\mathrm{cl}}(1 + \langle f_{\mathrm{cor}} \rangle), \tag{17.105}$$

where $\langle \cdots \rangle$ denotes a *canonical ensemble average*

$$\langle f_{\mathrm{cor}} \rangle = \frac{\int d\boldsymbol{p} \int d\boldsymbol{r} \, f_{\mathrm{cor}} \exp(-\beta H)}{\int d\boldsymbol{p} \int d\boldsymbol{r} \, \exp(-\beta H)}, \tag{17.106}$$

we see, using $\ln(1 + x) \approx x$, that

$$A = A^{\mathrm{cl}} - k_{\mathrm{B}}T \langle f_{\mathrm{cor}} \rangle. \tag{17.107}$$

By partial integration the second derivative of V can be rewritten as:

$$\langle \nabla_j^2 V \rangle = \beta \langle (\nabla_j V)^2 \rangle. \tag{17.108}$$

The $\hbar^2$ correction now reads

$$A = A^{\mathrm{cl}} + \frac{\hbar^2}{24(k_{\mathrm{B}}T)^2} \sum_j \frac{1}{m_j} \langle (\nabla_j V)^2 \rangle. \tag{17.109}$$

The averaged quantity is the sum of the *squared forces* on the particles.

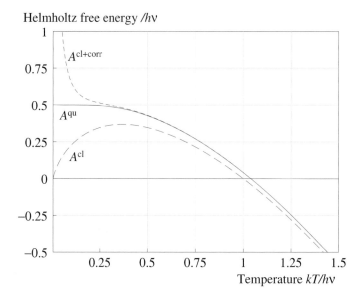

Figure 17.5 The classical Helmholtz free energy of the harmonic oscillator (long-dashed), the $\hbar^2$-correction added to the classical free energy (short-dashed) and the exact quantum free energy (solid).

The use of this $\hbar^2$-Wigner correction is described in Section 3.5 on page 70. Quantum corrections for other thermodynamic functions follow from A. As an example, we give the classical, quantum-corrected classical, and exact quantum free energy for the harmonic oscillator in Fig. 17.5. The improvement is substantial.

It is possible to include the $\hbar^2$-term into the Hamiltonian as an extra potential term:

$$V^{\mathrm{cor}} = -k_{\mathrm{B}}T f^{\mathrm{cor}}. \tag{17.110}$$

If done in this fashion, calculation of the force on particle i then requires a double summation over particles j and k, i.e., a three-body interaction.[19] The inclusion in a dynamics simulation would be cumbersome and time consuming, while not even being dynamically correct. However, there are several ways to devise effective interparticle interactions that will lead to the correct $\hbar^2$-correction to the free energy when applied to equilibrium simulations. An intuitively very appealing approach is to consider each particle as a Gaussian distribution. The width of such a distribution can be derived from Feynman's path integrals (see Section 3.3 on page 44) and leads to the Feynman–Hibbs potential, treated in Section 3.5 on page 70.

[19] See, e.g., Huang (1987)

Next we consider the exchange term, i.e., the last line of (17.102). We drop the first-order term in β, which is zero for an ideal gas and for high temperatures; it may however reach values of the order 1 for condensed phases. It is now most convenient to express the effect in an extra correction potential:

$$V^{\text{cor}} = -k_{\text{B}}T \sum_{i<j} \ln \left(1 \mp e^{-mk_{\text{B}}Tr_{ij}^2/\hbar^2}\right) \approx \pm k_{\text{B}}T \sum_{i<j} e^{-mk_{\text{B}}Tr_{ij}^2/\hbar^2}.$$

(17.111)

The first form comes from the equations derived by Uhlenbeck and Gropper (1932) (see also Huang, 1987); the second form is an approximation that is invalid for very short distances. This is an interesting result, as it indicates that fermions effectively repel each other at short distance, while bosons attract each other. This leads to a higher pressure for fermion gases and a lower pressure for boson gases, as was already derived earlier (Eq. (17.68) on page 469). The interparticle correction potential can be written in terms of the de Broglie wave length Λ (see (17.69) on page 469):

$$V_{ij}^{\text{cor}} = -k_{\text{B}}T \sum_{i<j} \ln \left(1 \mp \exp\left[-2\pi \left(\frac{r_{ij}}{\Lambda}\right)^2\right]\right) \approx \pm k_{\text{B}}T \exp\left[-2\pi \left(\frac{r_{ij}}{\Lambda}\right)^2\right].$$

(17.112)

This *exchange potential* is not a true, but an *effective* potential with the effect of correcting equilibrium ensembles to first-order for exchange effects. The effective potential "acts" only between identical particles. Figure 17.6 shows the form and size of the exchange potential. When compared to the interaction potential for a light atom (helium-4), it is clear that the effects of exchange are completely negligible for temperatures above 15 K. In "normal" molecular systems at "normal" temperatures exchange plays no role at all.

17.7 Pressure and virial

There are two definitions of pressure:[20] one stems from (continuum) mechanics and equals the normal component of the force exerted on a surface per unit area; the other stems from the first law of thermodynamics (16.24) and equals minus the partial derivative of the internal energy with respect to the volume at constant entropy, i.e., without exchange of heat. These definitions are equivalent for a system with homogeneous isotropic pressure: if a surface with area S encloses a volume V and a force $p\,dS$ acts on every

---

[20] The author is indebted to Dr Peter Ahlström and Dr Henk Bekker for many discussions on pressure in the course of preparing a review that remained unpublished. Some of the text on continuum mechanics originates from Henk Bekker.

potential energy $/k_{\mathrm{B}}T$

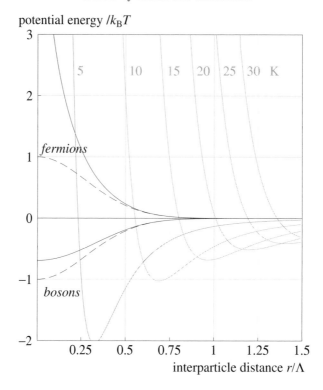

Figure 17.6 Effective exchange potential between fermions or bosons. The solid black curves are the first form of (17.112); the dashed black curves are the approximate second form. The distance is expressed in units of the de Broglie wavelength $\Lambda = h/\sqrt{2\pi mk_{\mathrm{B}}T}$. For comparison, the Lennard–Jones interaction for ^{4}He atoms, also expressed as a function of r/Λ, is drawn in gray for temperatures of 5, 10, 15, 20, 25 and 30 K.

surface element dS, moving the surface an infinitesimal distance δ inwards, then an amount of work $p\,S\delta = -p\,dV$ is done on the system, increasing its internal energy. But these definitions are not equivalent in the sense that the mechanical pressure can be defined locally and can have a tensorial character, while the thermodynamic pressure is a global equilibrium quantity. In statistical mechanics we try to average a detailed mechanical quantity (based on an atomic description) over an ensemble to obtain a thermodynamic quantity. The question to be asked is whether and how a mechanical pressure can be locally defined on an atomic basis. After that we can average over ensembles. So let us first look at the mechanical definition in more detail.

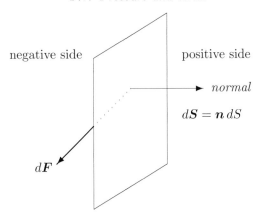

Figure 17.7 A force $d\boldsymbol{F} = \boldsymbol{\sigma} \cdot d\boldsymbol{S}$ acts *on* the negative side of a surface element.

17.7.1 The mechanical pressure and its localization

In a continuous medium a quantity related to to the local pressure, called *stress*, is given by a second-rank tensor $\boldsymbol{\sigma}(\boldsymbol{r})$, defined through the following relation (see Fig. 17.7): the force exerted by the material lying on the positive side of a static surface element dS with normal $\boldsymbol{n}$ (which points from negative to positive side), on the material lying on the negative side of dS, is given by

$$d\boldsymbol{F} = \boldsymbol{\sigma}(\boldsymbol{r}) \cdot \boldsymbol{n}\, dS = \boldsymbol{\sigma}(\boldsymbol{r}) \cdot d\boldsymbol{S}, \qquad (17.113)$$

$$dF_\alpha = \sum_\beta \sigma_{\alpha\beta}\, dS_\beta. \qquad (17.114)$$

The stress tensor is often decomposed into a diagonal tensor, the *normal stress*, and the *shear stress* $\boldsymbol{\tau}$ which contains the off-diagonal elements. The force acting on a body transfers momentum into that body, according to Newton's law. However, the stress tensor should be distinguished from the *momentum flux tensor* $\boldsymbol{\Pi}$, because the actual transport of particles also contributes to the momentum flux. Also the sign differs because the momentum flux is defined positive in the direction of the surface element (from inside to outside).[21] The momentum flux tensor is defined as

$$\Pi_{\alpha\beta} \stackrel{\text{def}}{=} -\sigma_{\alpha\beta} + \rho v_\alpha v_\beta, \qquad (17.115)$$

or

$$\boldsymbol{\Pi} = -\boldsymbol{\sigma} + \boldsymbol{v}\boldsymbol{J}, \qquad (17.116)$$

---

[21] There is no sign consistency in the literature. We follow the convention of Landau and Lifschitz (1987).

where ρ is the mass density and $\boldsymbol{v}\boldsymbol{J}$ is the dyadic vector product of the velocity $\boldsymbol{v}$ and the mass flux $\boldsymbol{J} = \rho \boldsymbol{v}$ through the surface element.

It is this momentum flux tensor that can be identified with the *pressure tensor*, which is a generalization of the pressure. If $\boldsymbol{\Pi}$ is isotropic, $\Pi_{\alpha\beta} = p\,\delta_{\alpha\beta}$, the force on a surface element of an impenetrable wall, acting from inside to outside, is normal to the surface and equals $p\,d\boldsymbol{S}$.

Is the pressure tensor, as defined in (17.116) unique? No, it is not. The stress tensor itself is not a physical observable, but is observable only through the action of the force resulting from a stress. From (17.113) we see that the force $\boldsymbol{F}_V$ acting on a closed volume V, as exerted by the surrounding material, is given by the integral over the surface S of the volume

$$\boldsymbol{F}_V = \int_S \boldsymbol{\sigma} \cdot d\boldsymbol{S}. \tag{17.117}$$

In differential form this means that the force density $\boldsymbol{f}(\boldsymbol{r})$, i.e., the force acting per unit volume, is given by the divergence of the stress tensor:

$$\boldsymbol{f}(\boldsymbol{r}) = \boldsymbol{\nabla} \cdot \boldsymbol{\sigma}(\boldsymbol{r}). \tag{17.118}$$

Thus only the divergence of the stress tensor leads to observables, and we are free to add any divergence-free tensor field $\boldsymbol{\sigma}_0(\boldsymbol{r})$ to the stress tensor without changing the physics. The same is true for the pressure tensor $\boldsymbol{\Pi}$. Without further proof we note that, although the local pressure tensor is *not* unique, its integral over a finite confined system, *is* unique. The same is true for a periodic system by cancelation of surface terms. Therefore the *average* pressure over a confined or periodic system is indeed unique.

Turning from continuum mechanics to systems of interacting particles, we ask the question how the pressure tensor can be computed from particle positions and forces. The particle flux component $\boldsymbol{v}\boldsymbol{J}$ of the pressure tensor is straightforward because we can count the number of particles passing over a defined surface area and know their velocities. For the stress tensor part all we have is a set of internal forces $\boldsymbol{F}_i$, acting on particles at positions $\boldsymbol{r}_i$.[22] From that we wish to construct a stress tensor such that

$$\boldsymbol{\nabla} \cdot \boldsymbol{\sigma}(\boldsymbol{r}) = \sum_i \boldsymbol{F}_i \delta(\boldsymbol{r} - \boldsymbol{r}_i). \tag{17.119}$$

Of course this construction cannot be unique. Let us first remark that a solution where $\boldsymbol{\sigma}$ is localized on the interacting particles is not possible for the simple reason that $\boldsymbol{\sigma}$ cannot vanish over a closed surface containing a

---

[22] Here we restrict the pressure as resulting from internal forces, arising from interactions within the system. If there are external forces, e.g., external electric or gravitational fields, such forces are separately added.

particle on which a force is acting, because the divergence inside the closed surface is not zero. As shown by Schofield and Henderson (1982), however, it is possible to localize the stress tensor on arbitrary *line contours* C_{0i} running from a reference point r_0 to the particle positions r_i:

$$\sigma_{\alpha\beta}(r) = -\sum_i F_{i,\alpha} \int_{C_{0i}} \delta(r - r_c)(dr_c)_\beta. \tag{17.120}$$

For each particle this function is localized on, and equally distributed over, the contour C_{0i}. Taking the divergence of σ we can show that (17.119) is recovered:

$$\begin{aligned}
\nabla \cdot \sigma(r) &= -\sum_i F_i \nabla \cdot \int_{C_{0i}} \delta(r - r_c)\, dr_c \\
&= \sum_i F_i[\delta(r_i) - \delta(r_0)] \\
&= \sum_i F_i \delta(r_i). \tag{17.121}
\end{aligned}$$

Proof The first step follows from the three-dimensional generalization of

$$\frac{d}{d\xi} \int_a^b f(\xi - x)\, dx = -\int_a^b \frac{d}{dx} f(\xi - x)\, dx = -f(\xi - b) + f(\xi - a), \tag{17.122}$$

the last step uses the fact that $\sum_i F_i = 0$ for internal forces. $\square$

If we integrate the stress tensor over the whole (confined) volume of the system of particles, only the end points in the line integral survive and we obtain the sum of the dyadic products of forces and positions:

$$\int_V \sigma\, d^3r = -\sum_i F_i(r_i - r_0) = -\sum_i F_i r_i, \tag{17.123}$$

which is independent of the choice of reference position r_0.

The introduction of a reference point is undesirable as it may localize the stress tensor far away from the interacting particles. When the forces are pair-additive, the stress tensor is the sum over pairs; for each pair i, j the two contours from the reference position can be replaced by one contour between the particles, and the reference position cancels out.

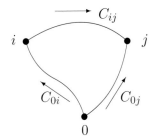

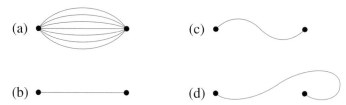

Figure 17.8 Four different contours to localize the stress tensor due to interaction between two particles. Contour (b) is the minimal path with optimal localization.

The result is

$$\sigma_{\alpha\beta} = \sum_{i<j} F_{ij\alpha} \int_{C_{ij}} \delta(\mathbf{r} - \mathbf{r}_c)\,(d\mathbf{r}_c)_\beta, \qquad (17.124)$$

with integral

$$\int_V \sigma_{\alpha\beta}\, d^3 r = F_{i\alpha}(r_{j\beta} - r_{i\beta}) = -F_{i\alpha}r_{i\beta} - F_{j\alpha}r_{j\beta}, \qquad (17.125)$$

consistent with (17.123). Here $F_{ij\alpha}$ is the α component of the force acting on i due to the interaction with j. The path between the interacting particle pair is irrelevant for the pressure: Fig. 17.8 gives a few examples including distribution over a collection of paths (a), the minimal path, i.e., the straight line connecting the particles (b), and curved paths (c, d) that do not confine the localization to the minimal path. Irving and Kirkwood (1950) chose the straight line connecting the two particles as a contour, and we recommend that choice.[23]

17.7.2 *The statistical mechanical pressure*

Accepting the mechanical definition of the pressure tensor as the momentum flux of (17.116), we ask what the average pressure over an ensemble is. First we average over the system, and then over the ensemble of systems. The average over the whole system (we assume our system is confined in space) is given by the integral over the volume, divided by the volume (using dyadic vector products):

$$\langle \mathbf{\Pi} \rangle_V = -\frac{1}{V} \int_V \sigma(\mathbf{r}) d^3 r + \frac{1}{V} \sum_i m_i \mathbf{v}_i \mathbf{v}_i, \qquad (17.126)$$

[23] This choice is logical but not unique, although Wajnryb *et al.* (1995) argue that additional conditions make this choice unique. The choice has been challenged by Lovett and Baus (1997; see also Marechal *et al.*, 1997) on the basis of a local thermodynamic pressure definition, but further discussion on a physically irrelevant choice seems pointless (Rowlinson, 1993). For another local definition see Zimmerman *et al.* (2004).

or, with (17.123):

$$\langle \Pi \rangle_V V = \sum_i \boldsymbol{F}_i \boldsymbol{r}_i + \sum_i m_i \boldsymbol{v}_i \boldsymbol{v}_i. \tag{17.127}$$

In a dynamic system this instantaneous volume-averaged pressure is a fluctuating function of time. We remark that (in contrast to the local tensor) this averaged tensor is symmetric, because in the first term the difference between an off-diagonal element and its transpose is a component of the total torque on the system, which is always zero in the absence of external forces. The second term is symmetric by definition. Finally, the *thermodynamic pressure tensor* $\boldsymbol{P}$ which we define as the ensemble average of the volume-averaged pressure tensor, is given by

$$\boldsymbol{P} V = \sum_i \langle \boldsymbol{F}_i \boldsymbol{r}_i \rangle + \sum_i m_i \langle \boldsymbol{v}_i \boldsymbol{v}_i \rangle, \tag{17.128}$$

where the angular brackets are equilibrium ensemble averages, or because of the ergodic theorem, time averages over a dynamic equilibrium system. $\boldsymbol{P}$ is a "sharp" symmetric tensor. For an isotropic system the pressure tensor is diagonal and its diagonal elements are the isotropic pressure p:

$$pV = \frac{1}{3} \operatorname{tr} \boldsymbol{P} V = \frac{1}{3} \sum_i \boldsymbol{F}_i \cdot \boldsymbol{r}_i + \frac{2}{3} \langle E_{\mathrm{kin}} \rangle. \tag{17.129}$$

The first term on the right-hand side relates to *the virial of the force*, already defined by Clausius (see, e.g., Hirschfelder *et al.*, 1954) and valid for a confined (i.e., bounded) system:

$$\Xi \stackrel{\mathrm{def}}{=} -\frac{1}{2} \sum_i \langle \boldsymbol{F}_i \cdot \boldsymbol{r}_i \rangle. \tag{17.130}$$

Here $\boldsymbol{F}_i$ is the total force on particle i, including external forces. The resulting virial Ξ is the total virial, which can be decomposed into an internal virial due to the internal forces and an external virial, for example caused by external forces acting on the boundary of the system in order to maintain the pressure. Written with the Clausius virial, (17.129) becomes

$$pV = \frac{2}{3} (\langle E_{\mathrm{kin}} \rangle - \Xi_{\mathrm{int}}). \tag{17.131}$$

This relation follows also directly from the classical *virial theorem*:

$$\Xi_{\mathrm{tot}} = \langle E_{\mathrm{kin}} \rangle \tag{17.132}$$

which is valid for a bounded system.[24] This virial theorem follows also from the generalized equipartition theorem, treated in Section 17.10 on page 503. Since the external virial due to an external force $-p\,d\boldsymbol{S}$ acting on any surface element dS equals

$$\Xi_{\text{ext}} = \frac{1}{2}p\oint_S \boldsymbol{r}\cdot d\boldsymbol{S} = \frac{3}{2}pV, \qquad (17.133)$$

the virial theorem immediately yields (17.131) (see Exercise 17.6).

Periodic boundary conditions

The virial expression and the pressure equations given above are valid for a bounded system, but not for an infinite system with periodic boundary conditions, as so often used in simulations. Under periodic boundary conditions the force on every particle may be zero while the system has a non-zero pressure: imagine the simple case of one particle in a cubic box, interacting symmetrically with six images. Then $\sum_i \boldsymbol{F}_i\boldsymbol{r}_i$ is obviously zero. For forces that can be written as a sum of pair interactions this problem is easily remedied by replacing $\sum_i \boldsymbol{F}_i\boldsymbol{r}_i$ by a sum over pairs:[25]

$$\sum_i \boldsymbol{F}_i\boldsymbol{r}_i = \sum_{i<j} \boldsymbol{F}_{ij}\boldsymbol{r}_{ij}, \qquad (17.134)$$

where $\boldsymbol{F}_{ij}$ is the force on i due to j and $\boldsymbol{r}_{ij} = \boldsymbol{r}_i - \boldsymbol{r}_j$. This sum can be taken over all minimum-image pairs if no more than minimum images are involved in the interaction. For interactions extending beyond minimum images, the pressure tensor can be evaluated over the volume of a unit cell, according to Irving and Kirkwood's distribution over a straight line, using the fraction of the line lying within the unit cell. Consider the simple case, mentioned above, with a single particle in a cubic unit cell of size $a \times a \times a$.

---

[24] See, e.g., Hirschfelder *et al.* (1954). A very extensive review on the virial theorem, including a discussion on the quantum-mechanical form, has been published by Marc and McMillan (1985).
[25] Erpenbeck and Wood (1977).

Assume that each of its six images exert a force F on the particle, with a zero vector sum. Each contribution Fa to the sum $\sum_i \boldsymbol{F}_i \cdot \boldsymbol{r}_i$ counts for 0.5 since the interaction line lies for 50% in the unit cell, so the total sum is $3Fa$, and the virial contribution to the pressure (17.129) equals $Fa/V = F/a^2$. This is correct, as we can see that this pressure, if acting externally on one side with area a^2, just equals the force acting "through" that plane.

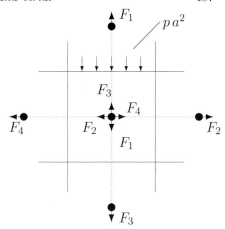

Now consider the interaction between two particles i and j (Fig. 17.9). In a lattice sum interactions between all image pairs of both i and j are included; note that there are always sets of equivalent pairs, e.g., in the case of a single shift with $\mathbf{n} = (010)$ (Fig. 17.9b):

$$\boldsymbol{r}_i - (\boldsymbol{r}_j + \mathbf{Tn}) = (\boldsymbol{r}_i - \mathbf{Tn}) - \boldsymbol{r}_j, \tag{17.135}$$

where $\mathbf{T}$ is the transformation matrix from relative coordinates in the unit cell to cartesian coordinates (see (6.3) on page 143), i.e., a matrix of which the columns are the cartesian base vectors of the unit cell, and $\mathbf{n} \in \mathbb{Z}^3$. Figure 17.9 shows three examples of image pairs, with one, two and three equivalent pairs, respectively. If we add up the fractions of the interaction lines that run through the unit cell, we obtain just one full interaction line, so the contribution of that set of pairs to the sum $\sum \boldsymbol{F}_{ij} \cdot \boldsymbol{r}_{ij}$ is given by $\boldsymbol{F}_{ij\mathbf{n}} \cdot (\boldsymbol{r}_i - \boldsymbol{r}_j - \mathbf{Tn})$. Note that each set of equivalent pairs contributes *only once* to the total energy, to the force on i, to the force on j and to the virial contribution to the pressure. Replacing the dot product by a dyadic product, the scalar contribution is generalized to a tensorial contribution. Summarizing, for a lattice sum of isotropic pair interactions $v_{ij}(r)$, the total potential energy, the force on particle i and the instantaneous pressure tensor (see (17.127)) are given by

$$E_{\text{pot}} = \frac{1}{2} \sideset{}{'}\sum_{i,j,\mathbf{n}} v_{ij}(r_{ij\mathbf{n}}), \tag{17.136}$$

$$\boldsymbol{F}_i = \sum_{j,\mathbf{n}} \boldsymbol{F}_{ij\mathbf{n}}, \quad \boldsymbol{F}_{ij\mathbf{n}} = \frac{dv_{ij}}{dr} \frac{\boldsymbol{r}_{ij\mathbf{n}}}{|\boldsymbol{r}_{ij\mathbf{n}}|}, \tag{17.137}$$

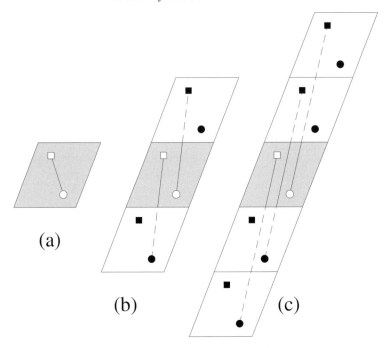

Figure 17.9 Three examples of "equivalent image pairs" of two interacting particles (open circle and open square in the shaded unit cell; images are filled symbols). For each image pair the fraction of the dashed interaction line lying in the unit cell, which contributes to the pressure, is drawn as a solid line.

$$\Pi V = \frac{1}{2} \sum_{i,j,\mathbf{n}}{}^{'} \mathbf{F}_{ij\mathbf{n}} \mathbf{r}_{ij\mathbf{n}} + \sum_{i} m_i \mathbf{v}_i \mathbf{v}_i. \qquad (17.138)$$

Here we use the notation

$$r_{ij\mathbf{n}} = |\mathbf{r}_{ij\mathbf{n}}|, \qquad (17.139)$$

$$\mathbf{r}_{ij\mathbf{n}} = \mathbf{r}_i - \mathbf{r}_j - \mathbf{Tn}. \qquad (17.140)$$

Note that the volume V is equal to $\det \mathbf{T}$. The sum is taken over all particles i and all particles j in the unit cell, and over all sets of three integers $\mathbf{n} : \mathbf{n} \in \mathbb{Z}^3$. This includes i and its images; the prime in the sum means that $j = i$ is excluded when $\mathbf{n} = \mathbf{0}$. The factor $\frac{1}{2}$ prevents double counting of pairs, but of course summation over i, j; $i \neq j$ can be restricted to all i with all $j > i$ because i and j can be interchanged while replacing $\mathbf{n}$ by $-\mathbf{n}$. The factor $\frac{1}{2}$ must be maintained in the $ii\mathbf{n}$ summation. The summation over images may well be conditionally convergent, as is the case for Coulomb interactions. This requires a specified summation order or special long-range summation techniques, as discussed in Section 13.10 on page 362.

The equation for the pressure (17.138) is usually derived for the canonical ensemble from the equation $p = -k_B T (\partial \ln Q / \partial V)_T$ or a tensorial variant that implies differentiating to the components of the transformation tensor **T**. The volume dependence in the partition function is then handled by transforming to scaled coordinates ρ ($\mathbf{r} = \mathbf{T}\rho$) which concentrates the volume dependence on **T**. The volume change is "coupled" to all particles in the system, rather then to particles on the surface as a volume change in a real experiment would do. This, of course, is also a *choice* that influences the instantaneous pressure, but not the ensemble-averaged pressure. Equation (17.138) is obtained.[26] It is interesting that this coupling to all particles is equivalent to Irving and Kirkwood's choice for the local pressure definition. Explicit equations for use with Ewald summation have been given by Nosé and Klein (1983) and for use with the Particle Mesh Ewald method by Essmann *et al.* (1995).

Pressure from center-of-mass attributes

Thus far we have considered detailed atomic motion and forces on atoms to determine the pressure. However, pressure is a result of translational motion and forces causing translational motion. For a system consisting of molecules it is therefore possible to consider only the center-of-mass (c.o.m.) velocities and the forces acting on the c.o.m. Equation (17.128) is equally valid when $\boldsymbol{F}_i$ are the forces acting on the c.o.m. and $\boldsymbol{r}_i$ and $\boldsymbol{v}_i$ denote c.o.m coordinates and velocities. Somehow the extra "intramolecular" virial should just cancel the intramolecular kinetic energy. Can we see why that is so?[27]

Consider a system of molecules with c.o.m. position $\boldsymbol{R}_i$, each consisting of atoms with positions $\boldsymbol{r}_k^i$ (see Fig. 17.10). A total force $\boldsymbol{F}_k^i$ is acting on this atom. Denoting the mass of the molecule as $M_i = \sum_k m_k^i$, the c.o.m. coordinate is given by

$$M_i \boldsymbol{R}_i = \sum_k m_k^i \boldsymbol{r}_k^i. \qquad (17.141)$$

Now we define intramolecular coordinates $\boldsymbol{s}_k^i$ of each atom with respect to

---

[26] Pressure calculations on the basis of Hamiltonian derivatives and their use in constant pressure algorithms have been pioneered by Andersen (1980) for isotropic pressure and by Parrinello and Rahman (1980, 1981) for the pressure tensor. A good summary is to be found in Nosé and Klein (1983) and an extensive treatment of pressure in systems with constraints has been given by Ciccotti and Ryckaert (1986).

[27] We roughly follow the arguments given in an appendix by Ciccotti and Ryckaert (1986).

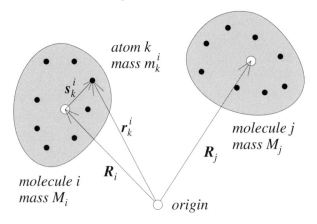

Figure 17.10 Definition of atoms clustered in molecules for discussion of the c.o.m. virial.

the c.o.m.:

$$s_k^i \stackrel{\text{def}}{=} r_k^i - R_i, \tag{17.142}$$

with the obvious relation

$$\sum_k m_k^i s_k^i = 0. \tag{17.143}$$

The total virial (see (17.130)) on an atomic basis can be split into a c.o.m. and an intramolecular part:

$$\begin{aligned}
\Xi_{\text{tot}} &= -\frac{1}{2}\sum_i\sum_k \langle F_k^i r_k^i \rangle = -\frac{1}{2}\sum_i\sum_k \langle F_k^i(R_i + s_k^i)\rangle \\
&= -\frac{1}{2}\langle F_i R_i \rangle - \frac{1}{2}\sum_i\left(\sum_k \langle F_k^i s_k^i\rangle\right) = \Xi_{\text{tot}}^{\text{com}} + \Xi_{\text{tot}}^{\text{intra}}. \tag{17.144}
\end{aligned}$$

The forces are the total forces acting on the atom. Likewise we can split the kinetic energy:

$$\begin{aligned}
E_{\text{kin}} &= \frac{1}{2}\sum_i\sum_k m_k^i \dot{r}_k^i \dot{r}_k^i = \frac{1}{2}\sum_i\sum_k m_k^i(\dot{R}_i + \dot{s}_k^i)(\dot{R}_i + \dot{s}_k^i) \\
&= \frac{1}{2}\sum_i M_i \dot{R}_i \dot{R}_i + \frac{1}{2}\sum_i\left(\sum_k m_k^i \dot{s}_k^i \dot{s}_k^i\right) \\
&= E_{\text{kin}}^{\text{com}} + E_{\text{kin}}^{\text{intra}}. \tag{17.145}
\end{aligned}$$

If we can prove that $\Xi_{\text{tot}}^{\text{intra}} = \langle E_{\text{kin}}^{\text{intra}}\rangle$, then we have proved that the pressure computed from c.o.m. forces and velocities equals the atom-based pressure.

The proof is simple and rests on the fact that neither the internal coordinates nor the internal velocities can grow to infinity with increasing time. First realize that $F_k^i = m_k^i \ddot{r}_k^i$; then it follows that for every molecule (we drop the superscripts i and use (17.143))

$$-\sum_k F_k s_k = -\sum_k m_k \ddot{s}_k s_k. \tag{17.146}$$

Ensemble-averaging can be replaced by time averaging:

$$-\sum_k \langle F_k s_k \rangle = -\lim_{T \to \infty} \frac{1}{T} \int_0^T \sum_k m_k \ddot{s}_k s_k \, dt$$

$$= \sum_k m_k \langle \dot{s}_k \dot{s}_k \rangle - \lim_{T \to \infty} \frac{1}{T} [s_k \dot{s}_k(T) - s_k \dot{s}_k(0)]. \tag{17.147}$$

Since the last term is a fluctuating but bounded quantity divided by T, the limit $T \to \infty$ is zero and we are left with the equality of intramolecular virial and kinetic energy, if averaged over an equilibrium ensemble. The reasoning is equivalent to the proof of the virial theorem (17.132) (Hirschfelder *et al.*, 1954). Note that the "molecule" can be any cluster of particles that does not fall apart in the course of time; there is no requirement that the cluster should be a rigid body.

The subtlety of virial-kinetic energy compensation is nicely illustrated by the simple example of an ideal gas of classical homonuclear diatomic molecules (bond length d) with an internal quadratic potential with force constant k. We can calculate the pressure from the total kinetic energy minus the internal atomic virial, but also from the kinetic energy of the c.o.m. minus the molecular virial. So the virial of the internal forces, which can only be due to a force acting in the bond direction, must be compensated by the intramolecular kinetic energy. Of the total of six degrees of freedom three are internal: the bond vibration and two rotational degrees of freedom, together good for an average of $\frac{3}{2}k_B T$ kinetic energy. The harmonic bond vibrational degree of freedom has an average kinetic energy of $\frac{1}{2}k_B T$, which is equal to the average potential energy $\langle \frac{1}{2}k(d - d_0)^2 \rangle$. Therefore the contribution to the virial is $\frac{1}{2}\langle Fd \rangle = \frac{1}{2}k\langle d(d-d_0) \rangle = \frac{1}{2}k\langle (d-d_0)^2 \rangle = \frac{1}{2}k_B T$, which exactly cancels the average kinetic energy of the bond vibration. But how about the contribution of the two rotational degrees of freedom? They have an average kinetic energy of $k_B T$, but where is the virial compensation? The answer is that rotation involves centrifugal forces on the bond, which is then stretched, causing a compensating elastic force in the bond direction. That force causes a contribution to the intramolecular virial that exactly compensates the rotational kinetic energy (see exercise 17.8).

With that problem solved, the next question is what happens if the bond length is treated as a constraint? In that case there is no kinetic energy in the bond vibration and their is no contribution to the virial due to vibrational motion. But there still is rotational kinetic energy and that is still compensated by the contribution to the virial of the *constraint force*. So when constraints are used in an atomic description, the constraint forces must be computed and accounted for in the atomic virial. Constraint forces are the forces that compensate components of other forces in the constraint directions; they follow directly from the constraint computation (see Section 15.8 on page 417).

17.8 Liouville equations in phase space

A classical system of particles that evolves under Hamiltonian equations of motion (see Section 15.3 on page 399) follows a *trajectory* in the $2n$-dimensional space of all its (generalized) coordinates q_i, $i = 1, \ldots n$ and conjugate momenta p_i $i = 1, \ldots n$, called *phase space*. Here n is the number of degrees of freedom, which equals $3\times$ the number of particles minus the number of *constrained* degrees of freedom, if any. A particular configuration of all coordinates and momenta at time t defines a *point* in phase space. Often it is convenient to denote the $2n$-dimensional vector $(q_1, \ldots q_n, p_1, \ldots p_n)^\mathsf{T}$ by the vector $\mathbf{z}$ ($\mathbf{z} \in \mathbb{R}^{2n}$).[28] We consider for the time being systems that obey a first-order differential equation, so that the evolution of a point in phase space is a deterministic initial value problem.

Consider evolution under the Hamilton equations (15.9):

$$\dot{q}_i = \frac{\partial \mathcal{H}(q, p)}{\partial p_i}, \tag{17.148}$$

$$\dot{p}_i = -\frac{\partial \mathcal{H}(q, p)}{\partial q_i}. \tag{17.149}$$

In symplectic notation, and introducing the matrix

$$\mathbf{L}_0 = \begin{pmatrix} \mathbf{0} & \mathbf{1} \\ -\mathbf{1} & \mathbf{0} \end{pmatrix}, \tag{17.150}$$

where $\mathbf{0}$ is a $n \times n$ all-zero matrix, and $\mathbf{1}$ is a $n \times n$ diagonal unit matrix,

---

[28] This notation is referred to as the *symplectic notation*. A *symplectic mapping* is an area (and volume) conserving mapping, such as the transformation of z for systems obeying Hamilton's equations. A *symplectic algorithm* is an algorithm to solve the time evolution that conserves area and volume in phase space.

Hamilton's equations can be written as

$$\dot{\mathbf{z}} = \mathbf{L}_0 \frac{\partial \mathcal{H}}{\partial \mathbf{z}}. \tag{17.151}$$

Another notation is writing the operation on $\mathbf{z}$ on the right-hand side of (17.151) as an *operator*:

$$\dot{\mathbf{z}} = i\hat{\mathcal{L}}\mathbf{z}. \tag{17.152}$$

$\hat{\mathcal{L}}$ is the *Liouville* operator, which we will define below in the context of the rate of change of density in phase space.

The rate equation (17.152) is more general than its Hamiltonian form of (17.151), and may for example contain the effects of external time-dependent fields. The operator is then called the *generalized Liouville operator*.

The rate equation can be formally integrated to yield a time evolution equation for the point $\mathbf{z}$ in phase space:

$$\mathbf{z}(t) = \exp[\int_0^t i\hat{\mathcal{L}}(\tau)\,d\tau]\,\mathbf{z}(0), \tag{17.153}$$

which for time-independent operators reduces to

$$\mathbf{z}(t) = e^{i\hat{\mathcal{L}}t}\mathbf{z}(0). \tag{17.154}$$

These equations are formally elegant but do not help solving the equations of motion. Exponential operators and the integral over a time-dependent operator must all be carefully defined to be meaningful (see Section 14.6 on exponential operators on page 385).

In statistical mechanics we deal not with single trajectories, but with *distribution functions* $f(\mathbf{z}, t)$ in phase space. We are interested in the time evolution of such distribution functions and in the equilibrium distributions corresponding to the various ensembles. When we know the distribution function, we can determine the *observable ensemble average* of any variable that is known as a function of the point in phase space:

$$\langle A \rangle(t) = \int A(\mathbf{z})\,f(\mathbf{z}, t)\,d\mathbf{z}. \tag{17.155}$$

In order to find a differential equation for the time evolution of $f(\mathbf{z}, t)$ we try to find the time derivative $\partial f/\partial t$. Since f concerns a probability distribution, its integrated density must be conserved, and any change in density must result from in- or outgoing flux $\mathbf{J} = f\dot{\mathbf{z}}$. The conservation law in $2n$ dimensions is similar to the continuity equation in fluid dynamics (see (9.3) on page 282):

$$\frac{\partial f(\mathbf{z}, t)}{\partial t} = -\boldsymbol{\nabla} \cdot \mathbf{J} = -\boldsymbol{\nabla} \cdot (f\dot{\mathbf{z}}) = -\dot{\mathbf{z}} \cdot \boldsymbol{\nabla} f - f\boldsymbol{\nabla} \cdot \dot{\mathbf{z}}, \tag{17.156}$$

where $\boldsymbol{\nabla}$ stands for the gradient operator in $\mathbf{z}$: $(\partial/\partial z_1, \ldots, \partial/\partial z_{2n})$. Writing the time derivative as the *material* or *Lagrangian* derivative D/Dt, i.e., the derivative seen when traveling with the flow (see page 282), we see that

$$\frac{Df}{Dt} \stackrel{\text{def}}{=} \frac{\partial f}{\partial t} + \dot{\mathbf{z}} \cdot \boldsymbol{\nabla} f = -f\boldsymbol{\nabla} \cdot \dot{\mathbf{z}}. \tag{17.157}$$

This equation is often referred to as the *generalized Liouville equation*. For a Hamiltonian system the term on the right-hand side is zero, as we see using (17.151):

$$\boldsymbol{\nabla} \cdot \dot{\mathbf{z}} = \boldsymbol{\nabla} \cdot \left[\mathbf{L}_0 \frac{\partial \mathcal{H}}{\partial \mathbf{z}} \right] = \sum_{i,j=1}^{2n} L_{0ij} \frac{\partial^2 \mathcal{H}}{\partial z_i \partial z_j}$$

$$= \sum_{i=1}^{n} \left(\frac{\partial^2 \mathcal{H}}{\partial q_i \partial p_j} - \frac{\partial^2 \mathcal{H}}{\partial p_i \partial q_j} \right) = 0. \tag{17.158}$$

This is the proper *Liouville equation*, which is very important in statistical mechanics. It states that for a Hamiltonian system the (probability) density in phase space does not change with time. This also means that a *volume* in phase space does not change with time: if one follows a bundle of trajectories that start in an initial region of phase space, then at a later time these trajectories will occupy a region of phase space with the same volume as the initial region.[29] This is also expressed by saying that the Hamiltonian probability flow in phase space is *incompressible*.

If the volume does not change, neither will a *volume element* used for integration over phase space. We have to be careful what we call a volume element. Normally we write the volume element somewhat loosely by a product $d\mathbf{z}$ or $d^{2n}\mathbf{z}$ or $dz_1 \ldots dz_{2n}$ or $\Pi_{i=1}^{2n} dz_i$, while we really mean the volume spanned by the local displacement vectors corresponding to the increments dz_i. These displacement vectors are proportional, but not necessarily equal to z_i. The volume spanned by the local displacement vectors only equals their product when these vectors are orthogonal; in general the volume is equal to the determinant of the matrix formed by the set of displacement vectors. The proper name for such a volume is the *wedge product*, but we

---

[29] We skip the intricate discussion on the possibility to define such regions, which relates to the fact that different trajectories can never cross each other. Even more intricate is the discussion on the possible *chaotic* behavior of Hamiltonian dynamical systems that destroys the notion of conserved volume.

shall not use the corresponding wedge notation. The volume element is written as

$$dV = \sqrt{g}dz_1 \cdots dz_{2n}, \qquad (17.159)$$

where g is the determinant of the *metric tensor* g_{ij}, which defines the *metric* of the space: the square of the *length* of the displacement ds caused by $dz_1, \ldots, dz_{2n}$ is determined by $(ds)^2 = \sum_{i,j} g_{ij}dz_i\, dz_j$.[30]

When coordinates are transformed from $\mathbf{z}(0)$ to $\mathbf{z}(t)$, the transformation is characterized by a transformation matrix $\mathbf{J}$ and the volume element transforms with the *Jacobian* J of the transformation, which is the determinant of $\mathbf{J}$:

$$\sqrt{g(t)}\, dz_1(t) \cdots dz_{2n}(t) = \sqrt{g(0)}\, dz_1(0) \cdots dz_{2n}(0), \qquad (17.160)$$

$$dz_1(t) \cdots dz_{2n}(t) = J\, dz_1(0) \cdots dz_{2n}(0), \qquad (17.161)$$

$$d\mathbf{z}(t) = \mathbf{J}\, d\mathbf{z}(0), \qquad (17.162)$$

$$J = det\, \mathbf{J}, \qquad (17.163)$$

$$J = \sqrt{g(t)/g(o)}. \qquad (17.164)$$

The Liouville equation (17.158) implies that the Jacobian of the transformation from $\mathbf{z}(0)$ to $\mathbf{z}(t)$ equals 1 for Hamiltonian systems. Hence the volume element in phase space, which is denoted by the product $d\mathbf{z} = dz_1(t) \cdots dz_{2n}(t)$, is invariant under a Hamiltonian or *canonical* transformation. A canonical transformation is *symplectic*, meaning that the so-called *two-form* $dq \wedge dp = \sum_i dq_i \wedge dp_i$ of any two vectors dq and dp, spanning a two-dimensional surface S, is invariant under the transformation.[31] This property is important in order to retain the notion of probability density in phase space $f(\mathbf{z})\, d\mathbf{z}$.

In practice, time evolutions are not always Hamiltonian and the probability flow could well loose its incompressibility. The question how the Jacobian (or the metric tensor) develops in such cases and influences the distribution functions has been addressed by Tuckerman *et al.* (1999). We'll summarize their results. First define the *phase space compressibility* κ:

$$\kappa(\mathbf{z}) \overset{\text{def}}{=} \boldsymbol{\nabla} \cdot \dot{\mathbf{z}} = \sum_i \frac{\partial \dot{z}_i}{\partial z_i}, \qquad (17.165)$$

[30] Consider polar coordinates (r, θ, ϕ) of a point in 3D space: changing θ to $\theta + d\theta$ causes a displacement vector of length $r\, d\theta$. Changing ϕ to $\phi + d\phi$ causes a displacement vector of length $r \sin \theta\, d\phi$. What is the metric tensor for polar coordinates and what is the square root of its determinant and hence the proper volume element?

[31] The two-form is the sum of areas of projections of the two-dimensional surface S onto the $q_i - p_i$ planes. See, e.g., Arnold (1975).

which is the essential factor on the right-hand side of the generalized Liouville equation (17.157). As we have seen, $\kappa = 0$ for incompressible Hamiltonian flow. Tuckerman *et al.* then proceed to show that the Jacobian J of the transformation from $\mathbf{z}(0)$ to $\mathbf{z}(t)$ obeys the differential equation

$$\frac{dJ}{dt} = J\,\kappa(\mathbf{z}),\qquad(17.166)$$

with solution

$$J(t) = \exp\left(\int_0^t \kappa[\mathbf{z}(\tau)]\,d\tau\right).\qquad(17.167)$$

If a function $w(\mathbf{z})$ is defined for which $\dot{w} = \kappa$, then

$$J(t) = e^{w(\mathbf{z}(t))-w(\mathbf{z}(0))},\qquad(17.168)$$

and

$$e^{-w(\mathbf{z}(t))}dz_1(t)\cdots dz_{2n}(t) = e^{-w(\mathbf{z}(0))}dz_1(0)\cdots dz_{2n}(0).\qquad(17.169)$$

Hence this modified volume element, with $\sqrt{g} = e^{-w(\mathbf{z})}$, is invariant under the non-Hamiltonian equations of motion. This enables us to compute equilibrium distribution functions generated by the non-Hamiltonian dynamics. Examples are given in Section 6.5 on page 194.

Let us return to Hamiltonian systems for which the Liouville equation (17.158) is valid. The time derivative of f, measured at a stationary point in phase space, is

$$\frac{\partial f}{\partial t} = -\dot{\mathbf{z}}\cdot\boldsymbol{\nabla} f = -i\hat{\mathcal{L}}f,\qquad(17.170)$$

where the *Liouville operator* is defined as[32]

$$i\hat{\mathcal{L}} \stackrel{\text{def}}{=} \dot{\mathbf{z}}\cdot\boldsymbol{\nabla} = \sum_{i=1}^{2n} \dot{z}_i\frac{\partial}{\partial z_i} = \sum_{i,j=1}^{2n} L_{0ij}\frac{\partial\mathcal{H}}{\partial z_j}\frac{\partial}{\partial z_i}$$

$$= \sum_{i,j=1}^{n}\left(\frac{\partial\mathcal{H}}{\partial p_j}\frac{\partial}{\partial q_i} - \frac{\partial\mathcal{H}}{\partial q_j}\frac{\partial}{\partial p_i}\right).\qquad(17.171)$$

This sum is called a *Poisson bracket*; if applied to a function f, it is written as $\{\mathcal{H}, f\}$. We shall not use this notation. Assuming a Hamiltonian that does not explicitly depend on time, the formal solution is

$$f(\mathbf{z}, t) = e^{-it\hat{\mathcal{L}}}f(\mathbf{z}, 0).\qquad(17.172)$$

---

[32] The convention to write the operator as $i\hat{\mathcal{L}}$ and not simply $\hat{\mathcal{L}}$ is that there is a corresponding operator in the quantum-mechanical evolution of the density matrix and the operator now is hermitian.

Note that the time-differential operator (17.170) for the space phase density has a sign opposite to that of the time-differential operator (17.152) for a point in phase space.

17.9 Canonical distribution functions

In this section we shall consider the classical distribution functions for the most important ensemble, the *canonical ensemble*, for various cases. The cases concern the distributions in phase space and in configuration space, both for cartesian and generalized coordinates and we shall consider what happens if internal constraints are applied.

In general phase space the canonical (NVT) equilibrium ensemble of a Hamiltonian system of n degrees of freedom ($= 3N$ for a system without constraints) corresponds to a density $f(\mathbf{z})$ in phase space proportional to $\exp(-\beta\mathcal{H})$:

$$f(\mathbf{z}) = \frac{\exp[-\beta\mathcal{H}(\mathbf{z})]}{\int \exp[-\beta\mathcal{H}(\mathbf{z})]\, d^{2n}\mathbf{z}}. \tag{17.173}$$

The classical canonical partition function for N particles (for simplicity taken as identical; if not, the $N!$ must be modified to a product of factorials for each of the identical types) is

$$Q = \frac{1}{h^{3N}N!} \int \exp[-\beta\mathcal{H}(\mathbf{z})]\, d^{2n}\mathbf{z}. \tag{17.174}$$

The Hamiltonian is the sum of kinetic and potential energy.

17.9.1 Canonical distribution in cartesian coordinates

We now write $\mathbf{r}_i, \mathbf{p}_i = m_i\dot{\mathbf{r}}_i$ ($i = 1, \ldots, N$) for the phase-space coordinates $\mathbf{z}$. The Hamiltonian is given by

$$\mathcal{H} = \sum_{i=1}^{N} \frac{\mathbf{p}_i^2}{2m_i} + V(\mathbf{r}). \tag{17.175}$$

The distribution function (17.173) can now be separately integrated over momentum space, yielding a *configurational canonical distribution function*

$$f(\mathbf{r}) = \frac{\exp[-\beta V(\mathbf{r})]}{\int_V \exp[-\beta V(\mathbf{r})]\, d^N\mathbf{r}}. \tag{17.176}$$

while the classical canonical partition function is given by integration of (17.174) over momenta:

$$Q = \frac{1}{N!} \left(\frac{2\pi k_B T}{h^2} \right)^{3N/2} \Pi_{i=1}^N m_i^{3/2} \int_V e^{-\beta V(\boldsymbol{r})} \, d^N \boldsymbol{r}. \tag{17.177}$$

Using the definition of the *de Broglie wavelength* Λ_i (17.69):

$$\Lambda_i = \frac{h}{\sqrt{2\pi m_i k_B T}},$$

the partition function can also be written as

$$Q = \frac{1}{N!} \Pi_{i=1}^N \Lambda_i^{-3} \int_V e^{-\beta V(\boldsymbol{r})} \, d^N \boldsymbol{r}. \tag{17.178}$$

17.9.2 Canonical distribution in generalized coordinates

In generalized coordinates (q, p) the kinetic energy is a function of the coordinates, even if the potential is conservative, i.e., a function of coordinates only. The Hamiltonian now reads (see 15.24) on page 402):

$$\mathcal{H} = \mathbf{p}^{\mathsf{T}} \mathbf{M}^{-1}(q)\mathbf{p} + V(q), \tag{17.179}$$

where $\mathbf{M}$ is the *mass tensor* (see (15.16) on page 401)

$$M_{kl} = \sum_{i=1}^N m_i \frac{\partial \boldsymbol{r}_i}{\partial q_k} \cdot \frac{\partial \boldsymbol{r}_i}{\partial q_l}. \tag{17.180}$$

In cartesian coordinates the mass tensor is diagonal with the masses m_i on the diagonal (each repeated three times) and the integration over momenta can be carried out separately (see above). In generalized coordinates the integration over momenta yields a q-dependent form that cannot be taken out of the integral:

$$Q = \frac{1}{N!} \left(\frac{2\pi k_B T}{h^2} \right)^{3N/2} \int_V |\mathbf{M}|^{1/2} e^{-\beta V(q)} \, d^n \mathbf{q}, \tag{17.181}$$

where we use the notation $|\mathbf{A}|$ for the determinant of $\mathbf{A}$. So, expressed as integral over generalized configurational space, there is a weight factor $(\det \mathbf{M})^{1/2}$ in the integrand. The integration over momenta is obtained by transforming the momenta with an orthogonal transformation so as to obtain a diagonal inverse mass tensor; integration then yields the square root of the product of diagonal elements, which equals the square root of the determinant of the original inverse mass matrix. It is also possible to arrive at this equation by first integrating over momenta in cartesian coordinates and then transforming from cartesian $(x_1, \ldots x_n)$ to generalized $(q_1, \ldots, q_n)$ coordinates by a transformation $\mathbf{J}$:

$$J_{ik} = \frac{\partial x_i}{\partial q_k} \tag{17.182}$$

with Jacobian $J(q) = |\mathbf{J}|$, yielding

$$Q = \frac{1}{N!} \left(\frac{2\pi k_{\mathrm{B}} T}{h^2} \right)^{3N/2} \Pi_{i=1}^N m_i^{3/2} \int_V J(q)\, e^{-\beta V(q)}\, d^n \mathbf{q}. \tag{17.183}$$

Apparently,

$$|\mathbf{M}|^{1/2} = \Pi_{i=1}^N m_i^{3/2} J, \tag{17.184}$$

as follows immediately from the relation between mass tensor (17.180) and transformation matrix:

$$\mathbf{M} = \mathbf{J}^{\mathrm{T}} \mathbf{M}_{\mathrm{cart}} \mathbf{J}, \tag{17.185}$$

where $\mathbf{M}_{\mathrm{cart}}$ is the diagonal matrix of masses, so that

$$|\mathbf{M}| = |\mathbf{J}|^2 \Pi_{i=1}^N m_i^3. \tag{17.186}$$

The result is that the canonical ensemble average of a variable $A(q)$ is given by

$$\langle A \rangle = \frac{\int A(q) |\mathbf{M}|^{1/2} \exp[-\beta V(q)]\, dq}{\int |\mathbf{M}|^{1/2} \exp[-\beta V(q)]\, dq}. \tag{17.187}$$

17.9.3 Metric tensor effects from constraints

The question addressed here is what canonical equilibrium distributions are generated in systems with constraints, and consequently, how averages should be taken to derive observables. Such systems are usually simulated in cartesian coordinates with the addition of Lagrange multipliers that force the system to remain on the constraint hypersurface in phase space; the dynamics is equivalent to that in a reduced phase space of non-constrained generalized coordinates. The result will be that there is an extra weight factor in the distribution function. This result has been obtained several times in the literature, for example see Frenkel and Smit (1996) or Ciccotti and Ryckaert (1986).

Consider generalized coordinates $q_1, \ldots, q_{3N} = (q'q'')$ that are chosen in such a way that the last n_c coordinates q'' are to be constrained, leaving the first $n = 3N - n_c$ coordinates q' free. The system is then restricted to n degrees of freedom q'. We first consider the fully, i.e., mathematically, constrained case, where $q'' = c$, c being a set of constants, and without any kinetic energy in the constrained coordinates. Writing the mass tensor in four parts corresponding to q' and q'':

$$\mathbf{M} = \begin{pmatrix} \mathbf{F} & \mathbf{D} \\ \mathbf{D}^\mathsf{T} & \mathbf{C} \end{pmatrix}, \tag{17.188}$$

the kinetic energy now equals

$$K = \frac{1}{2}\dot{\mathbf{q}}'^\mathsf{T}\mathbf{F}\dot{\mathbf{q}}', \tag{17.189}$$

leading to conjugate momenta

$$\mathbf{p}' = \mathbf{F}\dot{\mathbf{q}}', \tag{17.190}$$

which differ from the momenta in full space. The canonical average of a variable $A(q')$ obtained in the constrained system is

$$\langle A \rangle_c = \frac{\int A(q')|\mathbf{F}|^{1/2}\exp[-\beta V(q')]\,dq'}{\int |\mathbf{F}|^{1/2}\exp[-\beta V(q')]\,dq'}. \tag{17.191}$$

The same average is obtained from constrained dynamics in cartesian coordinates:

$$\langle A \rangle_c = \frac{\int A(\boldsymbol{r})\Pi_s\delta(\sigma_s(\boldsymbol{r}))\exp[-\beta V(\boldsymbol{r})]\,dr}{\int \Pi_s\delta(\sigma_s(\boldsymbol{r}))\exp[-\beta V(\boldsymbol{r})]\,dr}, \tag{17.192}$$

where $\sigma_s(\boldsymbol{r}) = 0$, $s = 1,\ldots,n_c$, are the constraint equations that remain satisfied by the algorithm.

Compare this with a classical physical system where q'' are *near constraints* that only negligibly deviate from constants, for example restrained by stiff oscillators. The difference with mathematical constraints is that the near constraints do contribute to the kinetic energy and have an additional potential energy $V_c(q'')$ as well. The latter is a harmonic-like potential with a sharp minimum. In order to obtain averages we need the full configuration space, but within the integrand we can integrate over q'':

$$Q'' = \int \exp[-\beta V_c(q'')]\,dq''. \tag{17.193}$$

The average over the near-constraint canonical ensemble is

$$\langle A \rangle_{nc} = \frac{\int A(q'')|M(q',c)|^{1/2}Q''\exp[-\beta V(q')\,dq']}{\int |M(q',c)|^{1/2}Q''\exp[-\beta V(q')\,dq']}. \tag{17.194}$$

Q'' may depend on q': for example, for harmonic oscillators Q'' depends on the force constants which may depend on q'. But this dependence is weak and often negligible. In that case Q'' drops out of the equation and the weight factor is simply equal to $|\mathbf{M}|^{1/2}$. Since $|F| \neq |M|$, the weight factor in the constraint ensemble is not equal to the weight factor in the classical physical near-constraint case. Therefore the constraint ensemble should be corrected by an extra weight factor $(|\mathbf{M}|/|\mathbf{F}|)^{1/2}$. This extra weight factor can also be expressed as $\exp[-\beta v_c(q')]$ with an extra *potential*

$$v_c(q') = -\frac{1}{2} k_{\mathrm{B}} T \ln \frac{|\mathbf{M}|}{|\mathbf{F}|}. \tag{17.195}$$

The ratio $|\mathbf{M}|/|\mathbf{F}|$ seems difficult to evaluate since both $\mathbf{M}$ and $\mathbf{F}$ are complicated large-dimensional matrices, even when there are only a few constraints. But a famous theorem by Fixman (1979) saves the day:

$$|\mathbf{M}|\,|\mathbf{Z}| = |\mathbf{F}|, \tag{17.196}$$

where $\mathbf{Z}$ is the $(q''q'')$ part of the matrix $\mathbf{M}^{-1}$:

$$\mathbf{M}^{-1} = \begin{pmatrix} \mathbf{X} & \mathbf{Y} \\ \mathbf{Y}^{\mathsf{T}} & \mathbf{Z} \end{pmatrix}, \quad Z_{st}(q') = \sum_i \frac{1}{m_i} \frac{\partial q_s''}{\partial \boldsymbol{r}_i} \cdot \frac{\partial q_t''}{\partial \boldsymbol{r}_i}. \tag{17.197}$$

$\mathbf{Z}$ is a low-dimensional matrix which is generally easy to compute. We find for the extra weight factor $|Z|^{-1/2}$, or

$$v_c(q') = \frac{1}{2} k_{\mathrm{B}} T \ln |\mathbf{Z}|.. \tag{17.198}$$

The corrected constrained ensemble average of an observable A can be expressed in cartesian coordinates as

$$\langle A \rangle = \frac{\int |Z|^{-1/2} A(\boldsymbol{r}) \Pi_s \delta(\sigma_s(\boldsymbol{r})) \exp[-\beta V(\boldsymbol{r})]\, d\boldsymbol{r}}{\int |Z|^{-1/2} \Pi_s \delta(\sigma_s(\boldsymbol{r})) \exp[-\beta V(\boldsymbol{r})]\, d\boldsymbol{r}}. \tag{17.199}$$

For completeness we give the ingenious proof of Fixman's theorem.

Proof From

$$\mathbf{M}^{-1}\mathbf{M} = \begin{pmatrix} \mathbf{X} & \mathbf{Y} \\ \mathbf{Y}^{\mathsf{T}} & \mathbf{Z} \end{pmatrix} \begin{pmatrix} \mathbf{F} & \mathbf{D} \\ \mathbf{D}^{\mathsf{T}} & \mathbf{C} \end{pmatrix} = 1,$$

we see that

$$\mathbf{X}\mathbf{F} + \mathbf{Y}\mathbf{D}^{\mathsf{T}} = 1,$$

$$\mathbf{Y}^{\mathsf{T}}\mathbf{F} + \mathbf{Z}\mathbf{D}^{\mathsf{T}} = 0.$$

Consider

$$\begin{pmatrix} \mathbf{F} & 0 \\ \mathbf{D}^{\mathsf{T}} & 1 \end{pmatrix} = \mathbf{M}\mathbf{M}^{-1} \begin{pmatrix} \mathbf{F} & 0 \\ \mathbf{D}^{\mathsf{T}} & 1 \end{pmatrix} = \mathbf{M} \begin{pmatrix} \mathbf{X} & \mathbf{Y} \\ \mathbf{Y}^{\mathsf{T}} & \mathbf{Z} \end{pmatrix} \begin{pmatrix} \mathbf{F} & 0 \\ \mathbf{D}^{\mathsf{T}} & 1 \end{pmatrix}$$

$$= \mathbf{M} \begin{pmatrix} \mathbf{X}\mathbf{F} + \mathbf{Y}\mathbf{D}^{\mathsf{T}} & \mathbf{Y} \\ \mathbf{Y}^{\mathsf{T}}\mathbf{F} + \mathbf{Z}\mathbf{D}^{\mathsf{T}} & \mathbf{Z} \end{pmatrix} = \mathbf{M} \begin{pmatrix} 1 & \mathbf{Y} \\ 0 & \mathbf{Z} \end{pmatrix}. \tag{17.200}$$

Hence $|\mathbf{F}| = |\mathbf{M}|\,|\mathbf{Z}|$. □

This effect is often referred to as the *metric tensor effect*, which is not a correct name, since it is not the metric tensor proper, but the mass(-metric) tensor that is involved.

The effect of the mass tensor is often zero or negligible. Let us consider a few examples:

- A *single distance constraint* between two particles: $q'' = r_{12} = |\mathbf{r}_1 - \mathbf{r}_2|$. The matrix $\mathbf{Z}$ has only one component Z_{11}:

$$Z_{11} = \frac{1}{m_1}\frac{\partial r_{12}}{\partial \mathbf{r}_1}\cdot\frac{\partial r_{12}}{\partial \mathbf{r}_1} + \frac{1}{m_2}\frac{\partial r_{12}}{\partial \mathbf{r}_2}\cdot\frac{\partial r_{12}}{\partial \mathbf{r}_2} = \frac{1}{m_1} + \frac{1}{m_2}. \tag{17.201}$$

  This is a constant, so there is no effect on the distribution function.
- A *single generalized distance constraint* that can be written as $q'' = R = |\sum_i \alpha_i \mathbf{r}_i|$, with α_i being constants. For this case $Z_{11} = \sum_i \alpha_i^2/m_i$ is also a constant with no effect.
- *Two distance constraints* r_{12} and r_{32} for a triatomic molecule with an angle ϕ between $\mathbf{r}_{12}$ and $\mathbf{r}_{32}$. The matrix $\mathbf{Z}$ now is a 2×2 matrix for which the determinant appears to be (see Exercise 17.9):

$$|\mathbf{Z}| = \left(\frac{1}{m_1} + \frac{1}{m_1}\right)\left(\frac{1}{m_2} + \frac{1}{m_3}\right) - \frac{1}{m_2^2}\cos^2\phi. \tag{17.202}$$

  This is a nonzero case, but the weight factor is almost constant when the bond angle is nearly constant.

More serious effects, but still with potentials not much larger than $k_{\mathrm{B}}T$, can be expected for bond length and bond angle constraints in molecular chains with low-barrier dihedral angle functions. It seems not serious to neglect the mass tensor effects in practice (as is usually done). It is, moreover, likely that the correction is not valid for the common case that constrained degrees of freedom correspond to high-frequency quantum oscillators in their ground state. That case is more complicated as one should really use flexible constraints to separate the quantum degrees of freedom rather than holonomic constraints (see page (v).)

17.10 The generalized equipartition theorem

For several applications it is useful to know the correlations between the fluctuations of coordinates and momenta in equilibrium systems. Statements like "the average kinetic energy equals $k_{\mathrm{B}}T$ per degree of freedom" (the equipartition theorem) or "velocities of different particles are not correlated" or the virial theorem itself, are special cases of the powerful *generalized equipartition theorem*, which states that

$$\left\langle z_i \frac{\partial H}{\partial z_j} \right\rangle = k_{\mathrm{B}} T \delta_{ij}, \tag{17.203}$$

where z_i, $i = 1, \dots, 2n$, stands, as usual, for any of the canonical variables $\{q, p\}$. We prove this theorem for bounded systems in the canonical ensemble, but it is also valid for other ensembles. Huang (1987) gives a proof for the microcanonical ensemble.

Proof We wish to prove

$$\frac{\int z_i \frac{\partial H}{\partial z_j} \exp[-\beta H] \, dz}{\int \exp[-\beta H] \, dz} = k_{\mathrm{B}} T \delta_{ij}.$$

Consider the following equality

$$-\beta z_i \frac{\partial H}{\partial z_j} e^{-\beta H} = z_i \frac{\partial e^{-\beta H}}{\partial z_j} = \frac{\partial}{\partial z_j} \left(z_i e^{-\beta H} \right) - \delta_{ij} e^{-\beta H}.$$

Integrating over phase space dz, the first term on the r.h.s. drops out after partial integration, as the integrand vanishes at the boundaries. Hence

$$-\beta \langle z_i \frac{\partial H}{\partial z_j} \rangle = -\delta_{ij},$$

which is what we wish to prove. □

Applying this theorem to $z_i = p_i = (\mathbf{M}\dot{\mathbf{q}})_i$; $z_j = p_j$ we find

$$\langle (\mathbf{M}\dot{\mathbf{q}})_i \dot{q}_j \rangle = \delta_{ij}, \tag{17.204}$$

or, in matrix notation:

$$\langle \dot{\mathbf{q}} \dot{\mathbf{q}}^{\mathsf{T}} \rangle = \mathbf{M}^{-1} k_{\mathrm{B}} T. \tag{17.205}$$

This is the classical equipartition: for cartesian coordinates (diagonal mass tensor) the average kinetic energy per degree of freedom equals $\frac{1}{2} k_{\mathrm{B}} T$ and velocities of different particles are uncorrelated. For generalized coordinates the kinetic energy per degree of freedom $\frac{1}{2} p_i \dot{q}_i$ still averages to $\frac{1}{2} k_{\mathrm{B}} T$, and the velocity of one degree of freedom is uncorrelated with the momentum of any other degree of freedom.

Another interesting special case is obtained when we apply the theorem to $z_i = q_i$; $z_j = q_j$:

$$\langle q_i \dot{p}_j \rangle = -\langle q_i F_j \rangle = k_{\mathrm{B}} T \delta_{ij}. \tag{17.206}$$

For $j = i$ this recovers the virial theorem (see (17.132) on page 485):

$$\Xi = -\frac{1}{2}\sum_{i=1}^{n}\langle q_i F_i\rangle = \frac{1}{2}nk_B T = \langle E_{\text{kin}}\rangle. \tag{17.207}$$

Including the cases $j \neq i$ means that the virial *tensor* is diagonal in equilibrium, and the average pressure is isotropic.

Exercises

17.1 Show for the canonical ensemble, where Q is a function of V and β, that $U = -\partial \ln Q/\partial \beta$. Show that this equation is equivalent to the Gibbs–Helmholtz equation, (16.32).

17.2 If the pressure p is defined as the ensemble average of $\partial E_i/\partial V$, then show that for the canonical ensemble $p = \beta^{-1}\partial \ln Q/\partial V$.

17.3 Derive (17.63) by considering how many points there are in **k**-space in a spherical shell between k and $k+dk$. Transform this to a function of ε.

17.4 Derive the quantum expressions for energy, entropy and heat capacity of the harmonic oscillator. Plot the heat capacity for the quantum and classical case.

17.5 Show that the quantum correction to the free energy of a harmonic oscillator (17.109) equals the first term in the expansion of the exact $A^{\text{qu}} - A^{\text{cl}}$ in powers of ξ.

17.6 See (17.133). Show that $\int_S \mathbf{r}\, d\mathbf{S} = 3V$, with the integral taken over a closed surface enclosing a volume V. Transform from a surface to a volume integral over the divergence of $\mathbf{r}$.

17.7 Carry out the partial differentiation of $A(V,T) = -k_B T \ln Q$ with respect to volume to obtain the isotropic form of (17.138). Assume a cubic $L \times L \times L$ lattice and use scaled coordinates $\mathbf{r}/L$.

17.8 Prove that the rotational kinetic energy of a harmonic homonuclear diatomic molecule $(2 \times \frac{1}{2}mv_{\text{rot}}^2)$ equals the contribution to the virial of the centripetal harmonic force $(2 \times mv_{\text{rot}}^2/(\frac{1}{2}d))$.

17.9 Compute the matrix $\mathbf{Z}$ for a triatomic molecule with constrained bond lengths.

18

Linear response theory

18.1 Introduction

There are many cases of interest where the relevant question we wish to answer by simulation is "what is the response of the (complex) system to an external disturbance?" Such responses can be related to experimental results and thus be used not only to predict material properties, but also to validate the simulation model. Responses can either be *static*, after a prolonged constant external disturbance that drives the system into a non-equilibrium *steady state*, or *dynamic*, as a reaction to a time-dependent external disturbance. Examples of the former are *transport properties* such as the heat flow resulting from an imposed constant temperature gradient, or the stress (momentum flow) resulting from an imposed velocity gradient. Examples of the latter are the optical response to a specific sequence of laser pulses, or the time-dependent induced polarization or absorption following the application of a time-dependent external electric field.

In general, responses can be expected to relate in a *non-linear* fashion to the applied disturbance. For example, the dielectric response (i.e., the polarization) of a dipolar fluid to an external electric field will level off at high field strengths when the dipoles tend to orient fully in the electric field. The optical response to two laser pulses, 100 fs apart, will not equal the sum of the responses to each of the pulses separately. In such cases there will not be much choice other than mimicking the external disturbance in the simulated system and "observing" the response. For time-dependent responses such simulations should be repeated with an ensemble of different starting configurations, chosen from an equilibrium distribution, in order to obtain statistically significant results that can be compared to experiment.

In this chapter we will concentrate on the very important class of *linear responses* with the property that the response to the sum of two disturbances

505

equals the sum of responses to each of the disturbances. To this class belong all responses to small disturbances in the *linear regime*; these are then proportional to the amplitude of the disturbance. The proportionality constant determines transport coefficients such as viscosity, thermal conductivity and diffusion constant, but also dielectric constant, refractive index, conductivity and optical absorption. Since the decay of a small perturbation, caused by an external disturbance, is governed by the same equations of motion that determine the thermal fluctuations in the equilibrium system, there is a relation between the decay function of an observable of the system after perturbation and the time-correlation function of spontaneous fluctuations of a related variable. In the next sections we shall elaborate on these relations.

In Section 18.2 the general relations between an external disturbance and the resulting linear response will be considered both in the time and frequency domain, without reference to the processes in the system that cause the response. In Section 18.3 the relation between response functions and the time correlation function of spontaneously fluctuating quantities will be considered for a *classical* system of particles that interact according to Hamilton's equations of motion.

18.2 Linear response relations

In this section we consider our system as a black box, responding to a disturbance $X(t)$ with a response $Y(t)$ (Fig. 18.1). The disturbance is an external force or field acting on the system, such as an electric field $E(t)$, and the response is an observable of the system, for example, a current density $j(t)$ resulting from the disturbance $E(t)$. Both X and Y may be vectorial quantities, in which case their relations are specified by tensors, but for simplicity of notation we shall stick to scalars here.

Exactly how the interactions between particles lead to a specific response does not concern us in this section. The only *principles* we assume the system to obey are:

  (i) (**causality**) the response never precedes its cause;
 (ii) (**relaxation**) the response will, after termination of the disturbance, in due time return to its equilibrium value.

Without loss of generality we will assume that the equilibrium value of Y is zero. So, when $X = 0$, $Y(t)$ will decay to zero.

A crucial role in the description of linear responses is played by the *delta-*

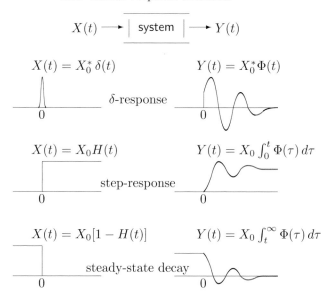

Figure 18.1 Black-box response to a small perturbation. Responses to a delta-disturbance, to a step disturbance (Heaviside function) and following a terminated steady state disturbance are sketched.

response $\Phi(t)$ (Fig. 18.1):

$$\text{if } X(t) = X_0^* \delta(t), \text{ then } Y(t) = X_0^* \Phi(t), \tag{18.1}$$

where X_0^* is the amplitude of the driving disturbance, taken small enough for the system response to remain linear. Note that we indicate this δ-function amplitude with a star, to remind us that X_0^* is not just a special value of X, but that it is a different kind of quantity with a dimension equal to the dimension of X, multiplied by time. Our two principles assure that

$$\Phi(t) = 0 \text{ for } t < 0, \tag{18.2}$$

$$\lim_{t \to \infty} \Phi(t) = 0. \tag{18.3}$$

The shape of $\Phi(t)$ is determined by the interactions between particles that govern the time evolution of Y. We note that $\Phi(t)$ is the result of a macroscopic experiment and therefore is an *ensemble average*: $\Phi(t)$ is the average result of delta-disturbances applied to many configurations that are representative for an equilibrium distribution of the system.

Because of the linearity of the response, once we know $\Phi(t)$, we know the response to an arbitrary disturbance $X(t)$, as the latter can be reconstructed from a sequence of δ-pulses. So the response to $X(t)$ is given by

$$Y(t) = \int_0^\infty X(t-\tau)\Phi(\tau)\,d\tau. \tag{18.4}$$

A special case is the response to a *step disturbance*, which is zero for $t < 0$ and constant for $t \geq 0$ (i.e., the disturbance is proportional to the Heaviside function $H(t)$ which is defined as 0 for $t < 0$ and 1 for $t \geq 0$). The response is then proportional to the integral of the δ-response function. Similarly, the response after suddenly switching-off a constant disturbance (leaving the system to relax from a steady state), is given by the integral from t to ∞ of the δ-response function. See Fig. 18.1.

Another special case is a *periodic* disturbance

$$X(t) = \Re\left[X(\omega)e^{i\omega t}\right], \tag{18.5}$$

which is a cosine function if $X_{(\omega)}$ is real, and a sine function if $X_{(\omega)}$ is purely imaginary. Inserting this into (18.4) we obtain a response in the frequency domain, equal to the one-sided Fourier transform of the delta-response:

$$Y(t) = \Re\left[X_{(\omega)}e^{i\omega t}\int_0^\infty \Phi(\tau)e^{-i\omega\tau}\,d\tau\right]. \tag{18.6}$$

Writing simply

$$X(t) = X(\omega)e^{i\omega t} \text{ and } Y(t) = Y(\omega)e^{i\omega t}, \tag{18.7}$$

with the understanding that the observables are the real part of those complex quantities, (18.6) can be rewritten in terms of the complex *frequency response* $\chi(\omega)$:

$$Y(\omega) = \chi(\omega)X(\omega), \tag{18.8}$$

with

$$\chi(\omega) = \int_0^\infty \Phi(\tau)e^{-i\omega\tau}\,d\tau. \tag{18.9}$$

The frequency response $\chi(\omega)$ is a *generalized susceptibility*, indicating how Y responds to X. It can be split into a real and imaginary part:

$$\chi(\omega) = \chi'(\omega) - i\chi''(\omega) \tag{18.10}$$

$$\chi'(\omega) = \int_0^\infty \Phi(\tau)\cos\omega\tau\,d\tau \tag{18.11}$$

$$\chi''(\omega) = \int_0^\infty \Phi(\tau)\sin\omega\tau\,d\tau \tag{18.12}$$

Note that the zero-frequency value of χ (which is real) equals the steady-state response to a constant disturbance:

$$\chi(0) = \int_0^\infty \Phi(\tau)\,d\tau, \tag{18.13}$$

$$X(t) = X_0 H(t) \rightarrow Y(\infty) = X_0 \chi(0). \tag{18.14}$$

For the case that X is an electric field and Y a current density, χ is the specific conductance σ. Its real part determines the current component *in phase* with the periodic field (which is dissipative), while its imaginary part is the current component 90° out of phase with the field. The latter component does not involve energy absorption from the field and is usually indicated with the term *dispersion*. Instead of the current density, we could also consider the *induced dipole density* P as the response; P is related to j since $j = dP/dt$. With $Y = P$, χ becomes ε_0 times the electrical susceptibility χ_e (see Chapter 13) and σ becomes indistinguishable from $i\omega\varepsilon_0\chi_e$. Thus the real part of the electrical susceptibility (or the dielectric constant, or the square root of the refractive index) corresponds to the non-dissipative dispersion, while the imaginary part is a dissipative absorption.

The Kramers–Kronig relations

There are interesting relations between the real and imaginary parts of a frequency response function $\chi(\omega)$, resulting from the causality principle. These are the famous *Kramers–Kronig relations*:[1]

$$\chi'(\omega) = \frac{2}{\pi} \int_0^\infty \frac{\omega' \chi''(\omega')}{\omega'^2 - \omega^2}\,d\omega', \tag{18.15}$$

$$\chi''(\omega) = \frac{2}{\pi} \int_0^\infty \frac{\omega \chi'(\omega')}{\omega^2 - \omega'^2}\,d\omega'. \tag{18.16}$$

Since the integrands diverge when ω' approaches ω, the integrals are not well-defined. They must be interpreted as the *principal value*: exclude an interval $\omega - \varepsilon$ to $\omega + \varepsilon$ from the integration and then take the limit $\varepsilon \rightarrow 0$.

The relations arise from the fact that both χ' and χ'' follow from the same delta-response function through equations (18.11) and (18.12). The proof of (18.16) is given below. The reader is challenged to prove (18.15).

Proof Start from (18.9). This is a one-sided Fourier (or Fourier–Laplace)

---

[1] These relations were first formulated by Kronig (1926) and Kramers (1927) and can be found in many textbooks, e.g., McQuarrie (1976).

transform, but taking into account that $\Phi(\tau) = 0$ for $\tau < 0$, the integral can be taken from $-\infty$ instead of zero. Thus we obtain

$$\chi(\omega) = \chi' - i\chi'' = \int_{-\infty}^{\infty} \Phi(\tau)e^{-i\omega\tau}\, d\tau,$$

with inverse transform

$$\Phi(\tau) = \frac{1}{2\pi}\int_{-\infty}^{\infty} \chi(\omega)e^{i\omega\tau}\, d\omega.$$

Using the symmetry properties of χ' and χ'' (see (18.11) and (18.12)):

$$\chi'(-\omega) = \chi'(\omega) \text{ and } \chi''(-\omega) = -\chi''(\omega),$$

we can write

$$\Phi(\tau) = \frac{1}{\pi}\int_0^{\infty} \chi'(\omega)\cos\omega\tau\, d\omega + \frac{1}{\pi}\int_0^{\infty} \chi''(\omega)\sin\omega\tau\, d\omega.$$

Now, using the fact that $\Phi(\tau) = 0$ for $\tau < 0$, we see that both integrals on the r.h.s. must be equal (for negative τ the last integral changes sign and the total must vanish). Therefore

$$\int_0^{\infty} \chi'(\omega)\cos\omega\tau\, d\omega = \int_0^{\infty} \chi''(\omega)\sin\omega\tau\, d\omega = \frac{\pi}{2}\Phi(\tau).$$

The equality is a direct result of the causality principle. Now insert the first expression for $\Phi(\tau)$ into (18.12) and obtain

$$\begin{aligned}
\chi''(\omega) &= \frac{2}{\pi}\int_0^{\infty} d\tau\, \sin\omega\tau \int_0^{\infty} d\omega'\, \chi'(\omega')\cos\omega'\tau \\
&= \frac{2}{\pi}\int_0^{\infty} d\omega'\chi'(\omega') \int_0^{\infty} d\tau\, \sin\omega\tau \cos\omega'\tau.
\end{aligned}$$

Note that we have changed the integration variable to ω' to avoid confusion with the independent variable ω. After rewriting the product of sin and cos as a sum of sines, the last integral can be evaluated to

$$\frac{1}{2}\left(\frac{1}{\omega + \omega'} + \frac{1}{\omega - \omega'}\right) = \frac{\omega}{\omega^2 - \omega'^2}$$

Here the primitive function has been assumed to vanish at the limit $\tau \to \infty$ because of the infinitely rapid oscillation of the cosines (if this does not satisfy you, multiply the integrand by a damping function $\exp(-\varepsilon\tau)$, evaluate the integral and then take the limit $\varepsilon \to 0$). Equation 18.16 results.   □

18.3 Relation to time correlation functions

In this section we consider our black box as a gray box containing a system of mutually interacting particles. We assume that the system obeys the classical Hamilton equations and is – in the absence of a disturbance – in an equilibrium state. In the linear regime the perturbation is so small that the system deviates only slightly from equilibrium. When after a delta-disturbance some observable Y deviates slightly from its equilibrium value, it will relax back to the equilibrium value through the same intra-system interactions that cause *spontaneous* thermal fluctuations of Y to relax. Therefore we expect that the time course of the relaxation of Y after a small perturbation (i.e., the time course of $\Phi(t)$) is directly related to the time correlation function of Y.

Let us be more precise. Consider the time correlation function of Y: $\langle Y(0)Y(t)\rangle$. The triangular brackets stand for an ensemble average. For a system in equilibrium, presumed to be ergodic, the ensemble average is also a time average over the initial time, here taken as the origin of the time scale. If the probability of Y in the equilibrium ensemble is indicated by $P_{\mathrm{eq}}(Y)$, we can write

$$\langle Y(0)Y(t)\rangle = \int Y_0 \, \langle Y(t)|Y(0) = Y_0\rangle \, P_{\mathrm{eq}}(Y_0) \, dY_0, \qquad (18.17)$$

where $\langle Y(t)|Y(0) = Y_0\rangle$ is the conditional ensemble-averaged value of Y at time t, given the occurrence of Y_0 at time 0. But that is exactly the response function after an initial disturbance of Y to Y_0:

$$\langle Y(t)|Y(0) = Y_0\rangle = Y_0 \frac{\Phi(t)}{\Phi(0)}. \qquad (18.18)$$

Here $\Phi(0)$ is introduced to normalize Φ. Inserting this into (18.17), we arrive at the equality:

$$\frac{\Phi(t)}{\Phi(0)} = \frac{\langle Y(0)Y(t)\rangle}{\langle Y^2\rangle}. \qquad (18.19)$$

This relation is often called the *first fluctuation–dissipation theorem* (Kubo, 1966); see for a further discussion page 258.

The response $Y(0)$ *after* a delta-disturbance $X(t) = X_0^*\delta(t)$ equals $X_0^*\Phi(0)$ (see (18.1)). Therefore:

$$\Phi(0) = Y(0)/X_0^*. \qquad (18.20)$$

This ratio can normally be computed without knowledge of the details of the intra-system interactions, as the latter have no time to develop during a delta-disturbance. The value of $\langle Y^2\rangle$ follows from statistical mechanical considerations and appears to be related to the delta-response. This relation between $Y(0)$ and $\langle Y^2 \rangle$, which we shall now develop, forms the basis of the Green-Kubo formula that relates the integral of time correlation functions to transport coefficients.

Following a delta-disturbance $X_0^* \delta(t)$, the point in phase space $\mathbf{z}$ (we use symplectic notation, see Section 17.8 on page 492) will shift to $\mathbf{z} + \Delta \mathbf{z}$. The shift $\Delta \mathbf{z}$ is proportional to X_0^*. For example, when the perturbation is an electric field $\boldsymbol{E}_0^* \delta(t)$, the ith particle with (partial) charge q_i will be subjected to a force $q_i \boldsymbol{E}_0^* \delta(t)$, leading to a shift in momentum $\Delta \boldsymbol{p}_i = q_i \boldsymbol{E}_0^*$. The phase-point shift leads to a first-order shift in the response function Y, which is simply a property of the system determined by the point in phase space:

$$\Delta Y = \sum_{i=1}^{2n} \Delta z_i \frac{\partial Y(\mathbf{z})}{\partial z_i}. \tag{18.21}$$

The delta-response $Y(0) = X_0^* \Phi(0)$ is the *ensemble average* of ΔY:

$$Y(0) = \langle \Delta Y \rangle = \int d\mathbf{z} \, e^{-\beta H(\mathbf{z})} \sum_{i=1}^{2n} \Delta z_i \frac{\partial Y(\mathbf{z})}{\partial z_i}. \tag{18.22}$$

By partial integration it can been shown (see Exercise 18.4) that

$$Y(0) = \beta \langle Y \Delta H \rangle, \tag{18.23}$$

where

$$\Delta H = \sum_{i=1}^{2n} \Delta z_i \frac{\partial H(\mathbf{z})}{\partial z_i}. \tag{18.24}$$

Combining (18.19), (18.20) and (18.24), we find a relation between the delta-response $\Phi(t)$ and the autocorrelation function of Y:

$$\Phi(t) = \frac{\beta}{X_0^*} \frac{\langle Y \Delta H \rangle}{\langle Y^2 \rangle} \langle Y(0) Y(t) \rangle. \tag{18.25}$$

This relation is only simple if $\langle Y \Delta H \rangle$ is proportional to $\langle Y^2 \rangle$. This is the case if

$$\Delta H \propto X_0^* Y, \tag{18.26}$$

imposing certain conditions on X and Y.

An example will clarify these conditions. Consider a system of (partially) charged particles with volume V, subjected to a homogeneous electric field $E(t) = E_0^* \delta(t)$. For simplicity we consider one dimension here; the extension to a 3D vector is trivial. Each particle with charge q_i will experience a force $q_i E(t)$ and will be accelerated during the delta disturbance. After the disturbance, the ith particle will have changed its velocity with $\Delta v_i = (q_i/m_i)E_0^*$. The total Hamiltonian will change as a result of the change in kinetic energy:

$$\Delta H = \Delta \sum_i \frac{1}{2} m_i v_i^2 = \sum_i m_i v_i \Delta v_i = E_0^* \sum_i q_i v_i. \tag{18.27}$$

Thus ΔH is proportional to the *current density* j:

$$j = \frac{1}{V} \sum_i q_i v_i. \tag{18.28}$$

So, if we take $Y = j$, then $\Delta H = E_0^* V j$ and (18.25) becomes

$$\Phi(t) = \frac{V}{k_B T} \langle j(0) j(t) \rangle. \tag{18.29}$$

Note that we have considered one dimension and thus j is the current density in one direction. In general j is a vector and Φ is a tensor; the relation then is

$$\Phi_{\alpha\beta}(t) = \frac{V}{k_B T} \langle j_\alpha(0) j_\beta(t) \rangle. \tag{18.30}$$

In isotropic materials $\mathbf{\Phi}$ will be a diagonal tensor and $\Phi = \frac{1}{3} \operatorname{tr} \mathbf{\Phi}$ is given by

$$\Phi(t) = \frac{V}{3k_B T} \langle \boldsymbol{j}(0) \cdot \boldsymbol{j}(t) \rangle. \tag{18.31}$$

Equation (18.31) relates the correlation function of the equilibrium current density fluctuation with the response function of the *specific conductance* σ, which is the ratio between current density and electric field:

$$\boldsymbol{j} = \sigma \boldsymbol{E}. \tag{18.32}$$

Using (18.9), we can express the frequency-dependent specific conductance in terms of current density fluctuations:

$$\sigma(\omega) = \frac{V}{3k_B T} \int_0^\infty \langle \boldsymbol{j}(0) \cdot \boldsymbol{j}(\tau) \rangle e^{-i\omega\tau} \, d\tau, \tag{18.33}$$

with the special case for $\omega = 0$:

$$\sigma_0 = \frac{V}{3k_B T} \int_0^\infty \langle \boldsymbol{j}(0) \cdot \boldsymbol{j}(\tau) \rangle \, d\tau. \tag{18.34}$$

Note that in these equations the average product of two current *densities*, multiplied by the volume, occurs:

$$V \langle j(o)j(t) \rangle = \frac{1}{V} \left(\sum_i q_i v_i(0) \right) \left(\sum_i q_i v_i(t) \right), \qquad (18.35)$$

which is indeed a statistically stationary quantity when the total volume of the system is much larger than the local volume over which the velocities are correlated.

Equation (18.31) is an example of a *Kubo formula* (Kubo *et al.*, 1985, p. 155), relating a time correlation function to a response function. Equation (18.34) is an example of a *Green–Kubo* formula (Green, 1954; Kubo, 1957), relating the integral of a time correlation function to a transport coefficient. There are many such equations for different transport properties.

In the conjugate disturbance $\boldsymbol{E}$ and current density $\boldsymbol{j}$, which obey the simple relation (18.26), we recognize the generalized force and flux of the thermodynamics of irreversible processes (see (16.98) in Section 16.10 on page 446). The product of force and flux is an energy dissipation that leads to an irreversible entropy production. Kubo relations also exist for other force–flux pairs that are similarly conjugated.

In this section we have considered the specific conductance as example of the Kubo and Green–Kubo formula. In the following sections other transport properties will be considered.

18.3.1 Dielectric properties

When the material is non-conducting and does not contain free charge carriers, the current density $j = (1/V) \sum q_i v_i$ is caused by the time derivative $\dot{P}$ of the dipole density $P = (1/V) \sum q_i x_i$. In this case there is no steady-state current and the zero-frequency conductivity vanishes. But we can connect to the conductivity case by realizing that the following relations exist between time correlation functions of $j = \dot{P}$ and P:

$$\frac{d}{dt} \langle P(0)P(t) \rangle = \langle P(0)\dot{P}(t) \rangle = -\langle \dot{P}(0)P(t) \rangle, \qquad (18.36)$$

$$\frac{d^2}{dt^2} \langle P(0)P(t) \rangle = \langle P(0)\ddot{P}(t) \rangle = -\langle \dot{P}(0)\dot{P}(t) \rangle = \langle \ddot{P}(0)P(t) \rangle. \quad (18.37)$$

These relations are easily derived when we realize that in an equilibrium system the time axis in correlation functions may be shifted: $\langle A(0)B(t) \rangle =$

$\langle A(-t)B(0)\rangle$. One of the relations in (18.37) is particularly useful:

$$\frac{d^2}{dt^2}\langle P(0)P(t)\rangle = -\langle j(0)j(t)\rangle, \tag{18.38}$$

as it allows us to translate the current fluctuations into polarization fluctuations. The equivalence of the Kubo formula (18.31) for the polarization response Φ_{P}, which is the integral of the current density response $\Phi(t)$, is:

$$\Phi_{\mathrm{P}}(t) = -\frac{V}{3k_{\mathrm{B}}T}\frac{d}{dt}\langle \boldsymbol{P}(0)\cdot\boldsymbol{P}(t)\rangle. \tag{18.39}$$

Realizing that $P(\omega) = \varepsilon_0[\varepsilon_r(\omega) - 1]E(\omega)$ (see Section 13.2 on page 336), we find for the frequency-dependent dielectric constant:

$$\varepsilon_r(\omega) = 1 - \frac{V}{3k_{\mathrm{B}}T}\int_0^\infty \left(\frac{d}{d\tau}\langle \boldsymbol{P}(0)\cdot\boldsymbol{P}(\tau)\rangle\right)e^{-i\omega\tau}\,d\tau. \tag{18.40}$$

At zero frequency the static dielectric constant is obtained:

$$\varepsilon_r(0) = 1 + \frac{V}{3k_{\mathrm{B}}T}\langle \boldsymbol{P}^2\rangle. \tag{18.41}$$

In simulations one monitors the total dipole moment $\boldsymbol{M} = \sum_i \boldsymbol{\mu}_i = V\boldsymbol{P}$ and computes $(1/V)\langle \boldsymbol{M}(0)\cdot\boldsymbol{M}(t)\rangle$. This is generally not a trivial calculation because $\boldsymbol{M}$ is a single quantity that fluctuates slowly and long simulations are needed to obtain accurate converged values for the correlation function.

Matters are somewhat more complicated than sketched above.[2] In fact, (18.40) and (18.41) can only be trusted for very dilute systems in which mutual dipole interactions can be ignored. The reason is that the *local* electric field, to which the molecules respond, includes the fields due to other dipoles and the reaction field introduced by boundary conditions. Without derivation we give the correct result (Neumann and Steinhauser, 1983) for the relation between the frequency-dependent dielectric constant and the correlation function of the total dipole moment $\boldsymbol{M}(t)$ for the case that a *reaction field* is employed in the simulation (see Section 6.3.5 on page 164 and (13.82) on page 347 for a description of reaction fields):

$$[\varepsilon_r(\omega) - 1]\frac{2\varepsilon_{\mathrm{RF}} + 1}{2\varepsilon_{\mathrm{RF}} + \varepsilon_r(\omega)} = \frac{1}{3\varepsilon_0 V k_{\mathrm{B}}T}\int_0^\infty \left(-\frac{d}{dt}\langle \boldsymbol{M}(0)\cdot\boldsymbol{M}(t)\rangle\right)e^{-i\omega t}\,dt, \tag{18.42}$$

---

[2] The theory relating dipole fluctuations with dielectric constants goes back to Kirkwood (1939). The theory for deriving the dielectric constant from dipole fluctuations in simulations with various boundary conditions is most clearly given by Neumann (1983), with extension to the frequency-dependent case by Neumann and Steinhauser (1983). The theory was tested on a Stockmayer fluid (Lennard–Jones particles with dipole moments) by Neumann *et al.* (1984).

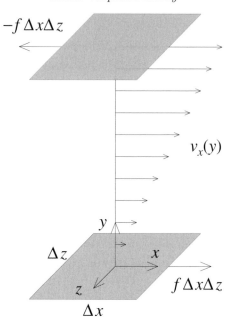

Figure 18.2 Two planes, moving relative to each other in a fluid, experience a viscous drag force proportional to the velocity gradient.

where $\varepsilon_{\mathrm{RF}}$ is the relative dielectric constant used for the reaction field. For the static dielectric constant $\varepsilon_s = \varepsilon_r(0)$ it follows that

$$(\varepsilon_s - 1)\frac{2\varepsilon_{\mathrm{RF}} + 1}{2\varepsilon_{\mathrm{RF}} + \varepsilon_r(\omega)} = \frac{1}{3\varepsilon_0 V k_{\mathrm{B}}T}\langle \boldsymbol{M}^2 \rangle. \tag{18.43}$$

Equations (18.42) and (18.43) are implicit equations for ε_r; they reduce to the simpler (18.40) and (18.41) when $\varepsilon_{\mathrm{RF}} = \infty$, i.e., for *conducting boundary conditions*. These are also valid for the use of complete lattice sums with dipole correction (see the discussion of tin-foil boundary conditions on page 373). For simulations with $\varepsilon_{\mathrm{RF}} = 1$, i.e., using a cutoff radius for Coulomb interactions, the dipole fluctuations are quenched and become rather insensitive to the the value of the dielectric constant. Such simulations are therefore unsuitable to derive dielectric properties.

18.3.2 *Viscosity*

Consider an isotropic fluid between two plates (each in the xz-plane and separated in the y-direction) that move with respect to each other in the x-direction (Fig. 18.2), causing a laminar flow with velocity gradient dv_x/dy.

On each plate the fluid will exert a *drag force* f per unit surface of the xz-plane, proportional to the velocity gradient. The proportionality constant is the *viscosity coefficient* η:

$$f = \eta \frac{\partial v_x}{\partial y}. \tag{18.44}$$

According to (17.113) on page 481 (see also Fig. 17.7), the force per unit surface in a continuum is determined by the stress tensor $\boldsymbol{\sigma}$:

$$d\boldsymbol{F} = \boldsymbol{\sigma} \cdot d\boldsymbol{S}. \tag{18.45}$$

For example, on the lower plane in Fig. 18.2, with $d\boldsymbol{S}$ in the y-direction, $F_x = \sigma_{xy} S_y = \sigma_{xy} \Delta x \Delta z$. This phenomenological definition agrees with the definition of η as given in the derivation of the Navier–Stokes equation in Section 9.2, where η connects the *off-diagonal* elements of the stress tensor with the velocity gradient as follows (see (9.11) and (9.12)):

$$\sigma_{\alpha\beta} = \eta \left(\frac{\partial u_\alpha}{\partial x_\beta} + \frac{\partial u_\beta}{\partial x_\alpha} \right) \quad (\alpha \neq \beta). \tag{18.46}$$

This is the basic equation defining η. How the stress tensor can be determined from simulations is explained in Section 17.7.2 on page 484. The off-diagonal elements of the *average* stress tensor over a given volume V are equal to the negative off-diagonal elements of the *pressure tensor*,[3] which is "measured" by the virial:

$$P_{\alpha\beta} = \frac{1}{V} \sum_i F_{i\alpha} x_{i\beta} = -\frac{1}{V} \int_V \sigma_{\alpha\beta}(\boldsymbol{r}) \, d^3 \boldsymbol{r}, \quad \alpha \neq \beta. \tag{18.47}$$

The (Green–)Kubo relations for the viscosity coefficient can be found by following the standard series of steps. We consider the xy-component of σ without loss of generality.

(i) Apply a delta-disturbance $g^* \delta(t)$ to $\partial u_x / \partial y$ by imposing an additional velocity $v_{ix} = g^* y_i \delta(t)$ to each particle in the considered volume. After the delta pulse the ith particle is *displaced* in the x-direction by $\Delta x_i = g^* y_i$.

(ii) Compute ΔH as a result of the disturbance, according to (18.24) and using (18.47):

$$\Delta H = \sum_i \Delta x_i \frac{\partial H}{\partial x_i} = -g^* \sum_i F_{xi} y_i = g^* V \sigma_{xy}. \tag{18.48}$$

---

[3] The pressure tensor also contains a momentum transfer part due to particle velocities, but that part is diagonal.

(iii) Define the response Y such that Y is proportional to ΔH. This is fulfilled for

$$Y = \sigma_{xy} = -\frac{1}{V} \sum_i F_{xi} y_i, \tag{18.49}$$

for which $\Delta H = g^* V Y$.

(iv) Find the δ-response function $\Phi(t)$ for Y from (18.25):

$$\Phi(t) = \frac{V}{k_{\mathrm{B}}T} \langle \sigma_{xy}(0)\sigma_{xy}(t) \rangle. \tag{18.50}$$

The end result is

$$\eta(\omega) = \frac{V}{k_{\mathrm{B}}T} \int_0^\infty \langle \sigma_{xy}(0)\sigma_{xy}(\tau) \rangle e^{-i\omega\tau} \, d\tau, \tag{18.51}$$

with the Kubo–Green relation for $\omega = 0$:

$$\eta_0 = \frac{V}{k_{\mathrm{B}}T} \int_0^\infty \langle \sigma_{xy}(0)\sigma_{xy}(\tau) \rangle \, d\tau. \tag{18.52}$$

In an isotropic fluid all six off-diagonal elements have the same correlation function and one can best use the average of the correlation functions of all off-diagonal elements of $\boldsymbol{\sigma}$.

The determination of viscosity through the Green-Kubo relation requires accurate determination of the (integral) of the correlation function of a heavily fluctuating quantity. Hess (2002a,b) concluded that it is more efficient to use non-equilibrium molecular dynamics (NEMD) to determine viscosity coefficients (see Section 18.5).

18.4 The Einstein relation

All Green-Kubo relations contain the integral of an autocorrelation function of a fluctuating observable $f(t)$:

$$\int_0^\infty \langle f(0)f(\tau) \rangle \, d\tau. \tag{18.53}$$

Numerical evaluations of such integrals are difficult if no knowledge on the analytical form of the tail of the correlation function is available. The statistics on the tail and – consequently – on the integral, is often poor.

An alternative is to monitor not $f(t)$, but its integral $F(t)$:

$$F(t) = \int_0^t f(t') \, dt', \tag{18.54}$$

and observe the behavior of $\langle F^2(t) \rangle$ for large t. The following *Einstein relation* is valid:

$$\lim_{t \to \infty} \frac{d}{dt} \langle F^2(t) \rangle = 2 \int_0^\infty \langle f(0)f(\tau) \rangle \, d\tau. \qquad (18.55)$$

This means that $\langle F^2(t) \rangle$, plotted versus t, should approach a straight line.

A common application is the determination of the single-particle diffusion constant

$$D = \int_0^\infty \langle v(0)v(\tau) \rangle \, d\tau, \qquad (18.56)$$

by observing the mean-squared displacement $\langle x^2(t) \rangle$, which approaches $2Dt$ for times much longer than the correlation time of the velocity.

Proof We prove (18.55). Consider

$$\frac{d}{dt} \langle F^2(t) \rangle = 2 \langle F(t)f(t) \rangle = 2 \int_0^t \langle f(t')f(t) \rangle \, dt'.$$

By substituting $\tau = t - t'$ the right-hand side rewrites to

$$2 \int_0^t \langle f(t - \tau)f(t) \rangle \, d\tau.$$

Since the ensemble average does not depend on the time origin, the integrand is a function of τ only and is equal to the autocorrelation function of $f(t)$. Obviously, the limit for $t \to \infty$ yields (18.55). $\qquad \square$

18.5 Non-equilibrium molecular dynamics

When (small) external "forces" are artificially exerted on the particles in a molecular-dynamics simulation, the system is brought (slightly) out of equilibrium. Following an initial relaxation the system will reach a steady state in which the response to the disturbance can be measured. In such *non-equilibrium molecular dynamics* (NEMD) methods the "forces" are chosen to represent the gradient that is appropriate for the transport property of interest. When the system under study is periodic, it is consistent to apply a gradient with the same periodicity. This implies that only spatial Fourier components at wave vectors which are integer multiples of the reciprocal basic vectors $(2\pi/l_x, 2\pi/l_y, 2\pi/l_z)$ can be applied. This limitation, of course, is a consequence of periodicity and the long-wavelength limit must be obtained by extrapolation of the observed box-size dependence. We now consider a few examples (Berendsen, 1991b).

18.5.1 Viscosity

In order to measure viscosity, we wish to impose a sinusoidal shear rate over the system, i.e., we wish to exert an *acceleration* on each particle. This is accomplished by adding at every time step Δt a velocity increment Δv_x to every particle

$$\Delta v_{ix} = A\Delta t \cos ky_i. \tag{18.57}$$

Here, A is a (small) amplitude of the acceleration and $k = 2\pi/l_y$ is the smallest wave vector fitting in the box. Any multiple of the smallest wave vector can also be used. When m is the mass of each particle and ρ is the number density, the external *force* per unit volume will be:

$$f_x^{\text{ext}}(y) = m\rho A \cos ky. \tag{18.58}$$

Since there is no pressure gradient in the x-direction, the system will react according to the Navier–Stokes equation:

$$m\rho\frac{\partial u_x}{\partial t} = \eta\frac{\partial^2 u_x}{\partial y^2} + m\rho A \cos ky. \tag{18.59}$$

The steady-state solution, which is approached exponentially with a time constant equal to $m\rho/\eta k^2$ (Hess, 2002b), is

$$u_x(y) = \frac{m\rho}{\eta k^2}A \cos ky. \tag{18.60}$$

Thus, the viscosity coefficient η is found by monitoring the gradient in the y-direction of the velocities v_x of the particles. Velocity gradients in non-equilibrium molecular dynamics simulations are determined by a least-squares fit of the gradient to the particle velocities. A *periodic* gradient can be measured by Fourier analysis of the velocity distribution.

18.5.2 Diffusion

Self-diffusion coefficients can easily be measured from an equilibrium simulation by monitoring the mean-square displacement of the particle as a function of time and applying the Einstein relation. The diffusion coefficient measured this way corresponds to the special case of a *tracer* diffusion coefficient of a single *tagged* particle that has the same interactions as the other particles. In general, the tagged particles can be of a different type and can occur in any mole fraction in the system. If not dilute, the diffusion flux is influenced by the hydrodynamic interaction between the moving particles.

Consider a binary mixture of two particle types 1 and 2, with mole fractions $x_1 = x$ and $x_2 = 1 - x$. Assume that the mixture behaves ideally:

$$\mu_i = \mu_i^0 + RT \ln x_i. \tag{18.61}$$

Now we can derive the following equation:

$$\boldsymbol{u}_1 - \boldsymbol{u}_2 = -\frac{D}{x(1-x)} \boldsymbol{\nabla} x. \tag{18.62}$$

In an NEMD simulation we apply two accelerations a_1 and a_2 to each of the particles of type 1 and 2, respectively. This is done by increasing the velocities (in a given direction) every step by $a_1 \Delta t$ for species 1 and by $a_2 \Delta t$ for species 2. The total force on the system must be kept zero in order to avoid acceleration of the center of mass:

$$M_1 x a_1 + M_2 (1 - x) a_2 = 0, \tag{18.63}$$

where M_1 and M_2 are the molar masses of species 1 and 2, respectively. The balance between driving force and frictional force is reached when

$$M_2 a_2 = -\frac{RT}{D} x (u_1 - u_2), \tag{18.64}$$

or, equivalently,

$$M_1 a_1 = \frac{RT}{D} (1 - x)(u_1 - u_2). \tag{18.65}$$

When $u_1 - u_2$ is monitored after a steady state has been reached, the diffusion coefficient is easily found from either of these equations. The steady state is reached exponentially with a time constant equal to MD/RT. The amplitudes a_1 and a_2 should be chosen large enough for a measurable effect and small enough for a negligible disturbance of the system; in practice the imposed velocity differences should not exceed about 10% of the thermal velocities.

18.5.3 Thermal conductivity

The thermal conductivity coefficient Λ can be measured by imposing a thermal flux J_q with the system's periodicity:

$$J_q(y) = A \cos ky. \tag{18.66}$$

This is accomplished by *scaling* the velocities of all particles every time step by a factor λ. The kinetic energy E_{kin} per unit volume changes by a factor

λ^2, such that

$$J_q = \frac{\Delta E_{\text{kin}}}{\Delta t} = \frac{(\lambda^2 - 1)}{\Delta t} \frac{3}{2} \rho k_{\text{B}} T. \tag{18.67}$$

The external heat flow causes a temperature change; the temperature $T(y)$ will obey the following equation:

$$\rho c_v \frac{\partial T}{\partial t} = J_q + \Lambda \frac{\partial^2 T}{\partial y^2}. \tag{18.68}$$

The steady-state solution then is

$$T(y) = T_0 + \frac{A}{\Lambda k^2} \cos ky. \tag{18.69}$$

Thus, by monitoring the Fourier coefficient of the temperature at wave vector k in the y-direction, the thermal conductivity coefficient Λ is found. Also here, the amplitude of the heat flux should be chosen large enough for a measurable effect and small enough for a negligible disturbance. In practice a temperature amplitude of 10 K is appropriate.

Exercises

18.1   When the delta response of a linear system equals an exponential decay with time constant τ_c, compute the response to a step function and the frequency response (18.9) of this system.

18.2   Verify the validity of he Kramers–Kronig relations for the frequency response of the previous exercise.

18.3   Prove (18.15) by following the same reasoning as in the proof given for (18.16).

18.4   Prove (18.23) by showing through partial integration that

$$\int e^{-\beta H(\mathbf{z})} \Delta z_i \frac{\partial Y}{\partial z_i} \, d\mathbf{z} = \beta \int e^{-\beta H(\mathbf{z})} Y \Delta z_i \frac{\partial H}{\partial z_i} \, d\mathbf{z}. \tag{E18.1}$$

18.5   When the total dipole fluctuation $\langle \mathbf{M}(0) \cdot \mathbf{M}(t) \rangle$ appears to decay exponentially to zero in a simulation with "tin-foil" boundary conditions, how then do the real and imaginary parts of the dielectric constant depend on ω? Consider $\varepsilon_r(\omega) = \varepsilon_r' - i\varepsilon_r''$. Plot ε_r'' versus ε_r' and show that this *Cole–Cole plot* is a semicircle.

18.6   Derive (18.62).

19

Splines for everything

19.1 Introduction

In numerical simulations one often encounters functions that are only given at discrete points, while values and derivatives are required for other values of the argument. Examples are:

(i) the reconstruction of an interaction curve based on discrete points, for example, obtained from extensive quantum calculations;

(ii) recovery of function values and derivatives for arbitrary arguments from tabulated values, for example, for potentials and forces in MD simulations;

(iii) the estimation of a definite or indefinite integral of a function based on a discrete set of derivatives, for example, the free energy in thermodynamic integration methods;

(iv) the construction of a density distribution from a number of discrete events, for example, a radial distribution function.

In all these cases one looks for an interpolation scheme to construct a complete curve from discrete points or *nodes*. Considerations that influence the construction process are (i) the *smoothness* of the curve, (ii) the *accuracy* of the data points, and (iii) the *complexity* of the construction process.

Smoothness is not a well-defined property, but it has to do with two aspects: the *number of continuous derivatives* (continuous at the nodes), and the *integrated curvature C*, which can be defined as the integral over the square of the second derivative over the relevant interval:

$$C[f] = \int_a^b \{f''(x)\}^2 \, dx, \qquad (19.1)$$

523

or, in the multidimensional case:

$$C[f] = \int_V \{\nabla^2 f(\boldsymbol{r})\}^2 \, d\boldsymbol{r}. \qquad (19.2)$$

The term "curvature" is used very loosely here.[1] If data points are not in-finitely accurate, there is no reason why the curve should go exactly through the data points. Any curve from which the data points deviate in a statis-tically acceptable manner, is acceptable from the point of view of fitting to the data. In order to choose from the – generally infinite number of – ac-ceptable solutions, one has to apply additional criteria, as compliance with additional theoretical requirements, minimal curvature, minimal complexity of the curve specification, or maximum "uncertainty" ("entropy") from an information-theoretical point of view. Finally, one should always choose the *simplest* procedure within the range of acceptable methods.

The use we wish to make of the constructed function may prescribe the number of derivatives that are continuous at the nodes. For example, in an MD simulation using an algorithm as the Verlet or leap-frog scheme, the first error term in the prediction of the coordinate is of the order of $(\Delta t)^4$. The accuracy depends on the cancellation of terms in $(\Delta t)^3$, which involve the first derivatives of the forces. For use in such algorithms we wish the derivative of the force to be continuous, implying that the second deriva-tive of the potential function should be continuous. Thus, if potential and forces are to be derived from tabulated functions, the interpolation proce-dure should not only yield continuous potentials, but also continuous first and second derivatives of the potential. This, in fact, is accomplished by *cubic spline interpolation*. Using cubic spline interpolation, far less tabu-lated values are needed for the same accuracy than when a simple linear interpolation scheme would have been used.

We first restrict our considerations to the one-dimensional case of *polyno-mial* splines, which consist of piecewise polynomials for each interval. These *local* polynomials are by far to be preferred to global polynomial fits, which are ill-behaved and tend to produce oscillating solutions. The polynomial splines cannot be used when the function is not single-valued or when the x-coordinates of the data points cannot be ordered ($x_0 \le x_1 \le \cdots \le x_n$); one then needs to use *parametricn* splines, where both x and y (and any

---

[1] It would be better to call the curvature, as defined here, *the total energy of curvature*. Curvature is defined as the change of tangential angle per unit of length along the curve, which is the inverse of the radius of the circle that fits the local curved line segment. This curvature is a local property, invariant for orientation of the line segment. For a function $y(x)$ the curvature can be expressed as $y''(1+y')^{-3/2}$, which can be approximated by $y''(x)$. If an elastic rod is bent, the elastic energy per unit of rod length is proportional to the square of the curvature. See Bronstein and Semendjajew (1989) for definitions of curvature.

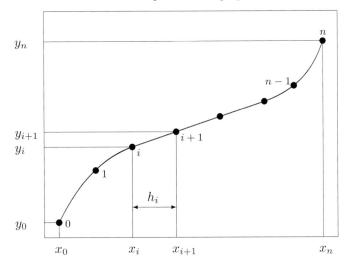

Figure 19.1 Cubic spline interpolation of $n + 1$ points in n intervals. The i-th interval has a width of h_i and runs from x_i to x_{i+1}. First and second derivatives are continuous at all nodes.

further coordinates in higher-dimensional spaces) are polynomial functions of one (or more, in higher-dimensional spaces) parameter(s). To this category belong B-splines and Bezier curves and surfaces. B-splines are treated in Section 19.7; they are often used to construct smooth surfaces and multi-dimensional interpolations, for example in the *smooth-particle mesh-Ewald* (SPME) method for computing long-range interactions in periodic systems (see Section 13.10.6 in Chapter 13 on page 373). For Bezier curves we refer to the literature. For many more details, algorithms, programs in C, and variations on this theme, the reader is referred to Engeln-Müllges and Uhlig (1996). Programs including higher-dimensional cases are also given in Späth (1973). A detailed textbook is de Boor (1978), while *Numerical Recipes* (Press *et al.*, 1992) is a practical reference for cubic splines.

In the following sections the spline methods are introduced, with emphasis on the very useful cubic splines. A general Python program to compute one-dimensional cubic splines is provided in Section 19.6. Section 19.7 treats B-splines.

19.2 Cubic splines through points

Consider $n+1$ points x_i, y_i, $i = 0, \ldots, n$ (Fig. 19.1). We wish to construct a function $f(x)$ consisting of piece-wise functions $f_i(x - x_i)$, $i = 0, \ldots, n-1$, defined on the interval $[0, h_i = x_{i+1} - x_i]$, with the following properties:

  (i) each $f_i(\xi)$ is a polynomial of the third degree;
 (ii) $f_i(0) = y_i$, $\quad i = 0, \ldots, n-1$;
(iii) $f_{n-1}(h_{n-1}) = y_n$;
 (iv) $f_i(h_i) = f_{i+1}(0)$, $\quad i = 0, \ldots, n-2$;
  (v) $f_i'(h_i) = f_{i+1}'(0)$, $\quad i = 0, \ldots, n-2$;
 (vi) $f_i''(h_i) = f_{i+1}''(0)$, $\quad i = 0, \ldots, n-2$.

There are n intervals with $4n$ parameters to describe the n piecewise functions; the properties provide $(n+1)+3(n-1) = 4n-2$ equations. In order to solve for the unknown parameters, two extra conditions are needed. These are provided by specification of two further derivatives at the end nodes, choosing from the first, second or third derivatives of either or both end nodes.[2] In the case of *periodic cubic splines*, with period $x_n - x_0$, for which $y_n = y_0$, it is specified that the first and second derivative at the first point are equal to the first and second derivative at the last point. If it is specified that the second derivatives at both end points vanish, *natural* splines result. It is clear that *quadratic splines* will be obtained when the function and its first derivative are continuous (one extra condition being required), and that quartic and higher-order splines can be defined as well.

The function in the ith interval is given by

$$f(x_i + \xi) = f_i(\xi) = f_i + g_i\xi + p_i\xi^2 + q_i\xi^3 \quad (0 \leq \xi \leq h_i). \tag{19.3}$$

Here f_i, g_i, p_i and q_i are the four parameters to be determined from the conditions given above. We see immediately that the values of the function and its first derivative in point x_i are

$$f_i = f(x_i) = y_i, \tag{19.4}$$

and

$$g_i = f'(x_i) = f_i', \tag{19.5}$$

respectively, while $2p_i$ is the second derivative at x_i and $6q_i$ is the third derivative, which is constant throughout the i-th interval, and discontinuous at the nodes.

---

[2] It is wise to specify one condition at each end node; specifying two conditions at one end node may lead to accumulating errors giving erratic oscillations.

The continuity conditions lead to the following equations:

$$f_i(h_i) = f_i + g_i h_i + p_i h^2 + q_i h^3 = f_{i+1}, \tag{19.6}$$

$$f_i'(h_i) = g_i + 2p_i h_i + 3q_i h_i^2 = g_{i+1} = f_{i+1}', \tag{19.7}$$

$$f_i''(h_i) = 2p_i + 6q_i h_i = 2p_{i+1} = f_{i+1}'', \tag{19.8}$$

valid for $i = 0, \ldots, n - 2$.

Four parameters suffice for the reconstruction of $f(x)$ in each interval. It is convenient to use the function values and the first derivatives in the nodes for reconstruction. Thus, for the i-th interval we obtain by solving p_i and q_i from (19.6) and (19.7):

$$p_i = \frac{1}{h_i} \left[\frac{3(f_{i+1} - f_i)}{h_i} - 2g_i - g_{i+1}, \right] \tag{19.9}$$

$$q_i = \frac{1}{h_i^2} \left[\frac{-2(f_{i+1} - f_i)}{h_i} + g_i + g_{i+1} \right]. \tag{19.10}$$

Any value of $f(x)$ within the i-th interval is easily found from (19.3) when both the function and derivative values are given at the nodes of that interval. So the interpolation is practically solved when the first derivatives are known, and the task of constructing a cubic spline from a set of data points is reduced to the task of finding the first derivatives in all points (including the end nodes).[3]

The continuity of the second derivatives is given by (19.8). Written in terms of f and g, this condition gives $n - 1$ equations for $n + 1$ unknowns $g_0, \ldots, g_n$:

$$\frac{1}{h_i} g_i + 2 \left(\frac{1}{h_i} + \frac{1}{h_{i+1}} \right) g_{i+1} + \frac{1}{h_{i+1}} g_{i+2}$$
$$= 3 \left(\frac{f_{i+1} - f_i}{h_i^2} + \frac{f_{i+2} - f_{i+1}}{h_{i+1}^2} \right), \quad i = 0, \ldots, n - 2. \tag{19.11}$$

These equations must be augmented by the two additional conditions, and – depending on the exact conditions – will lead to a matrix equation with a symmetric, tridiagonal matrix. For example, if the first derivatives g_0 and g_n at the end points are known, they can be removed from the unknown vector and lead to a matrix equation with $\mathbf{g} = [g_1, \ldots g_{n-1}]$ as the unknown vector:

$$\mathbf{Ag} = \mathbf{b}, \tag{19.12}$$

---

[3] We could have chosen to solve for the second derivatives at all nodes, as is often done in the literature, since the functions in each interval are also easily constructed from knowledge of the values of the function and its second derivatives at the nodes. The computational effort is similar in both cases.

$$
\begin{pmatrix}
d_1 & s_1 & & & & \\
s_1 & d_2 & s_2 & & & \\
 & \ddots & \ddots & \ddots & & \\
 & & \ddots & d_{n-2} & s_{n-2} \\
 & & & s_{n-2} & d_{n-1}
\end{pmatrix}
\begin{pmatrix}
g_1 \\
g_2 \\
\vdots \\
g_{n-2} \\
g_{n-1}
\end{pmatrix}
=
\begin{pmatrix}
b_1 - g_0/h_0 \\
b_2 \\
\vdots \\
b_{n-2} \\
b_{n-1} - g_n/h_{n-1}
\end{pmatrix},
$$

where we use the general notation

$$
s_i = h_i^{-1}, \quad i = 1, \ldots, n - 1, \tag{19.13}
$$

and

$$
d_i = 2(s_{i-1} + s_i), \quad i = 1, \ldots, n - 1 \tag{19.14}
$$
$$
b_i = 3s_{i-1}^2(f_i - f_{i-1}) + 3s_i^2(f_{i+1} - f_i), \quad i = 1, \ldots, n - 1. \tag{19.15}
$$

In case the second derivative is given at an end node, say at $x_0 : f_0'' = 2p_0$, the extra condition reads

$$
h_0 p_0 = \frac{3}{h_0}(f_1 - f_0) - 2g_0 - g_1. \tag{19.16}
$$

Now g_0 cannot be eliminated and an extra row and column must be added to the matrix:

$$
\begin{pmatrix}
d_0 & s_0 & & & & \\
s_0 & d_1 & s_1 & & & \\
 & s_1 & d_2 & s_2 & & \\
 & & \ddots & \ddots & \ddots & \\
 & & & \ddots & d_{n-2} & s_{n-2} \\
 & & & & s_{n-2} & d_{n-1}
\end{pmatrix}
\begin{pmatrix}
g_0 \\
g_1 \\
g_2 \\
\vdots \\
g_{n-2} \\
g_{n-1}
\end{pmatrix}
=
\begin{pmatrix}
b_0 \\
b_1 \\
b_2 \\
\vdots \\
b_{n-2} \\
b_{n-1} - g_n s_{n-1}
\end{pmatrix},
$$
$$\tag{19.17}$$

with

$$
d_0 = 2s_0, \tag{19.18}
$$
$$
b_0 = 3(f_1 - f_0)s_0^2 - \frac{1}{2}f_0''. \tag{19.19}
$$

Again, a symmetric tridiagonal matrix is obtained. Similarly, a given second derivative at the end point x_n can be handled., yielding an extra row at the bottom and column at the right with

$$
d_n = 2s_{n-1}, \tag{19.20}
$$
$$
b_n = 3(f_n - f_{n-1})s_{n-1}^2 - \frac{1}{2}f_n''. \tag{19.21}
$$

If the third derivative f_0''' is specified (at x_0), the same matrix is obtained as in the previous case, but with values

$$d_0 = s_0, \tag{19.22}$$

$$b_0 = 2(f_1 - f_0)s_0^2 + \frac{1}{6}f_0''' h_0. \tag{19.23}$$

For the third derivative f_n''' specified at x_n, the extra elements are

$$d_n = s_{n-1}, \tag{19.24}$$

$$b_n = 2(f_n - f_{n-1})s_{n-1}^2 + \frac{1}{6}f_n''' h_{n-1}. \tag{19.25}$$

For *periodic* splines, $f(x_n+\xi) = f(x_0+\xi)$, and the function and its first two derivatives are continuous at x_n. Thus there are n unknowns $g_0, \ldots, g_{n-1}$, and the matrix equation now involves a symmetric tridiagonal *cyclic* matrix:

$$\mathbf{Ag} = \mathbf{b}, \tag{19.26}$$

$$
\begin{pmatrix}
d_0 & s_0 & & & s_{n-1} \\
s_0 & d_1 & s_1 & & \\
& \ddots & \ddots & \ddots & \\
& & \ddots & d_{n-2} & s_{n-2} \\
s_{n-1} & & & s_{n-2} & d_{n-1}
\end{pmatrix}
\begin{pmatrix}
g_0 \\
g_1 \\
\vdots \\
g_{n-2} \\
g_{n-1}
\end{pmatrix}
=
\begin{pmatrix}
b_0 \\
b_1 \\
\vdots \\
b_{n-2} \\
b_{n-1}
\end{pmatrix},
$$

where the matrix elements are given by (19.14), (19.13) and (19.15), with additional elements

$$d_0 = 2(s_0 + 2s_{n-1}), \tag{19.27}$$

$$b_0 = 3(f_1 - f_0)s_0^2 + 3(f_0 - f_{n-1})s_{n-1}^2, \tag{19.28}$$

$$b_{n-1} = \text{as in (19.15) with } f_n = f_0. \tag{19.29}$$

As the matrices are well-behaved (diagonally dominant, positive definite), the equations can be simply solved; algorithms and Python programs are given in Section 19.6.

Cubic splines have some interesting properties, related to the curvature, as defined in (19.1). We refer to Kreyszig (1993) for the proofs.

(i) Of all functions (continuous and with continuous first and second derivatives) that pass through n given points, and have given first derivatives at the end points, the cubic spline function has the smallest curvature. The cubic spline solution is unique; all other functions with these properties have a larger curvature. One may say that the cubic spline is the *smoothest* curve through the points.[4]

(ii) Of all functions (continuous and with continuous first and second derivatives) that pass through n given points, the function with smallest curvature is a *natural* cubic spline, i.e., with zero second derivatives at both ends.

So, if for some good reason, you look for the function with smallest curvature through a number of given points, splines are *the* functions of choice.

Figure 19.2 shows the cubic (periodic) spline solution using as x, y data just the following points, which sample a sine function:

$$x_i = 0, \frac{\pi}{2}, \pi, \frac{3\pi}{2}, 2\pi, \quad y_i = 0, 1, 0, -1, 0.$$

The spline function (dotted line) is almost indistinguishable from the sine itself (solid curve), with a largest deviation of about 2%. Its first derivatives at $x = 0$ and $x = \pi$ differ somewhat from the ideal cosine values. The cubic interpolation using exact derivatives $(1, 0, -1, 0, 1)$ gives a somewhat better fit (dashed curve), but with slight discontinuities of the second derivatives at $x = 0$ and $x = \pi$.

19.3 Fitting splines

There is one good reason *not* to draw spline functions through a number of given points. That is if the points represent *inaccurate* data. Let us assume that the inaccuracy is a result of statistical random fluctuations.[5] The data points y_i then are random deviations from values $f_i = f(x_i)$ of a function $f(x)$ that we much desire to discover. The value of $d_i = y_i - f_i$ is a random sample from a *distribution function* $p_i(d_i)$ of the random variable. That is, if the data points are statistically independent; if they are not, the whole set of deviations is a sample from a *multivariate* probability distribution. We need at least *some* knowledge of these distribution functions, best obtained from separate observations or simulations:

---

[4] This is the rationale for the name *spline*, borrowed from the name of the thin elastic rods used by construction engineers to fit between pairs of nails on a board, in order to be able to draw smooth outlines for shaping construction parts. The elastic deformation energy in the rod is proportional to the integral of the square of the second derivative, at least for small deviations from linearity. The rod will assume the shape that minimizes its elastic energy. If the ends of the rods are left free, natural splines result. Lasers and automated cutting machines have made hardware splines obsolete.

[5] Be aware of, and check for, experimental errors or programming errors, and – using simulations – for insufficient sampling and inadequate equilibration. An observable may appear to be randomly fluctuating, but still only sample a limited domain. This is a problem of ergodicity that cannot be solved by statistical methods alone.

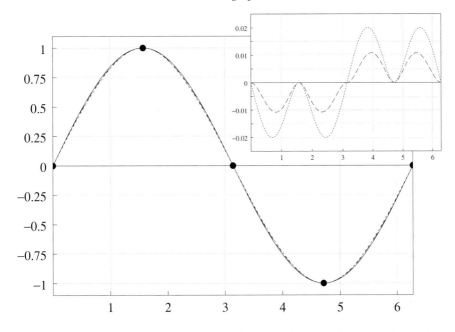

Figure 19.2 A cubic periodic spline (dotted) fitted through five points sampling a sine wave (solid curve). Dashed curve: cubic interpolation using function values and first derivatives at the sample points. The inset shows the differences with the sine function.

(i) their *expectation* (or "expectation value"), defined as the average over the distribution function, must be assumed to be zero. If not, there is a *bias* in the data that – if known – can be removed.

$$\int_{\infty}^{\infty} x p_i(x)\, dx = 0, \qquad (19.30)$$

(ii) their *variances* σ_i^2, defined as the expectation of the square of the variable over the unbiased distribution function

$$\sigma_i^2 = \int_{\infty}^{\infty} x^2 p_i(x)\, dx. \qquad (19.31)$$

For our purposes it is sufficient to have an estimate of σ_i. It enables us to determine the sum of weighted residuals, usually indicated by chi-square:

$$\chi^2 = \sum_{i=0}^{n} \frac{(y_i - f_i)^2}{\sigma_i^2}. \qquad (19.32)$$

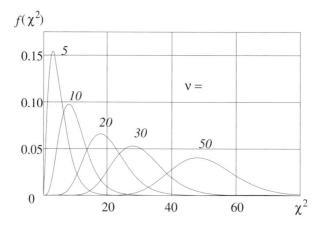

Figure 19.3 The chi-square probability distribution for various values of ν, the number of degrees of freedom.

If the weighted deviations $(y_i - f_i)/\sigma_i$ are samples from normal distributions, χ^2 will be distributed according to the *chi-square distribution*

$$f(\chi^2|\nu) = [2^{\nu/2}\Gamma(\frac{\nu}{2})]^{-1}(\chi^2)^{\nu/2-1} \exp(-\frac{\chi^2}{2})\,d\chi^2, \tag{19.33}$$

as depicted in Fig. 19.3. The parameter ν is the number of *degrees of freedom*; if the deviations are uncorrelated, $\nu = n$ for our purposes. The expectation (average value over the distribution) of χ^2 is equal to ν, and for large ν the χ^2-distribution tends to a Gaussian distribution with variance 2ν. From the cumulative χ^2-distribution confidence intervals can be computed: these indicate the range in which, say, 90% of the samples are expected to occur, or, say, the value that will be exceeded in 1% of the cases. Note that the median value for χ^2, expected to be exceeded in 50% of the cases, is close to ν.

Chi-square tables can be used to decide what deviation from a fitted curve is considered acceptable. The logical choice is to take the value at the *expectation* of χ^2, which is the 50% value, for which $\chi^2 = \nu$ is a sufficient approximation. If a small value is set, the fitted curve may follow the noisy data too closely; if a large value is allowed, details may be missed that are real.

Extensive tables and equations can be found in Beyer (1991) and in Abramowitz and Stegun (1965), as well as in most textbooks on statistics. In Table 19.1 an excerpt is given of values that are not exceeded by χ^2/n in a given fraction of the cases. The 50% value is the most neutral expectation, which statistically will be exceeded in half of the cases. The

Table 19.1 *Values that are not exceeded by χ^2/n in the fraction of cases F mentioned in the header row. Vertical is the number of degrees of freedom ν.*

$F =$	0.01	0.05	0.10	0.50	0.90	0.95	0.99
ν							
1	0.000	0.004	0.016	0.455	2.706	3.841	6.635
2	0.005	0.051	0.105	0.693	2.303	2.996	4.605
3	0.038	0.117	0.195	0.789	2.084	2.605	3.782
4	0.074	0.178	0.266	0.839	1.945	2.374	3.319
5	0.111	0.229	0.322	0.870	1.847	2.214	3.017
6	0.145	0.273	0.367	0.891	1.774	2.099	2.802
7	0.177	0.310	0.405	0.907	1.717	2.010	2.639
8	0.206	0.342	0.436	0.918	1.670	1.938	2.511
9	0.232	0.369	0.463	0.927	1.632	1.880	2.407
10	0.256	0.394	0.487	0.934	1.599	1.831	2.321
11	0.278	0.416	0.507	0.940	1.570	1.789	2.248
12	0.298	0.436	0.525	0.945	1.546	1.752	2.185
13	0.316	0.453	0.542	0.949	1.524	1.720	2.130
14	0.333	0.469	0.556	0.953	1.505	1.692	2.082
15	0.349	0.484	0.570	0.956	1.487	1.666	2.039
20	0.413	0.543	0.622	0.967	1.421	1.571	1.878
25	0.461	0.584	0.659	0.973	1.375	1.506	1.773
30	0.498	0.616	0.687	0.978	1.342	1.459	1.696
40	0.554	0.663	0.726	0.983	1.295	1.394	1.592
50	0.594	0.695	0.754	0.987	1.263	1.350	1.523
60	0.625	0.720	0.774	0.989	1.240	1.318	1.473
70	0.649	0.739	0.790	0.990	1.222	1.293	1.435
80	0.669	0.755	0.803	0.992	1.207	1.273	1.404
90	0.686	0.768	0.814	0.993	1.195	1.257	1.379
100	0.701	0.779	0.824	0.993	1.185	1.243	1.358
∞	$1-c$	$1-b$	$1-a$	1	$1+a$	$1+b$	$1+c$
	$a = 1.81/\sqrt{\nu}$		$b = 2.32/\sqrt{\nu}$			$c = 3.29/\sqrt{\nu}$	

number of degrees of freedom ν is not necessarily equal to the number of data points over which the square deviations are summed; correlation between the points and the "use" of data points to determine parameters in a function to which the data are fitted, will decrease the number of effective degrees of freedom. For example, if a straight line (two parameters) is fitted to two points, there are no deviations and no degrees of freedom; if a straight line is fitted to n independent data points, there are only $n - 2$

degrees of freedom. If we would duplicate every data point, the values of χ^2 and of n would both double, but χ^2/n would not change, and neither would the number of degrees of freedom change. The effective number of degrees of freedom equals the number of data points divided by a *correlation length* of the data and reduced by the number of (independent) parameters in the fitted function.

It is clear that a criterium based on acceptance of a χ^2-value does not suffice to choose the parameters of a model. In the context of splines we choose the additional criterium to *minimize the curvature* of the constructed function. We choose natural splines, as these have the least curvature among all splines through the same points.[6] The total curvature C of a cubic spline is given by

$$C = \int_{x_0}^{x_n} (f'')^2 \, dx = f'' f \big|_{x_0}^{x_n} - \int_{x_0}^{x_n} f' f''' \, dx = -\sum_{i=0}^{n-1} \int_0^{h_i} f_i' f_i''' \, d\xi$$

$$= -6 \sum_{i=0}^{n-1} q_i \int_0^{h_i} f_i' \, d\xi = -6 \sum_{i=0}^{n-1} q_i (f_{i+1} - f_i). \tag{19.34}$$

This can be rewritten as

$$C = 6 \left[q_0 f_0 + \sum_{i=1}^{n-1} (q_i - q_{i-1}) f_i - q_{n-1} f_n \right] = 6 \sum_{i=0}^{n} \delta_i f_i, \tag{19.35}$$

where

$$\delta_i = q_i - q_{i-1}, \quad i = 1, \ldots, n-1,$$

$$\delta_0 = q_0, \quad \delta_n = -q_{n-1}. \tag{19.36}$$

In order to minimize C while χ^2 is to be constrained at a given value, we use a *Lagrange multiplier*[7] λ and minimize the function $C + \lambda \chi^2$:

$$\text{Minimize} \ \sum_0^n \delta_i f_i + \lambda \sum \frac{(y_i - f_i)^2}{\sigma_i^2}. \tag{19.37}$$

Here we have absorbed the constant 6 in the Lagrange multiplier. Note that the variables are $f_i, i = 0, \ldots, n$, while the δ_i's follow from the spline through f_i; y_i and σ_i are constants. For λ an initial value is chosen, and the

---

[6] There are circumstances that external knowledge would imply other boundary conditions than vanishing second derivatives, e.g. in the case of periodic conditions. It is possible to construct programs for other boundary conditions, but in general one may choose some extra points at, or just beyond, the boundaries such that certain prescribed conditions will be fulfilled.

[7] This is the standard way to optimize a function in the presence of a constraint. For its use see Chapter 17, footnote on page 456.

procedure is iterated until the desired value of χ^2 is obtained. Since χ^2 is a statistical quantity, it may well happen that the desired value of χ^2 is not even reached with the smoothest possible curve (a straight line with $C = 0$); in that case the straight line is an acceptable least squares fit to a linear function.

In order to solve (19.37), we solve the set of equations obtained by setting the derivatives to all f_k equal to zero:

$$\delta_k + \sum_{i=0}^{n} f_i \frac{\partial \delta_i}{\partial f_k} - 2\lambda \frac{y_k - f_k}{\sigma_k^2} = 0, \quad k = 1, \ldots, n. \tag{19.38}$$

Now, it is easy to show that δ is a homogeneous linear function of f: $\delta = \mathbf{M}f$. The matrix $\mathbf{M}$ appears to be symmetric, as can be shown by a straightforward but cumbersome inspection of the matrices $\mathbf{M}$ is composed of. We leave the proof to the interested reader. Since $\partial \delta_i / \partial f_k = M_{ik} = M_{ki}$, the second term in (19.38) equals $\sum M_{ki} f_i = \delta_k$. Thus

$$\lambda \frac{y_k - f_k}{\sigma_k^2} = \delta_k, \quad i = 0, \ldots, n, \tag{19.39}$$

is the condition to be imposed on the spline parameters to obtain the smoothest curve (in terms of least curvature) with a constrained value of χ^2. By varying λ, the smoothest curve that is statistically compatible with the data is obtained.

Equation (19.39), together with (19.10), expressing q in f and g, and the continuity conditions that lead to the matrix equation (19.12) or (19.26), yields a matrix equation from which the spline parameters can be solved. The equation is now five-diagonal but symmetric; complete expressions can be found in Engeln-Müllges and Uhlig (1996), who also give expressions for different boundary conditions. A Python program is given in Section 19.6.

Figure 19.4 shows fitting splines through 11 points that were generated as normal-distributed random deviations from the parabola $y = 0.2x^2$. Three fitting splines were generated with different overall weight factors for the data points: $w = 1000$ yields a spline through the points; $w = 0.2$ yields a value of χ^2/n approximately equal to the mean expectation 1; $w = 0.001$ yields a least squares fit to an (almost) straight line. Note that the "best" fit (dashed line) is *not* equivalent to a least squares fit to a polynomial of the second degree. If *knowledge* is available from other sources about the functional form that is to be expected, a direct least-squares fit to such a functional form is to be preferred above a fitting spline. The latter will always choose the smoothest curve, statistically consistent with the data.

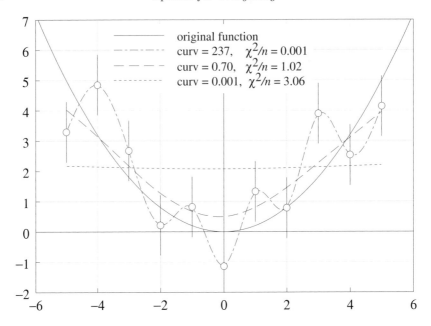

Figure 19.4 Fitting splines approximating noisy data points. Three curves were generated with different weighting factors (1000; 0.2; 0.001). The results range from a spline through the points to an almost straight line. The "best" fit has an average chi-square deviation close to 1.

What we are looking for is the smoothest curve consistent not only with the data, but with all the knowledge we may possess about the expected curve.

19.4 Fitting distribution functions

Suppose we wish to generate a probability distribution function $\rho(x)$ from a number of statistical samples from that distribution. For example, we wish to generate a radial distribution function between two specified species of atoms, or an angular distribution of relative orientations of molecular dipoles. If there are sufficient independent data, this task simply reduces to constructing a histogram by sorting the data into *bins* $x_0, \ldots, x_n$. The number of samples n_i in each bin follows *multinomial* statistics: when the probability to find a sample in the i-th bin is p_i, then the joint probability to find the distribution $n_1, \ldots, n_n$, given the total number of samples N, is given by

$$P(n_1, \ldots, n_n) = N! \, \Pi_{i=1}^{n} \frac{p_i^{n_i}}{n_i!}. \tag{19.40}$$

The (marginal) probability to find exactly n_i samples in the i-th bin, irrespective of the occupancy of other bins, follows *binomial* statistics:

$$P(n_i|N) = \frac{N!}{n_i!(N-n_i)!} p_i^{n_i} (1-p_i)^{N-n_i}. \tag{19.41}$$

For a sufficient number of bins, the probability to find a sample in one particular bin becomes small. In the limit of $p_i \to 0$ (but keeping the expected number in the bin $\mu_i = Np_i$ constant), the binomial distribution becomes a Poisson distribution:

$$P(n_i|\mu_i) = \frac{\mu^{n_i} e^{-\mu}}{n_i!}. \tag{19.42}$$

The expected value μ_i in bin (x_i, x_{i+1}) equals

$$\mu_i = N \int_{x_i}^{x_{i+1}} \rho(x) \, dx. \tag{19.43}$$

The observed number can be written as

$$n_i = \mu_i + \varepsilon_i, \tag{19.44}$$

with ε_i a random variable with average zero and variance μ_i. Thus the observed number n_i has a standard deviation approximately equal to the square root of n_i itself. When the number of samples in each bin is large, the histogram itself is a good representation of the required distribution function.

When the available number of samples is small, it is not possible to construct a histogram that has both a high resolution and a high precision. Can we construct a smooth distribution function that is compatible with the sampled data? There is one difficulty: the number in each bin is not a sample of the distribution function at one point (say, the center of the bin), but rather a sample of the integral of the distribution function over the bin. This difficulty is resolved if we start with the *cumulative* distribution N_i:

$$N_i = \sum_{j=1}^{i} n_j, \tag{19.45}$$

which is a sample of $(N\times)$ the *cumulative distribution function* $F(x_i)$:

$$F(x) = \int_{-\infty}^{x} \rho(x) \, dx. \tag{19.46}$$

So we fit a spline to the observed numbers N_i/N, knowing that the function must start at 0 and end at 1. We also know that the derivative of the

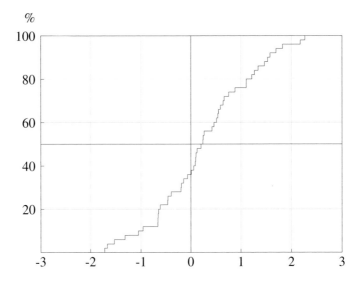

Figure 19.5 Cumulative distribution of 50 samples taken from a normal distribution.

function, being a density, must always be non-negative. But we must also know the standard deviations of N_i/N.

Since

$$N_i = E[N_i] + \sum_{j=1}^{i} \varepsilon_j, \qquad (19.47)$$

we can approximate by the first term in a Taylor expansion

$$\frac{N_i}{N} \approx F(x_i)\left(1 + \frac{1}{E[N_i]} \sum_{j=1}^{i} \varepsilon_j - \frac{1}{E[N]} \sum_{j=1}^{n} \varepsilon_j\right) = F(x_i) + \eta_i. \qquad (19.48)$$

The variance of the relative error η_i can be easily computed, using the facts that the variance of ε_j equals μ_j (expectation of n_j) and that ε_i and ε_j are uncorrelated for $i \neq j$. We find

$$\sigma_i^2 = \frac{1}{N} F_i(1 - F_i) \approx \frac{N_i}{N}\left(1 - \frac{N_i}{N}\right). \qquad (19.49)$$

The standard deviations are zero at both ends, which are therefore forced to 0 and 1, and in the middle reaches the value of $0.5/\sqrt{N}$.

We now consider an example: the construction of the distribution function of 50 samples from a normal distribution. Figure 19.5 shows the cumulative distribution of the data. In Fig. 19.6 the cumulative distribution is put into bins of width 0.5, and a fitting spline is constructed with such a weight that

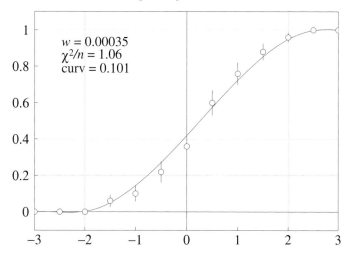

Figure 19.6 A fitting spline approximating the cumulative distribution.

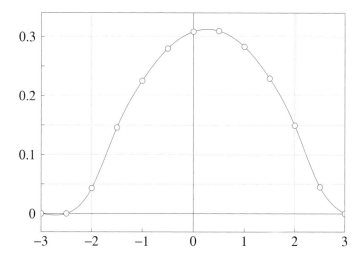

Figure 19.7 Distribution constructed (by a spline) to the first derivative of the cumulative fit.

$\chi^2/n = 1$. Finally, Fig. 19.7 shows a spline fit through the derivative values of Fig. 19.6. The distribution is still some way from being Gaussian, but it is the smoothest (cumulative) distribution compatible with the data.

19.5 Splines for tabulation

Splines can be used to recover forces and energies from tabulated values. Assume we have a centrosymmetric potential $V(r)$ between a pair of particles, defined for $0 < r \le r_c$, with r_c a *cutoff distance* beyond which the potential is identically zero. The particles have cartesian coordinates $\mathbf{r}_i$ and $\mathbf{r}_j$; we use the convention $\mathbf{r} = \mathbf{r}_i - \mathbf{r}_j$ and $r = |\mathbf{r}|$. The force $\mathbf{F} = \mathbf{F}_i$ on particle i, which is equal and opposite to the force acting on particle j, is given by

$$\mathbf{F}(\mathbf{r}) = -\frac{dV}{dr}\frac{\mathbf{r}}{r}. \tag{19.50}$$

We wish to tabulate values of V such that:

- V and dV/dr are recovered simultaneously from the table;
- the derivative of the force is continuous – this is accomplished by using cubic spline interpolation for V;
- the interpolation should be fast – this requires storage of not only the values of the potential, but (at least) also of its derivative, allowing local cubic interpolation;
- a search procedure to find the table index for an arbitrary argument should be avoided – the table index should follow directly from the argument;
- the computationally expensive square root evaluation should be avoided – this means that table indices should be derived from r^2 rather than r.

These requirements lead to the following procedures.

Construction of the table

(i) Decide on the number of table entries $n + 1$ (the table index runs from 0 to n).
(ii) Decide on the cutoff radius r_c.
(iii) Define a scaled variable x, proportional to r^2, running from 0 to n.

$$x = \frac{nr^2}{r_c^2}. \tag{19.51}$$

The table index i runs from 0 to n.

(iv) Construct a table with function values f_i and derivatives g_i. The values should be the required potential and its derivative. Defining

$$r_i = r_c\sqrt{\frac{i}{n}}, \tag{19.52}$$

the values are

$$f_i = V(r_i), \tag{19.53}$$

$$g_i = \frac{dV}{dx} = \frac{r_c^2}{2nr_i}\frac{dV}{dr}(r_i). \tag{19.54}$$

The values g_i can also be determined from a spline fit to f_i. In that case the second derivatives are strictly continuous; if the g_i's are determined from analytical derivatives, the interpolated values approach the analytical function more accurately, but there will be slight discontinuities in the second derivatives at the nodes.

For fast recovery from memory f_i and g_i should be contiguously stored, for example in a two-dimensional array, rather than in separate arrays.[8]

Recovery from the table

We wish to obtain $V(r)$ and $\boldsymbol{F}(r)$ for a given value of r^2, where

$$r^2 = (x_i - x_j)^2 + (y_i - y_j)^2 + (z_i - z_j)^2. \tag{19.55}$$

(i) Determine $x = nr^2/r_c^2$, $i = \text{int}(x)$ and $\xi = x - i$.
(ii) Fetch $f_i, g_i, f_{i+1}, g_{i+1}$ from the table.
(iii) Determine

$$p = 3f_{i+1} - 3f_i - 2g_i - g_{i+1}, \tag{19.56}$$
$$q = -2f_{i+1} + 2f_i + g_i + g_{i+1}. \tag{19.57}$$

(iv) Determine V and dV/dx:

$$V = f_i + \xi g_i + \xi^2 p + \xi^3 q, \tag{19.58}$$
$$\frac{dV}{dx} = g_i + 2\xi p + 3\xi^2 q. \tag{19.59}$$

(v) Reconstruct the force:

$$\boldsymbol{F} = -\frac{2n}{r_c^2} \frac{dV}{dx} \boldsymbol{r}. \tag{19.60}$$

We note that the determination of p and q takes six floating point operations (flops),[9] while the actual interpolation takes ten flops. One may choose to store also p_i and q_i, thus saving 6 out of 16 flops at the expense of doubling the required memory. The total number of flops is comparable to that required to compute Lennard–Jones potential and forces; for almost any other kind of interaction tables are more efficient than analytical evaluation.

Using r^2 instead of r causes a rather inappropriate mapping of r onto the table index, the spacing in r being much finer for r close to r_c than for values close to the potential minimum, where high accuracy is required. A more

---

[8] This requirement depends on the architecture of the computer used; it is certainly true for vector processors that will fetch several words from memory at once. Processors that will hold all tables in cache memory are less demanding in this respect.

[9] For example, as follows: $u = f_{i+1} - f_i$; $q = -2u + g_i + g_{i+1}$; $p = -q + u - g_i$.

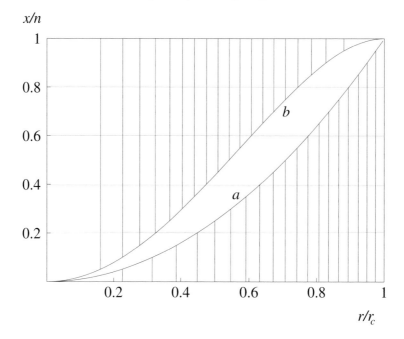

Figure 19.8 Two mappings from distance r to table index x, both avoiding square-root operations: a shows the quadratic mapping of (19.51), b shows the mapping according to (19.61). For both mappings the intervals in r are indicated when the range of x is divided into 20 equal intervals.

appropriate mapping that still avoids square-root operations, but requires additional floating point operations, is

$$x = \frac{2nr^2}{r_c^2}\left(1 - \frac{r^2}{2r_c^2}\right). \tag{19.61}$$

Both mappings are represented in Fig. 19.8, showing the better distribution of intervals for the mapping of (19.61) than for the quadratic mapping of (19.51).

19.6 Algorithms for spline interpolation

In this section the relevant programs for computing, interpolating and integrating cubic spline functions are given. These programs can be downloaded from the book's website.[10]

The general structure is as follows:

- Data points are given as list or array **data=[x, y]**, containing two lists or

[10] http://www.cambridge.org/9780521835275 or http://www.hjcb.nl.

arrays x and y of abscissa and ordinates. The x-values must be ordered: $x_i \leq x_{i+1}$. For example, a sampled sine function:

```
[[0.,0.5*pi,pi,1.5*pi,2.*pi],[0.,1.,0.,-1.,0.]].
```

- Data ready for interpolation contain the function values f, as well as the first derivatives g: `data=[x,f,g]`. f is equal to the data ordinates y for splines through the points; for fitting splines y_i does not lie on the curve, but f_i does. Example (a sampled sine):

```
[[0.,0.5*pi,pi,1.5*pi,2.*pi],
 [0.,1.,0.,-1.,0.],[1.,0.,-1.,0.,1.]].
```

- The function `cinterpol(`x`,data)` returns the cubic-spline interpolated value $f(x)$ or – if x is an array – an array of interpolated values; the spline is specified by `data=[x,f,g]`.

- The function `cspline(data[,cond])` returns the list `[x,f,g]` that specifies the spline through `data=`$[x, y]$. The optional parameter `cond=[[`*first, sec, third*`]`, `[`*first, sec, third*`]]` specifies the extra conditions at the two end nodes: On both the first and the last node the values of *either* the first, the second, or the third derivative must be given; the two other values must be given as `None`. If all values are "None", a periodic spline is assumed. The default is the *natural spline*, with:

```
cond=[[None,0.,None],[None,0.,None]].
```

- The function `cintegral(data)` returns the values of the integral of the spline function at the nodes. The first value is zero; the last value represents the total integral.

- The function `fitspline(data,weight)` produces a natural fitting spline approximating the data points. It uses the argument `weight` as a general weighting factor for the importance of the minimization of χ^2 relative to the minimization of the curvature. A large weight selects a spline through the points, while a small weight tends to perform a least-squares fit to a straight line. The weight is multiplied by the individual weights $1/\sigma_i^2$; the standard deviations must be contained in the data. The function returns not only the spline data but also the curvature and the value of χ^2. The iteration to obtain the required deviation is left to the user, who may wish to graphically inspect intermediate results.

PYTHON PROGRAM 19.1 **cinterpol(x,data)**
Cubic interpolation of spline data

```
01 def cinterpol(x,data):
02 # returns interpolated cubic spline function for argument x
03 # x can b scalar or array
04 # data=[x-values, function values, derivative values] (lists or arrays)
05     xx=array(data[0])
06     n=alen(xx)
07     if (len(data[1])!=n or len(data[2])!=n):
08         print ' ERROR: data arrays of unequal length'
09         print ' cinterpol aborted'
10         return None
11     nx=alen(x)
12     def fun(x):
13         i=argmin(abs(x-xx))
14         if x<xx[i]: i=i-1
15         x0=xx[i]; x1=xx[i+1]
16         f0=data[1][i]; f1=data[1][i+1]
17         g0=data[2][i]; g1=data[2][i+1]
18         h=x1-x0; g=(f1-f0)/h
19         dx=x-x0
20         xsqh=dx*dx/h
21         return f0+g0*dx+(3.*g-2.*g0-g1)*xsqh+(-2.*g+g0+g1)*xsqh*dx/h
22     if nx==1:
23         if (x<xx[0] or x>xx[n-1]):
24             print ' ERROR: x outside range'
25             print ' cinterpol aborted'
26             return None
27         return fun(x)
28     f=empty_like(x)
29     for i in range(nx):
30         z=x[i]
31         if (z<xx[0] or z>xx[n-1]):
32             print ' ERROR: x[%d] outside range' % i
33             f[i]=None
34         else: f[i]=fun(z)
35     return f
```

Comments

It is assumed that scipy has been loaded (see page xiii). Lines 06–10 check for equal length of both input arrays. Lines 12–21 define the interpolated value for a scalar argument; lines 28–34 apply the function for all elements of the input array x.

PYTHON PROGRAM 19.2 **cintegral(data)**
Integral of spline data

```
01 def cintegral(data):
02 # returns array of integral of cubic spline function at nodes
03 # data=[x-values, function values, derivative values] (lists or arrays)
04   x=data[0]; f=data[1]; g=data[2]
05   n=len(x)
06   if (len(f)!=n or len(g)!=n):
07       print ' ERROR: data arrays of unequal length'
08       print ' cintegral aborted'
09       return None
10   s=arange(n,dtype=float); s[0]=0.
11   for i in range(n-1):
12       h=x[i+1]-x[i]
13       s[i+1]=s[i]+0.5*(f[i]+f[i+1]+h*(g[i]-g[i+1])/6.)
14   return s
```

PYTHON PROGRAM 19.3 **cspline(data[,cond])**
Generates first derivatives for cubic spline through set of data

---

```
01 def cspline(data, cond=[[None,0.,None],[None,0.,None]]):
02 # appends derivative array to data=[x,y] using cubic splines through y
03 # x,y can be lists or arrays
04 # returns [x,f,g], f=y (array), g=first derivative array
05 # cond prescribes first, second or third derivatives at end points
06 # default condition is natural splines
07     from linalg import solve
08     x=array(data[0]); f=array(data[1])
09     n=alen(x)-1
10     if (alen(f)!=n+1):
11         print ' ERROR: unequal length of x and y in data'
12         print ' cspline aborted'
13         return None
14     xdif=diff(x)
15     if xdif.min()<=0.:
16         print ' ERROR: nonincreasing x values'
17         print ' cspline aborted'
18         return None
19     s=1./xdif
20     d=arange(n+1,dtype=float) # predefine d
21     for i in range(1,n): d[i]=2.*(s[i-1]+s[i])
22     b=arange(n+1,dtype=float) # predefine b
23     for i in range(1,n):
24         b[i]=3.*((f[i]-f[i-1])*s[i-1]**2 + (f[i+1]-f[i])*s[i]**2)
25     if (cond==[[None,None,None],[None,None,None]]): # periodic
26         if (f[0]!=f[-1]):
27             print ' ERROR: periodic spline with',
28             print ' unequal first and last values'
29             print ' cspline aborted'
30             return None
31         else:
32             d[0]=2.*(s[0]+s[n-1])
33             d=d[:-1]
34             A=diag(d)
35             for i in range(n-1): A[i,i+1]=A[i+1,i]=s[i]
36             A[0,n-1]=A[n-1,0]=s[n-1]
37             b[0]=3.*((f[1]-f[0])*s[0]**2 + (f[n]-f[n-1])*s[n-1]**2)
38             b=b[:-1]
39             g=solve(A,b)
40             g=concatenate((g,[g[0]]))
41             return [x,f,g]
42     head=True # extra first row and column
43     tail=True # extra last row and column
44     if (cond[0][0]!=None):
45         b[1]=b[1]-cond[0][0]*s[0]
46         head=False
47     elif (cond[0][1]!=None):
48         d[0]=2.*s[0]
49         b[0]=3.*(f[1]-f[0])*s[0]**2-0.5*cond[0][1]
50     else:
51         d[0]=s[0]
52         b[0]=2.*(f[1]-f[0])*s[0]**2+cond[0][2]/(6.*s[0])
53     if (cond[1][0]!=None):
```

```
54          b[n-1]=b[n-1]-cond[1][0]*s[n-1]
55          tail=False
56      elif (cond[1][1]!=None):
58          d[n]=2.*s[n-1]
59          b[n]=3.*(f[n]-f[n-1])*s[n-1]**2-0.5*cond[1][1]
60      else:
61          d[n]=s[n-1]
62          b[n]=2.*(f[n]-f[n-1])*s[n-1]**2+cond[1][2]/(6.*s[n-1])
63      if (not head):
64          s=s[1:]; d=d[1:]; b=b[1:]
65      if (not tail):
66          s=s[:-1]; d=d[:-1]; b=b[:-1]
67      A=diag(d)
68      for i in range(alen(d)-1): A[i,i+1]=A[i+1,i]=s[i]
69      g=solve(A,b)
70      if (not head): g=concatenate(([cond[0][0]],g))
71      if (not tail): g=concatenate((g,[cond[1][0]]))
72      return [x,f,g]
```

Comments

It is assumed that `scipy` has been loaded (see page xiii). Line 07 imports the scipy package `linalg`, which contains the linear equation solver `solve`, used in lines 39 and 69. Although the matrices A are tridiagonal, no use is made of more sophisticated linear solvers for sparse matrices in scipy subpackage `linsolve`, as the latter are still under development.

PYTHON PROGRAM 19.4 **fitspline(data,weight)**

Generates natural cubic fitting spline, approximating a set of data, given with standard deviations (algorithm adapted from Späth, 1973).

```
01 def fitspline(data,weight):
02 # computes cubic spline fitting to data
03 # data = [x,y,sig] sig = estimated s.d. of y
04 # x,y,sig can be lists or arrays
05 # weight = overall weight factor attached to data
06 # returns[[x,f,g],[curvature, chisquare deviation]]
07     x=array(data[0]); y=array(data[1]); sig=array(data[2])
08     n=alen(x)-1
09     np1=n+1; n1=n-1; n2=n-2; n3=n-3
10     if (alen(y)!=np1 or alen(sig)!=np1):
11         print ' ERROR: unequal lengths in data'
12         print ' FitSpline aborted'
13         return None
14     q=arange(np1,dtype=float); p=arange(np1,dtype=float)
15     g=arange(np1,dtype=float)
16     y2=arange(np1,dtype=float); y2[0]=y2[n]=0.
17     w=weight/sig**2
18     f=concatenate((diff(x),[0.]))
19     q1=q2=q[0]=p[0]=p[1]=0.
20     j1=0
21     w1=1./w[0]; w2=1./w[1]
22     h=0.; h1=f[0]; h2=f[1]
23     r1=(y[1]-y[0])*h1
24     for k in range(n1):
25         k1=k+1; k2=k+2
26         if (k>0): h=p1-q1*q[j1]
27         h3=f[k2]
28         w3=1./w[k2]
29         s=h1+h2
```

```
30          t=2./h1+2./h2+6.*(h1*h1*w1+s*s*w2+h2*h2*w3)
31          r2=(y[k2]-y[k1])*h2
32          p2=1./h2-6.*h2*(w2*s+w3*(h2+h3))
33          q3=6.*w3*h2*h3
34          z=1./(t-q1*p[k]-h*q[k])
35          if (k<n2): q[k1]=z*(p2-h*p[k1])
36          if (k<n3): p[k2]=z*q3
37          r=6.*(r2-r1)
38          if (k>0): r=r-h*g[j1]
39          if (k>1): r=r-q1*g[j2]
40          g[k]=z*r
41          j2=j1; j1=k
42          q1=q2; q2=q3
43          p1=p2
44          h1=h2; h2=h3
45          w1=w2; w2=w3
46          r1=r2
47      y2[n1]=g[n2]
48      if (n2!=0):
49          y2[n2]=g[n3]-q[n2]*y2[n1]
50          if (n3!=0):
51              for j in range(n3):
52                  k=n3-j
53                  y2[k]=g[k-1]-q[k]*y2[k+1]-p[k+1]*y2[k+2]
54      h1=0
55      for k in range(n):
56          j2=k+1
57          g[k]=f[k]
58          h2=f[k]*(y2[j2]-y2[k])
59          q[k]=h2/6.
60          f[k]=y[k]-(h2-h1)/w[k]
61          p[k]=0.5*y2[k]
62          h1=h2
63      f[n]=y[n]+h1/w[n]
64      for k in range(n):
65          j2=k+1
66          h=g[k]
67          g[k]=(f[j2]-f[k])*h-(y2[j2]+2.*y2[k])/(6.*h)
68      g[n]=(f[n]-f[n1])*h+(2.*y2[n]+y2[n1])/(6.*h)
69      curv=0. # compute curvature
70      for k in range(n1): curv=curv+q[k]*(f[k+1]-f[k])
71      curv=-6.*curv
72      print 'curvature = ', curv, '    ',
73      chisq=0. # compute chisquare sum
74      for k in range(np1): chisq=chisq+((y[k]-f[k])/sig[k])**2
75      print 'chisquare = ', chisq
76      return [[x,f,g],[curv,chisq]]
```

Comments

This program is adapted from Späth (1973) and contains an explicit linear solver; it does not require the package `linalg`. A trial weight must be given; the program returns both the curvature and the chi-square sum. The weight should be adjusted such that the chi-square sum is statistically acceptable, i.e., near the number of data points.

19.7 B-splines

B-spline functions[11] $N_i^{(p)}(t)$ are local smooth functions of a parameter t (p is the *order* of the spline), which are used to construct smooth B-spline curves in one- or multidimensional space by linear superposition, given a set of *control points* (also called *de Boor points*) $\boldsymbol{d}_i$ at the *nodes* t_i:

$$\boldsymbol{F}(t) = \sum_i N_i^{(p)}(t)\boldsymbol{d}_i. \tag{19.62}$$

Each B-spline function is a polynomial with *local support*, i.e., it has (positive) values only on a limited local interval and is zero outside this interval. The support depends on the order p of the spline and is equal to the interval $[t_i, t_{i+p}]$. The higher the order p, the smoother the curve, but the larger its support. If the control points are vectors in 3D space, a curve in 3D space results; the control points may also be real or complex scalars that reconstruct a (real or complex) function $f(x)$. B-spline *surfaces* result from a product of spline functions with two parameters u and v and a two-dimensional set of control points:

$$\boldsymbol{F}(u, v) = \sum_{i,j} N_i^{(p)}(u)N_j^{(p)}(v)\boldsymbol{d}_{ij}. \tag{19.63}$$

The B-spline functions are *normalized* if they integrate to unity.

The normalized B-spline function $N_i^{(p)}$ is defined recursively:

$$p = 1:\ N_i^{(1)}(t) = 1 \text{ for } t_i \le t < t_{i+1};\ 0 \text{ elsewhere}, \tag{19.64}$$

$$p \ge 2:\ N_i^{(p)}(t) = \frac{t - t_i}{t_{i+p-1} - t_i}N_i^{(p-1)}(t) + \frac{t_{i+p} - t}{t_{i+p} - t_{i+1}}N_{i+1}^{(p-1)}(t). \tag{19.65}$$

Most common are the *cardinal* B-splines, which have equidistant intervals $t_{i+1} - t_i$, and especially with distance 1 ("normalized nodes"). The nodes (or knots) are then a set of consecutive integers. They produce *uniform* B-spline curves. These cardinal B-spline functions are thus defined as

$$p = 1:\ N_i^{(1)}(t) = 1 \text{ for } i \le t < i + 1;\ 0 \text{ elsewhere}, \tag{19.66}$$

---

[11] See Engeln-Müllges and Uhlig (1996). B-splines are also introduced in Essmann *et al.* (1995). The name derives from B for "basis". Although in principle already known by Laplace and Euler, B-spline functions and their applications in approximation theory were developed in the 1960s and 1970s. The basic mathematical theory can be found in Schumaker (1981), but a more practical guide is a published lecture series by Schoenberg (1973). A book by de Boor (1978) provides Fortran programs. A relatively more modern description can be found in Chapter 4 of Chui (1992). There is no universally agreed notation for B-splines functions. Our notation N is most common for the *normalized* cardinal spline functions, although the meaning of sub- and superscripts may differ. Schoenberg (1973) denotes non-normalized B-spline functions by Q and central B-spline functions, which are shifted to have the origin at the maximum of the function, by M. The terms *knots* and *nodes* are both used for the control points. One should distinguish between B-splines (curves or surfaces) and B-spline functions. The former are linear combinations of the latter and could be called *B-spline series*.

$$p \geq 2: \quad N_i^{(p)}(t) = \frac{t-i}{p-1} N_i^{(p-1)}(t) + \frac{i+p-t}{p-1} N_{i+1}^{(p-1)}(t). \quad (19.67)$$

Clearly N_i is simply translated over i units with respect to N_0:

$$N_i^{(p)}(t) = N_0^{(p)}(t-i); \quad \text{support of } N_i^{(p)}(t) : \ i \leq t < i+p. \quad (19.68)$$

We may omit the subscript 0 and thus define the basic B-spline functions $N^{(p)}(t)$ for the support $0 \leq t < p$, from which all other functions $N_i^{(p)}(t)$ are derived by translation, as

$$p = 1: \quad N^{(1)}(t) = 1 \text{ for } 0 \leq t < 1; \ 0 \text{ elsewhere}, \quad (19.69)$$

$$p \geq 2: \quad N^{(p)}(t) = \frac{t}{p-1} N^{(p-1)}(t) + \frac{p-t}{p-1} N^{(p-1)}(t-1). \quad (19.70)$$

The functions are symmetric around $t = p/2$, where they attain a maximum. Some texts define *centralized* B-spline functions by taking the origin at $p/2$; for odd p this has the disadvantage that the knots do not occur at integer arguments. The B-spline functions defined here are also called *forward* B-spline functions (Silliman, 1974).

Figure 19.9 shows the first five normalized cardinal splines for $i = 0$. The functions $N^{(p)}(t)$ do not only integrate to 1, but the sum of their values at the nodes equals unity as well:

$$\sum_{j=i}^{i+p} N_i^{(p)}(j) = 1. \quad (19.71)$$

Thus they, so to say, distribute the number 1 among points close to the center $p/2$.

An interesting property of cardinal B-spline functions is that they represent the distribution generated by the sum of p independent random numbers η, each homogeneously distributed in the interval $0 \leq \eta < 1$. As p becomes larger, according to the central limit theorem, this distribution tends to a normal distribution with mean $p/2$ and variance $p/12$. Therefore

$$\lim_{p \to \infty} N^{(p)}(t) = \frac{1}{\sigma \sqrt{2\pi}} \exp\left[-\frac{(t-p/2)^2}{2\sigma^2}\right], \quad \sigma^2 = \frac{p}{12}. \quad (19.72)$$

This property is a direct consequence of a general convolution theorem for B-spline functions: each $N^{(p)}$ is the *convolution* of $N^{(p-q)}$ $(1 \leq q < p)$ with $N^{(q)}$. For the special case $q = 1$ $N^{(q)}$ is the unit pulse function and the convolution becomes another recurrent relation

$$N^{(p)}(t) = (N^{(p-1)} * N^{(1)})(t) = \int_0^1 N^{(p-1)}(t-t')\, dt' = \int_{t-1}^t N^{(p-1)}(t')\, dt'. \quad (19.73)$$

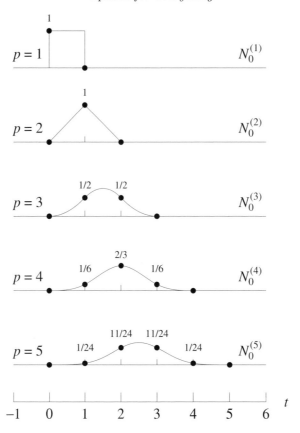

Figure 19.9 The first five normalized cardinal B-spline functions $N_0^{(p)}(t)$.

The convolution theorem implies that the Fourier transform of $N^{(p)}$ is the product of the Fourier transforms of $N^{(q)}$ and $N^{(p-q)}$. Since the Fourier transform of the unit pulse $N^{(1)}$ is proportional to $\sin \frac{1}{2}\omega/\omega$ (ω is the conjugate variable in reciprocal space),[12] the Fourier transform of $N^{(p)}$ is proportional to the p-th power of that function, disregarding a phase factor due to the shifted origin. This transform decays faster in reciprocal space for higher-order splines.

A B-spline function of order p consists of piecewise polynomials of degree $p-1$ and is $p-2$ times continuously differentiable; thus $N^{(4)}(t)$ is twice differentiable, which makes it (and higher-order functions) suitable to express energy functions: the first derivative (needed for forces) is sufficiently smooth as it has a continuous derivative. For the derivatives the following

---

[12] This applies to the central function; the Fourier transform of the function $N^{(1)}$ contains a phase factor $\exp(\frac{1}{2}i\omega)$ because the origin is shifted with respect to the symmetry axis.

relation holds:

$$\frac{dN_0^{(p)}(t)}{dt} = N_0^{(p-1)}(t) - N_1^{(p-1)}(t). \tag{19.74}$$

B-spline curves are not interpolating, as the control points are not equal to the nodal values. Each nodal value is composed of nearby control points according to (19.62). One may construct interpolating curves by deriving the control points from a corresponding set of equations. For *closed*, i.e., periodic, B-spline curves, function values are periodic on $[0, n)$ and there are exactly n nodes. One therefore needs n known values to solve for the coefficients c_i. When the known values are the nodal function values, an interpolating periodic spline is obtained:

$$f(t) = \sum_{i=0}^{n-1} c_i \sum_{k \in \mathbb{Z}} N_{i+kn}^{(p)}(t). \tag{19.75}$$

The sum over k takes care of the periodicity, but only a few values are included because of the compact support of the functions. For example, the interpolating fourth-order B-spline for a sine function $\sin \frac{1}{2}\pi t$ on the periodic interval $[0, 4)$ (Fig. 19.10), given the nodal values $(0, 1, 0, -1)$ at the nodes $(0,1,2,3)$ is

$$f(t) = \frac{3}{2}(N_{-1} + N_3) - \frac{3}{2}(N_1 + N_5). \tag{19.76}$$

These coefficients are the solution of the following set of equations:

$$\frac{1}{6}\begin{pmatrix} 0 & 1 & 4 & 1 \\ 1 & 0 & 1 & 4 \\ 4 & 1 & 0 & 1 \\ 1 & 4 & 1 & 0 \end{pmatrix}\begin{pmatrix} c_0 \\ c_1 \\ c_2 \\ c_3 \end{pmatrix} = \begin{pmatrix} 0 \\ 1 \\ 0 \\ -1 \end{pmatrix}, \tag{19.77}$$

and have the values $(0, -3/2, 0, 3/2)$. The supporting B-spline functions are depicted in Fig. 19.10. Not surprising this B-spline is identical to the cubic interpolating spline shown in Fig. 19.2 on page 531.

The recurrence relation (19.67) allows the following simple algorithm to compute spline values (Griebel *et al.*, 2003). The value of p must be an integer not smaller than 2. While a recurrent program is elegant, a non-recurrent version would be more efficient.

PYTHON PROGRAM 19.5 **Bspline**

Computes the value of the cardinal B-spline function $N^{(p)}(t)$ of real t and integer order $p \leq 2$.

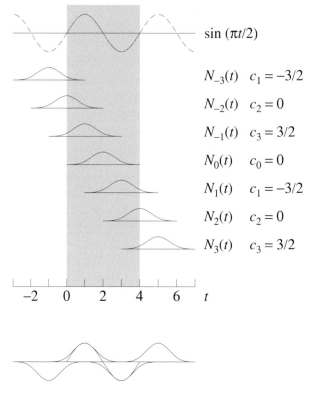

Figure 19.10 The sine function, periodic on the interval $[0, 4)$, approximated by a fourth-order B-spline.

```
01 def Bspline(t,p):
02    if (t<=0. or t>=p): return 0.
03    if (p==2): return 1.-abs(t-1.)
04    return (t*Bspline(t,p-1) + (p-t)*Bspline(t-1.,p-1))/(p-1.)
```

Comments

Recurrence is pursued until the order $p = 2$ is reached, which is the "hat-function" specified in line 2.

For non-periodic (*open*) B-splines that are to be fitted on a function known at $n+1$ points $t = 0, \ldots, n$, one needs coefficients for all the B-spline functions that share their support with the interval $[0, n)$, which are $n + p - 1$ in number. This means that, in addition to the $n + 1$ nodal values, $(p - 2)$ extra conditions must be specified. These can be points outside the interval or first or second derivative specifications. Instead of function values at the nodes, values at arbitrary (e.g., random) arguments may be specified, as long as they provide a sufficient number of independent data to solve for the coefficients.

While in general a set of equations must be solved to find the spline coefficients, there is an exception for *exponential functions*. Functions of the type $f(t) = z^t$, among which the important class of exponentials $f(t) = \exp(i\omega t)$, can be approximated by a B-spline series

$$f(t) \approx \sum_{n=-\infty}^{\infty} c_n N_n^{(p)}(t). \tag{19.78}$$

These are called *Euler exponential splines* (Schoenberg, 1973; Silliman, 1974). We wish to find the coefficients c_n such that the spline series equals $f(t) = z^t$ exactly at all integer values of t. Now consider $f(t+1)$. On the one hand we may expand the shifted series as follows:

$$f(t+1) \approx \sum_{n=-\infty}^{\infty} c_n N_n^{(p)}(t+1) = \sum_{n=-\infty}^{\infty} c_n N_{n-1}^{(p)}(t) = \sum_{n=-\infty}^{\infty} c_{n+1} N_n^{(p)}(t), \tag{19.79}$$

but, on the other hand:

$$f(t+1) = z f(t). \tag{19.80}$$

Therefore there is a relation between coefficients:

$$c_{n+1} = z c_n. \tag{19.81}$$

Denoting c_0 by c, it follows that

$$c_n = z^n c, \tag{19.82}$$

and

$$f(t) \approx c \sum_{n=-\infty}^{\infty} z^n N_n^{(p)}(t). \tag{19.83}$$

The infinite sum takes care of the periodicity. The constant c follows from the requirement that the approximation is exact at the nodes $t = k$, $k \in \mathbb{Z}$:

$$z^k = c \sum_{n=-\infty}^{\infty} z^n N_n^{(p)}(k). \tag{19.84}$$

This can be transformed by replacing $n - k$ by n and shifting to the spline function $N = N_0$, yielding

$$c = \left[\sum_{n=1}^{p-1} z^{-n} N^{(p)}(n) \right]^{-1}. \tag{19.85}$$

Thus in (19.83) and (19.85) we have obtained the solution for an interpolating exponential spline.

These exponential splines are particularly useful to approximate terms in a

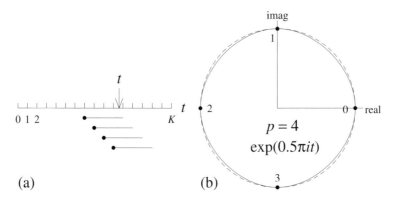

Figure 19.11 (a) the t-axis discretized in $K = 16$ intervals of unit length for use in FFT. The function to be transformed is periodic on $[0, K)$. The supports of the four fourth-order interpolating spline functions, needed to approximate $\exp(2\pi imt/K)$, is indicated by four lines. (b) Fourth-order spline interpolation (drawn curve) for $\exp(0.5\pi it)$ (dashed circle).

Fourier sum, such as $\exp(i\omega t)$ (for arbitrary t), when fast Fourier transforms (FFT) are used, defined only at discrete values of ω and t. Take for example a periodic t-axis divided into K (K is an even integer) intervals of unit length. In FFT (see page 324) the angular frequencies are discretized in K steps

$$\omega_m = \frac{2\pi m}{K}; \quad m = -\frac{K}{2}, -\frac{K}{2} + 1, \ldots, \frac{K}{2} - 1, \tag{19.86}$$

and therefore one needs to evaluate $\exp(2\pi imt/K)$ for $|m| \leq K/2$. Only the values $t = n, \; n = 0, \ldots K - 1$ can be handled by the FFT, and thus one wishes to express $\exp(2\pi imt/K)$ in an interpolating spline series with known values at integer values of t. This is accomplished by (19.83) and (19.85):

$$\exp\left(\frac{2\pi im}{K}t\right) \approx c \sum_{n=-\infty}^{\infty} \exp\left(\frac{2\pi imn}{K}\right) N_n^{(p)}(t), \tag{19.87}$$

$$c = \left[\sum_{n=1}^{p-1} \exp\left(-\frac{2\pi imn}{K}\right) N^{(p)}(n)\right]^{-1}. \tag{19.88}$$

Figure 19.11(a) shows the t axis for the example $K = 16$ and the supports for the four 4-th order spline functions needed for the indicated value of t. The quantity to be Fourier transformed is thus distributed over the four preceding integer values. In Fig. 19.11(b) the fourth-order spline approximation (drawn curve) of the function $\exp(0.5\pi it)$ (providing four points per period) is compared with the exact function (dashed circle) to show the accuracy of the interpolation (better than 0.03).

The absolute error in the approximation equals approximately $(|\omega|/\pi)^p$ (Schoenberg, 1973; Silliman, 1974). This makes the approximation quite imprecise for the highest frequency $\omega_{max} = \pi$ ($m = K/2$) that occurs in the FFT. In fact, for the highest frequency the exponential $\exp(i\pi n)$ is alternating between $+1$ and -1, and the resulting spline is fully real. A fourth-order spline does reproduce the real part $\cos \pi t$ of $\exp(\pi t)$ fairly well (within 0.02) but fails completely to reproduce the imaginary part. For *odd* values of the spline order p there is no solution since the sum evaluates to $0/0$ (e.g., for $p = 5$ the sum would be $2/24 - 22/24 + 22/24 - 2/24 = 0$). This combination of values must be excluded. In usual applications the grid is chosen sufficiently fine for the highest frequency terms to be of vanishing importance.

References

Abraham, F. F. (2003). How fast can cracks move? A research adventure in materials failure using millions of atoms and big computers. *Adv. Phys.*, **52**, 727–90.

Abraham, F. F., Walkup, R., Gao. H. *et al.* (2002). Simulating materials failure by using up to one billion atoms and the world's fastest computer: work-hardening. *Proc. Natl. Acad. Sci.*, **99**, 5783–7.

Abramowitz, M. and Stegun, I. A. (1965). *Handbook of Mathematical Functions*. New York, Dover Publ.

Adelman, S. A. and Doll, J. D. (1974). Generalized Langevin equation approach for atom/solid-surface scattering: Collinear atom/atomic chain model. *J. Chem. Phys.*, **61**, 4242–5.

Ahlström, P., Wallqvist, A., Engström, S. and Jönsson, B. (1989). A molecular dynamics study of polarizable water. *Mol. Phys.*, **68**, 563–81.

Allen, M. P. and Tildesley, D. J. (1987). *Computer Simulation of Liquids*. Oxford, Clarendon Press.

Altman, S. L. (1986). *Rotations, Quaternions and Double Groups*. Oxford, Clarendon Press.

Amadei, A., Linssen, A. B. M. and Berendsen, H. J. C. (1993). Essential dynamics of proteins. *Proteins*, **17**, 412–25.

Amadei,A., Chillemi, G., Ceruso, M., Grottesi, A. and Di Nola, A. (2000). Molecular dynamics simulations with constrained roto-translational motions: theoretical basis and statistical mechanical consistency. *J. Chem. Phys.*, **112**, 9–23.

Andersen, H. C. (1980). Molecular dynamics simulations at constant pressure and/or temperature. *J. Chem. Phys.*, **72**, 2384–93.

Andersen, H. C. (1983). Rattle: a "velocity" version of the Shake algorithm for molecular dynamics calculations. *J. Comput. Phys.*, **52**, 24–34.

Anderson, J. B. (1975). A random-walk simulation of the Schrödinger equation: H_3^+. *J. Chem. Phys.*, **63**, 1499–1503.

Anderson, J. B. (1976). Quantum chemistry by random walk: H ^{2}P, H_3^+ D_{3h} $^1A_1'$, H_2 $^3\Sigma_u^+$, H_4 $^1\Sigma_g^+$, Be ^{1}S. *J. Chem. Phys.*, **65**, 4121–7.

Anderson, J. B. (1995). Exact quantum chemistry by Monte Carlo methods. In *Quantum Mechanical Electronic Structure Calculations with Chemical Accuracy*, ed. S.R. Langhoff. Dordrecht, Kluwer, pp. 1–45.

Anderson, J. B., Traynor, C.vA. and Boghosian, B. M. (1993). An exact quantum Monte Carlo calculation of the helium-helium intermolecular potential. *J. Chem. Phys.*, **99**, 345–51.

Appel, A. W. (1985). An efficient program for many-body simulation. *SIAM J. Sci Statist. Comput.*, **6**. 85–103.

Applequist, J., Carl, J. R. and Fung, K.-K. (1972). An atom dipole interaction model for molecular polarizability. Application to polyatomic molecules and determination of atom polarizabilities. *J. Amer. Chem. Soc.*, **94**, 2952–60.

Arnold, V. I. (1975). *Mathematical Methods of Classical Mechanics*. Graduate Texts in Math, **60**. New York, Springer–Verlag.

Aspuru-Guzik,, A., Salomón-Ferrer, R., Austin, B. *et al.* (2005). Zori 1.0: a parallel quantum Monte Carlo electronic structure package. *J. Comput. Chem.*, **26**, 856–62.

Assaraf, R., Caffarel, M. and Khelif, A. (2000). Diffusion Monte Carlo methods with a fixed number of walkers. *Phys. Rev. E*, **61**, 4566–75.

Athènes, M. (2004). A path-sampling scheme for computing thermodynamic properties of a many-body system in a generalized ensemble. *Eur. Phys. J. B (Cond. Matt.)*, **38**,651–64.

Auffinger, P. and Beveridge, D. L. (1995). A simple test for evaluating the truncation effects in simulations of systems involving charged groups. *Chem. Phys. Lett.*, **234**, 413–5.

Axilrod, B. M. and Teller, E. (1948). Interaction of the van der Waals type between three atoms. *J. Chem. Phys.*, **11**, 299-300.

Bakowies, D. and Thiel, W. (1996). Hybrid models for combined quantum mechanical and molecular mechanical approaches. *J. Phys. Chem.*, **100**, 10580–94.

Bala, P., Lesyng, B. and McCammon, J. A. (1994). Applications of quantum–classical and quantum–stochastic molecular dynamics simulations for proton transfer processes. *Chem. Phys.*, **180**, 271–85.

Banik, S. K., Chaudhuri, J. R. and Ray, D. S. (2000). The generalized Kramers theory for nonequilibrium open one-dimensional systems. *J. Chem. Phys.*, **112**, 8330–7.

Barnes, J. E. and Hut, P. (1986). A hierarchical $\mathcal{O}(N \log N)$ force-calculation algorithm. *Nature* **324**, 446–9.

Becke, A. D. (1988). Density-functional exchange-energy approximation with correct asymptotic behaviour. *Phys. Rev.*, **A38**, 3098–100.

Becke, A. D. (1992). Density-functional thermochemistry I. The effect of the exchange-only gradient correction. *J. Chem. Phys.* **96**, 2155–60.

Becke, A. D. (1993). Density-functional thermochemistry III. The role of exact exchange. *J. Chem. Phys.*, **98**, 5648–52.

Beeman, D. (1976). Some multistep methods for use in molecular dynamics calculations. *J. Comput. Phys.*, **20**, 130–9.

Bekker, H. (1996). Molecular dynamics simulation methods revised. Ph.D. thesis, University of Groningen, June 1996. (Electronic version available through http://dissertations.rug.nl/) .

Bekker, H. (1997). Unification of box shapes in molecular simulations. *J. Comput. Chem.*, **18**, 1930–42.

Bekker, H., Van den Berg, J. P. and Wassenaar, T. A. (2004). A method to obtain a near-minimal-volume molecular simulation of a macromolecule, using periodic boundary conditions and rotational constraints. *J. Comput. Chem.*, **25**, 1037–46.

Bell, J. S. (1976). The theory of local beables. *Epistemological Lett.*, **9**, 11; reprinted in *Dialectica*, **39** (1985), 85–96 and in Bell (1987).

Bell, J. S. (1987). *Speakable and Unspeakable in Quantum Mechanics*. Cambridge, UK, Cambridge University Press.

Ben-Nun, M. and Martinez, T. J. (1998). Nonadiabatic molecular dynamics: validation of the multiple spawning method for a multidimensional problem. *J. Chem. Phys.*, **108**, 7244–57.

Berendsen, H. J. C. (1991a). Incomplete equilibration: a source of error in free energy calculations. In *Proteins, Structure, Dynamics, Design*, Ed. V. Renugopalakrishnan *et al.*, Leiden, Escom, pp. 384–92.

Berendsen, H. J. C. (1991b). Transport properties computed by linear response through weak coupling to a bath. In *Computer Simulation in Material Science. Interatomic Potentials, Simulation Techniques and Applications*, Ed. M. Meyer and V. Pontikis, NATO ASI Series **E, 205**, Dordrecht, Kluwer, pp. 139–55.

Berendsen, H. J. C. (1993). Electrostatic interactions. In *Computer Simulation of Biomolecular Systems*, vol 2, ed. W. F. van Gunsteren, P. K. Weiner and A. J. Wilkinson. Leiden, ESCOM, pp. 161–81.

Berendsen, H. J. C. (1996). Bio-molecular dynamics comes of age. *Science* **271**, 954.

Berendsen, H. J. C. (1998). A glimpse of the Holy Grail? (Perspectives: Protein Folding). *Science* **282**, 642–3.

Berendsen, H. J. C. and Mavri, J. (1993). Quantum simulation of reaction dynamics by density matrix evolution. *J. Phys. Chem.*, **97**, 13464–68.

Berendsen, H. J. C. and Mavri, J. (1997). Simulating proton transfer processes: Quantum dynamics embedded in a classical environment. In *Theoretical Treatment of Hydrogen Bonding*, Chapter 6, ed. Hadži, D. Chicester, Wiley, pp. 119–41.

Berendsen, H. J. C. and van der Velde, G. A. (1972). In *MD and MC on Water*. CECAM Workshop report, ed. H. J. C. Berendsen. Orsay, France, Centre Européen de Calcul Atomique et Moléculaire, p. 63.

Berendsen, H. J. C. and van Gunsteren, W. F. (1986). Practical algorithms for dynamic simulations. In *Molecular Dynamics Simulation of Statistical Mechanical Systems*, ed. G. Ciccotti and W. Hoover. Amsterdam, North Holland.

Berendsen, H. J. C., Postma, J. P. M., van Gunsteren, W. F. and Hermans, J. (1981). Interaction models for water in relation to protein hydration. In: *Intermolecular Forces: Proceedings of the Fourteenth Jerusalem Symposium on Quantum Chemistry and Biochemistry*, ed. B. Pullman. Dordrecht, Reidel Publ. Cy., pp. 331–42.

Berendsen, H. J. C., Postma, J. P. M., van Gunsteren, W. F., DiNola, A. and Haak, J. R. (1984). Molecular dynamics with coupling to an external bath. *J. Chem. Phys.*, **81**, 3684–90.

Berendsen, H. J. C., Postma, J. P. M. and van Gunsteren, W. F. (1985). Statistical mechanics of molecular dynamics: the calculation of free energy. In *Molecular Dynamics and Protein Structure*, Ed. J. Hermans, Western Springs, Ill., Polycrystal Book Service, pp. 43–6.

Berendsen, H. J. C., Grigera, J. R. and Straatsma, T. P. (1987). The missing term in effective pair potentials. *J. Phys. Chem.*, **91**, 6269–71.

Berens, P. H., Mackay, D. H. J., White, G. M. and Wilson, K. R. (1983). Thermodynamics and quantum corrections from molecular dynamics for liquid water. *J. Chem. Phys.*, **79**, 2375–89.

Berg, B. A. and Neuhaus, T. (1991). Multicanonical algorithms for first order phase transitions. *Phys. Lett. B*, **267**, 249–53.

Bergdorf, M., Peter, C. and Hünenberger, P. H. (2003). Influence of cut-off truncation and artificial periodicity of electrostatic interactions in molecular simulations of solvated ions: a continuum electrostatics study. *J. Chem. Phys.*, **119**, 9129–44.

Berne, B. J. (1999). Molecular dynamics in systems with multiple time scales: Reference System Propagator Algorithms. In *Computational Molecular Dynamics: Challenges, Methods, Ideas*, ed. P. Deuflhard, J. Hermans, B. Leimkuhler *et al.* Berlin, Springer–Verlag, pp. 297–317.

Berne, B. J. and Thirumalai, D. (1986). On the simulation of quantum systems: path integral methods. *Ann. Rev. Phys. Chem.*, **37**, 401–24.

Beyer, W. H. (1990). *CRC Standard Probability and Statistics, Tables and Formulae*. Boca Raton, USA, CRC Press.

Billing, G. D. (1994). Mean-field molecular dynamics with surface hopping. *Int. Revs Phys. Chem.*, **13**, 309–36.

Binder, K. and Heermann, D. W. (2002). *Monte Carlo Simulation in Statistical Physics. An Introduction*. Fourth edn. Berlin, Springer–Verlag.

Board, J. A. and Schulten, K. (2000). The fast multipole algorithm. *Computing in Science and Engineering*, **2**, 76-9.

Boek, E. S., Coveney, P. V., Lekkerkerker, H. N. W. and van der Schoot, P. (1997). Simulating the rheology of dense colloidal suspensions using dissipative particle dynamics. *Phys. Rev. E*, **55**, 3124–33.

Bohm, D. (1952a). A suggested interpretation of the quantum theory in terms of "hidden" variables. I. *Phys. Rev.*, **85**, 166-79.

Bohm, D. (1952b). A suggested interpretation of the quantum theory in terms of "hidden" variables. II. *Phys. Rev.*, **85**, 180-93.

Bohr, N. (1928). Das Quantenpostulat und die neuere Entwicklung der Atomistik. *Naturwiss.*, **16**, 245–57.

Borgis, D. and Hynes, J. T. (1991). Molecular-dynamics simulation for a model nonadiabatic proton transfer reaction in solution. *J. Chem. Phys.*, **94**, 3619–28.

Borgis, D., Lee, S. and Hynes, J. T. (1989). A dynamical theory of nonadiabatic proton and hydrogen atom transfer reaction rates in solution. *Chem. Phys. Lett.*, **162**, 19–26.

Born, M. (1920). Volumen und Hydratationswärme der Ionen. *Z. Physik*, **1**, 45–8.

Born, M. and Oppenheimer, R. (1927). Zur Quantentheorie der Molekeln. *Ann. Physik*, **84**, 457–84.

Bossis, G., Quentrec, B. and Boon, J. P. (1982). Brownian dynamics and the fluctuation-dissipation theorem. *Mol. Phys.,*, **45**, 191–6.

Braly, L. B., Cruzan, J. D., Liu, K., Fellers, R. S. and Saykally, R. J. (2000a). Terahertz laser spectroscopy of the water dimer intermolecular vibrations. I. D_2O. *J. Chem. Phys.*, **112**, 10293–313.

Braly, L. B., Liu, K., Brown, M.G. *et al.* (2000b). Terahertz laser spectroscopy of the water dimer intermolecular vibrations. II. H_2O. *J. Chem. Phys.*, **112**, 10314–26.

Bright Wilson, E. (1968). DFT explained. *Structural Chemistry and Biology*, 753–60.

Broglie, L. de (1927). La mécanique ondulatoire et la structure atomique de la matière et du rayonnement. *J. de Physique*, **8**, 225–41.

Bronstein, I. N. and Semendjajew, K. A. (1989). *Taschenbuch der Mathematik*. Frankfurt/Main, Verlag Harri Deutsch.

Brooks III, C. L., Pettitt, B. M. and Karplus, M. (1985). Structural and energetic effects of truncating long-ranged interactions in ionic and polar fluids. *J. Chem. Phys.*, **83**, 5897–908.

Brown, L. M., ed. (2000). *Selected Papers of Richard Feynman*. Singapore, World Scientific.

Burnham, C. J., Li, J., Xantheas, S. S. and Leslie, M. (1999). The parametrization of a Thole-type all-atom polarizable model from first principles and its application to the study of water clusters ($n = 2 - 21$) and the phonon spectrum of ice Ih. *J. Chem. Phys.*, **110**, 4566–81.

Caillol, J. -M. (1994). Comments on the numerical simulations of electrolytes in periodic boundary conditions. *J. Chem. Phys.*, **101**, 6080–90.

Caldwell, J., Dang, L. X. and Kollman, P. A. (1990). Implementation of non-additive intermolecular potentials by use of molecular dynamics: development of a water-water potential and water-ion cluster interactions. *J. Amer. Chem. Soc.*, **112**, 9144–7.

Cao, J. and Berne, B. J. (1990). Low-temperature variational approximation for the Feynman quantum propagator and its application to the simulation of the quantum system. *J. Chem. Phys.*, **92**, 7531–9.

Car, R. and Parrinello, M. (1985). Unified approach for molecular dynamics and density-functional theory. *Phys. Rev. Lett.*, **55**, 2471–4.

Carrillo-Trip, M., Saint-Martin, H. and Ortega-Blake, I. (2003). A comparative study of the hydration of Na^+ and K^+ with refined polarizable model potentials. *J. Chem. Phys.*, **118**, 7062–73.

Ceperley, D. M. and Alder, B. J. (1980). Ground state of the electron gas by a stochastic method. *Phys. Rev. Lett.*, **45**, 566–9.

Ceperley, D. M. and Kalos, M. H. (1979). Quantum many-body problems. In *Monte Carlo Methods in Statistical Physics*, ed. K. Binder. Berlin, Springer–Verlag, pp. 145–94.

Chandler, D. (1987). *Introduction to Modern Statistical Mechanics*. Oxford, Oxford University Press.

Chandler, D. and Wolynes, P. G. (1981). Exploiting the isomorphism between quantum theory and classical statistical mechanics of polyatomic fluids. *J. Chem. Phys.*, **74**, 4078–95.

Chandra Singh, U. and Kollman, P. A. (1986). A combined *ab initio* quantum mechanical and molecular mechanical method for carrying out simulations on complex molecular systems: Applications to the $CH_3Cl + Cl^-$ exchange reaction and gas phase protonation of polyethers. *J. Comput. Chem.*, **7**, 718–30.

Chen, B., Xing, J. and Siepmann, J. I. (2000). Development of polarizable water force fields for phase equilibrium calculations. *J. Phys. Chem. B*, **104**, 2391–401.

Chialvo, A. A. and Cummings, P. T. (1996). Engineering a simple polarizable model for the molecular simulation of water applicable over wide ranges of state conditions. *J. Chem. Phys.*, **105**, 8274–81.

Christen, M., Hünenberger, P. H., Bakowies, D. *et al.* (2005). The GROMOS software for biomolecular simulation: GROMOS05. *J. Comput. Chem.*, **26** 1719–51.

Chui, C. K. (1992). *An Introduction to Wavelets*. Boston, Academic Press.

Ciccotti, G. and Ryckaert, J. -P. (1981). On the derivation of the generalized Langevin equation for interacting Brownian particles. *J. Stat. Phys.*, **26**, 73–82.

Ciccotti, G. and Ryckaert, J. -P. (1986). Molecular dynamics simulations of rigid molecules. *Comput. Phys. Rep.*,, **4**, 345–92.

Cieplack, P., Kollman, P. and Lybrand, T. (1990). A new water potential including polarization: Application to gas-phase, liquid and crystal properties of water. *J. Chem. Phys.*, **92**, 6755–60.

Clementi, E. and Roetti, C. (1974). . *Atomic Data and Nuclear Data Tables*, **14**, 177–478.

Cohen, E. G. D. and Mauzerall, D. (2004). A note on the Jarzynski equality. *J. Stat. Mech.*, **2004**, P07006.

Cohen, M. H. (2002). Classical Langevin dynamics for model Hamiltonians. *arXiv:cond-mat/0211002*, http://arxiv.org/ (6 pp.).

Cooley, J. W. and Tukey, J. W. (1965). An algorithm for the machine calculation of complex Fourier series. *Mathematics of Computation* **19**, 297–301.

Cramer, C. J. (2004). *Essentials of Computational Chemistry, Theories and Models*. 2nd edn. Chicester, UK, Wiley.

Crooks, G. E. (2000). Path-ensemble averages in systems driven far from equilibrium. *Phys. Rev. E*, **61**, 2361–6.

D'Alessandro, M., Tenebaum, A. and Amadei, A. (2002). Dynamical and statistical mechanical characterization of temperature coupling algorithms. *J. Phys. Chem. B.*, **106**, 5050–7.

Dang, L. X. (1992). The nonadditive intermolecular potential for water revised. *J. Chem. Phys.*, **97**, 2659–60.

Dang, L. X. and Chang, T. -M. (1997). Molecular dynamics study of water clusters, liquid, and liquid-vapor interface with many-body potentials. *J. Chem. Phys.*, **106**, 8149–59.

Darden, T., York, D. and Pedersen, L. (1993). Particle mesh Ewald: An $n \log n$ method for Ewald sums in large systems. *J. Chem. Phys.*, **98**, 10089–92.

de Boor (1978). *Practical Guide to Splines*. New York, Springer–Verlag.

de Groot, S. R. and Mazur, P. (1962). *Nonequilibrium Thermodynamics*. Amsterdam, North Holland.

de Leeuw, S. W., Perram, J. W. and Smith, E. R. (1980). Simulation of electrostatic systems in periodic boundary conditions. I. Lattice sums and dielsctric constants. *Proc. Roy. Soc. London*, **A 373**, 26–56.

de Leeuw, S. W., Perram, J. W. and Petersen, H. G. (1990). Hamilton's equations for constrained dynamical systems. *J. Stat. Phys.*, **61**, 1203–22.

den Otter, W. K. (2000). Thermodynamic integration of the free energy along a reaction coordinate in Cartesian coordinates. *J. Chem. Phys.*, **112**, 7283–92.

den Otter, W. K. and Briels, W. J. (1998). The calculation of free-energy differences by constrained molecular-dynamics simulations. *J. Chem. Phys.*, **109**, 4139–46.

de Pablo, J. J., Prausnitz, J. M., Strauch, H. J. and Cummings, P. T. (1990). Molecular simulation of water along the liquid–vapor coexistence curve from 25 °C to the critical point. *J. Chem. Phys.*, **93**, 7355–9.

de Raedt, H. (1987). Product formula algorithms for solving the time-dependent Schrödinger equation. *Comput. Phys. Reports*, **7**, 1–72.

de Raedt, H. (1996). Computer simulations of quantum phenomena in nano-scale devices. *Ann. Revs Comput. Phys.*, **IV**, 107–46.

de Raedt, B., Sprik, M. and Klein, M. L. (1984). Computer simulation of muonium in water. *J. Chem. Phys.*, **80**, 5719–24.

Deserno, M. and Holm, C. (1998a). How to mesh up Ewald sums. I. A theoretical and numerical comparison of various mesh routines. *J. Chem. Phys.*, **109**, 7678–93.

Deserno, M. and Holm, C. (1998b). How to mesh up Ewald sums. II. An accurate error estimate for the particleparticleparticle–mesh algorithm. *J. Chem. Phys.*, **109**, 7694–701.

Dewar, M. J. S., Zoebisch, E. G., Healy, E. F. and Stewart, J. J. P. (1985). Development and use of quantum mechanical molecular models. 76. AM1: a new general purpose quantum mechanical molecular model. *J. Am. Chem. Soc.*, **107**, 3902–9.

Dick Jr, B. G. and Overhauser, A. W. (1958). Theory of the dielectric constant of alkali halide crystals. *Phys. Rev.*, **112**, 90–103.

DiNola, A., Berendsen, H. J. C. and Edholm, O. (1984). Free energy determination of polypeptide conformations generated by molecular dynamics. *Macromol.*, **17**, 2044–50.

Dirac, P. A. M. (1958). *The Principles of Quantum Mechanics*. Oxford, Clarendon Press.

Doi, M. and Edwards, S. F. (1986). *The Theory of Polymer Dynamics*. Oxford, Clarendon Press.

Doob, J. L. (1942). The Brownian movement and stochastic processes. *Ann. Math.*, **43**, 351–69; also reproduced in Wax (1954).

Edberg, R., Evans, D. J. and Morris, G. P. (1986). Constrained molecular dynamics: Simulations of liquid alkanes with a new algorithm. *J. Chem. Phys.*, **84**, 6933–9.

Edholm, O. and Berendsen, H. J. C. (1984). Entropy estimation from simulations of non-diffusive systems. *Mol. Phys.*, **51**, 1011–28.

Edholm, O., Berendsen, H. J. C. and van der Ploeg, P. (1983). Conformational entropy of a bilayer membrane derived from dynamics simulation. *Mol. Phys.*, **48**, 379–88.

Ehrenfest, P. (1927). Bemerkung über die angenäherte Gültigkeit der klassischen Mechanik innerhalb der Quantenmechanik. *Z. Physik*, **45**, 455–7.

Engeln-Müllges, G. and Uhlig, F. (1996). *Numerical Algorithms with C*. Berlin, Springer–Verlag.

Erpenbeck, J. J. and Wood, W. W. (1977). Molecular dynamics techniques for hard-core systems. In *Statistical Mechanics, Part B: Time-Dependent Processes*, ed. B.J. Berne. New York, Plenum Press, pp. 1–40.

Español, P. (1995). Hydrodynamics for dissipative particle dynamics. *Phys. Rev. E*, **52**, 1734–42.

Español, P. (1998). Fluid particle model. *Phys. Rev. E*, **57**, 2930–48.

Español, P. and Revenga, M. (2003). Smoothed dissipative particle dynamics. *Phys. Rev. E*, **67**, 26705 (12 pp.).

Español, P. and Warren, P. B. (1995). Statistical mechanics of dissipative particle dynamics. *Europhys. Lett.*, **30**, 191–6.

Español, P., Serrano, M. and Öttinger, H. C. (1999). Thermodynamically admissible form for discrete hydrodynamics. *Phys. Rev. Lett.*, **83**, 4542–5.

Essex, J. W. and Jorgensen, W. L. (1995). An empirical boundary potential for water droplet simulations. *J. Comput. Chem.*, **16**, 951–72.

Essmann, U., Perera, L., Berkowitz, M., Darden, T., Lee, H. and Pedersen, L. (1995). A smooth particle mesh Ewald method. *J. Chem. Phys.*, **103**, 8577–93.

Evans, D. J. (1977). On the representation of orientation space. *Mol. Phys.*, **34** 317–25.

Evans, D. J. (1983). Computer experiment for non-linear thermodynamics of Couette flow. *J. Chem. Phys.*, **78**, 3297–302.

Evans, D. J. and Morriss, G. P. (1983a). The isothermal isobaric molecular dynamics ensemble. *Phys. Lett.*, **98A**, 433–6.

Evans, D. J. and Morriss, G. P. (1983b). Isothermal isobaric molecular dynamics. *Chem. Phys.*, **77**, 63–6.

Evans, D. J. and Morriss, G. P. (1984). Non-Newtonian molecular dynamics. *Phys. Rep.*, **1**, 297–344.

Ewald, P. P. (1921). Die Berechnung optischer und elektrostatischer Gitterpotentiale. *Ann. Physik*, **64**, 253–87.

Fan, X. J., Phan-Thien, N., Ng, T. Y., Wu, X. H. and Xu, D. (2003). Microchannel flow of a macromolecular suspension. *Phys. Fluids*, **15**, 11–21.

Feenstra, K. A., Hess, B. and Berendsen, H. J. C. (1999). Improving efficiency of large time-scale molecular dynamics simulations of hydrogen-rich systems. *J. Comput. Chem.*, **20**, 786–98.

Fellers, R. S., Braly, L. B., Saykally, R. J. and Leforestier, C. (1999). Fully coupled six-dimensional calculations of the water dimer vibration-rotation-tunneling states with split Wigner pseudospectral approach. II. Improvements and tests of additional potentials. *J. Chem. Phys.*, **110**, 6306–18.

Fényes, I. (1952). Eine wahrscheinlichkeitstheoretische Begründung und Interpretation der Quantenmechanik. *Z. Physik*, **132**, 81–106.

Feynman, R. P. (1948). Space-time approach to non-relativistic quantum mechanics. *Rev. Mod. Phys.*, **20**, 367–87; also reprinted in Brown (2000), pp. 177-97.

Feynman, R. P. and Hibbs, A. R. (1965). *Quantum mechanics and path integrals*. New York, McGraw-Hill.

Field, M. J., Bash, P. A. and Karplus, M. (1990). A combined quantum mechanical and molecular mechanical potential for molecular dynamics simulations. *J. Comput. Chem.*, **11**, 700–33.

Figueirido, F., Levy, R. M., Zhou, R. and Berne, B. J. (1997). Large scale simulation of macromolecules in solution: Combining the periodic fast multipole method with multiple time step integrators. *J. Chem. Phys.*, **106**, 9835–49; Erratum: *ibid.*, **107** (1997), 7002.

Fincham, D. (1992). Leapfrog rotational algorithms. *Mol. Simul.*, **8**, 165–78.

Fixman, M. (1974). Classical statistical mechanics of constraints: a theorem and application to polymers. *Proc. Natl. Acad. Sci.*, **71**, 3050–3.

Flekkøy, E. G. and Coveney, P. V. (1999). From molecular to dissipative particle dynamics. *Phys. Rev. Lett.*, **83**, 1775–8.

Flekkøy, E. G., Coveney, P. V. and Fabritiis, G. D. (2000). Foundations of dissipative particle dynamics. *Phys. Rev. E*, **62**, 2140–57.

Flügge, S. (1974). *Practical Quantum Mechanics*. New York, Springer–Verlag.

Fosdick, L. D. (1962). Numerical estimation of the partition function in quantum statistics. *J. Math. Phys.*, **3**, 1251–64.

Foulkes, W. M. C., Mitas, L., Needs, R. J. and Rajagopal, G. (2001). Quantum Monte Carlo simulation of solids. *Revs Modern Phys.*, **73**, 33–83.

Fraaije, J. G. E. M. (1993). Dynamic density functional theory for microphase separation kinetics of block copolymer melts. *J. Chem. Phys.*, **99**, 9202–12.

Fraaije, J. G. E. M., van Vlimmeren, B. A. C., Maurits, N. M. *et al.* (1997). The dynamic mean-field density functional method and its application to the mesoscopic dynamics of quenched block copolymer melts. *J. Chem. Phys.*, **106**, 4260–9.

Frenkel, D. and Smit, B. (2002). *Understanding Molecular Simulation*. 2nd edn. San Diego and London, Academic Press.

Friedman, H. A. (1975). Image Approximation to the Reaction Field. *Mol. Phys.*, **29**, 1533–43.

Gao, J. and Thompson, M. A., eds (1998). Combines Quantum Mechanical and Molecular Mechanical Methods. *ACS Symposium Series*, **712**, Amer. Chem. Soc.

Gardiner, F. W. (1990). *Handbook of Stochastic Methods for Physics, Chemistry and the Natural Sciences*. 2nd edn. Berlin, Springer–Verlag.

Gasiorowicz, S. (2003). *Quantum Physics*. 3rd edn. New York, Wiley.

Gear, C. W. (1971). *Numerical Initial Value Problems in Ordinary Differential Equations*. Englewood Cliffs, N.J., USA.

Gibbs, J. W. (1957). *The Collected Works of J. Willard Gibbs*. New Haven, Yale University Press.

Giese, T. J. and York, D. M. (2004). Many-body force field models based solely on pairwise Coulomb screening do not simultaneously reproduce correct gas-phase and condensed-phase polarizability limits. *J. Chem. Phys.*, **120**, 9903–6.

Gillan, M. J. (1988). The quantum simulation of hydrogen in metals. *Phil. Mag. A*, **58**, 257–83.

Gindensperger, E., Meier, C. and Beswick, J. A. (2000). Mixing quantum and classical dynamics using Bohmian trajectories. *J. Chem. Phys.*, **113**, 9369–72.

Gindensperger, E., Meier, C. and Beswick, J. A. (2002). Quantum-classical dynamics including continuum states using quantum trajectories. *J. Chem. Phys.*, **116**, 8–13.

Gindensperger, E., Meier, C. and Beswick, J. A. (2004). Hybrid quantum/classical dynamics using Bohmian trajectories. *Adv. Quantum Chem.*, **47**, 331–46.

Glen, R. C. (1994). A fast empirical method for the calculation of molecular polarizability. *J. Comput.-Aided Mol. Design*, **8**, 457–66.

Goldstein, H. (1980). *Classical Mechanics*. 2nd edn. Reading, Addison–Wesley.

Goldstein, H., Poole, C. and Safko, J. (2002). *Classical Mechanics*. 3rd edn. San Francisco, Addison-Wesley.

Grassberger, P. (2002). Go with the winners: a general Monte Carlo strategy. *Comput. Phys. Comm.*, **147**, 64–70.

Green, M. S. (1954). Markoff random processes and the statistical mechanics of time-dependent phenomena. II. Irreversible processes in fluids. *J. Chem. Phys.*, **22**, 398–413.

Greengard, L. and Rokhlin, V. (1987). A fast algorithm for particle simulation. *J. Comput. Phys.*, **73**, 325–48.

Griebel, M., Knapek, S., Zumbusch, G. and Caglar, A. (2003). *Numerische Simulation in der Moleküldynamik. Numerik, Algorithmen, Parallelisierung, Anwendungen*. Berlin, Springer–Verlag.

Grimm, R. C. and Storer, R. G. (1971). Monte-Carlo solution of Schrödinger's equation. *J. Comput. Phys.*, **7**, 134–56.

Grmela, M. and Öttinger, H. C. (1997). Dynamics and thermodynamics of complex fluids. I Development of a general formalism. *Phys. Rev. E*, **56**, 6620–32.

Groenhof, G. *et al.* (2004). Photoactivation of the Photoactive Yellow Protein: why photon absorption triggers a trans-to-cis isomerization of the chromophore in the protein. *J. Amer. Chem. Soc.*, **126**, 4228–33.

Groot, R. D. and Warren, P. B. (1997). Dissipative particle dynamics: bridging the gap between atomistic and mesoscopic simulation. *J. Chem. Phys.*, **107**, 4423–35.

Grossman, J. C. and Mitas, L. (2005). Efficient quantum Monte Carlo energies for molecular dynamics simulations. *Phys. Rev. Lett.*, **94**, 056403.

Grubmüller, H., Heymann, B. and Tavan, P. (1996). Ligand binding and molecular mechanics calculation of the streptavidin-biotin rupure force. *Science*, **271**, 997–9.

Guerra, F. (1981). Structural aspects of stochastic mechanics and stochastic field theory. *Phys. Reports*, **77**, 263–312.

Guillot, B. (2002). A reappraisal of what we have learnt during three decades of computer simulations on water. *J. Mol. Liquids*, **101**, 219–60.

Guillot, B. and Guissani, Y. (1998). Quantum effects in simulated water by the Feynman–Hibbs approach. *J. Chem. Phys.*, **108**, 10162–74.

Guissani, Y. and Guillot, B. (1993). A computer simulation study of the liquid-vapor coexistence curve of water. *J. Chem. Phys.*, **98**, 8221–35.

Hahn, B. and Stock, G. (2000). Quantum-mechanical modeling of the femtosecond isomerization in rhodopsin. *J. Phys. Chem. B*, **104**, 1146–9.

Hamilton, W. R. (1844). On quaternions; or on a new system of imaginaries in algebra. *Phil. Mag.*, 3rd ser. **25**, 489–95.

Hammes-Schiffer, S. (1996). Multiconfigurational molecular dynamics with quantum transitions: Multiple proton transfer reactions. *J. Chem. Phys.*, **105**, 2236–46.

Handy, N. C. (1996). Density functional theory. In *Quantum Mechanical Simulation Methods for Studying Biological Systems*, ed. D. Bicout and M. Field. Berlin, Springer–Verlag, pp. 1–35.

Handy, N. C. and Lee, A. M. (1996). The adiabatic approximation. *Chem. Phys. Lett.*, **252**, 425–30.

Hansmann, U. H. E. and Okamoto, Y. (1993). Prediction of peptide conformation by multicanonical algorithm: new approach to the multiple-minima problem. *J. Comput. Chem.*, **14**, 1333–8.

Hawkins, G. D., Cramer, C. J. and Truhlar, D. G. (1996). Parametrized models of aqueous free energies of solvation based on pairwise descreening of solute atomic charges from a dielectric medium. *J. Phys. Chem.*, **100**, 19824–39.

Heisenberg, W. (1927). Über den anschaulichen Inhalt der quantentheoretischen Kinematik und Mechanik. *Z. Physik*, **43**, 172–98.

Heisenberg, W. and Bohr, N. (1963). Die Kopenhagener Deutung der Quantentheorie. *Dokumente der Naturwissenschaft* **4**, Stuttgart, Ernst Battenberg Verlag.

Hernandez, R. (1999). The projection of a mechanical system onto the irreversible generalized Langevin equation. *J. Chem. Phys.*, **111**, 7701–4.

Hernández-Cobos, J., Saint-Martin, H., Mackie, A. D., Vega, L. F. and Ortega-Blake, I. (2005). Water liquid-vapor equilibria predicted by refined *ab initio* derived potentials. *J. Chem. Phys.*, **123**, 044506 (8 pp.).

Hess, B. (2002a). Stochastic concepts in molecular simulation. Ph.D. thesis, University of Groningen, the Netherlands.
(electronic version available through http://dissertations.rug.nl/).

Hess, B. (2002b). Determining the shear viscosity of model liquids from molecular dynamics simulations. *J. Chem. Phys.*, **116**, 209–17.

Hess, B., Bekker, H., Berendsen, H. J. C. and Fraaije, J. G. E. M. (1997). LINCS: A linear constraint solver for molecular simulations. *J. Comput. Chem.*, **18**, 1463–72.

Hess, B., Saint-Martin, H. and Berendsen, H. J. C. (2002). Flexible constraints: An adiabatic treatment of quantum degrees of freedom, with application to the flexible and polarizable mobile charge densities in harmonic oscillators model for water. *J. Chem. Phys.*, **116**, 9602-10.

Hetherington, J. H. (1984). Observations on the statistical iteration of matrices. *Phys. Rev.*, **A 30**, 2713–9.

Hill, T. L. (1956). *Statistical Mechanics*. New York, McGraw-Hill.

Hiller, M. (1983). *Mechanische Systeme, Eine Einführung in die analytische Mechanik und Systemdynamik*. Berlin, Springer–Verlag.

Hirschfelder, J. O, Curtiss, C. F and Bird, R. B. (1954). *Molecular Theory of Gases and Liquids*. New York, Wiley.

Hockney, R. and Eastwood, J. (1988). *Computer Simulation using Particles*. London, Institute of Physics Publishing.

Hohenberg, P. and Kohn, W. (1964). Inhomogeneous electron gas. *Phys. Rev.*, **B 136**, 864–71.

Holland, P. R. (1993). *The Quantum Theory of Motion. An Account of the de Broglie-Bohm Causal Interpretation of Quantum Mechanics*. Cambridge, UK, Cambridge University Press.

Hoogerbrugge, P. J. and Koelman, J. M. V. A. (1992). Simulating microscopic hydrodynamic phenomena with dissipative particle dynamics. *Europhys. Lett.*, **19**, 155–60.

Hoover, W. G. (1985). Canonical dynamics: Equilibrium phase-space distributions. *Phys. Rev.*, **A31**, 1696–7.

Hoover, W. G., Ladd, A. J. C. and Mran, B. (1982). High strain rate plastic flow studied via non-equilibrium molecular dynamics. *Phys. Rev. Lett.*, **48**, 1818–20.

Huang, K. (1987). *Statistical Mechanics*. 2nd edn. New York, Wiley.

Huber, T., Torda, A. E. and van Gunsteren, W. F. (1997). Structure optimization combining sof-core interaction functions, the diffusion equation method, and molecular dynamics. *J. Phys. Chem. A*, **101**, 5926–30.

Hummer, G. (2001). Fast-growth thermodynamic integration: error and efficiency analysis. *J. Chem. Phys.*, **114** 7330–7.

Hummer, G. and Szabo, A. (2001). Free energy reconstruction from nonequilibrium single-molecule pulling experiments. *Proc. Natl Acad. Sci.*, **298**, 3658–61.

Hummer, G., Pratt, L. R. and Garcia, A. E. (1996). Free energy of ionic hydration. *J. Phys. Chem.*, **100**, 1206–15.

Iancu, F. O. (2005). Droplet dynamics in a fluid environment. A mesoscopic simulation study. Ph. D. thesis, Technical University, Delft, the Netherlands.

Irving, J. H. and Kirkwood, J. G. (1950). The statistical mechanical theory of transport processes. IV. The equations of hydrodynamics. *J. Chem. Phys.*, **18**, 817–29.

Isralewitz, B., Gao, M and Schulten, K. (2001). Steered molecular dynamics and mechanical functions of proteins. *Curr. Opinion Struct. Biol.*, **11**, 224–30.

Izrailev, S., Stepaniants, S., Balsera, M., Oono, Y. and Schulten, K. (1997). Molecular dynamics study of unbinding of the avidin–biotin complex. *Biophys. J.*, **72**, 1568–81.

Jahnke, E. and Emde, F. (1945). *Tables of Functions with Formulae and Curves*. New York, Dover Publications.

Jarzynski, C. (1997a). Nonequilibrium equality for free energy differences. *Phys. Rev. Lett.*, **78**, 2690–3.

Jarzynski, C. (1997b). Equilibrium free-energy differences from nonequilibrium measurements: a master equation approach. *Phys. Rev. E*, **56**, 5018–35.

Jarzynski, C. (2004). Response to Cohen and Mauzerall. *arXiv*:cond-mat/07340. http://arxiv.org/.

Jaynes, E. T. (1957a). Information theory and statistical mechanics I. *Phys. Rev.*, **106**, 620–30.

Jaynes, E. T. (1957b). Information theory and statistical mechanics II. *Phys. Rev.*, **108**, 171–90.

Jedlovszky, P. and Richardi, J. (1999). Comparison of different water models from ambient to supercritical conditions: A Monte Carlo simulation and molecular Ornstein-Zernike study. *J. Chem. Phys.*, **110**, 8019–31.

Jensen, F. (2006). *Introduction to Computational Chemistry*. 2nd edn. New York, Wiley.

Johnson, J. K., Zollweg, J. A. and Gubbins, K. E. (1993). The Lennard–Jones equation of state revisited. *Mol. Phys.*, **78**, 591–618.

Jones, H. F. (1990). *Groups, Representations, and Physics*. Bristol, Institute of Physics Publ.

Jordan, P. C., van Maaren, P. J., Mavri, J., van der Spoel, D. and Berendsen, H. J. C. (1995). Towards phase transferable potential functions: Methodology and application yo nitrogen. *J. Chem. Phys.*, **103**, 2272–85.

Jorgensen, W. L. (1981). Transferable intermolecular potential function for water, alcohols, and ether. Application to liquid water. *J. Amer. Chem. Soc.*, **103**, 335–40.

Jorgensen, W. L., Chandrasekhar, J., Madura, J. D., Impey, R. W. and Klein, M. L. (1983). Comparison of simple potential functions for simulating liquid water. *J. Chem. Phys.*, **79**, 926–35.

Juffer, A. H., Botta, E. F. F, Van Keulen, B. A. M., Van der Ploeg, A. and Berendsen, H. J. C. (1991). The electric potential of a macromolecule in a solvent: A fundamental approach. *J. Comput. Phys.*, **97**, 144–71.

Kalos, M. H. (1962). Monte Carlo calculations of the ground state of three- and four-body nuclei. *Phys Rev.*, **128**, 1791–5.

Kapral, R. and Ciccotti, G. (1999). Mixed quantum–classical dynamics. *J. Chem. Phys.*, **110**, 8919–29.

Karplus, M. and Kushick, J. N. (1981). Method for estimating the configurational entropy of macromolecules. *Macromol.*, **14**, 325–32.

Kästner, J. and Thiel, W. (2005). Bridging the gap between thermodynamic integration and umbrella samplng provides a novel analysis method: "umbrella integration". *J. Chem. Phys.*, **123**, 144104/1–5.

Kershaw, D. (1964). Theory of hidden variables. *Phys. Rev.*, **136**, B1850–6.

King, G. and Warshel, A. (1989). A surface-constrained all-atom solvent model for effective simulations of polar solutions. *J. Chem. Phys.*, **91**, 3647–61.

Kirkpatrick, S., Gelatt, C. D. and Vecchi, M. P. (1983). Optimization by simulated annealing. *Science*, **220**, 671–80.

Kirkwood, J. G. (1933). Quantum statistics of almost classical ensembles. *Phys. Rev.*, **44**, 31–35. A correction was published in *Phys. Rev.*, **45** (1934), 116–7.

Kirkwood, J. G. (1939). The dielectric polarization of polar liquids. *J. Chem. Phys.*, **7**, 911–19.

Koelman, J. M. V. and Hoogerbrugge, P. J. (1993). Dynamic simulations of hard-sphere suspensions under steady shear. *Europhys. Lett.*, **21**, 363–8.

Kohn, W. and Sham, L. J. (1965). Self-consistent equations including exchange and correlation effects. *Phys. Rev.*, **A 140**, 1133–8.

Kramers, H. A. (1927). Diffusion of light by atoms. Atti Congr. Internat. dei Fisici, Como, p. 545.

Kramers, H. A. (1940). Brownian motion in a field of force and the diffusion model of chemical reactions. *Physica*, **7**, 284–304.

Kranenburg, M., Venturoli, M. and Smit, B. (2003). Phase behavior and induced interdigitation in bilayers studied with dissipative particle dynamics. *J. Phys. Chem. B*, **107**, 11491-501.

Kreyszig, E. (1993). *Advanced Engineering Mathematics*. 7th edn. New York, Wiley.

Kronig, R. de L. (1926). On the theory of dispersion of X-rays. *J. Opt. Soc. Amer. and Rev. Sci. Instr.*, **12**, 547–57.

Kubo, R. (1957). Statistical-mechanical theory of irreversible processes. I. General theory and simple application to magnetic and conduction problems. *J. Phys. Soc. Japan*, **12**, 570–86.

Kubo, R. (1966). The fluctuation–dissipation theorem and Brownian motion. In

Many-Body Theory, ed. R. Kubo. Tokyo–New York, Syokabo–Benjamin, pp. 1–16.

Kubo, R., Toda, M. and Hashitsume, N. (1985). *Statistical Physics. II. Nonequilibrium Statistical Mechanics*. Berlin, Springer–Verlag.

Kuharski, R. A. and Rossky, P. J. (1984). Quantum mechanical contributions to the structure of liquid water. *Chem. Phys. Lett.*, **103**, 357–62.

Kusalik, P. G., Liden, F. and Svishchev, I. M. (1995). Calculation of the third virial coefficient for water. *J. Chem. Phys.*, **103**, 10169–75.

Kyrala, A (1967). *Theoretical Physics: Applications of vectors, matrices, tensors and quaternions*. Philadelphia, W.B. Saunders Cy.

Laasonen, K. and Nieminen, R. M. (1990). Molecular dynamics using the tight-binding approximation. *J. Phys.: Condens. Matter*, **2**, 1509–20.

Lamoureux, G., MacKerell Jr, A. D. and Roux, B. (2003). A simple polarizable model of water based on classical Drude oscillators. *J. Chem. Phys.*, **119**, 5185–97.

Landau, D. P. and Binder, K. (2005). *A Guide to Monte Carlo Simulations in Statistical Physics*. 2nd edn. Cambridge, UK, Cambridge University Press.

Landau, L. D. and Lifshitz, E. M. (1981). *Quantum Mechanics (Non-relativistic Theory)*. 3rd edn. Oxford, Butterworth–Heinemann.

Landau, L. D. and Lifshitz, E. M. (1982). *Mechanics*. 3rd edn. Oxford, Butterworth–Heinemann.

Landau, L. D. and Lifshitz, E. M. (1987). *Fluid Mechanics*. 2nd edn. Oxford, Butterworth–Heinemann.

Landau, L. D. and Lifshitz, E. M. (1996). *Statistical Physics*. 3rd edn. Oxford, Butterworth–Heinemann.

Laria, D., Ciccotti, G., Ferrario, M. and Kapral, R. (1992). Molecular-dynamics study of adiabatic proton-transfer reactions in solution. *J. Chem. Phys.*, **97**, 378–88.

Leach, A. R. (2001). *Molecular Modelling. Principles and Applications*. 2nd edn. Harlow, UK, Pearson Education.

Lee, C., Yang, W. and Parr, R. G. (1988). Development of the Colle-Salvetti correlation energy formula into a functional of the electron density. *Phys. Rev.*, **B 37**, 785–9.

Lee, T. S., York, D. M. and Yang, W. (1996). Linear-scaling semiempirical quantum calculations for macromolecules. *J. Chem. Phys.*, **105**, 2744–50.

Leforestier, C., Braly, L. B., Liu, K., Elrod, M. J. and Saykally, R. J. (1997). Fully coupled six-dimensional calculations of the water dimer vibration–rotation–tunneling states with a split Wigner pseudo spectral approach. *J. Chem. Phys.*, **106**, 8527–44.

Leimkuhler, B. J. (1999). Comparison of geometric integrators for rigid body simulation. In *Computational Molecular Dynamics: Challenges, Methods, Ideas*, ed. P. Deuflhard, J. Hermans, B. Leimkuhler *et al.*, Berlin, Springer–Verlag, pp. 349–62.

Leimkuhler, B. and Reich, S. (1994). Symplectic integration of constrained Hamiltonian systems. *Math. Comput.*, **63**, 589–605.

Leimkuhler, B. J. and Skeel, R. D. (1994). Symplectic numerical integrators in constrained Hamiltonian systems. *J. Comput. Phys.*, **112**, 117–25.

Lepreore, C. L. and Wyatt, R. E. (1999). Quantum wave packet dynamics with trajectories. *Phys. Re, Lett.*, **82**, 5190–3.

Levin, F. S. (2002). *An Introduction to Quantum Theory*. Cambridge, UK, Cambridge University Press.

Levitt, M. (1983). Protein folding by restrained energy minimization and molecular dynamics. *J. Mol. Biol.*, **170**, 723–64.

Lide, D. R. (1994). *CRC Handbook of Chemistry and Physics*. 75th edn. Boca Raton, Chemical Rubber Publishing Company.

Lie, S. and Engel, F. (1888). *Theorie der Transformationsgruppen*. Leipzig, Teubner.

Linssen, A. B. M. (1998). Molecular dynamics simulations of haloalkane dehalogenase. A statistical analysis of protein motions. Ph.D. thesis, University of Groningen, the Netherlands (electronic version available through http://dissertations.rug.nl/).

Liu, H., Mark, A. E. and van Gunsteren, W. F. (1996). Estimating the relative free energy of different molecular states with respect to a single reference state. *J. Phys. Chem.*, **100**, 9485–94.

Lovett, R. and Baus, M. (1997). A molecular theory of the Laplace relation and the local forces on a curved interface. *J. Chem. Phys.*, **106**, 635–44.

Lowe, C. P. (1999). An alternative approach to dissipative particle dynamics. *Europhys. Lett.*, **47**, 145–51.

Lu, H., Isralewitz, B., Krammer, A., Vogel, V. and Schulten, K. (1998). Unfolding of titin immunoglobulin domains by steered molecular dynamics. *Biophys. J.*, **75**, 662–71.

Lucy, L. B. (1977). A numerical approach to the testing of the fission hypothesis. *Astron. J.*, **82**, 1013–24.

Madelung, E. (1926). Quantentheorie in hydrodynamischer Form. *Z. Physik*, **40**, 322–6.

Mak, C. H. and Andersen, H. C. (1990). Low-temperature approximations for Feynman path integrals and their applications in quantum equilibrium and dynamical problems. *J. Chem. Phys.*, **92**, 2953–5.

Marc, G. and McMillan, W. G. (1985). The virial theorem. *Adv. Chem. Phys.*, **58**, 209–361.

Marechal, M., Baus, M. and Lovett, R. (1997). The local pressure in a cylindrical liquid-vapor interface: a simulation study. *J. Chem. Phys.*, **106**, 645–54.

Marinari, E. and Parisi, G. (1992). Simulated tempering: a new Monte Carlo scheme. *Europhys. Lett.*, **19**, 451–8.

Marrink, S. -J. and Mark, A. E. (2003a). The mechanism of vesicle fusion as revealed by molecular dynamics simulations. *J. Amer. Chem. Soc.*, **125**, 11144–5.

Marrink, S.- J. and Mark, A. E. (2003b). Molecular dynamics simulation of the formation, structure, and dynamics of small phospholipid vesicles. *J. Amer. Chem. Soc.*, **125**, 15233–42.

Marrink, S. -J. and Mark, A. E. (2004). Molecular view of hexagonal phase formation in phospholipid membranes. *Biophys. J.*, **87**, 3894–900.

Marrink, S. -J., de Vries, A. H. and Mark, A. E. (2004). Coarse grained model for semi-quantitative lipid simulations. *J. Phys. Chem. B*, **108**, 750-60.

Marrink, S. -J., Risselada, J. and Mark, A. E. (2005). Simulation of gel phase formation and melting in lipid bilayers using a coarse grained model. *Chem. Phys. Lipids*, **135**, 223–44.

Martyna, G. J., Tuckerman, M. E. and Klein, M. L. (1992). Nosé–Hoover chains: the canonical ensemble via continuous dynamics. *J. Chem. Phys.*, **97**, 2635–43.

Marx, D. and Hutter, J. (2000). *Ab Initio* molecular dynamics: theory and implementation. In *Modern Methods and Algorithms of Quantum Chemistry*, ed. J. Grotendorst, Forschungszentrum Jülich, NIC Series, Vol. **1**, pp. 301–449. Available from http://www.fz-juelich.de/nic-series/Volume1/.

Mathews, J. and Walker, R. L. (1970). *Mathematical Methods of Physics*. Reading, Addison–Wesley.

Maurits, N. M., Zvelindovsky, A. V., Sevink, G. J. A., van Vlimmeren, B. A. C. and Fraaije, J. G. E. M. (1998a). Hydrodynamic effects in three-dimensional microphase separation of block copolymers: dynamic mean-field density functional approach. *J. Chem. Phys.*, **108**, 9150-4.

Maurits, N. M., Zvelindovsky, A. V. and Fraaije, J. G. E. M. (1998b). Viscoelastic effects in threedimensional microphase separation of block copolymers: dynamic mean-field density functional approach. *J. Chem. Phys.*, **109**, 11032-42.

Maurits, N. M., Zvelindovsky, A. V. and Fraaije, J. G. E. M. (1999). Equation of state and stress tensor in inhomogeneous compressible copolymer melts: dynamic mean-field density functional approach. *J. Chem. Phys.*, **108**, 2638-50.

Mavri, J. (2000). Molecular dynamics with nonadiabatic transitions: a comparison of methods. *Mol. Simul.*, **23**, 389–411.

Mavri, J. and Berendsen, H. J. C. (1994). Dynamical simulation of a quantum harmonic oscillator in a noble-gas bath by density matrix evolution. *Phys. Rev. E*, **50**, 198-204.

Mavri, J. and Berendsen, H. J. C. (1995). Calculation of the proton transfer rate using density matrix evolution and molecular dynamics simulations: inclusion of the proton excited states. *J. Phys. Chem.*, **99** (1995) 12711–7.

Mavri, J. and Grdadolnik, J. (2001). Proton potential in acetylacetone. *J. Phys. Chem.*, **A 105**, 2039–44.

Mavri, J., Berendsen, H. J. C. and van Gunsteren, W. F. (1993). Influence of solvent on intramolecular proton transfer in hydrogen malonate. Molecular dynamics simulation study of tunneling by density matrix evolution and nonequilibrium solvation. *J. Phys. Chem.*,, **97**, 13469–76.

Mazur, A. K. (1997). Common molecular dynamics algorithms revisited: accuracy and optimal time steps of Störmer-leapfrog integrators. *J. Comput. Phys.*, **136**, 354–65.

McDowell, H. K. (2000). Quantum generalized Langevin equation: explicit inclusion of nonlinear system dynamics. *J. Chem. Phys.*, **112**, 6971–82.

McQuarrie, D. A. (1976). *Statistical Mechanics*. New York, Harper and Row.

McWeeny, R. (1992). *Methods of Molecular Quantum Mechanics*. Academic Press.

Metropolis, N., Metropolis, A. W., Rosenbluth, M. N., Teller, A. H. and Teller, E. (1953). Equation of state calculations by fast computing machines. *J. Chem. Phys.*, **21**, 1087–92.

Meiners, J.-C. and Quake, S. R. (1999). Direct measurement of hydrodynamic cross correlations between two particles in an external potential. *Phys. Rev. Lett.*, **82**, 2211–4.

Merzbacher, E. (1998). Quantum Mechanics. 3rd edn. Hoboken, New York, Wiley.

Meyer, H. D. and Miller, W. H. (1979). A classical analog for electronic degrees of freedom in nonadiabatic collision processes. *J. Chem. Phys.*, **70**, 3214–23.

Millot, C., Soetens, J. -C., Martins Costa, M. T. C., Hodges, M. P. and Stone, A. J. (1998). Revised anisotropic site potentials for the water dimer and calculated properties. *J. Phys. Chem. A*, **102**, 754–70.

Miyamoto, S. and Kollman, P. A. (1992). SETTLE: An analytical version of the SHAKE and RATTLE algorithms for rigid water molecules. *J. Comput. Chem.*, **13**, 952–62.

Mohr, P. J. and Taylor, B. N. (2005). CODATA recommended values of the fundamental physical constants: 2002. *Rev. Mod. Phys.*, **77**, 1–107.

Monaghan, J. (1988). An introduction to SPH. *Comput. Phys. Comm.*, **48**, 89–96.

Mori, H. (1965a). Transport, collective motion, and Brownian motion. *Progr. Theor. Phys.*, **33**, 423–55.

Mori, H. (1965b). A continued fraction representation of the time correlation functions. *Progr. Theor. Phys.*, **34**, 399–416.

Morishita, T. (2000). Fluctuation formulas in molecular-dynamics simulations with the weak coupling heat bath. *J. Chem. Phys.*, **113**, 2976–82.

Morozov, A. N., Zvelindovsky, A. V. and Fraaije, J. G. E. M. (2000). Orientational phase transitions in the hexagonal phase of a diblock copolymer melt under shear flow. *Phys. Rev. E*, **61**, 4125-32.

Morse, P. M. (1929). Diatomic molecules according to the wave mechanics. II Vibrational levels. *Phys. Rev.*, **34**, 57–64.

Müller, U. and Stock, G. (1998). Consistent treatment of quantum-mechanical and classical degrees of freedom in mixed quantum-classical simulations. *J. Chem. Phys.*, **108**, 7516–26.

Müller-Plathe, F. (2002). Coarse-graining in polymer simulation: From the atomistic to the mesoscopic scale and back. *ChemPhysChem.*, **3**, 754–69.

Nelson, E. (1966). Derivation of the Schrödinger equation fron Newtonian mechanics. *Phys. Rev.*, *150*, 1079–85.

Neumann, M. (1983). Dipole moment fluctuation formulas in computer simulations of polar systems. *Mol. Phys.*, **50**, 841–58.

Neumann, M. and Steinhauser, O. (1983). On the calculation of the frequency-dependent dielectric constant in computer simulations. *Chem. Phys. Lett.*, **102**, 508–13.

Neumann, M., Steinhauser, O. and Pawley, G. S. (1984). Consistent calculations of the static and frequency-dependent dielectric constant in computer simulations. *Mol. Phys.*, **52**, 97–113.

Nicholls, A. and Honig, B. (1991). A rapid finite difference algorithm, utilizing successive over relaxation to solve the Poisson–Boltzmann equation. *J. Comput. Chem.*, **12**, 435–45.

Nicolas, J. J., Gubbins, K. E., Streett, W. B. and Tildesley, D. J. (1979). Equation of state for the Lennard–Jones fluid. *Mol. Phys.*, **37**, 1429–54.

Niedermeier, C. and Taven, P. (1994). A structure adapted multipole method for electrostatic interactions in protein dynamics. *J. Chem. Phys.*, **101**, 734–48.

Nielsen, S., Kapral, R. an Ciccotti, G. (2000). Non-adiabatic dynamics in mixed quantum-classical systems. *J. Stat. Phys.*, **101**, 225–42.

Nielsen, S. O., Lopez, C. F., Srinivas, G. and Klein, M. L. (2003). A coarse grain model for n-alkanes parameterized from surface tension data. *J. Chem. Phys.*, **119**, 7043–9.

Norrby, L. J. (1991). Why is mercury liquid? Or, why do relativistic effects not get into chemistry textbooks?. *J. Chem. Education*, **68**, 110–3.

Nosé, S. (1984a). A unified formulation of the constant temperature molecular dynamics method. *J. Chem. Phys.*, **81**, 511–9.

Nosé, S. (1984b). A molecular dynamics method for simulations in the canonical ensemble. *Mol. Phys.*, **52**, 255–68.

Nosé, S. and Klein, M. L. (1983). Constant pressure molecular dynamics for molecular systems. *Mol. Phys.*, **50**, 1055–76.

Onsager, L. (1931a). Reciprocal relations in irreversible processes I. *Phys Rev.*, **37**, 405–26.

Onsager, L. (1931b). Reciprocal relations in irreversible processes II. **38**, 2265–79.

Onufriev, A., Bashford, D. and Case, D. A. (2004). Exploring protein native states and large-scale conformational changes with a modified generalized Born model. *Proteins*, **55**, 383–94.

Oostenbrink, C. and van Gunsteren, W. F. (2003). Single-step perturbation to calculate free energy differences from unphysical reference states. *J. Comput. Chem.*, **24**, 1730–9.

Oostenbrink, C. and van Gunsteren, W. F. (2006). Calculating zeros: non-equilibrium free energy calculations. *Chem Phys.*, **323**, 102–8.

Öttinger, H. C. (1998). General projection operator formalism for the dynamics and thermodynamics of complex fluids. *Phys. Rev. E*, **57**. 1416–20.

Öttinger, H. C. and Grmela, M. (1997). Dynamics and thermodynamics of complex fluids. II. Illustrations of a general formalism. *Phys. Rev. E*, **56**. 6633–55.

Pang, T. (2006). Computational Physics. 2nd edn. Cambridge, UK, Cambridge University Press.

Papoulis, A. (1965). *Probability, Random Variables and Stochastic Processes*, International Student edn. Tokyo, McGraw-Hill Kogakusha.

Paricaud, P., Předota, M., Chialvo, A. A. and Cummings, P. T. (2005). From dimer to condensed phases at extreme conditions: Accurate predictions of the properties of water by a Gaussian charge polarizable model. *J. Chem. Phys.*, **122**, 244511–4.

Park, S. and Schulten, K. (2004). Calculating potentials of mean force from steered molecular dynamics simulations. *J. Chem. Phys.*, **120**, 5946–61.

Park, S., Khalili-Araghi, F., Tajkhorshid, E. and Schulten, K. (2003). Free energy calculation from steered molecular dynamics simulations using Jarzynskis equality. *J. Chem. Phys.*, **119**, 3559–66.

Parr, R. G. and Yang, W. (1989). *Density Functional Theory*. Oxford, Oxford University Press.

Parrinello, M. and Rahman, A. (1980). Crystal structure and pair potentials: A molecular-dynamics study. *Phys. Rev. Lett.*, **45**, 1196–9.

Parrinello, M. and Rahman, A. (1981). Polymorphic transitions in single crystals: A new molecular dynamics method. *J. Appl. Phys.*, **52**, 7182–90.

Parrinello, M. and Rahman, A. (1984). Study of an F center in molten KCl. *J. Chem. Phys.*, **80**, 860–7.

Pearlman, D. A. and Kollman, P. A. (1989a). A new method for carrying out free energy perturbation calculations: dynamically modified windows. *J. Chem. Phys.*, **90**, 2460–70.

Pearlman, D. A. and Kollman, P. A. (1989b). The lag between the Hamiltonian and the system configuration in free energy perturbation calculations. *J. Chem. Phys.*, **91**, 7831–9.

Pechukas, P. (1969a). Time-dependent semiclassical scattering theory. I. Potential scattering. *Phys. Rev.*, **181**, 166–174.

Pechukas, P. (1969b). Time-dependent semiclassical scattering theory. II. Atomic collisions. *Phys. Rev.*, **181**, 174–185.

Perram, J. W., Petersen, H. G. and de Leeuw, S. W. (1988). An algorithm for the simulation of condensed matter which grows as the 3/2 power of the number of particles. *Mol. Phys.*, **65**, 875–89.

Pikkemaat, M. G., Linssen, A. B. M., Berendsen, H. J. C. and Janssen, D. B. (2002). Molecular dynamics simulations as a tool for improving protein stability. *Protein Eng.*, **15**, 185–92.

Postma, J. P. M. (1985). MD of H_2O. A Molecular Dynamics Study of Water.. *Ph.D. Thesis, University of Groningen*, April 1985.

Powles, J. G. and Rickayzen, G. (1979). Quantum corrections and the computer simulation of molecular fluids. *Mol. Phys.*, **38**, 1875–92.

Press, W. H., Teukolsky, S. A., Vettering, W. T. and Flannery, B. P. (1993). *Numerical Recipes, The Art of Scientific Computing*. 2nd edn. Cambridge, UK, Cambridge University Press.

Prezhdo, O. V. and Brooksby, C. (2001). Quantum backreaction through the Bohmian particle. *Phys. Rev. Lett.*, **86**, 3215–9.

Prezhdo, O. V. and Rossky, P. J. (1997a). Mean-field molecular dynamics with surface hopping. *J. Chem. Phys.*, **107**, 825–34.

Prezhdo, O. V. and Rossky, P. J. (1997b). Evaluation of quantum transition rates from quantum-classical molecular dynamics simulations. *J. Chem. Phys.*, **107**, 5863–77.

Prigogine, I. (1961). *Introduction to Thermodynamics of Irreversible Processes*. New York, Interscience.

Pyykkö, P. (1988). Relativistic effects in structural chemistry. *Chem. Rev.*, **88**, 563–94.

Rahman, A. and Stillinger, F. H. (1971). Molecular dynamics study of liquid water. *J. Chem. Phys.*, **55**, 3336–59.

Rapaport, D. (2004). *The Art of Molecular Dynamics Simulation*. 2nd edn. Cambridge, UK, Cambridge University Press.

Redfield, A. G. (1965). Relaxation theory: density matrix formulation. *Adv. Magn. Reson.*, **1**, 1–32.

Reynolds, P. J., Ceperley, D. M., Alder, B. and Lester Jr, W. A. (1982). Fixed-node quantum Monte carlo for molecules. *J. Chem. Phys.*, **77**, 5593–603.

Rice, S. O. (1954). Mathematical analysis of random noise. In *Selected Papers on Noise and Stochastic Processes*, ed. N. Wax. New York, Dover Publications, pp. 133–294.

Rick, S. W., Stuart, S. S. and Berne, B. J. (1994). Dynamical fluctuating charge force fields: Application to liquid water. *J. Chem. Phys.*, **101**, 6141–56.

Risken, H. (1989). *The Fokker–Planck equation*. Berlin, Springer–Verlag.

Rodrigues, O. (1840). Des lois géométriques qui régissent les déplacements d'un système solide dans l'espace, et de la variation des coordonnées provenant de ses déplacements considérés indépendammant des cause qui peuvent les produire. *J. de Mathématiques Pures et Appliquées*, **5**, 380–440.

Roos, B. O, Taylor, P. R. and Siegbahn, E. M. (1980). A complete active space SCF method (CASSCF) using a density matrix formulated super-CI approach. *Chem. Phys.*, **48**, 157–73.

Rowlinson, J. S. (1993). Thermodynamics of inhomogeneous systems. *Pure Appl. Chem.*, **65**, 873–82.

Roy, P. -N. and Voth, G. A. (1999). On the Feynman path centroid density for Bose–Einstein and Fermi–Dirac statistics. *J. Chem. Phys.*, **110**, 3647–52.

Ryckaert, J. P., Ciccotti, G. and Berendsen, H. J. C. (1977). Numerical integration of the cartesian equations of motion of a system with constraints: molecular dynamics of n-alkanes. *J. Comput. Phys.*, **23**, 327–41.

Saint-Martin, H., Hernández-Cobos, J., Bernal-Uruchurtu, M. I., Ortega-Blake, I. and Berendsen, H. J. C. (2000). A mobile charge densities in harmonic oscillators (MCDHO) molecular model for numerical simulations: the water–water interaction. *J. Chem. Phys.*, **113**, 10899–912.

Saint-Martin, H., Hess, B. and Berendsen, H. J. C. (2004). An application of flexible constraints in Monte Carlo simulations of the isobaric–isothermal ensemble of liquid water and ice Ih with the polarizable and flexible mobile charge densities in harmonic oscillators model. *J. Chem. Phys.*, **120**, 11133–43.

Saint-Martin, H., Hernández-Cobos, J. and Ortega-Blake, I. (2005). Water models based on a single potential energy surface and different molecular degrees of freedom. *J. Chem. Phys.*, **122**, 224509 (12 pp.).

Salcedo, L. L. (2003). Comment on "Quantum backreaction through the Bohmian particle". *Phys. Rev. Lett.* **90**, 118901.

Schäfer, H., Mark, A. E. and van Gunsteren, W. F. (2000). Absolute entropies from molecular dynamics simulation trajectories. *J. Chem. Phys.*, **113**, 7809–17.

Schaefer, M. and Karplus, M. (1996). A comprehensive analytical treatment of continuum electrostatics. *J. Phys. Chem.*, **100**, 1578–99.

Schenter, G. K. (2002). The development of effective classical potentials and the quantum-statistical mechanical second virial coefficient of water. *J. Chem. Phys.*, **117**, 6573–81.

Schiff, L. L. (1968). *Quantum Mechanics*. 3rd edn. New York, McGraw-Hill.

Schlichter, C. P. (1963). *Principles of Magnetic Resonance*. New York, Harper and Row.

Schlijper, A. G., Hoogerbrugge, P. J. and Manke, C. W. (1995). Computer simulation of dilute polymer solutions with the dissipative particle dynamics method. *J. Rheol.*, **39**, 567–79.

Schlitter, J. (1993). Estimation of absolute and relative entropies of macromolecules using the covariance matrix. *Chem. Phys. Lett.*, **215**, 617–21.

Schmidt, K. E. (1987). Variational and Green's function Monte Carlo calculations of few-body systems. In *Models and Methods in Few-Body Physics*, ed. L.S. Ferreira, A. Fonseca and L. Streit. Lecture Notes in Physics **273**. Berlin, Springer–Verlag, pp. 363–407.

Schmidt, K. E. and Ceperley, D. M. (1992). Monte Carlo techniques for quantum fluids, solids and droplets. In *The Monte Carlo Method in Condensed Matter Physics*, ed. K. Binder, Topics in Applied Physics, **71**. Berlin, Springer Verlag, pp. 205–48.

Schmidt, K. E. and Kalos, M. H. (1984). . In *Application of the Monte Carlo Method in Statistical Physics*, ed. K. Binder, Topics in Applied Physics, **36**. Berlin, Springer-Verlag, pp. 125–43.

Schneider, T. and Stoll, E. (1978). Molecular dynamics study of a three-dimensional ne-component model for distortive phase transitions. *Phys. Rev.*, **B 17**, 1302-22.

Schoenberg, I. J. (1973). *Cardinal Spline Interpolation*. Regional Conference Series in Applied Mathematics. Philadelphia, Penn, SIAM.

Schofield, P. and Henderson, J. R. (1982). Statistical mechanics of inhomogeneous fluids. *Proc. Roy. Soc. London* **A 379**, 231–46.

Schumaker, L. L. (1981). *Spline Functions: Basic Theory*. New York, Wiley.

Schurr, J. M. and Fujimoto, B. S. (2003). Equalities for the non-equilibrium work transferred from an external potential to a molecular system. Analysis of single-molecule extension experiments. *J. Phys. Chem.*, **B 107**, 14007–19.

Selloni, A., Carnevali, P., Car, R. and Parrinello, M. (1987). Localization, hopping, and diffusion of electrons in molten salts. *Phys. Rev. Lett.*, **59**, 823–6.

Serrano, M. and Español, P. (2001). Thermodynamically consistent mesoscopic fluid particle model. *Phys. Rev. E*, **64**, 46115 (18 pp.).

Serrano, M., De Fabritiis, G. Español, P., Flekkøy, E. G. and Coveney, P. V. (2002). Mesoscopic dynamics of Voronoi fluid particles. *J. Phys. A: Math. Gen.*, **35**, 1605–25.

Sesé, L. M. (1992). A quantum Monte Carlo study of liquid Lennard–Jones methane, path integral and effective potentials. *Mol. Phys.*, **76**, 1335–46.

Sesé, L. M. (1993). Feynman–Hibbs quantum effective potentials for Monte Carlo simulations of liquid neon. *Mol. Phys.*, **78**, 1167–77.

Sesé, L. M. (1994). Study of the Feynman–Hibbs effective potential against the path-integral formalism for Monte Carlo simulations of quantum many-body Lennard–Jones systems. *Mol. Phys.*, **81**, 1297–312.

Sesé, L. M. (1995). Feynman–Hibbs potentials and path integrals for quantum Lennard–Jones systems: theory and Monte Carlo simulations. *Mol. Phys.*, **85**, 931–47.

Sesé, L. M. (1996). Determination of the quantum static structure factor of liquid neon within the Feynman–Hibbs picture. *Mol. Phys.*, **89**, 1783–802.

Sevink, G. J. A., Zvelindovsky, A. V., van Vlimmeren, B. A. C., Maurits, N. M. and Fraaije, J. G. E. M. (1999). Dynamics of surface-directed mesophase formation in block copolymer melts. *J. Chem. Phys.*, **110**, 2250-6.

Shabana, A. A (1989). Dynamics of Multibody Systems. New York, Wiley.

Shannon, C. E. (1948). The mathematical theory of communication. *Bell Systems Techn. J.*, **27**, 379; 623. Reprinted In C.E. Shannon and W. Weaver, *The Mathematical theory of Communication*, Urbana, IL, University of Illinois Press (1949).

Sigfridsson, E. and Ryde, U. (1998). Comparison of methods for deriving atomic charges from the electrostatic potential and moments. *J. Comput. Chem.*, **19**, 377–95.

Silliman, S. D. (1974). The numerical evaluation by splines of Fourier transforms. *J. Approximation Theory*, **12**, 32–51.

Singh, A. K. and Sinha, S. K. (1987). Equilibrium properties of molecular fluids in the semiclassical limit. II Hydrogen molecules. *Phys. Rev. A*, **35**, 295–9.

Smit, B. (1992). Phase diagrams of Lennard–Jones fluids. *J. Chem. Phys.*, **96**, 8639–40.

Smith, W. (1982). Point multipoles in the Ewald summation. *CCP-5 Quarterly*, **4**, 13, Daresbury Lab, UK.

Sorella, S. (1998). Green function Monte Carlo with stochastic reconfiguration. *Phys. Rev. Lett.*, **80**, 4558–61.

Soto, P. and Mark, A. E. (2002). The effect of the neglect of electronic polarization in peptide folding simulations. *J. Phys. Chem. B*, **106**, 12830–3.

Spångberg, D. and Hermansson, K. (2004). Many-body potentials for aqueous Li^+,

Na^+, Mg^{++}, and Al^{+++}: Comparison of effective three-body potentials and polarizable models. *J. Chem. Phys.*, **120**, 4829–43.

Späth, H. (1973). *Spline-Algorithmen zur Konstruktion glatter Kurven und Flächen.* München, Wien, R. Oldenbourg Verlag.

Sprik, M. and Ciccotti, G. (1998). Free energy from constrained molecular dynamics. *J. Chem. Phys.*, **109**, 7737–44.

Stern, H. A., Rittner, F., Berne, B. J. and Friesner, R. A. (2001). Combined fluctuating charge and polarizable dipole models: Application to a five-site water potential function. *J. Chem. Phys.*, **115**, 2237–51.

Stewart, J. J. P. (1989a). Optimization of parameters for semiempirical methods. I Method. *J. Comput. Chem.*, **10**, 209–20.

Stewart, J. J. P. (1989b). Optimization of parameters for semiempirical methods. II Applications. *J. Comput. Chem.*, **10**, 221–64.

Still, W. C., Tempczyck, A., Hawley, R. C. and Hendrickson, T. (1990). Semianalytical treatment of solvation for molecular mechanics and dynamics. *J. Amer. Chem. Soc.*, **112**, 6127–9.

Stillinger, F. H. and Rahman, A. (1972). Molecular dynamics study of temperature effects on water structure and kinetics. *J. Chem. Phys.*, **57**, 1281–92.

Stillinger, F. H. and Rahman, A. (1974). Improved simulation of liquid water by molecular dynamics. *J. Chem. Phys.*, **60**, 1545–57.

Straatsma, T. P. and Berendsen, H. J. C. (1988). Free energy of ionic hydration: Analysis of a thermodynamic integration technique to evaluate free energy differences by molecular dynamics. *J. Chem. Phys.*, **89**, 5876–86.

Straatsma, T.P. and McCammon, J.A. (1990a). Molecular dynamics simulations with interaction potentials including polarization. Development of a noniterative method and application to water. *Mol. Simul.*, **5**, 181–92.

Straatsma, T. P. and McCammon, J. A. (1990b). Free energy thermodynamic integrations in molecular dynamics simulations using a noniterative method to include electronic polarization. *Chem. Phys. Lett.*, **167**, 252–54.

Straatsma, T. P. and McCammon, J. A. (1991). Free energy evaluation from molecular dynamics simulations using force fields including electronic polarization. *Chem. Phys. Lett.*, **177**, 433–40.

Sugita, Y. and Okamoto, Y. (1999).

. Replica-exchange molecular dynamics method for protein folding. *Chem. Phys. Lett.*, **314**, 141–51

Sun, X. and Miller, W.H. (1997). Mixed semiclassicalclassical approaches to the dynamics of complex molecular systems. *J. Chem. Phys.*, **106**, 916-27.

Suzuki, M. (1991). General theory of fractal path integrals with applications to many-body theories and statistical physics. *J. Math. Phys.*, **32**, 400–7.

Svensson, M., Humbel, S. and Froese, R. D. J. *et al.* (1996). ONION: A multilayered integral MO + MM method for geometry optimizations and single point energy predictions. A test for Diels-Alder reactions and $Pt(P(t\text{-}Bu)_3)_2$ + H_2 oxidative addition. *J. Phys. Chem.*, **100**, 19357–63.

Svishchev, I. M., Kusalik, P. G., Wang, J. and Boyd, R. J. (1996). Polarizable point-charge model for water: Results under normal and extreme conditions. *J. Chem. Phys.*, **105**, 4742–50.

Swart, M. (2003). AddRemove: A new link model for use in QM/MM studies. *Int. J. Quant. Chem.*, **91**, 177–83.

Swart, M., van Duijnen, P. Th. and Snijders, J. G. (2001). A charge analysis derived from an atomic multipole expansion. *J. Comput. Chem.*, **22**, 79-88.

Swendsen, R. H. and Wang, J.-S. (1986). Replica Monte Carlo simulation of spin-glasses. *Phys. Rev. Lett.*, **57**, 2607–9.

Swope, W. C., Andersen, H. C., Berens, P. H. and Wilson, K. R. (1982). A computer simulation method for the calculation of equilibrium constants for the formation of physical clusters of molecules: application to small water clusters. *J. Chem. Phys.*, **76**, 637–49.

Szabo, A. and Ostlund, N. S. (1982). *Modern Quantum Chemistry*. New York, McGraw-Hill.

Takabayasi, T. (1952). On the formulation of quantum mechanics associated with classical pictures. *Prog. Theor. Phys.*, **8**, 143–82.

Tanizaki, S., Mavri, J., Partridge, H. and Jordan, P. C. (1999). Unusual distributed charge models for water's electric potential. *Chem. Phys.*, **246**, 37–47.

Tappura, K., Lahtela–Kakkonen, M. and Teleman, O. (2000). A new soft-core potential function for molecular dynamics applied to the prediction of protein loop conformations. *J. Comput. Chem.*, **21**, 388–397.

Thirumalai, D., Hall, R. W. and Berne, B. J. (1984). Path integral Monte carlo study of liquid neon and the quantum effective pair potential. *J. Chem. Phys.*, **81**, 2523-7.

Thole, B. T. (1981). Molecular polarizabilities calculated with a modified dipole interaction. *Chem. Phys.*, **59**, 341–50.

Tironi, I. G., Sperb, R., Smith, P. E. and Van Gunsteren, W. F. (1995). A generalized reaction field method for molecular dynamics simulations. *J. Chem. Phys.*, **102**, 5451–9.

Toda, M., Kubo, R. and Saito, N. (1983). *Statistical Physics. I. Equilibrium Statistical Mechanics*. Berlin, Springer–Verlag.

Torrie, G. M. and Valleau, J. P. (1974). Monte Carlo free energy estimates using non-Boltzmann sampling: application to the sub-critical Lennard–Jones fluid. *Chem. Phys. Lett.*, **28**, 578–81.

Torrie, G. M. and Valleau, J. P. (1977). Non-physical sampling distributions in Monte Carlo free energy estimation: umbrella sampling. *J. Comput. Phys.*, **23**, 187–99.

Toukmaji, A., Sagui, C., Board, J. and Darden, T. (2000). Efficient particle–mesh Ewald based approach to fixed and induced dipolar interactions. *J. Chem. Phys.*, **113**, 10913–27.

Toxvaerd, S. and Olsen, O. H. (1990). Canonical molecular dynamics of molecules with internal degrees of freedom. *Ber. Bunsenges. phys. Chem.*, **94**, 274–8.

Trotter, H. F. (1959). On the product of semigroup of operators. *Proc. Amer. math. Soc.*, **10**, 545–51.

Tuckerman, M., Berne, B. J. and Martyna, G. J. (1992). Reversible multiple time scales molecular dynamics. *J. Chem. Phys.*, **97**, 1990–2001.

Tuckerman, M. E., Mundy, C. J. and Martyna, G. J. (1999). On the classical statistical mechanics of non-Hamiltonian systems. *Europhys. Lett.*, **45**, 149–55.

Tully, J. C. (1990). Molecular dynamics with electronic transitions. *J. Chem. Phys.*, **93**, 1061–71.

Tully, J. C. and Preston, R. K. (1971). Trajectory surface hopping approach to nonadiabatic molecular collisions: the reaction of H with D_2. *J. Chem. Phys.*, **55**, 562–72.

Uhlenbeck, G. E. and Gropper, L. (1932). The equation of state of a non-ideal Einstein-Bose or Fermi–Dirac gas. *Phys. Rev.*, **41**, 79–90.

Umrigar, C. J., Nightingale, M. P. and Runge, K. (1993). A diffusion Monte Carlo algorithm with very small time-step errors. *J. Chem. Phys.*, **99**, 2865–90.

van der Spoel, D., van Maaren, P. J. and Berendsen, H. J. C. (1998). A systematic study of water models for molecular simulation: derivation of water models optimized for use with a reaction field. *J. Chem. Phys.*, **108**, 10220–30.

van der Spoel, D., Lindahl, E., Hess, B. *et al.* (2005). GROMACS: fast, flexible, and free. *J. Comput. Chem.*, **26**, 1701–18.

van Duijnen, P. Th. and Swart, M. (1998). Molecular and atomic polarizabilities: Thole's model revisited. *J. Phys. Chem. A*, **102**, 2399–407.

van Gunsteren, W. F. and Berendsen, H. J. C. (1977). Algorithms for macromolecular dynamics and constraint dynamics. *Mol. Phys.*, **34**, 1311–27.

van Gunsteren, W. F., Beutler, T. C., Fraternali, F., King, P. M., Mark, A. E. *et al.* (1993). Computation of free energy in practice: choice of approximations and accuracy limiting factors. In: *Computer Simulation of Biomolecular Systems*, vol **2**, Eds W. F. van Gunsteren *et al.*, Leiden, Escom, pp 315–48.

van Kampen, N. G. (1981). *Stochastic Processes in Physics and Chemistry*. Amsterdam, North Holland.

van Maaren, P. J. and van der Spoel, D. (2001). Molecular dynamics simulations of water with novell shell-model potentials. *J. Phys. Chem. B*, **105**, 2618–26.

van Vlimmeren, B. A. C., Maurits, N. M., Zvelindovsky, A. V., Sevink, G. J. A., Fraaije, J. G. E. M. (1999). Simulation of 3D mesoscale structure formation in concentrated aqueous solution of the triblock polymer surfactants (ethylene oxide)13(propylene oxide)30(ethylene oxide)13 and (ethylene oxide)19(propylene oxide)33(ethylene oxide)19. Application of dynamic mean-field density functional theory. *Macromol.*, **32**, 646-56.

Verhoeven, J. and Dymanus, A. (1970). Magnetic properties and molecular quadrupole tensor of the water molecule by beam-maser Zeeman spectroscopy. *J. Chem. Phys.*, **52**, 3222–33.

Verlet, L. (1967). Computer "experiments" on classical fluids. I. Thermodynamical properties of Lennard–Jones molecules. *Phys. Rev.*, **159**, 98–103.

Vesely, F. J. (2001). *Computational Physics*. 2nd edn. New York, Kluwer Academic/ Plenum Publishing.

Vink, J. C. (1993). Quantum mechanics in terms of discrete beables. *Phys. Rev.*, **A 48**, 1808–18.

Wajnryb, E., Altenberger, A. R. and Dahler, J. S. (1995). Uniqueness of the microscopic stress tensor. *J. Chem. Phys.*, **103**, 9782–7.

Wallqvist. A. and Berne, B. J. (1985). Path-integral simulation of pure water. *Chem. Phys. Lett.*, **117**, 214–9.

Warshel, A. and Levitt, M. (1976). Theoretical studies of enzymic reactions: Dielectric, electrostatic and steric stabilization of the carbonium ion in the reaction of lysozyme. *J. Mol. Biol.*, **103**, 227–49.

Wassenaar, T. A. (2006). Molecular Dynamics of Sense and Sensibility in Processing and Analysis of Data. Ph. D. thesis, University of Groningen, the Netherlands (electronic version available from http://dissertations.rug.nl/.

Wassenaar, T. A. and Mark, A. E. (2006). The effect of box shape on the dynamic properties of proteins simulated under periodic boundary conditions. *J. Comput. Chem.*, **27**, 316–25.

Wax, N., ed. (1954). *Noise and Stochastic Processes*. New York, Dover Publ.

Webster, F., Rossky, P. J. and Friesner, R.A. (1991). Nonadiabatic processes in condensed matter: semi-classical theory and implementation. *Comp. Phys. Comm.*, **63**, 494–522.

Weisstein, E. W. (2005). Spherical Harmonic Addition Theorem. *MathWorld–A Wolfram Web Resource*, available at http://mathworld.wolfram.com.

Weizel, W. (1954). Ableitung der quantenmechanischen Wellengleichung des Mehrteilchensystems aus einem klassischen Modell. *Z. Physik*, **136**, 582–604.

Wesselingh, J. A. and Krishna, R. (1990). *Mass Transfer*. New York and London, Ellis Horwood.

Widom, B. (1963). Some topics in the theory of fluids. *J. Chem. Phys.*, **39**, 2808–12.

Widom, B. (2002). *Statistical Mechanics. A concise introduction for chemists*. Cambridge, UK, Cambridge University Press.

Wigner, E. (1932). On the quantum correction for thermodynamic equilibrium. *Phys. Rev.*, **40**, 749–59.

Wittenburg, J. (1977). *Dynamics of Systems of Rigid Bodies*. Stuttgart, Teubner.

Wolniewicz, L. (1993). Relativistic energies of the ground state of the hydrogen molecule. *J. Chem. Phys.*, **99**, 1851–68.

Wood, R. H. (1991). Estimation of errors in free energy calculations due to the lag between the Hamiltonian and the system configuration. *J. Phys. Chem.*, **95**, 4838–42.

Wood, R. H. (1995). Continuum electrostatics in a computational universe with finite cutoff radii and periodic boundary conditions: Correction to computed free energies of ionic solvation. *J. Chem. Phys.*, **103**, 6177–87.

Woolfson, M. M. and Pert, G. J. (1999). *An Introduction on Computer Simulation*. Oxford, Oxford University Press.

Wormer, P. E. S and van der Avoird, A. (2000). Intermolecular potentials, internal motions, and spectra of van der Waals and hydrogen-bonded complexes. *Chem. Rev.*, **100**, 4109–43.

Wu, Y. -S. M., Kuppermann, A. and Anderson, J. B. (1999). A very high accuracy potential energy surface for H_3. *Phys. Chem. Chem. Phys.*, **1**, 929–37.

Wyatt, R. E. (2005). *Quantum Dynamics with Trajectories. Introduction to Quantum Hydrodynamics*. New York, Springer–Verlag.

Yang, W. (1991a). Direct calculation of electron density in density-functional theory. *Phys. Rev. Lett.*, **66**, 1438–41.

Yang, W. (1991b). Direct calculation of electron density in density-functional theory: Implementation for benzene and a tetrapeptide. *Phys. Rev.*, **A 66**, 7823–6.

Yang, W. and Lee, T. S. (1995). A density-matrix divide-and-conquer approach for electronic structure calculations in large molecules. *J. Chem. Phys.*, **103**, 5674–8.

Yoneya, M., Berendsen, H. J. C. and Hirasawa, K. (1994). A non-iterative matrix method for constraint molecular dynamics simulation. *Mol. Simul.*, **13**, 395–405.

Ytreberg, F. M. and Zuckerman, D. M. (2004). Efficient use of non-equilibrium measurement to estimate free energy differences for molecular systems. *J. Comput. Chem.*, **25**, 1749–59.

Yu, H., Hansson, T. and van Gunsteren, W. F. (2003). Development of a simple, self-consistent polarizable model for liquid water. *J. Chem. Phys.*, **118**, 221–34.

Zeiss, G. D. and Meath, W. J. (1975). The H_2O–H_2O dispersion energy constant and the dispersion of the specific refractivity of diluted water vapour. *Mol. Phys.*, **30**, 161–9.

Zhang, Y. and Yang, W. (1999). A pseudobond approach to combining quantum mechanical and molecular mechanical methods. *J. Chem. Phys.*, **110**, 46–54.

Zhou, J., Reich, S. and Brooks, B. R. (2000). Elastic molecular dynamics with self-consistent flexible constraints. *J. Chem. Phys.*, **112**, 7919–29.

Zhu, S. -B., Singh, S. and Robinson, G. W. (1991). A new flexible/polarizable water model. *J. Chem. Phys.*, **95**, 2791–9.

Zimmerman. J. A., Web III, E. B., Hoyt, J. J. *et al.* (2004). Calculation of stress in atomistic simulation. *Modelling Simul. Mater. Sci. Eng.*, **12**, 319–32.

Zvelindovsky, A. V., Sevink, G. J. A., van Vlimmeren, B. A. C., Maurits, N. M. and Fraaije, J. G. E. M. (1998a). Three-dimensional mesoscale dynamics of block copolymers under shear: The dynamic density-functional approach. *Phys. Rev. E*, **57**, R4879-82.

Zvelindovsky, A. V., van Vlimmeren, B. A. C., Sevink, G. J. A., Maurits, N. M. and Fraaije, J. G. E. M. (1998b). Three-dimensional simulation of hexagonal phase of a specific polymer system under shear: The dynamic density functional approach. *J. Chem. Phys.*, **109**, 8751-4.

Zwanzig, R. W. (1954). High-temperature equation of state by a perturbation method. I. Nonpolar gases. *J. Chem. Phys.*, **22** 1420–6.

Zwanzig, R. (1960). Ensemble method in the theory of irreversibility. *J. Chem. Phys.*, **33**, 1338–41.

Zwanzig, R. (1961). Memory effects in irreversible thermodynamics. *Phys. Rev.*, **124**, 983–92.

Zwanzig, R. (1965). Time-correlation functions and transport coefficients in statistical mechanics. *Ann. Rev. Phys. Chem.*, **16**, 67–102.

Zwanzig, R. (1973). Nonlinear generalized Langevin equations. *J. Stat. Phys.*, **9**, 215–20.

Index

587